ments

Group 3	Group 4	Group 5	Group 6	Group 7	Group 0
					4 **He** Helium 2
11 **B** Boron 5	12 **C** Carbon 6	14 **N** Nitrogen 7	16 **O** Oxygen 8	19 **F** Fluorine 9	20 **Ne** Neon 10
27 **Al** Aluminium 13	28 **Si** Silicon 14	31 **P** Phosphorus 15	32 **S** Sulphur 16	35.5 **Cl** Chlorine 17	40 **Ar** Argon 18

Mass

59	59 **Ni** Nickel 28	64 **Cu** Copper 29	65 **Zn** Zinc 30	70 **Ga** Gallium 31	73 **Ge** Germanium 32	75 **As** Arsenic 33	79 **Se** Selenium 34	80 **Br** Bromine 35	84 **Kr** Krypton 36
103 m	106 **Pd** Palladium 46	108 **Ag** Silver 47	112 **Cd** Cadmium 48	115 **In** Indium 49	119 **Sn** Tin 50	122 **Sb** Antimony 51	128 **Te** Tellurium 52	127 **I** Iodine 53	131 **Xe** Xenon 54
192 m	195 **Pt** Platinum 78	197 **Au** Gold 79	201 **Hg** Mercury 80	204 **Tl** Thallium 81	207 **Pb** Lead 82	209 **Bi** Bismuth 83	210 **Po** Polonium 84	210 **At** Astatine 85	222 **Rn** Radon 86

152 u ium	157 **Gd** Gadolinium 64	159 **Tb** Terbium 65	162.5 **Dy** Dysprosium 66	165 **Ho** Holmium 67	167 **Er** Erbium 68	169 **Tm** Thulium 69	173 **Yb** Ytterbium 70	175 **Lu** Lutetium 71
243 n cium	247 **Cm** Curium 96	247 **Bk** Berkelium 97	251 **Cf** Californium 98	254 **Es** Einsteinium 99	253 **Fm** Fermium 100	256 **Md** Mendelevium 101	254 **No** Nobelium 102	257 **Lr** Lawrencium 103

HAMPSHIRE
EDUCATION COMMITTEE

St Edmunds School

NAME	Form	Date of Issue	* Condition	Teacher's Initials
EC	STAFF	Book		

*A: New. B: Good. C: Fair. D: Poor.

This book is the property of the Hampshire Education Committee, and the parent or guardian of the pupil to whom it is lent is responsible for any loss or damage.

880632

Investigating Chemistry

Investigating Chemistry

L. Davies, B.Sc., C.Chem.M.R.I.C.,
Head of Applied Chemistry, Whitehaven Grammar School

M. J. Denial, B.Sc., C.Chem.M.R.I.C.,
Head of the Science Department, King Edward VII School, Sheffield

A. W. Locke, B.Sc., Ph.D., C.Chem.M.R.I.C.
Head of Pure Chemistry, Whitehaven Grammar School

M. E. Reay, B.Sc.,
Formerly Whitehaven Grammar School

Heinemann Educational Books · London

Heinemann Educational Books Ltd
LONDON EDINBURGH MELBOURNE AUCKLAND TORONTO
HONG KONG SINGAPORE KUALA LUMPUR NEW DELHI
IBADAN NAIROBI LUSAKA JOHANNESBURG
KINGSTON/JAMAICA

ISBN 0 435 64164 6

First published 1973
Reprinted 1973, 1974
Reprinted with corrections 1976, 1978

Published by
Heinemann Educational Books Ltd
48 Charles Street, London W1X 8AH

Filmset by Keyspools Ltd, Golborne, Lancs
Printed and bound in Great Britain by
Cox & Wyman Ltd, London, Fakenham and Reading

Introduction

It is generally agreed that the Nuffield schemes have produced more stimulating changes in school chemistry than any other single factor in recent years. For this chemistry teachers are heavily in debt to the Nuffield Foundation. Most Examining Boards have since introduced modified 'O' level and C.S.E. syllabuses, and we have been provided with a wide choice of modern syllabuses and many 'new' experiments which can be successfully introduced into any scheme.

Inevitably opinions about modern chemistry courses differ widely. We feel that a purely heuristic approach cannot be used successfully with the majority of pupils. The lack of a textbook in such cases is a serious disadvantage; pupils find it difficult to know what to revise for examinations; they feel insecure when asked to revise only ideas, and absence causes many problems.

Investigating Chemistry is a modification of conventional schemes in the light of the Nuffield approach and we consider that it includes the best of the old and the new. It is thus suitable for schools following any of the 'O' level courses prescribed by the major examining boards. We hope that it will also prove useful for C.S.E. work.

Considerable emphasis has been placed throughout on structure and its relation to properties. Basic material such as atomic structure, bonding, and periodicity are introduced early in the course. We have included all topics such as energy changes, simple kinetics, detergency, organic chemistry and polymers. Instead of treating them in isolation, such topics have been incorporated into the text wherever they form part of a continuous story. Thus the extended treatment of a subject may follow in a later section, but in the same chapter as the basic work.

It is intended that, in the teaching sequence, this more advanced material is omitted in the elementary stages and then dealt with separately later in the course, but when the syllabus is completed and the book is being used for revision, all the references to one particular aspect of chemistry are found together. It is not intended that the material would be taught chapter by chapter and in the sequence presented. Sections, rather than chapters, should be regarded as the units of teaching material. A suggested teaching sequence is given later in the Introduction and is considered in more detail in *Notes for Teachers*, published separately.

It was our original intention to consider some topics in greater depth, and others were scheduled for inclusion. However, lack of space has forced us to confine the material mainly to that needed for modern 'O' level syllabuses. For those interested in the extension of some of the work beyond these rather narrow confines, we have produced *Further Topics in Investigating Chemistry* in which not only are some of the topics dealt with in the basic book taken further, e.g. the proton donor theory of acidity, further types of bonding and corrosion, but some additional topics such as social and economic factors in industry, radioactivity, fuel cells, organic nomenclature, and the structure of metals are introduced. This supplementary book should stimulate the more able pupils who always want to investigate further and will provide a useful transition from 'O' to 'A' level work as well as forming a basis for other types of post 'O' level science courses.

It is our experience that the bridge between 'O' level and more advanced work is not crossed by the great majority of textbooks, even when the gap is a comparatively narrow one. This is inevitable because textbooks are usually prepared to cater for specific courses. We feel that many of the sixth formers now staying on to study 'A' level sciences need a transition book which links some of the basic ideas and provides a platform for extension. We hope that *Investigating Chemistry* and *Further Topics in Investigating Chemistry* will meet these requirements.

The theory is interspersed with many experiments so that almost every aspect of the work is approached through the laboratory. In this respect we are indebted to all those in the teaching profession who have contributed to the evolution of chemistry teaching. We do not claim that many of the experiments are original. For example, some are derived from ideas in the Nuffield courses and others from the School Science Review or established texts, though we may have modified them slightly in the light of our own experience. We see chemistry teaching as an evolving interchange of ideas; teachers put old experiments into new situations, or use new interpretations. We are sure that many teachers using this book will further modify our ideas according to their own interests; this is what chemistry teaching is about.

We have suggested whether each experiment could be a demonstration, classwork, or both, but it is impossible to be dogmatic about this because such decisions depend upon the experience of the teacher, the equipment that is available, the age of the pupils, the ability of the pupils, and so on. If an experiment title is given 'one star' we consider that the experiment is suitable for classwork. Two stars indicate that the work could be either (or both) classwork or demonstration, and an experiment marked with three stars would normally be a demonstration. Teachers are responsible for the safety of pupils within a laboratory, and although the experiments used in the text have been conducted many times, it cannot be overemphasised, particularly to inexperienced colleagues, that even the most traditional or apparently trivial experiment is still a potential source of danger. It is essential that teachers familiarize themselves with reactions and manipulations before a lesson takes place, and that careful thought is given to laboratory management and potential hazards before *each* practical session. More details about safety factors and hints on practical work are given in *Notes for Teachers*.

Perhaps more important than the content of the book is the approach. Our aim has been to maintain a spirit of enquiry whilst providing a firm framework on which pupils can develop their own ideas. The work is designed to provoke thought and discussion; a considerable degree of guidance is given, but the investigational spirit is fostered by the withholding of the results and conclusions.

Every effort has been made to present chemistry as a living and developing science. Too often pupils have the idea that chemistry is just fact and that all has been discovered. We have tried to make it clear, wherever possible, that there are new frontiers to be crossed, loose ends to be tied, and that the chemistry pupil of today may become a research chemist of tomorrow. In any event, it is essential that in a technology-based democracy everyone should understand enough about chemistry to appreciate some of the vital decisions facing us in the future.

Modern nomenclature has been used throughout. Organic chemicals are given systematic names (unless they have very large molecules) followed by trivial names in brackets, but the trivial name is only used once in each piece of work. We are aware that this will not meet with everyone's approval, but we feel that it is important to avoid using a compromise between the two extremes which would result in some chemicals being given only the systematic name and others only the trivial name. Pupils brought up in this way certainly come to appreciate the advantages of a systematic nomenclature.

SI units have been used, but in some instances we have felt justified in making minor modifications. For example, reproductions of historical tables have retained their original units. Temperatures are quoted in K and °C except in the case of thermometer ranges specified under apparatus. Atmospheres have been retained in a few tables where the information is for interest or comparison and because we feel that in such cases the unit is more directly related to the experience of the pupils.

Notes for Teachers includes suggestions about using the book, notes on the teaching sequence that is outlined at the end of this introduction, comments on some of the experiments, and additional material which could be used, where time and circumstances permit. A † in front of an Experiment title indicates that more information about the experiment (e.g. safety factors) will be found in *Notes for Teachers.*

A SUGGESTED TEACHING SEQUENCE

No two people would ever teach the same syllabus in exactly the same way, and we have tried to present the work so that the order in which it is taught is flexible. The following scheme is one example of the way in which the material has been taught. It is a logical progression and the basic aim has been to introduce as many principles as possible early in the course, and then to use these in establishing the more factual concepts. In each topic we have tried to reserve a little 'depth' for later work, partly because it may be more easily appreciated with maturity and also to provide a final year containing a relatively small amount of new material, all of which, however, requires a revision of earlier work before it can be established. However, many such sequences are possible and these are likely to be further modified depending on the nature of elementary science courses, the time allocated to the teaching of chemistry, the resources and manpower available, etc. A more detailed consideration of the following scheme is in *Notes for Teachers.*

Topic 1 An Investigation into the Nature of Matter

1.1 Evidence for the Existence of Fundamental Particles
1.2 How Small are the Particles?
1.3 Particles in Motion. An Introduction to the Kinetic Theory

Topic 2 The Materials and Methods with which a Chemist Works

2.1 How Chemists obtain Pure Compounds
2.2 Criteria of Purity
2.3 Elements and Compounds
2.4 Compounds and Mixtures
2.5 An Investigation into the Nature of 'Liquid X'

Topic 3 Some Important Chemicals
3.1 Acids
3.2 Bases and Alkalis
3.3 Strong and Weak Acids and Alkalis. The pH Scale
3.4 Neutralization
3.5 Salts

Topic 4 An Investigation into the Nature of Matter (continued)
1.4 The Atomic Theory
(*Further Topics:* The Discovery of the Structure of the Atom)
1.5 A More Detailed Consideration of Atomic Structure. Atomic Mass and Molecular Mass. Isotopes
1.6 The Mole and Chemical Formulae.
1.7 Ionic and Covalent Bonding. Valency

Topic 5 Gases. Air, Oxygen, Hydrogen
4.1 The Preparation and Collection of Gases
4.2 The Air
4.3 Oxygen and the Oxides
4.4 Hydrogen

Topic 6 Some Electrochemistry
6.1 Electrolysis

Topic 7 The Periodic Classification
5.1 Historical Development
5.2 Electronic Structure and the Periodic Table
5.3 Trends Across a Typical Period
5.4 Trends Down a Metallic Group
5.5 Trends Down a Non-metallic Group

Topic 8 More Electrochemistry
6.2 The Electrochemical Series
6.3 Voltaic Cells and Rusting

Topic 9 Three-dimensional Chemistry
7.1 Some Basic Ideas
(*Further Topics:* Some Metallic and Other Giant Structures)
7.2 Some Giant Structures. Allotropy
7.3 Molecular Crystals
(*Further Topics:* The Prediction of Molecular Shapes)
7.4 The Shapes of Individual Molecules
7.5 Isomerism

Topic 10 Some Metal Chemistry
(*Further Topics:* A General Survey of the Properties of Elements)
10.1 General Preparations and Properties of Metallic Compounds
10.2 Some Reactive Metals. Sodium, Calcium, Magnesium and Aluminium

Topic 11 Rate of Change
14.1 Rate of Change
14.2 Reversible Reactions

Topic 12 Some Non-metal Chemistry
11.1 Carbon and Silicon
11.2 Nitrogen and Phosphorus

Topic 13 Oxidation and Reduction. Oxidation Number

Topic 14 Organic Chemistry
(*Further Topics:* The Different World of Organic Chemistry, Nomenclature, etc.)
12.1 Hydrocarbons, Alkanes, Alkenes and Alkynes
12.2 Alcohols

Topic 15 Energy in Chemistry
(*Further Topics:* More about Bonding)
13.1 What is Energy?
13.2 Some Determinations of Enthalpy Changes

Topic 16 A Further Study of Acids
(*Further Topics*)

Topic 17 Faraday's Laws
6.4 Faraday's Laws

Topic 18 Some of the Less Reactive Metals
10.3 Zinc, Iron, Lead and Copper

Topic 19 Water and Hydrogen Peroxide
9.1 Water, the Most Important Chemical
9.2 Hardness of Water. Soaps and Detergents
9.3 Hydrogen Peroxide

Topic 20 Some More Non-metal Chemistry
11.3 Oxygen and Sulphur
11.4 The Halogens, with Particular Reference to Chlorine

Topic 21 Large Molecules
12.3 Large Molecules

Topic 22 The Chemical Industry
(*Further Topics:* Social and Economic Factors in Industry)
14.3 Some Important Industrial Processes

Calculations, Chemical laws, etc.
We feel that these, and also some of the work in *Further Topics,* can be introduced in stages at almost any part of the course according to the discretion of the teacher and the interests of the class. For this reason they have not been included in the suggested sequence. Most of the quantitative chemistry required is in Chapter 15 and we prefer to leave the greater part of this until fairly late in the course so that the interpretation of equations etc. revises much earlier, basic work, and because the calculations then seem easier to the pupils.

ACKNOWLEDGEMENTS
Some of the questions at the end of each chapter are reproduced by kind permission of the following examining boards:

The Associated Examining Board (A.E.B.)
University of Cambridge Local Examinations Syndicate (C.)
The Joint Matriculation Board (J.M.B.)
The Welsh Joint Education Committee (W.J.E.C.)

We thank these examining bodies for allowing us to use their questions, and the origin of such questions is acknowledged separately in the text.

We should also like to thank the many firms, organizations and individuals who have so generously provided information and material. In particular, our thanks are due to Richard Imrie for the cover photograph, the Marchon Division of Albright and Wilson Ltd. (Whitehaven), Unilever Ltd., The British Steel Corporation (Workington), British Nuclear Fuels Ltd., United Kingdom Atomic Energy Authority, The British Petroleum Co. Ltd., The Shell Oil Company, Pilkington Brothers Ltd., Imperial Smelting Processes Ltd., and Imperial Chemical Industries Ltd. Photographs provided by these organizations have been acknowledged separately in the text.

If any names of organizations, individuals or books have been inadvertently omitted from the above list we extend our sincere apologies.

We would like to thank Martyn Berry and J. P. Chippendale for their detailed reading of the typescript and many helpful comments. We would also like to thank our Publishers, particularly Mr Hamish MacGibbon, for their advice and patience; and also the wives and husbands who have had to endure so much during the preparation of this book.

L.D.
M. J. D.
A. W. L.
M. E. R.

1973

Contents

INTRODUCTION v

Chapter 1 An Investigation into the Nature of Matter

1.1 Evidence for the existence of fundamental particles 1
1.2 How small are the particles? 5
1.3 Particles in motion. An introduction to the kinetic theory 9
1.4 The Atomic Theory 16
1.5 A more detailed consideration of atomic structure. Atomic mass and molecular mass. Isotopes 21
1.6 The mole and chemical formulae 24
1.7 Ionic and covalent bonding. Valency 30
Questions on Chapter 1 40

Chapter 2 The Materials and Methods with which a Chemist works

2.1 How chemists obtain 'pure compounds' 43
2.2 Criteria of purity 58
2.3 Elements and compounds 62
2.4 Compounds and mixtures 64
2.5 An investigation into the nature of 'liquid X' 66
Questions on Chapter 2 68

Chapter 3 Some Important Chemicals

3.1 Acids 70
3.2 Bases and alkalis 76
3.3 Strong and weak acids and alkalis. The pH scale 79
3.4 Neutralization 82
3.5 Salts 87
Questions on Chapter 3 98

Chapter 4 Gases. Air, Oxygen, Hydrogen

4.1 The preparation and collection of gases 100
4.2 The air 106
4.3 Oxygen and the oxides 114
4.4 Hydrogen 119
Questions on Chapter 4 125

Chapter 5 The Periodic Classification

5.1 Historical development 127
5.2 Electronic structure and the Periodic Table 132
5.3 Trends across a typical period 135
5.4 Trends down a metallic group 140
5.5 Trends down a non-metallic group 143
Questions on Chapter 5 148

Chapter 6 Electrochemistry

6.1 Electrolysis 151
6.2 The electrochemical series 167
6.3 Voltaic cells and rusting 177
6.4 Faraday's Laws 179
Questions on Chapter 6 183

Chapter 7 Three-dimensional Chemistry

7.1 Some basic ideas 185
7.2 Some giant structures. Allotropy 192
7.3 Molecular crystals 194
7.4 The shapes of individual molecules 198
7.5 Isomerism 200
Questions on Chapter 7 202

Chapter 8 Oxidation and Reduction. Oxidation Number 204

Questions on Chapter 8 211

Chapter 9 Water. Soaps and Detergents. Hydrogen Peroxide

9.1 Water, the most important chemical 213
9.2 Hardness of water. Soaps and detergents 216
9.3 Hydrogen peroxide 228
Questions on Chapter 9 230

Chapter 10 The Metals and their Compounds

10.1 General preparations and properties of metallic compounds 231
10.2 Some reactive metals. Sodium, calcium, magnesium and aluminium 245
10.3 Some of the less reactive metals. Zinc, iron, lead and copper 254
Questions on Chapter 10 266

Chapter 11 The Chemistry of some Important and Characteristic Non-metals

11.1 Carbon and silicon 268
11.2 Nitrogen and phosphorus 277
11.3 Oxygen and sulphur 288
11.4 The halogens, with particular reference to chlorine 297
Questions on Chapter 11 302

Chapter 12 An Introduction to Organic Chemistry
12.1 Hydrocarbons. Alkanes, alkenes and alkynes 305
12.2 Alcohols 321
12.3 Large molecules 329
Questions on Chapter 12 341

Chapter 13 Energy Changes in Chemistry
13.1 What is energy? 343
13.2 Some determinations of enthalpy changes 349
Questions on Chapter 13 358

Chapter 14 Rate of Change. The Chemical Industry
14.1 Rate of change 360
14.2 Reversible reactions. Equilibria 367
14.3 Some important industrial processes 381
Questions on Chapter 14 396

Chapter 15 Chemical Measurements. Volumes and Masses
15.1 Chemical formulae. How they are worked out and what we can learn from them 399
15.2 Volumetric analysis 403
15.3 Volume changes in gases 407
15.4 Chemical equations 417
Questions on Chapter 15 425

INDEX 429

1 An Investigation into the Nature of Matter

1.1 EVIDENCE FOR THE EXISTENCE OF FUNDAMENTAL PARTICLES

Introduction

'What is it made of?' This is a question asked by most small children almost as soon as they can talk and even before this they try to take to pieces anything they can lay hands on. This may be thought to be sheer destructiveness but is more likely to be natural curiosity. Down the ages it has been this curiosity which has led men to investigate the nature of the world around them. It has inspired great discoveries and led to striking advances in our knowledge.

young scientist investi-es the nature of matter!

Chemistry is primarily concerned with what things are made of and how various substances react with each other. One of the chemist's most important tasks is to find out as much as possible about the nature of matter. By matter we mean something which occupies space and has mass. If you are a chemistry student you are also a chemist and in this first chapter you will begin to learn something of the way in which a chemist works; how he experiments, records and deduces, how he works out a theory and then tries to prove it by producing as much evidence as possible in its favour, but is always ready to abandon one theory for another if he finds that it fits the facts more closely.

You may sometimes think as you read on that you have picked up a history book by mistake, but it is important for you to know how the knowledge of certain aspects of the subject has grown and developed, for in chemistry, as in other branches of knowledge, men advance by climbing on to the shoulders of their predecessors.

Much important work has been done in the past but during this century the pace of research and discovery has accelerated at a phenomenal rate. These are exciting times for the scientist and future generations of workers in the field of chemistry will find that there is much to be learnt and many new fields of research to choose from. Some time in the future you may use the knowledge and experience of some eminent present-day chemists to help you in a new branch of research, and you will perhaps have cause to be grateful for the work of those scientists who have gone before you.

Can We Prove the Existence of Atoms?

One of the main discoveries of the earlier scientists was that complex substances can be broken down into simpler ones. The substances which cannot be split up into anything simpler by chemical means are called elements. About ninety of these elements occur in nature and others have been man-made so that over a hundred have now been recognized.

For the moment we are not concerned with the elements, which are studied more fully in Chapter 2. We think that a chemistry course should start with fundamentals and so the early part of this book deals mainly with the tiny particles of which the elements themselves are composed. These particles are called atoms, as you probably already know.

It is almost impossible to have lived in this modern world without having heard of atoms, but when we begin to think about the problem of *proving* that matter is made up of atoms we find that we have quite a difficult task. We have to demonstrate that particles exist which are too small to be seen by even the most powerful light microscope, a problem which baffled scientists for many hundreds of years.

The idea that all matter is composed of tiny particles is not new. The word atom comes from the Greek word 'atomos' which means indivisible and the Greeks, as long ago as 500 BC, were arguing as to whether it was possible to go on dividing matter indefinitely or whether an end would finally be reached with a fundamental particle which could not be divided further. Because they were debating about something extremely small, no direct evidence could be produced by either side and the argument for and against the existence of fundamental particles has lasted down the centuries.

Even in a present-day school chemistry laboratory we have no direct way of proving the existence of atoms, but from the knowledge we already have of matter and its behaviour and from a few simple experiments we can produce quite a mass of evidence which supports the view that matter is made up of tiny particles. As your knowledge and experience of chemistry increases, the weight of evidence in favour of the particulate theory of matter will grow, and you will accept the idea as a practical concept and use it as a good working hypothesis.

You may have taken part in a competition in which you have had to guess what the object is inside a sealed box. By moving the object round and shaking the box you can determine whether the object is round or square, hard or soft, and soon you will have a fairly good picture of the unseen object. By accumulating evidence in this indirect way you will probably be able to identify the object as, for example, a rubber, a magnet, or a cork. In a similar manner the experiments you will do in this section will help to accumulate evidence in favour of the particulate theory of matter. Each experiment taken separately contributes very little to the argument, but the results as a whole can only be interpreted satisfactorily if we accept the idea that matter is made up of very small particles.

Some Introductory Experiments

Experiment 1.1
To Examine how a Dissolved Solid Behaves in a Solvent *

Apparatus
Litre flask, four test-tubes, white paper. Potassium permanganate.

Procedure
Half fill the litre flask with water and drop in two or three crystals of potassium

permanganate. Swirl the water round until the crystals dissolve. Fill two clean test-tubes with the liquid from the flask and view them against a sheet of white paper. Almost fill the flask with water, swirl the solution for a short time and repeat the experiment with the test-tubes.

Points for Discussion

1. Did the potassium permanganate spread evenly through the water?
2. If potassium permanganate is made up of tiny particles all of the same size, would you expect it to behave as it did in the water?
3. Can you explain the result you observed in any other way?

Experiment 1.2
† A Simple Experiment with Ethoxyethane ***

Apparatus
Watch-glass.
Ethoxyethane (diethyl ether).

Procedure
Pour a few drops of ethoxyethane on to a watch-glass and if possible go outside the room for a few minutes, leaving the watch-glass on the bench. When you return go to different parts of the room and notice if you can smell the ethoxyethane. Observe what has happened to the ethoxyethane on the watch-glass. Try to estimate the volume of the room in cubic centimetres.

Points for Discussion

1. Compare the starting volume of ethoxyethane with the volume it occupied after evaporation. By how many times had it 'spread out'?
2. We did nothing to the liquid to make it 'come apart' in this way. Perhaps it already consisted of tiny particles which, given the chance to do so, separate from each other and spread across the room. If the liquid was not already divided up in some way it is difficult to imagine how it could 'disintegrate' so evenly of its own accord.

Experiment 1.3
† Two Into One Will Go *

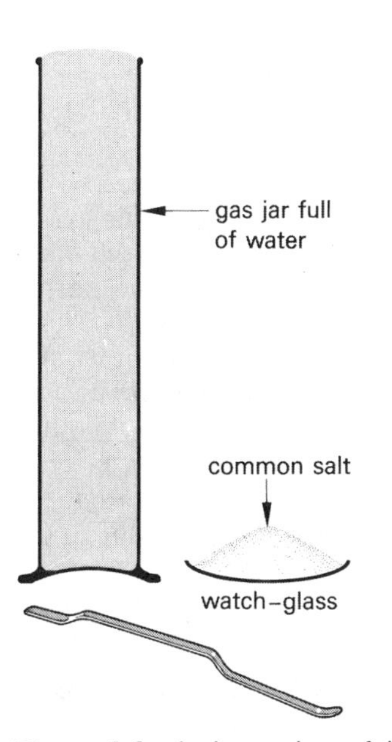

Figure 1.1 Is the gas jar as full as it appears?

Apparatus
Small gas jar, spatula.
Sodium chloride (common salt).

Procedure
Pour water into the gas jar until it is *completely* full. Add a spatula measure of salt. Continue to add salt to the contents of the gas jar by spatula measures until the liquid overflows (Figure 1.1).

Points for Discussion

1. Was the jar as 'full' as you thought?
2. Could you add grains of rice to a jar 'full' of marbles?
3. What spaces did the salt fill up?
4. Can you explain what happened in terms of salt and water particles?

Experiment 1.4
†A Surprising Change in Volume***

Apparatus
Burette with rubber bung.
Ethanol.

Procedure
(a) Pour water into the burette until it is nearly half full. Carefully add ethanol until the total volume is on the top mark.
(b) Close the burette mouth with a bung and invert the burette several times so that the liquids are thoroughly mixed.

Note the level of the mixture and compare it with the level before the liquids were mixed.

Points for Discussion
1. Try to explain what has happened.
2. Would it be correct to say that the ethanol particles have gone between the water particles? Give a reason for your answer.

3. The experiment can be explained if we imagine that chemicals consist of tiny particles; that these differ in size from one chemical to another, and that when chemicals are mixed their particles can pack more closely together to make use of the available space. As an extreme example it could be imagined that small particles of one chemical could fit into the spaces between the larger particles of a different chemical. Consider the addition of grains of rice to a jar full of marbles as an example; some if not all the rice would go into the spaces between the marbles. This does not prove that particles exist, but it is difficult to explain what happens in any other way.

Can the particle theory explain change of state?

One final experiment can be done before we leave this part of the subject. It is a very simple experiment that you will have done in one form or another many times before but this time you will be carrying out the work in the light of the particle theory.

Experiment 1.5
†Water, Ice, Steam*

Apparatus
100 cm^3 beaker, 250 cm^3 beaker, tripod, gauze, asbestos square, Bunsen burner, small round bottom flask, watch-glass. Ice, ether.

Procedure
Place the piece of ice in the small beaker. Stand this beaker in the larger beaker containing hot water. When the ice has melted, heat the water in the small beaker until it boils and condense the steam by holding a flask of cold water over the beaker. Catch some of the condensed water on a watch-glass and leave it in a refrigerator for a short time.

Points for Discussion
1. Compare what happened when the water turned to steam with the experiment you did earlier with the ethoxyethane.
2. Did you know that one kettleful of water will produce about 1700 kettlefuls of steam? Can you account for this?
3. Have you tried the effect of a drop of ethoxyethane on the back of your hand? Can you explain what happens?
4. What was needed to turn the ice to water and the water to steam?

Another Example of Change of State

You will have seen a big football match on television or in reality and will remember how the military band stands perfectly still and plays before the match, each bandsman in exactly the right place so that the whole band forms a solid square, yet is made up of individual soldiers. Then at a word

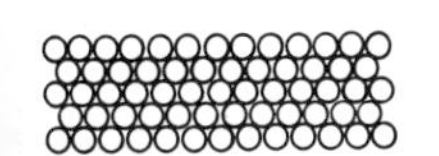

band stands and plays in positions
ticles in a SOLID)

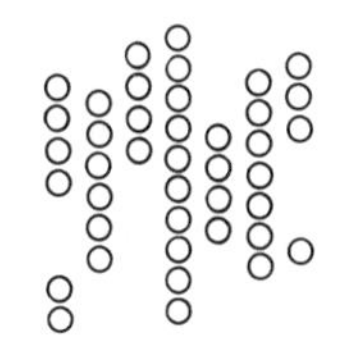

band starts marching
hin a fixed area
ticles in a LIQUID)

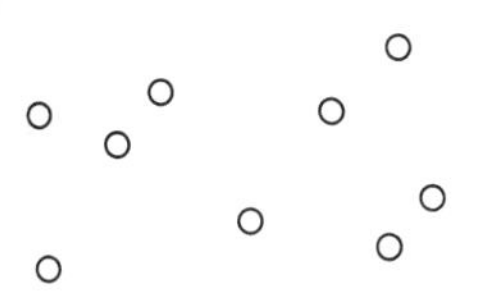

band disperses to various
ts of the field
rticles in a GAS)

ure 1.2 The 'states of natter' band

of command the band begins a complicated series of movements, crossing and recrossing, forming different patterns and shapes but still remaining a single entity. You will probably have recognized the comparison with ice and water, but when does the band resemble the behaviour of steam? Perhaps when they disperse to different parts of the field (Figure 1.2).

A Summary of the Evidence in Favour of the Particle Theory

You should now begin to understand how the accumulation of evidence in favour of a theory gives more and more support to its validity. The first experiment was meant to demonstrate that the potassium permanganate was spread out *evenly* through the water and this could only happen if it was made up of similar tiny particles. The ether was spread out in the air even more thinly than the potassium permanganate in the water. Can you suggest why?

The addition of salt to a completely 'full' jar of water can be explained if we suppose that the salt particles and water particles can pack together to make more use of the available space and a similar rearrangement of particles could have occurred with the ethanol and water.

Similarly the change of state of water into steam or ice is much easier to understand if we think of water as consisting of small particles which can spread out or come closer together according to the energy they possess.

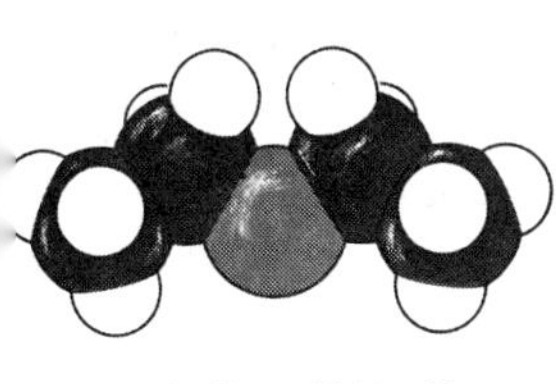

igure 1.3 A molecule of ethoxyethane

Matter is Made of Particles. A Working Hypothesis

We will now assume that the evidence we have produced has convinced you that all matter *is* made up of small particles. Scientists everywhere accept that this is so and we will use this concept throughout our work. At the beginning of the section we mentioned atoms. An atom is the smallest particle into which an element can be divided and still keep its individuality. You will probably have heard of another particle called a *molecule*. A molecule is a group or cluster of atoms. The smallest particles of the ethoxyethane vapour are molecules, each consisting of carbon, hydrogen, and oxygen atoms (Figure 1.3). Another small particle that you will be learning about is called an *ion*. An ion is an atom or group of atoms which possesses an electric charge.

If we assume that ice, liquid water, and steam are all made up of molecules, it is obvious that they are held most closely and tightly together in the ice and least tightly and furthest apart in the steam. Many substances can exist in turn in the solid, liquid, and vapour state and the same reasoning can be applied to these substances. So here is another point to think about: the tiny particles which make up all matter, whether they be atoms, molecules or ions, rarely exist as free particles but are held together in some way. If you refer to your answers to the last two questions in the previous Points for Discussion you will realize that considerable energy is needed to pull the particles apart. Until you become more familiar with the correct use of the terms *atoms*, *molecules* and *ions* it is better to use the general term *particles* when describing the fundamental units of matter. For example, it is quite wrong to call a particular particle an atom when it is really an ion or a molecule.

1.2 HOW SMALL ARE THE PARTICLES?

Although it is not possible to measure the size of individual particles accurately in a school laboratory, the following experiments are designed to give you some idea of the order of size of these minute fragments of matter.

Experiment 1.6
† The Divisibility of Matter ***

Apparatus
Burette and stand, measuring cylinder, graduated flasks.
1 litre of potassium permanganate solution containing 1 g $litre^{-1}$.

Procedure
Measure out 10 cm^3 of the solution. We will call it the dark purple solution. Pour it into a litre flask and add water until the total volume is 1 litre. Invert the flask several times to mix the solution. You have now diluted the deep purple solution 100 times. Pour 5 cm^3 of this solution into a graduated flask and make it up to 1 litre as before. This dilutes the solution (rose pink?) 200 times. It should now be pale pink. Pour some of this solution into a burette and allow it to run out a drop at a time. Count the number of drops in 1 cm^3 of the solution.

Can you work out the weight of potassium permanganate contained in one drop of the pale pink solution?

Your working could be something like this:

In 1000 cm^3 of the dark purple solution there is 1 g potassium permanganate.

In 1 cm^3 of the dark purple solution there is 1/1000 g potassium permanganate.

In 1 cm^3 of rose pink solution there is $1/1000 \times 1/100$ g potassium permanganate.

In 1 cm^3 of pale pink solution there is $1/1000 \times 1/100 \times 1/200$ g potassium permanganate.

Suppose you find that 1 cm^3 of the pale pink solution contains twenty-five drops. In one drop of this solution there is

$1/1000 \times 1/100 \times 1/200 \times 1/25$ g of potassium permanganate.

$= 1/500\,000\,000$ g of potassium permanganate.

The original gramme of potassium permanganate has been divided into at least 500 million parts (Figure 1.4).

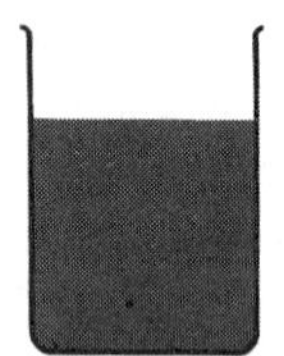

Dark purple solution one drop contains $\frac{1}{25\,000}$ g of potassium permanganate

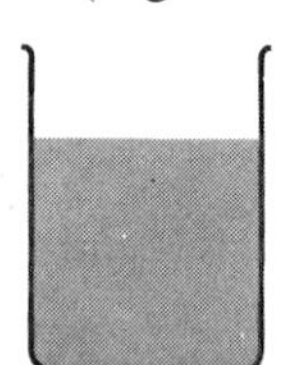

Rose pink solution one drop contains $\frac{1}{2\,500\,000}$ g of potassium permanganate

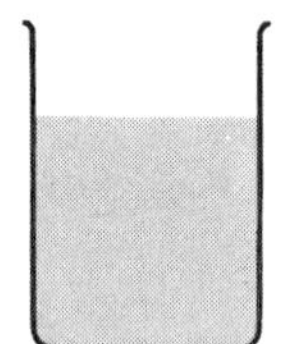

Pale pink solution one drop contains $\frac{1}{500\,000\,000}$ g of potassium permanganate

Figure 1.4 Dilution of a solution of potassium permanganate

Experiment 1.7
† How Far Will an Oil Drop Spread Out? *

Apparatus
Teat pipette, large bowl.
Oil (oleic acid dissolved in petroleum ethoxyethane), tin of talcum powder or flowers of sulphur.

Procedure
Half fill the bowl with water and sprinkle the surface lightly with talc or sulphur. Using the teat pipette allow one drop of the oil to fall on the centre of the water surface.

Points for Discussion

1. Why did the mixture spread out and then contract slightly?
2. Was the film left on the surface just oil, just ethoxyethane, or both?
3. What was the purpose of the talcum powder?
4. Why was the oil diluted with ethoxyethane?
5. Why does the oil drop stop spreading?
6. In terms of particles, what will be the thickness of the oil layer?

The oil drop leaves the pipette

It reaches the surface of the water and starts to spread out

It spreads further

It spreads to the limit
– one molecule thick

Figure 1.5 The spreading oil drop

Experiment 1.8

†To Find the Size of a Single Lead Shot*

Apparatus

Small rectangular tray, measuring cylinder. Lead shot.

Procedure

Spread the lead shot over the tray so that it is completely covered and nowhere more than one shot thick. Calculate the area of the base of the tray in cm². Pour the lead shot into a measuring cylinder and find its volume in cm³ (Figure 1.6).

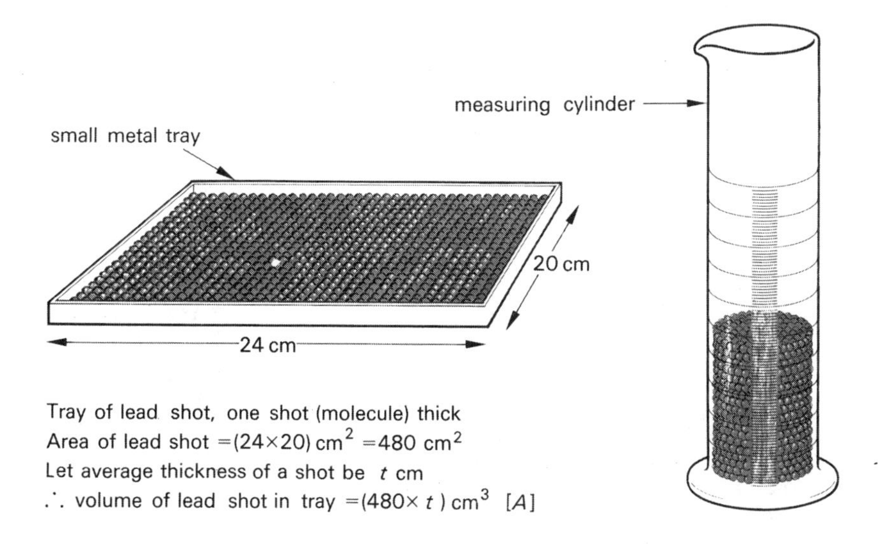

Figure 1.6 Finding the size of a single lead shot

We can imagine the lead shot in the tray to form a rectangular shape just one particle thick.

Points for Discussion

1. It may be possible to measure the size of a lead shot directly, but suppose you were not able to do this, i.e. that a lead shot was like a molecule in this respect. How is the thickness of the layer of shot in your tray related to its area and volume?
2. Use your results and your answer to point 1 to calculate the average size of a lead shot.

Experiment 1.9
† To Find the Size of an Oil Molecule*

Principle

From the last two experiments you have learnt that a layer of oil spreads out until it is one particle (molecule in this case) thick, and that you can find the thickness of a *very* thin layer if you know its area and the volume that it occupies.

The thickness of the film of oil is, therefore, the thickness of a molecule of oil. If we think of the molecule of oil as a tiny sphere, we can find its thickness in the same way as we found the thickness of a lead shot, although in this case we will be working with many millions of molecules, each one so tiny as to be invisible under the most powerful light microscope.

Apparatus

Bowl, teat pipette, burette, burette stand.
Talcum powder, a solution of oleic acid (the oil) in petroleum ethoxyethane, containing 0.1 cm^3 of oil per 1000 cm^3 of solution.

Procedure

You are going to repeat the experiment with the oil drop but this time you will need to know its exact volume.

(a) *Finding the volume of oil in one drop of oil–ether mixture.*

Nearly fill the burette with water and note the volume. Slowly add drops of water from the teat pipette to the water in the burette, counting the drops until you have transferred exactly 1 cm^3 of water. Make a note of the number of drops used. This will be very similar to the number of drops of oil in 1 cm^3.

(b) *Finding the area of a monomolecular layer of the oil which was present in one drop of the oil–ethoxyethane mixture.*

Wash out the pipette with the oil solution and refill it. Scatter a thin layer of talc over the surface of the water with which you have half filled the bowl. Using the pipette allow one drop of oil to fall on to the centre of the talc surface. The ethoxyethane evaporates leaving a monomolecular film of oil, and the area covered by this oil film will be clearly visible as it spreads out and pushes the talc aside.

Measure the dimensions of this clear space in centimetres and calculate its area in square centimetres. The shape of the space may not be very regular but you should be able to make a rough approximation of the total area (Figure 1.7) and this is all that is needed.

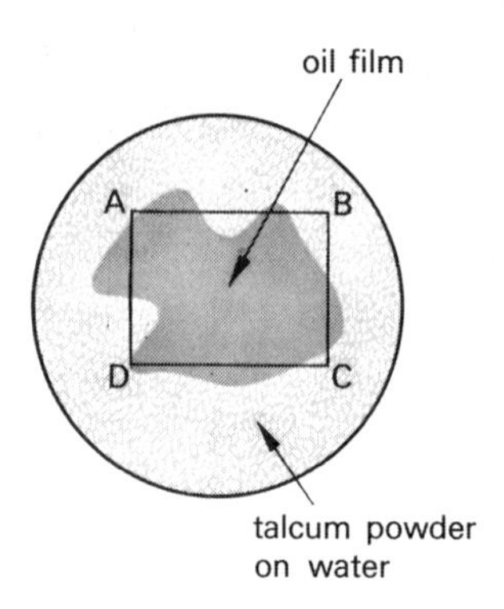

Figure 1.7 Estimating the area occupied by the oil film

(c) *The calculation*
You now have all the data needed to find the thickness of the oil layer (T cm).

Suppose your teat pipette delivers 25 drops to 1 cm^3.

Volume of oil in 1 litre of solution
$= 0.1\ cm^3$

Volume of oil in 1 cm^3 of solution
$= 0.1/1000\ cm^3$

Volume of oil in one drop of solution
$= (0.1/1000 \times 1/25)\ cm^3$

Using the formula:

thickness (T) = Volume/area (A)

$T = (0.1/1000 \times 1/25 \times 1/A)$ cm

use *your* results to calculate the size of an oil molecule.

Some Comparisons of Atomic Size with Everyday Objects

You were probably surprised that you were able to measure the size of such a very small particle as an oil molecule and perhaps even more surprised at how extremely small such molecules are. Yet a molecule of oleic acid is comparatively 'large' as it contains fifty-four atoms. Single atoms and simple molecules must obviously be very much smaller than an oleic acid molecule. The diameter of an atom is of the order of $1/10^8$ cm. You probably find it difficult to envisage such an extremely tiny particle but some of the following comparisons may help you to understand better the world of the very small.

If a drop of water could be magnified to the size of the earth, then on the same scale its constituent atoms would be as large as golf balls.

Two million hydrogen atoms would cover an average full stop.

In 1000 cm^3 of water there are as many atoms as there would be grains of sand if the whole of the earth's surface, continents and oceans, were covered with a layer of sand, one foot thick.

See how many more comparisons of this kind you can find.

1.3 PARTICLES IN MOTION. AN INTRODUCTION TO THE KINETIC THEORY

It will be clear from some of the previous experiments, such as the spreading out of the ethoxyethane and steam, that the particles, at least of a gas, are moving. The next set of experiments will help you to learn more about these moving particles.

Movement of Particles in Gases, Liquids, and Solids

Experiment 1.10
The Movement of Bromine Molecules in Air ***

Apparatus
Two gas jars, white paper or card, cover glass, teat pipette.
Bromine. (This chemical must only be used by the teacher. Gloves and goggles, should be used when working with bromine, and a fume cupboard is essential.)

Procedure
(a) Using the teat pipette place one drop of bromine at the bottom of a gas jar and immediately cover the jar. Bromine is volatile and quickly vaporizes. Watch the passage of the bromine up the jar by placing a piece of white paper behind it. How

long does the bromine take to reach the top of the jar (Figure 1.8(a))?
(b) Quickly remove the cover glass and invert a gas jar of air over the first gas jar. Watch and record the further progress of the bromine up the second gas jar (Figure 1.8(b)).

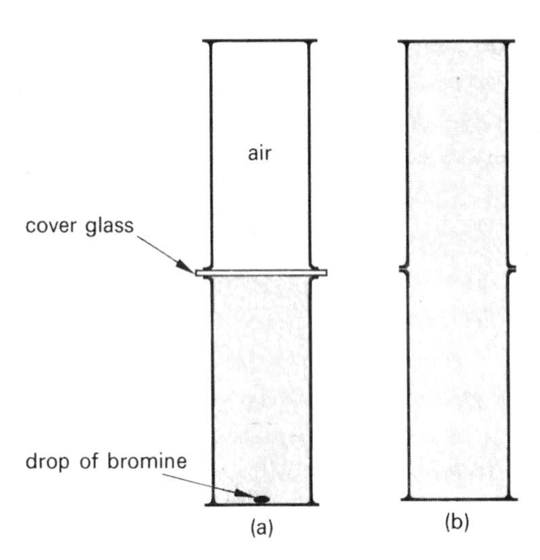

Figure 1.8 The movement of bromine particles in air (a) with cover glass separating gas jars and (b) five minutes after cover glass is removed

Points for Discussion

1. Why does the bromine vapour rise, even though it is denser than air? Why does the bromine not remain as a drop of liquid?

2. If the second jar had been completely empty (i.e. evacuated) would the bromine vapour have moved

A more slowly; B more quickly; C at the same rate

as when the jar was filled with air? Explain how you reached your answer.

3. It would appear that particles in liquids and gases are continuously on the move. Note that the unceasing movement of one set of particles within another produces uniform mixing of the liquids and gases. This movement of particles from a region of high concentration to one of lower concentration is known as *diffusion.* Use the word (or its derivative, e.g. diffused) to explain what happened in the experiment. Note that more than one substance diffused.

Experiment 1.11
The Movement of Copper(II) Sulphate Particles in a Liquid

Apparatus
Gas jar or measuring cylinder, pipette. Concentrated copper(II) sulphate solution, polystyrene sphere.

Procedure
Pour copper(II) sulphate solution into the gas jar or measuring cylinder until it is about two-thirds full. Float the polystyrene ball on the surface of the solution. By means of the pipette run water on to the polystyrene ball so that it forms an upper layer to the top of the jar. Remove the ball when the container is full and stand the jar or measuring cylinder on a shelf where it can remain undisturbed for as long as is required.

Points for Discussion

1. What happens to the 'blue substance' in the lower layer?

2. What general term is used for this movement?

3. What conclusion can you make regarding the movement of particles in a gas compared with particles in a liquid?

Experiment 1.12
†Movement of Particles in Solids? ***

Apparatus
Watch-glass, top pan balance.
Camphor block.

Procedure
Place the block of camphor on a watch-glass and find the total weight. Leave for about half an hour. Weigh the camphor and watch-glass again.

Points for Discussion
1. Compare the two weighings.
2. How can you explain any difference in terms of movement of the particles of camphor on the surface?
3. This is strictly *not* diffusion in solids, but diffusion of particles from the surface of the solid into the air. Diffusion of particles within the solid itself is extremely slow.

Brownian Movement

An interesting and historic experiment on the movement of particles was first carried out by the botanist Robert Brown in 1827. He made his discovery by accident when he was observing some pollen grains suspended in water under a microscope. He noticed that the pollen was moving about in an erratic manner. As he had not seen any similar movement when examining larger and heavier particles and as he was able to discount the effects of draughts and convection currents, Brown concluded that the movements 'belonged to the particles themselves'. Brown's observations were later explained by C. Wiener in 1863 as being due to bombardment of the pollen grains by the much smaller but rapidly moving water particles.

This movement of visible particles caused by smaller but invisible ones is known as Brownian movement (Figure 1.9). You may be shown Brownian movement in a liquid.

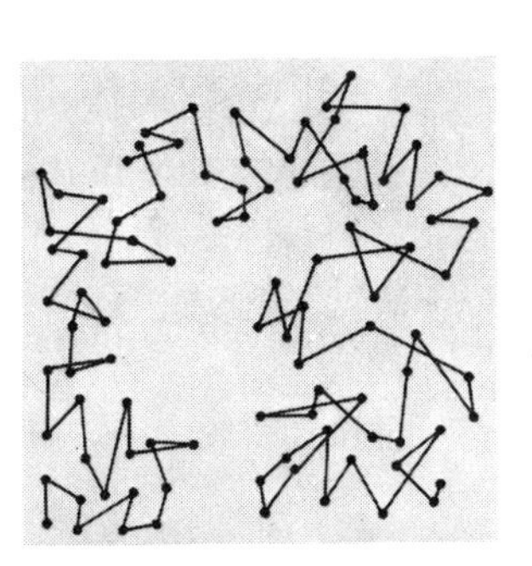

Figure 1.9 Brownian movement. This shows the path taken by a single particle suspended in water. The position of the particle was recorded at ten second intervals

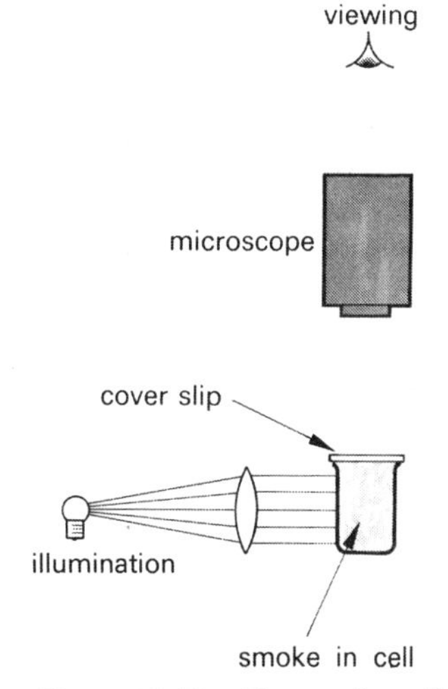

Figure 1.10 The smoke cell

Brownian Movement in Air

An excellent method of showing Brownian movement is by the 'Smoke Cell' (Figure 1.10). A specially constructed piece of apparatus allows smoke particles, illuminated from the side, to be viewed under a microscope. The light is reflected from the smoke particles (very small fragments of carbon) and numerous bright pin points of light can be seen as the smoke particles are pushed in all directions. Can you explain what causes them to move in this way?

The Movement of Molecules of Gases

You will have already decided from previous experiments that the movement of the molecules of a gas is much more rapid than that of particles in a liquid, and your next question could be, 'How fast?'. From the evidence that bromine vapour takes about three minutes to travel thirty centimetres you might conclude that the particles are moving at the rate of the traditional snail. The next experiment will help you to find more exactly how long it takes molecules of gases to move from one place to another.

Experiment 1.13
To Compare the Rates of Movement of Two Gases ***

Apparatus
Glass tube about 90 cm long and 2.5 cm diameter, supported horizontally by two retort stands, two rubber bungs to fit the tube (Figure 1.11), two test-tubes (150 × 25 mm), two test-tube stands, cotton wool.
Concentrated hydrochloric acid, concentrated ammonia solution.

Procedure
(a) Almost fill one test tube with the acid and stand it at one end of the bench. Make a small ball of cotton wool, and using tweezers dip it in the acid. Remove the cotton wool and hold it over the test tube so that surplus acid drips back into the tube. When it has stopped dripping, insert the cotton wool into one end of the long tube and stopper it with a bung.
(b) Repeat the procedure as quickly as possible with another ball of cotton wool, another pair of tweezers, but dipping the cotton wool this time into the *other* test tube which has been almost filled with ammonia solution. Insert the ammonia-soaked piece of cotton wool into the *other* end of the tube and stopper with another bung. (It is an advantage to insert both pieces of cotton wool at the same time, so that both gases start to diffuse at the same instant.)
(c) Leave the tube undisturbed for some time, but watch for any developments. Note the time taken for any change to occur.

Points for Discussion
1. The liquids themselves are not important in this experiment; it is the gases which come from them that we are concerned with. When ammonia gas (NH_3) meets hydrogen chloride gas (HCl), they join together to form the white solid, ammonium chloride (NH_4Cl). You will see this demonstrated

i.e. $NH_3(g) + HCl(g) \longrightarrow NH_4Cl(s)$

2. Where did the two gases meet? Have they moved at equal speeds?

3. Measure the distance that each gas has moved, and knowing the time it took them to meet, calculate the apparent speed of each gas in metres per second.

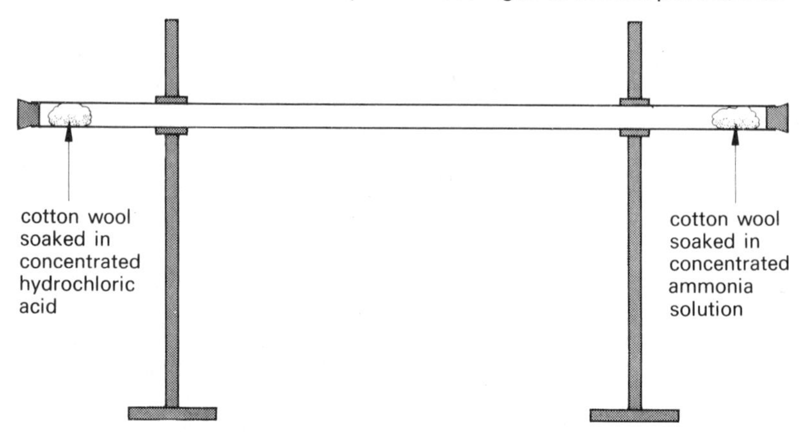

Figure 1.11 To compare the rate of movement of two gases

The Enormous Speeds of Gas Molecules

trying to get to you ...monia!

You probably found from your last experiment that the gases were moving along the tube at rates of approximately 1/800 metres per second and so you may be surprised to learn that the molecules of ammonia actually move at an average speed of 550 metres per second and those of hydrogen chloride at 450 metres per second. (Yes, you were correct in finding that the two gases move at different speeds.) How is it then, that a gas molecule, moving at a rate of 500 metres per second, takes twenty minutes to move one metre? The answer is, of course, that we are not dealing with one molecule in isolation but with countless millions, all moving in straight lines at enormous speeds and in all possible directions, continually bumping into each other and rebounding. Our ammonia molecule has to fight its way down the tube impeded by other ammonia molecules and also by the oxygen and nitrogen molecules from the air already there. This is rather like having to make your way down the subway of a tube station in the rush hour in the opposite direction to the majority of the crowd of people. No wonder the ammonia molecules and the hydrogen chloride molecules took so much time to reach each other. The actual speed of some gas molecules is summarized in Table 1.1.

Table 1.1 Actual speed of some gas molecules at 273 K (0 °C)

Hydrogen	1823 metres per second
Oxygen	460 metres per second
Nitrogen	493 metres per second
Carbon dioxide	390 metres per second

This means that the molecules in air at room temperature are travelling on average a quarter of a mile per second. In each second one molecule makes about 10^9 collisions; more than the ticks of a second hand of a watch in thirty years.

The Kinetic Theory

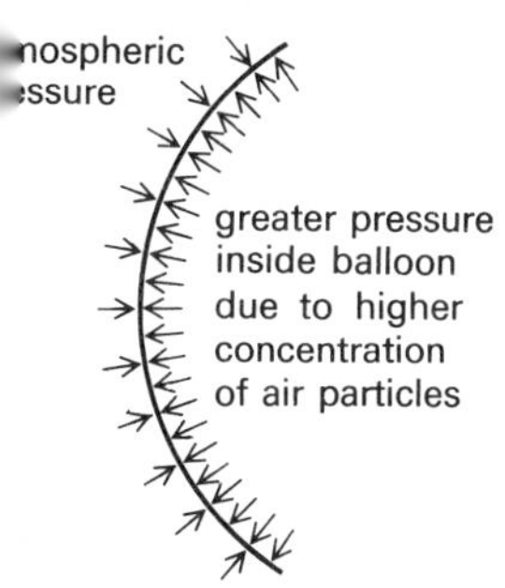

...gure 1.12 How air pressure keeps a balloon inflated

The energy possessed by particles of solids, liquids, and gases is called kinetic energy (energy of motion). So far we have not said much about the movement of the particles of a solid. This is because this movement is not apparent, as the particles do not change their relative positions, although they do rotate and vibrate. Particles in liquids and gases also rotate and vibrate and in addition move from place to place. The effects of these *translational movements* can be observed.

The theory which deals with the energy due to particle movement is the *kinetic theory*. It is still called a theory as we have no direct evidence of the movement of these submicroscopic particles. As the movement of molecules of gases has so much more effect on their properties than in the case of liquids and solids, the applications of the kinetic theory are, in the main, confined to gases.

Applications of the Kinetic Theory

Gas Pressure

You will have known since your early days that gases exert pressure. Your first balloon, your bicycle, the family car, will have made the fact familiar to you. You will now begin to understand what causes this pressure. It is the constant bombardment of the gas molecules against the inside of the balloon or tyre. The balloon is also pounded by air molecules on the outside but the hits per unit area are greater on the inside because there the gas is more concentrated (Figure 1.12).

We shall now see how, using the kinetic theory, we can predict how gases will behave under certain conditions and whether these predictions are borne out by experimental results.

The Effect of Change of Temperature on Gas Pressure

When we gave you a list of the speeds of various gases we were careful to mention that these speeds were at 273 K (0 °C). This is important because a change of temperature makes a considerable difference to the velocity

Table 1.2 The effect of temperature on the average velocity of molecules of gases

	273 K (0 °C)	*373 K (100 °C)*	*1093 K (820 °C)*
Hydrogen	1823	2430	3626
Oxygen	460	538	920
Nitrogen	493	576	986
Carbon dioxide	390	456	780

Velocity in metres per second

An increase in temperature causes an increase in pressure

of the molecules (Table 1.2). If the temperature of a gas is to be increased it must be given heat. Now heat is a form of energy and this increase in energy shows itself as an increase in the kinetic energy of the molecules. The average speed of the molecules, therefore, increases with rise in temperature. If the speed of the molecules increases then so will the number of collisions with the walls of the container. This increase in temperature should produce an increase in pressure and it can be shown by experiment that this is true. In fact you probably found this out for yourself without realizing it when you discovered what happened to a balloon which was too near the fire or you left a bicycle with fully blown-up tyres in the hot sun.

The law which tells us exactly how much the pressure increases with rise in temperature was first discovered in 1782 by J. A. Charles and bears his name. You will learn more about Charles' Law later in both your chemistry and physics lessons (see also page 408).

Effect on Gas Pressure of Change of Volume

Imagine a cylindrical vessel with a movable piston in the top (Figure 1.13(a)) containing a fixed mass of gas. If the piston is pushed down so that the volume of the gas is halved (Figure 1.13(b)) the molecules of gas will

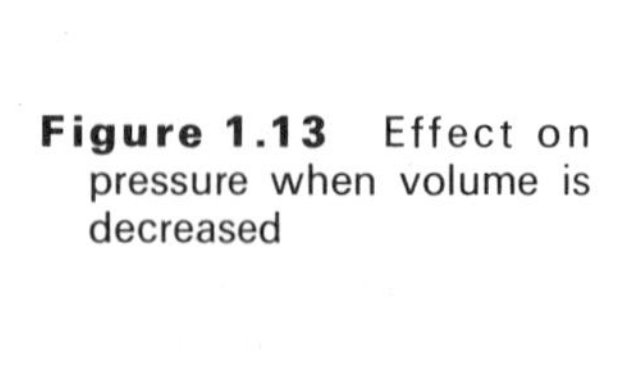

Figure 1.13 Effect on pressure when volume is decreased

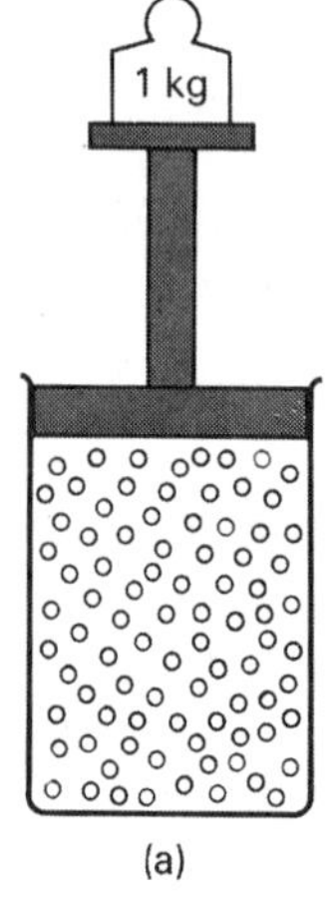

(a)

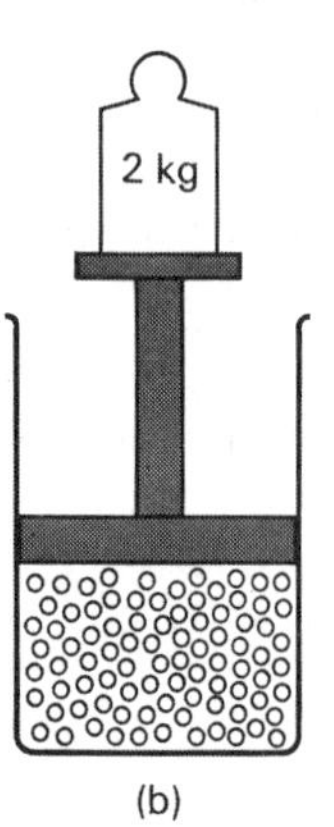

(b)

same number of molecules in half the volume

be more tightly packed and will make twice as many collisions as before, i.e. the pressure will be doubled. Thus as the volume decreases, the pressure increases and the converse is also true; as the volume increases the pressure decreases. This fact was first determined experimentally in 1662 by Robert Boyle. Boyle's Law is something you will hear more about in the future (see page 408).

Change of State

We have seen how, according to the kinetic theory, gas molecules move more quickly as the temperature is increased. If the temperature of a gas is gradually lowered the molecules will move more and more slowly, until eventually the weak attractive forces which exist between all molecules become effective and the molecules are pulled much more closely together. The gas condenses to a liquid.

Also, compression of a gas will squeeze the molecules closely together until again the gas condenses. Thus a combination of increase in pressure and decrease in temperature would seem to be the conditions necessary to liquefy a gas and this hypothesis again can be proved to be correct by experiment. In fact, all gases can be liquid under these conditions, including air (see page 394).

GAS
Molecules comparatively widely separated. Move at great speeds

LIQUID
Molecules packed close together in a random fashion. Free to move

SOLID
Molecules held in fixed pattern but vibrating and rotating

ure 1.14 Particles in lid, liquid, and vapour ates

Table 1.3 The effect of different pressures on the boiling points of gases

Gas	*Pressure (atm)*	*Boiling point K (°C)*	*Pressure (atm)*	*Boiling point K (°C)*
Ammonia	110	403 (130)	1.0	240 (−33)
Methane	45	191 (−82)	1.0	113 (−160)
Chlorine	76	417 (144)	6.0	283 (10)
Carbon dioxide	75	304 (31)	60	294 (21)

Table 1.3 shows the effect of the two variables, temperature and pressure, on the liquefaction of gases. You will see that gases will liquefy at different temperatures if the pressure is varied.

As we have seen in the diffusion experiments, the molecules in a liquid are still able to move about but much more slowly than in a gas. As the liquid is cooled this movement becomes slower still until, when the liquid solidifies, the molecules become fixed in a three-dimensional pattern. Each molecule still has sufficient energy to vibrate and rotate about a fixed point but less and less vigorously as the temperature drops and the energy decreases.

practically impossible to press a solid

The reverse process can also be explained by the kinetic theory in a similar manner. As the temperature rises the particles of a solid vibrate more quickly until they have sufficient energy to break out of the regular three-dimensional arrangement and the solid melts. Heating the liquid makes the molecules more and more energetic until they are able to overcome the weak forces holding them together and the liquid vaporizes.

Figure 1.14 shows diagrammatically the three states of matter and how the particles behave.

Summary

1. Matter is composed of tiny particles. Elements are made up of atoms or molecules and compounds are made up of ions or molecules (groups of atoms).
2. All the particles are constantly moving and because of this movement they possess kinetic energy.
3. Particles of a solid do not move from their relative positions but vibrate and rotate about fixed points.

4. Particles of a liquid move erratically and comparatively slowly, constantly colliding with each other.
5. Particles of a gas move at very high speeds with frequent collisions.
6. The kinetic theory can be used to explain gas pressure and how it varies with changes in temperature and volume. It also explains change of state.

Points for Discussion

1. Can you explain why:
   (a) liquids have no shape of their own but take the shape of their container;
   (b) gases completely fill their container?
2. Why are gases so easily compressible whereas it is practically impossible to compress a liquid or a solid?
3. Suppose a solid was cooled to a point where the molecules were absolutely at rest. Would it be possible to lower the temperature further? Explain your answer.

1.4 THE ATOMIC THEORY

John Dalton

We have mentioned how the concept of atoms as tiny indivisible particles of matter persisted down the ages, and you now understand something of the nature of matter and the behaviour of the ultimate particles of solids, liquids, and gases. It is now time to learn something of the work of John Dalton, who revived the age-old theory that matter consisted of atoms and went on to consider the masses of the atoms and how compounds were formed from them.

Figure 1.15 John Dalton

John Dalton (Figure 1.15) was born in the small village of Eaglesfield in Cumbria in 1766. His parents were poor and he was educated at the village school, but he showed early promise of brilliance in science and mathematics and in his early twenties he obtained a teaching post at Manchester College, where he spent the rest of his life doing scientific research.

In 1808 he published his Atomic Theory which was the outcome of many years of determined work and brilliant thinking. The main points of his theory were:

1. The elements are made up of tiny particles of matter called atoms.
2. Atoms are indivisible and indestructible.
3. Atoms of any one element are identical and have the same mass.
4. Atoms of different elements have differing masses.
5. When elements combine to form compounds, combination takes place between small whole numbers of atoms to form what Dalton called 'compound atoms'.

The two important new ideas in Dalton's theory of atoms were that each atom had its own individual mass and that chemical combination took place between atoms. This theory explained the laws of Constant and Multiple Proportions which had already been formulated from experimental work on chemical compounds, and stimulated research on the masses of atoms and how they combine, for the next fifty years.

Symbols for the Atoms

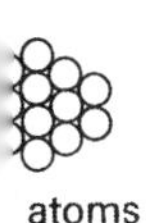

gure 1.16 Dalton's theory on combination between elements

From his statement that when compounds were formed, small whole numbers of atoms combined, Dalton worked out that a reaction between a small number of atoms could be taken as representing the whole bulk of the reacting substances (Figure 1.16). Thus when the element iron combined with the element sulphur to form iron sulphide, one atom of iron combined with one atom of sulphur to form a 'compound atom' of iron sulphide. This simple reaction between the two atoms was repeated many millions of times and was representative of the bulk behaviour of the two elements in the reaction, which could be written:

1 atom of iron + 1 atom of sulphur → 1 'compound atom' of iron sulphide

We have taken a very simple reaction as an example but you must understand that atoms do not always combine in such a simple ratio.

Dalton's next step was to write symbols for the atoms so that a reaction, such as the one we have used, could be represented by a concise kind of

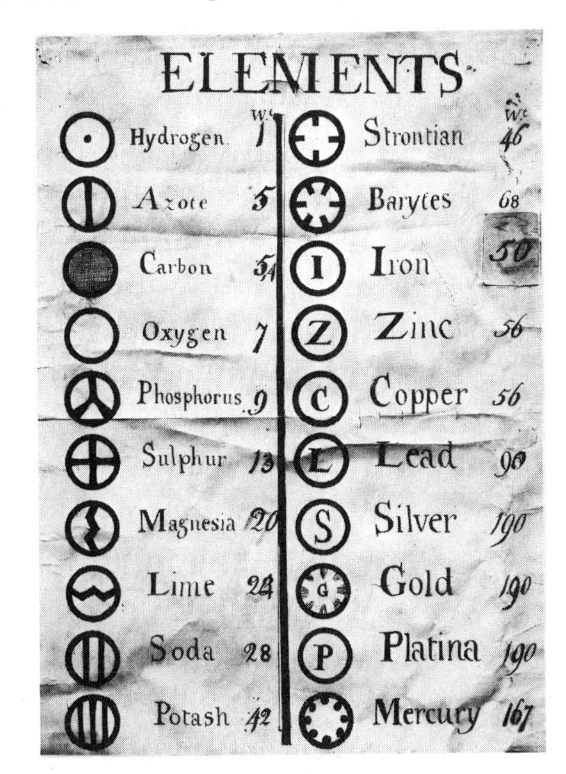

ELEMENTS

Element	W.t	Element	W.t
Hydrogen	1	Strontian	46
Azote	5	Barytes	68
Carbon	5.4	Iron	50
Oxygen	7	Zinc	56
Phosphorus	9	Copper	56
Sulphur	13	Lead	90
Magnesia	20	Silver	190
Lime	24	Gold	190
Soda	28	Platina	190
Potash	42	Mercury	167

Figure 1.17 Some of Dalton's atomic symbols. Do you know the modern name for azote? Why should lime not have been included in the list?

chemical shorthand. Dalton invented a set of symbols, some of which are shown in Figure 1.17. They were never generally used but were superseded by the system suggested by J. J. Berzelius in which an atom of an element is represented by the initial letter of the element, e.g. H stands for one atom of hydrogen, C for one atom of carbon, N for one atom of nitrogen etc. Where several elements have the same initial letter, a second letter is added. Co is the symbol for cobalt; Ca the symbol for calcium. As it was usual in the time of Berzelius (1811) to use Latin names for the elements, copper (cuprum) is represented by Cu, and iron (ferrum) by Fe.

Dalton's Influence on Chemistry in the Nineteenth Century

Dalton's Atomic Theory proved to be a tremendous stimulus to the scientists of his time and produced a new interest in chemistry. Dalton realized that atoms were far too small to be weighed or measured by any means then known, but he was able to work out the *relative* masses of some of the atoms, taking the mass of hydrogen as the standard. Other scientists continued this work and gradually tables of atomic masses were built up. You may well find it difficult to understand how the early scientists could find the relative masses of particles too small to be seen, but later in the course you will learn some of the methods by which relative atomic and molecular masses were determined. Compounds were analysed and the proportions in which the atoms combined were worked out. All this work led to the discovery of new substances and new reactions, and progress in chemical research was accelerated. In 1865 a method was discovered for measuring with some degree of accuracy the true masses of individual atoms and by the end of the century, the size and mass of the atoms of many of the elements were known. The scientists knew nothing about the actual make-up of the atoms, which were still generally thought of as tiny, homogeneous, indivisible spheres, rather like extremely minute billiard balls.

The Structure of the Atom

We owe our present day knowledge of atomic structure to the work of a number of scientists in the early years of the twentieth century. An atom is now thought to consist of a very small and extremely dense region, the *nucleus*, surrounded by a much larger volume of negative charge. The main sub-atomic particles are the following.

The proton is found in the nucleus of every atom and carries a unit positive charge. It has a mass approximately the same as that of the hydrogen atom, i.e. one atomic mass unit.

The neutron is uncharged and has about the same mass as the proton. It forms part of the nucleus of all atoms except hydrogen.

The electron carries unit negative charge and has a mass only about 1/1800 of the proton. Electrons can be considered to orbit the nucleus at great speeds; although this picture of orbiting electrons is not really satisfactory, it helps to explain a large proportion of basic chemistry.

Table 1.4 Data on atomic particles

Particle	*Mass (amu)*	*Approximate mass*	*Charge*
Proton	1.007 58	1.0	+1
Neutron	1.008 98	1.0	0
Electron	0.000 544 8	1/1840	−1

aw through his disguise
nediately—he only has
protons!'

As atoms are electrically neutral the number of protons and the number of electrons in any given atom must be the same. The neutrons play an important part in the stability of the nucleus as they reduce the inter-repulsive effects of the protons (like charges repel each other). If the nucleus consisted only of protons, the repulsive forces between these particles would cause it to disintegrate. In fact, even though many neutrons are present, all atoms containing a large number of protons (more than eighty-two) do tend to disintegrate spontaneously by radioactive decay.

The difference between atoms of one element and those of another is due to the differing numbers of electrons, protons, and neutrons they contain. An atom is therefore characterized by the number of electrons, protons and neutrons it contains and various terms are used to indicate the numbers of these sub-atomic particles present in any atom.

The Atomic Number

The atomic number of an atom is defined as the number of protons present in that atom. This number is usually written as a subscript in front of the symbol for the atom, e.g. a carbon atom contains six protons, therefore its atomic number is six and this is written $_{6}C$. What is the atomic number of oxygen? How would you write this?

The Mass Number

The mass number of an atom is defined as the sum of the numbers of protons and neutrons present in that atom. This number is usually written as a superscript in front of the symbol for the atom, e.g. a carbon atom contains six protons and six neutrons, therefore its mass number is twelve and this is written ^{12}C. Given that an atom of oxygen contains eight neutrons, write down the symbol for an atom of oxygen showing its mass number.

A combination of these two numbers (e.g. $^{12}_{6}C$, or $^{35}_{17}Cl$) gives us nearly all the information we need to know about a particular atom. From these numbers we can work out the number of electrons, protons, and neutrons in the atom.

Table 1.5 Number of protons, electrons and neutrons in atoms of the first twenty elements (most abundant isotopes only)

Atom and symbol		*Number of protons*	*Number of electrons*	*Number of neutrons*	*Mass number*
Hydrogen	H	1	1	0	1
Helium	He	2	2	2	4
Lithium	Li	3	3	4	7
Beryllium	Be	4	4	5	9
Boron	B	5	5	6	11
Carbon	C	6		6	
Nitrogen	N	7			14
Oxygen	O	8			16
Fluorine	F		9	10	
Neon	Ne		10		20
Sodium	Na		11	12	23
Magnesium	Mg	12		12	
Aluminium	Al	13			27
Silicon	Si		14	14	
Phosphorus	P		15		31
Sulphur	S			16	32
Chlorine	Cl			18	35
Argon	Ar		18	22	
Potassium	K	19		20	
Calcium	Ca		20	20	

Thus Atomic number = The number of protons or electrons
Mass number = The number of protons + neutrons
And Mass number − Atomic number = The number of neutrons

Draw Table 1.5 in your notebook, and working in a similar manner, fill in the spaces.

The Arrangement of the Electrons

Although it is impossible to plot the paths of the rapidly moving electrons, we do know the varying quantities of energy they possess and can arrange them in sets of energy levels. Thus the energy level (or shell) nearest to the nucleus can contain a maximum of two electrons, the next shell up to eight, and in the next energy level there can be as many as eighteen electrons.

The hydrogen atom, which has the lowest atomic number, has one electron in the lowest energy level; helium has two. The lithium atom has three electrons, two, the maximum number, in the first energy level and one in the second. Beryllium, atomic number four, has two electrons in the first shell and two in the second. This pattern continues until with neon, atomic number ten, the second energy level has the full complement of eight electrons. With the sodium atom, atomic number eleven, a new shell is started and this atom has two electrons in the first energy level, eight in the second and one in the third.

Diagrammatic Representation of the Atoms

The paths of the orbiting electrons round the nucleus cannot be shown with any degree of accuracy by diagrams or models. One reason for this is that the electrons are moving with such high speeds, another is that the electrons are on the border line between matter and energy and are now thought to have the characteristics of both waves and particles. Thus they can most accurately be pictured as a cloud of negative charge, denser in some areas than in others. Again it is not possible to draw a scale model of the atom because of the comparatively enormous distances of the electrons from the tiny nucleus. Atomic sizes are of the order of a hundred thousand times the size of the nucleus. This means that if the atom could be enlarged to the size of Wembley Stadium, the nucleus would be represented on the same scale by a hazelnut somewhere in the centre.

All the same, diagrammatic representations of atoms can be useful and by convention the various energy levels of the electrons are shown as concentric circles. Each circle corresponds to the region where an electron, if it was a particle, would be most likely to be found. The methods of depicting the various particles differ, so it is wise to give a key to any diagrams you make to show atomic structure.

Figure 1.18 shows representations of some of the lighter atoms. Using a similar method draw diagrams of atoms of fluorine $^{19}_{9}F$, potassium $^{39}_{19}K$ and argon $^{40}_{18}A$.

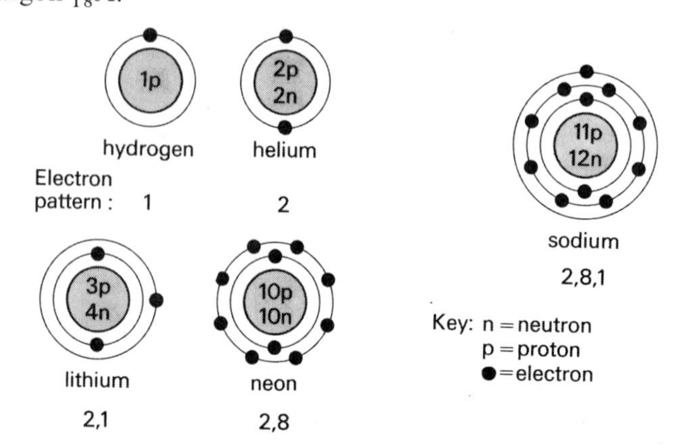

Figure 1.18 Diagrammatic representation of some of the lighter atoms

1.5 A MORE DETAILED CONSIDERATION OF ATOMIC STRUCTURE. ATOMIC MASS AND MOLECULAR MASS. ISOTOPES

Isotopes

Look again at Table 1.5. You should have found that the chlorine atom has seventeen protons, seventeen electrons, and eighteen neutrons. Some chlorine atoms however, consist of seventeen protons, seventeen electrons, and twenty neutrons, i.e.

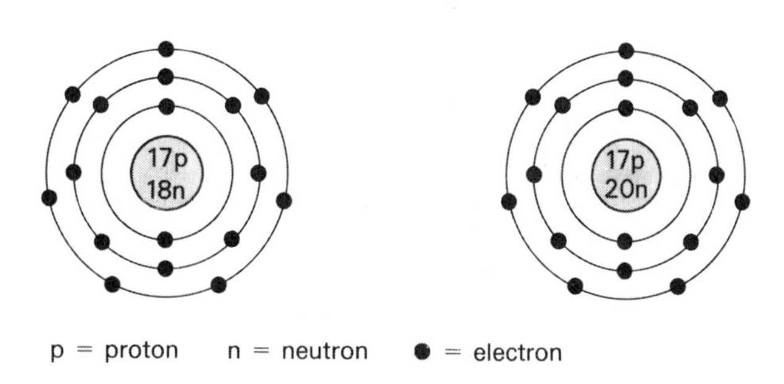

Atoms of the same element which contain different numbers of neutrons are called isotopes. Isotopic atoms each contain the same number of protons and the same number of electrons, as they must if they are to be atoms of the same element, but have differing numbers of neutrons. Thus, chlorine consists of two isotopes $^{35}_{17}Cl$ and $^{37}_{17}Cl$. These isotopes are present in the approximate ratio of three to one respectively, but it is rare for such large proportions of both isotopes to occur in a natural sample of an element. In most cases where an element is composed of isotopic atoms, one of the isotopes predominates and the others are present to a much lesser extent. For example, carbon consists of 98.89 per cent $^{12}_{6}C$ and 1.11 per cent $^{13}_{6}C$* and oxygen of 99.76 per cent $^{16}_{8}O$, 0.04 per cent $^{17}_{8}O$ and 0.20 per cent $^{18}_{8}O$. The percentage of each isotope present in any naturally occurring sample of an element is always the same (although a few exceptions are now known to occur when the element has been produced from a radioactive substance).

Isotopes always have the same chemical properties. This is because they have the same number of electrons and you will learn in Section 5.2 that it is the number of electrons present in the outer shell of an atom that determines its chemical properties. The only effect of the extra neutrons is to increase the mass of the atom, and this has a negligible effect on the chemical properties.

ome elements have a large umber of isotopes

Most elements occur as isotopic mixtures and some elements have a large number of isotopes (e.g. tin has ten). However, a few elements (aluminium, cobalt, fluorine, iodine, manganese, phosphorus, and sodium) have no natural isotopes, although it is now possible to produce radioactive isotopes of any element including those which have no naturally occurring forms.

*Plus minute traces of the radioactive isotope $^{14}_{6}C$.

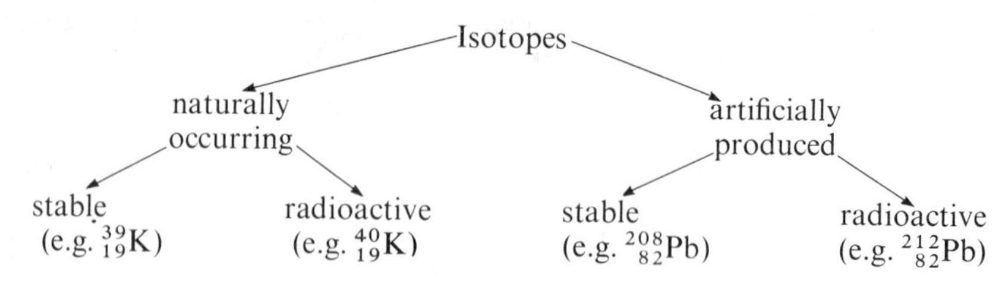

Nowadays an enormous number of isotopes is known, both naturally occurring and artificially produced.

Atomic Mass

Dalton introduced the idea that atoms of the same element all had the same mass and differed in mass from atoms of other elements. It was obviously impossible to weigh individual atoms, and so it was decided to fix a scale for comparing the mass of atoms. Hydrogen was the lightest element and the mass of its atom was arbitrarily fixed as one unit (i.e. H = 1). The masses of other atoms were then found by comparing their masses with that of a hydrogen atom.

The mass of any atom compared with that of a hydrogen atom was called the *atomic mass*. Thus the atomic mass of an oxygen atom is sixteen, and this means that one atom of oxygen weighs sixteen times as much as one atom of hydrogen. Similarly, the atomic mass of nitrogen is fourteen and this means that one atom of nitrogen is fourteen times heavier than one atom of hydrogen.

As work on determining atomic masses progressed it was found that hydrogen was an unsuitable standard as it did not combine with many elements, and in the early twentieth century the standard was changed from hydrogen = 1 to oxygen = 16. Oxygen was chosen as the new standard because it combined with many more elements than did hydrogen. However, this scale led to a problem when it was discovered that isotopes of oxygen existed, i.e. $^{17}_{8}O$, $^{18}_{8}O$ (see page 21). Physicists then redefined their standard making the commonest isotope of oxygen, $^{16}_{8}O$, equal to 16.000. On the other hand, the chemists took the average mass of all the oxygen atoms present in a normal sample of pure oxygen (i.e. including isotopes) to be equal to 16. This meant that two scales for measuring atomic masses were in use, the Physical Atomic Mass Scale and the Chemical Atomic Mass Scale. As the standards used for each scale differed slightly, so the atomic masses measured on each scale differed slightly.

An amicable agreement

This confused situation was clarified in 1961 when, at the recommendation of the International Union of Pure and Applied Chemistry, the standard for comparison was again changed, this time to $^{12}_{6}C$ = 12. That is, an atom of carbon–12, $^{12}_{6}C$, was taken to weigh exactly twelve units. The choice of carbon–12 as the standard was made for practical reasons.

The atomic mass of an element is now defined as, *the mass of an 'average atom' of the element relative to that of an atom of $^{12}_{6}C$, the mass of which is taken as 12 units*. An 'average atom' of an element is the weighted mean of the masses of all the atoms present in the normal isotopic mixture of the element.

We now know that an atom has mass because its constituent particles, i.e. the protons, electrons, and neutrons have mass (see Table 1.4, page 18). If you look at the list of atomic masses you will observe that, for example, the atomic mass of helium is almost four times that of hydrogen. We can now see why this is the case by using the approximate masses of the sub-atomic particles listed in Table 1.4, i.e. a proton = 1 amu, a neutron = 1 amu, and an electron = 0 amu. A hydrogen atom consists of a proton and electron and will therefore have a mass of $1+0 = 1$ amu, whereas a helium atom consists of two protons, two electrons and two neutrons and will therefore have a mass of $2+0+2 = 4$ amu. Similarly an atom of lithium containing three protons, three electrons and four neutrons would have an atomic mass of $3+0+4 = 7$ amu.

On this basis we would expect all atomic masses to be whole numbers, or almost so if we allowed for our slight approximations in the masses of the protons, electrons, and neutrons. However, some elements have atomic masses which differ widely from a whole number, for example the approximate atomic mass of chlorine is 35.5. If we consider a chlorine atom as containing seventeen protons, seventeen electrons, and eighteen neutrons, then its mass would be $17+0+18 = 35$ amu. Why then is the mass 35.5 amu? We cannot have half of a proton or neutron. The reason is that the atomic mass of an element is an average mass of all the atoms making up that element and so we have to consider the mass and proportion of each isotope present in a normal sample of the element.

In the case of chlorine, the most common isotope is $^{35}_{17}Cl$ which has a mass of approximately 35 amu, and in addition there is also a $^{37}_{17}Cl$ isotope which has a mass of approximately 37 amu. About three-quarters of chlorine atoms are the $^{35}_{17}Cl$ isotope, the other quarter being the $^{37}_{17}Cl$ isotope. This means that out of every four atoms of the element chlorine, three atoms have a mass of 35 amu and one atom has a mass of 37 amu. Thus the average mass of a chlorine atom is

$$\frac{(3\times 35)+(1\times 37)}{4} = 35.5 \text{ amu.}$$

Can you work out the approximate atomic mass of bromine, given that it is composed of approximately equal amounts of the two isotopes $^{79}_{35}Br$ and $^{81}_{35}Br$ which have masses of approximately 79 amu and 81 amu respectively? What would be the new atomic mass, if the isotopes happened to exist in a 2:1 ratio, the lightest isotope being in excess?

Table 1.6 Differences between the atomic mass and the mass number for the element chlorine

Natural chlorine	*Atomic mass of natural chlorine*	*Mass number of natural chlorine*
$^{35}_{17}Cl$ $^{35}_{17}Cl$ $^{37}_{17}Cl$ $^{35}_{17}Cl$	This is the mass of an 'average atom' of chlorine, based on the carbon–12 scale and = 35.5, i.e. $\frac{(3\times 35)+(37)}{4}$ The atomic mass, (a) is not a whole number due to the existence of isotopes; (b) does not refer to any one atom of the element.	Chlorine has two mass numbers, 35 and 37. These mass numbers, (a) are whole numbers because they refer to a total *number* of protons and neutrons; (b) refer to a particular atom only;

Thus if we make allowance for the proportion of isotopes present in a given element, it should be possible to calculate its atomic mass very precisely by using the accurate masses of the proton, electron, and neutron.

An instrument called a mass spectrometer is used to measure very accurately the atomic masses and proportions of all the isotopes present in a particular element. It is calibrated on the $^{12}_{6}C = 12$ scale and the technique is called mass spectrometry. A detailed discussion of mass spectrometry is outside the scope of this book.

It is usual in school work, in order to simplify calculations, to take the atomic mass of an element as the nearest whole number, and this usually means using the mass number of the most abundant isotope. The differences between the mass number and the atomic mass are summarized in Table 1.6 (page 23), using chlorine as an example.

Molecular Mass

Molecular mass is found by adding together the atomic masses of all the atoms present in a molecule. For example, a water molecule (H_2O) would have a molecular mass of $1+1+16 = 18$, using the approximate atomic masses of hydrogen and oxygen. *Molecular mass* is defined as *the mass of an 'average molecule' of an element or compound relative to that of the mass of an atom of* $^{12}_{6}C$ *which is taken as 12 units*. The term molecule which we shall discuss further in Section 1.7, in addition to covering true molecular cases (e.g. oxygen molecules, O_2; water molecules, H_2O) is often wrongly used for compounds which may be totally or partially ionic, e.g. sodium chloride, NaCl, and copper(II) sulphate, $CuSO_4$. If the compound is made up of ions we should not really talk about its 'molecular mass', and another term has been introduced to overcome this difficulty. This is the *formula mass* which is defined as *the combined mass of all the atoms making up the formula of a compound compared with the mass of an atom of* $^{12}_{6}C$.

1.6 THE MOLE AND CHEMICAL FORMULAE

The Mole

In the previous section we considered atomic masses. Refer to the table of atomic masses and compare the masses of a few individual atoms. Notice that a helium atom is about four times as heavy as a hydrogen atom and a carbon atom is about three times heavier than a helium atom. How much heavier is an atom of sulphur than an atom of oxygen, or an atom of silicon than an atom of lithium?

Imagine we could weigh out one gramme of hydrogen, then one gramme of any other element will contain fewer atoms, because the atoms of these other elements are all heavier than those of hydrogen. As an example, suppose we measure out one gramme of hydrogen and this is made up of x atoms of hydrogen (each of which has a mass of about 1 amu, remember). How many atoms of helium would we have to measure out to obtain one gramme of helium? As each helium atom (mass approximately 4 amu) is four times heavier than a hydrogen atom, then we would only need one-quarter of the number of helium atoms to make up one gramme of helium, i.e. $\frac{1}{4}$x atoms. Can you work out how many atoms of (a) carbon (mass 12 amu) and (b) oxygen (mass 16 amu) would be needed to make up one gramme? In these examples we have chosen to keep the mass constant and calculate the relative number of atoms which would be equivalent to this mass. However, a chemist usually prefers to work with a fixed number

Equal masses do not mean equal numbers!

of atoms of a substance, rather than a fixed mass of the substance. For example, if he wishes to compare certain chemical properties of say, three metals, then it is of little use taking one gramme of each metal. To get a true indication of their relative reactivities he must use an equal number of atoms, because it is the atoms that take part in chemical reactions and there is not an equal number of atoms in, for example, one gramme of iron or copper or lead.

Let us now consider the previous examples 'the other way round'. We found that

1 g of hydrogen contained x atoms
1 g of helium contained $\frac{1}{4}$x atoms
1 g of carbon contained $\frac{1}{12}$x atoms
1 g of oxygen contained $\frac{1}{16}$x atoms

Now, if we fix the number of atoms as x atoms, then

x atoms of hydrogen have a mass of 1 g
x atoms of helium have a mass of 4 g
x atoms of carbon have a mass of 12 g
x atoms of oxygen have a mass of 16 g

Avogadro's number!

Do you notice anything about the number of grammes that make up x atoms of each element? Where have you seen these numbers before? They are, of course, the atomic masses of the appropriate elements. This means that if we measure out elements in the same proportions as their atomic masses, we will always be dealing with the same number of atoms, no matter which element we use. Thus a chemist is really able to 'count the number of atoms by weighing'. Therefore, although we cannot see atoms, and do not need to know their individual masses, we can still control the number of atoms by weighing.

Nowadays not only do we know that 1 g of hydrogen, 12 g of carbon etc. contain an equal number of atoms (x atoms), we also know what this number is. It has been calculated as 6.02×10^{23} and is named in honour of the Italian scientist, Amedeo Avogadro, and called the *Avogadro Number*. You will see that it is the number of atoms in 1 g of hydrogen, 4 g of helium, 12 g of carbon etc., or expressing it another way,

6.02×10^{23} atoms of hydrogen have a mass of 1 g
6.02×10^{23} atoms of helium have a mass of 4 g
6.02×10^{23} atoms of carbon have a mass of 12 g
6.02×10^{23} atoms of oxygen have a mass of 16 g etc.

The amount of substance containing this number of particles (6.02×10^{23}) is a basic scientific unit called THE MOLE.

The essential feature of the mole is that it is an *amount of substance* usually expressed as a mass (in grammes).

For example, as seen earlier,

12 grammes of carbon contain 6.02×10^{23} atoms.

Thus

a mole of carbon atoms (C) has a mass of 12 grammes.
a mole of hydrogen atoms (H) has a mass of 1 gramme.
a mole of oxygen atoms (O) has a mass of 16 grammes.
a mole of chlorine atoms (Cl) has a mass of 35.5 grammes.

How many grammes of (a) magnesuim, (b) copper, (c) zinc, would you have to measure out in order to have a mole of atoms in each case? You might be allowed to do this practically by weighing out the powdered form of each element separately into three similar beakers or jam jars. Remember that a mole of any element contains the same number of atoms, i.e. the Avogadro Number.

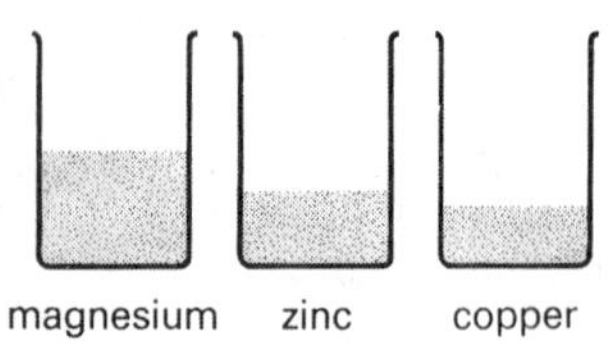

Figure 1.19 Relative volumes occupied by gramme-atoms of magnesium, zinc, and copper

In view of this you may be surprised to see that magnesium, which has the smallest atoms of the three elements, occupies the largest volume. (In case you were unable to measure out the elements, the relative amounts in the beakers are illustrated in Figure 1.19.) Don't worry, you have not made a great error and measured out more atoms than for the other two elements. The explanation is that atoms of different elements are packed together in different ways and the atoms of copper and zinc are more closely packed than those of magnesium.

If possible you should have practice in working out and weighing various proportions of moles. For example, work out and weigh:

(a) half a mole of sulphur atoms
(b) one-tenth of a mole of iron atoms (use iron filings)
(c) $\frac{1}{4}$ mole of magnesium atoms (use magnesium powder).

You could also weigh out,

(a) 3.01×10^{23} atoms of zinc
(b) 12.04×10^{23} atoms of carbon
(c) 6.02×10^{23} atoms of sulphur.

Which of the following contains the greatest number of atoms?

(a) 18 g of carbon or 2 g of hydrogen
(b) 7 g of oxygen or 7 g of nitrogen
(c) 16 g of sulphur or 8 g of oxygen.

By calculating the masses required and actually weighing out these amounts you will get used to thinking in terms of moles and using the idea in your work. As you will see, the concept enables you to weigh out a given number of atoms of any particular element.

Some of the questions at the end of this chapter are designed to further help you to understand the idea of the mole. It is a most important idea which is used in many later chapters, so make sure that you understand the principles.

Moles of Compounds

So far we have applied the mole idea to elements only, but most matter exists in the form of compounds. Some of these compounds are built up of molecules, some of ions and others have more complicated structures. We can still refer to a *mole of compound* but this time we will be referring to 6.02×10^{23} molecules or formula units rather than atoms, because these are the particles of which the compounds are constructed. For example,

A mole of water (H_2O) has a mass of 18 grammes (1+1+16) and contains 6.02×10^{23} water *molecules*.

A mole of carbon dioxide (CO_2) has a mass of 44 grammes (12+16+16) and contains 6.02×10^{23} *molecules* of carbon dioxide.

A mole of sodium hydroxide (NaOH) has a mass of 40 grammes (23+16+1) and contains 6.02×10^{23} *formula units* of sodium hydroxide.

A mole of copper(II) sulphate ($CuSO_4$) has a mass of 160 grammes (64+32+16+16+16+16) and it contains 6.02×10^{23} *formula units* of copper(II) sulphate.

The term '*formula unit*' (or sometimes just called '*unit*') can be applied to *all compounds* whatever their structure, i.e. whether they are made up of molecules, ions, or have more complicated structures. However, it is only correct to refer to a *mole of molecules* when the compound is actually made up of *molecules*.

The term *mole of ions* can also be used but it refers to 6.02×10^{23} ions of the *same type*, e.g. a mole of sodium ions (Na^+), and thus is *not* used to describe an ionic compound which is made up of at least two kinds of ions. The term is useful when considering *solutions* of ionic compounds in which ions separate from each other. In the solid ionic compound, e.g. sodium chloride, we refer to a mole of sodium chloride units which in this case is made up of a mole of sodium ions (Na^+) *and* a mole of chloride ions (Cl^-) joined together. When in solution these ions separate from each other and we can then refer to a mole of the individual ions. Similarly if a mole of magnesium chloride units ($MgCl_2$) is dissolved in water, we can obviously say that the solution contains *a mole of dissolved magnesium chloride units*; it is also equally true to say that the solution contains a *mole of magnesium ions* and *two moles of chloride ions* and that there are three moles of ions altogether. Think about this carefully.

The importance of always stating the *type* of particle being considered cannot be overemphasized. For example, just stating 'a mole of oxygen' could mean a mole of oxygen *atoms* (O) or a mole of oxygen *molecules* (O_2) or a mole of oxygen (oxide) ions (O^{2-}). This important point also crops up when considering the molarity of solutions of ionic compounds where it is essential to state on which ion the molarity calculations are based.

Note: Sometimes chemists use special terms which automatically indicate the type of particle being referred to. The *mole* can be subdivided into a number of such terms:

(i) a '*gramme-atom*' is the mass (in grammes) of 6.02×10^{23} atoms of the same kind. It is equivalent to the atomic mass of the element expressed in grammes.

(ii) a '*gramme-molecule*' is the mass (in grammes) of 6.02×10^{23} *molecules* of the same kind. It is equivalent to the molecular mass of the element or compound expressed in grammes.

(iii) a '*gramme-formula*' is the mass (in grammes) of 6.02×10^{23} *formula units* of the same kind. It is equivalent to the formula mass of the compound expressed in grammes.

(iv) a '*gramme-ion*' is the mass (in grammes) of 6.02×10^{23} *ions* of the same kind. It is equivalent to the ionic mass of the ion expressed in grammes.

Molar Solutions

A molar solution contains one mole of dissolved substance per litre of solution. For example, a molar solution of a substance A contains 6.02×10^{23} particles of A per litre of solution. Thus when 58.5 grammes of sodium chloride, or 40 grammes of sodium hydroxide etc. are dissolved in water and made up with water to one litre of solution, a molar solution of that compound results and is recorded as 1M sodium chloride or 1M NaCl etc. Thus 106 grammes of anhydrous sodium carbonate (Na_2CO_3) dissolved in a litre of solution can be described as a *molar (1M) solution of sodium carbonate*. It is equally correct to say that the solution is *one molar (1M) with respect to carbonate ions* (i.e. it contains one mole of carbonate ions per litre of solution), but it is *two molar (2M) with respect to sodium ions* (i.e. it contains two moles of sodium ions per litre of solution).

Using Moles to Find Formulae

A particular compound, if it is pure, consists of the same elements combined in the same proportions by mass no matter how it has been made (Section 2.4). Nowadays it is possible to look up the formula of any compound in a reference book, but remember that at some time an experiment had to be carried out using the compound in order to find its formula. We shall now attempt an experiment to find the formula of magnesium oxide, i.e. the compound formed when magnesium reacts with oxygen, and you will see how a knowledge of the mole enables this to be done.

Experiment 1.14
†Determination of the Formula of Magnesium Oxide *

Aim

1. To find the masses of magnesium and oxygen which combine together when the compound magnesium oxide is formed.
2. To convert these masses into moles so that we can compare the *numbers* of atoms of magnesium and oxygen which react together.
3. Hence to calculate the simplest ratio of the numbers of magnesium and oxygen atoms which combine together. This gives us the empirical formula of the compound, i.e. the simplest ratio of the atoms in the compound. For example, is it MgO, Mg_2O, Mg_3O, MgO_2, Mg_2O_3 etc.?

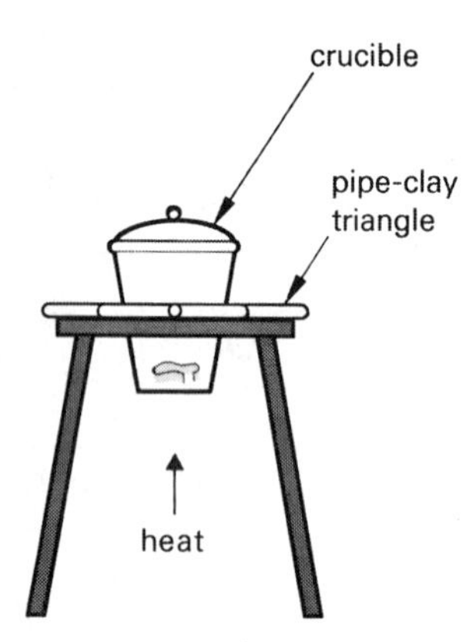

Figure 1.20 Heating magnesium in a crucible

Apparatus

Crucible and lid, tripod, Bunsen burner, pipe-clay triangle, asbestos square, tongs, balance.
Magnesium ribbon.

Procedure

(a) Weigh a crucible and lid on the balance and record this mass.

(b) Scrape the ribbon, if necessary, to remove any oxide film. Coil about 15 cm of the ribbon around a pencil and place the ribbon in the crucible.

(c) Replace the lid on the crucible and reweigh.

(d) Place the crucible containing the magnesium ribbon on a pipe-clay triangle supported by a tripod, and heat using a Bunsen burner as in Figure 1.20. Make sure the flame is low at first and then gradually increase it. Heat strongly for a few minutes. Remove the Bunsen burner and slightly lift the crucible lid by means of the tongs. Quickly replace the lid taking care to lose as little magnesium oxide 'smoke' as possible. Repeat this process until the magnesium ceases to flare up. When this stage has been reached remove the crucible lid and heat strongly to make sure that the combustion is complete.

(e) Remove the flame and allow the crucible to cool. When cool, replace the lid and reweigh the whole.

(f) Reheat the crucible and contents, cool and reweigh. If the mass is not the same as at the end of part (e), repeat the heatings until consecutive final mass measurements agree (i.e. the mass remains constant).

Points for Discussion

1. Why was the lid kept on at first?
2. Why was the lid *slightly* lifted from time to time?
3. Why is it necessary to repeat the heatings until the mass of the crucible and contents becomes constant?

Results

Record the results of *your* experiment as follows:

Mass of crucible and lid	= (23.1 g)
Mass of crucible, lid, and coil of magnesium ribbon	= (25.5 g)
Mass of crucible, lid, and magnesium oxide	= (27.1 g)
∴ Mass of magnesium	= (2.4 g)
Mass of oxygen combining with this mass of magnesium	= (1.6 g)

A theoretical set of experimental results are quoted in brackets and these will be used to show you how to work out the empirical formula.

Calculation

Make reference to the aims of this experiment and you will observe that we first of all wish to find:

1. The masses of magnesium and oxygen which combine together. These have been given in the results table; can you see how they have been worked out? 2.4 g of magnesium combine with 1.6 g of oxygen.
2. The next step is to convert these masses to moles. (Refer to page 26.)
$\frac{2.4}{24}$ moles of magnesium atoms combine with $\frac{1.6}{16}$ moles of oxygen i.e. $\frac{1}{10}$ mole of magnesium combine with $\frac{1}{10}$ mole of oxygen or 1 mole of magnesium combines with 1 mole of oxygen
$\therefore$ 6.02×10^{23} atoms of magnesium combine with 6.02×10^{23} atoms of oxygen or 1 atom of magnesium combines with 1 atom of oxygen.
3. Thus the simplest ratio of combination is one atom of magnesium with one atom of oxygen, and therefore the empirical formula of magnesium oxide is *MgO*.

The *empirical* formula is that found by experiment and represents the simplest ratio of the combining atoms. The true formula is $(MgO)_n$ where n is an integer (i.e. any whole number). For example if n = 1 then the formula would be MgO, if n = 2 the formula would be Mg_2O_2 etc. Further experimental work is necessary in order to find the value of n (Section 15.1), but this need not concern us here as you can check which formula is correct by looking in a book of data.

Work out your experimental results in a similar manner and see if they agree with this formula we have just worked out.

Points for Discussion

1. If the experimental results had shown that 6 g of magnesium combined with 4 g of oxygen, show that these figures could also be used to calculate the same empirical formula for magnesium oxide.
2. As you will observe, what we really wish to find first of all is the masses of the two elements which combine together. This can be done either by making the compound as we have done in the magnesium–oxygen example, or by breaking down the compound as illustrated in the following example.

Example 1.1

Hydrogen was passed over a heated sample of pure copper(II) oxide, as shown in Figure 1.21. The hydrogen removed the oxygen from the copper(II) oxide to form water vapour (steam) and left only copper metal in the tube. As this experiment requires the use of a stream of hydrogen in heated apparatus, it is too dangerous for you to perform, so we have given you a typical set of results.

Mass of porcelain boat empty = 14.4 g
Mass of porcelain boat and copper(II) oxide = 18.4 g

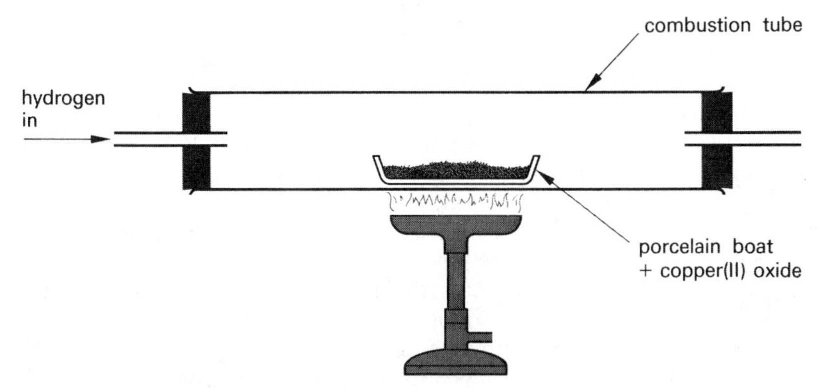

Figure 1.21 The reduction of copper(II) oxide using hydrogen

Mass of porcelain boat and copper	= 17.6	g
∴ Mass of copper	=	g
Mass of oxygen combining with this mass of copper	=	g

Complete this table and then calculate the empirical formula of copper(II) oxide in exactly the same way as you did for magnesium oxide. (Take the atomic masses of copper and oxygen as 64 and 16 respectively.)

3. If another experiment showed that 2 g of calcium combined with 3.55 g of chlorine, what would be the empirical formula of calcium chloride? (Take the atomic masses of calcium and chlorine as 40 and 35.5 respectively.)

1.7 IONIC AND COVALENT BONDING. VALENCY

Introduction

In the last section you performed an experiment which showed that the simplest formula of magnesium oxide was MgO, that is, one atom of magnesium combined with one atom of oxygen. We must now ask ourselves such questions as (a) Why do magnesium and oxygen combine together, rather than exist separately as magnesium and oxygen atoms? (b) Why is it that they combine in the ratio 1:1 and not say 2:1, 1:2, 2:3 etc.? (c) What holds the magnesium and oxygen atoms together once they have combined?

The ideas developed in this section will help you to answer these questions and you will be able to apply these ideas to other compounds.

Structure of the Noble Gases

In Section 1.4 we considered the electronic configurations of the first twenty elements and you will remember that the electrons were placed in definite shells. By referring to the diagrams of electronic structures on page 19 and to your own drawings you will see that some atoms have a complete outer shell of electrons. These atoms are helium (2.), neon (2, 8.), and argon (2, 8, 8.). Such elements belong to a group known as the *noble gases*. The noble gases all have one characteristic feature, can you find out what this is? Have you heard of neon oxide, helium chloride, argon sulphate etc.? See if you can find such compounds in a book of data. It may also help you to know that this group of elements used to be called the *inert gases*.

Stability of the Noble Gases

The noble gases are very stable elements, showing very little chemical activity. This stability must obviously be linked with their electronic structures and the fact that each shell has its full quota of electrons.

If a system reacts so that it achieves greater stability, energy is released. Nearly all spontaneous changes that occur in nature take place with release of energy so that the final state is more stable than the initial one. A stone balanced on the edge of a cliff is in an unstable state. As it drops it loses energy and when it comes to rest at the bottom of the cliff it is once more in a stable state. The water at the top of a waterfall is less stable than that in the pool below. The energy lost by the falling water is in some cases converted into electric power.

He never joins with us

In the case of atoms, the most stable states are those in which the atom has the electronic configuration of a noble gas. Thus atoms of all the other elements could become more stable if they could achieve this configuration. How can the less stable atoms attain these stable electronic structures? Free atoms are rarely found in nature because of their tendency to become more stable. Can you think which types of atoms *are* found free in nature?

All other atoms join together in an effort to become more stable, even if in some cases it means that they have to join to another atom of the same kind. For example chlorine exists as chlorine molecules (Cl_2) and not just as single chlorine atoms (Cl).

Ionic Bonding

Let us consider an atom of sodium and an atom of chlorine. Draw their electronic structures in your notebook, and answer the following questions:

(a) Which is the *nearest* noble gas to sodium? Draw its electronic structure in your notebook.
(b) How does its electronic configuration differ from that of a sodium atom?
(c) What then must an atom of sodium do to attain this structure?

Answer these same three questions with respect to the chlorine atom instead of the sodium atom.

The Sodium Ion

You should have found that the sodium atom must *lose* one electron in order to attain the electronic configuration of a neon atom.

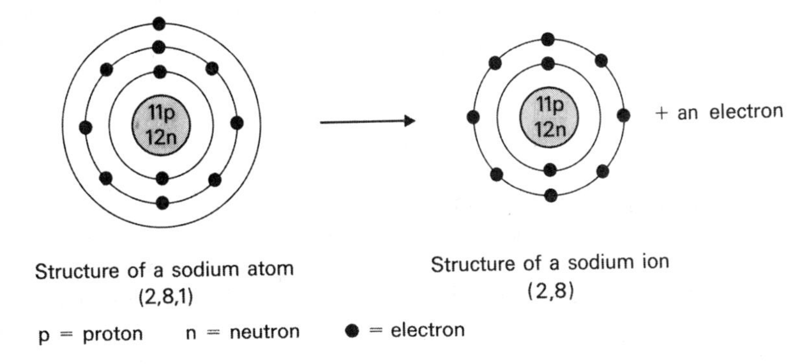

You will notice that when a sodium atom loses an electron, the resultant structure is a 'mixture'. The electronic structure is that of a neon atom, but the nucleus is that of a sodium atom. As the number of protons is unaltered, the atomic number of the resultant structure is still eleven so it is a derivative of sodium. [The atomic number = the number of protons and is only equal to the number of electrons when we consider a 'neutral atom'.] The resulting structure is called a *sodium ion.*

An ion is formed when an atom or radical (page 39) has either lost or gained one or more electrons. In this case the sodium atom has lost one electron and formed a sodium ion. Is there an equal number of electrons and protons in the ion? Is the structure still neutral? As the number of electrons has decreased by one, the ion produced has eleven protons (each carrying a unit positive charge) and ten electrons (each carrying a unit negative charge), so there is an overall surplus charge of one positive unit. This is usually written as a superscript after the symbol for the element, i.e. Na^+.

The number of electrons possessed by any atom or ion determines its chemical properties as we shall discuss more fully in Section 5.2. Thus the sodium *ion,* which has ten electrons arranged in the neon configuration, will be unreactive and will therefore show different chemical properties from an *atom* of sodium which has eleven electrons. For example, we eat sodium ions (Na^+) which are present in common salt (Na^+Cl^-), but you can imagine what would happen if we tried to eat the very reactive sodium metal (Na).

The Chloride Ion

Now consider the chlorine atom. You should have worked out that it must *gain* one electron in order to attain the electronic configuration of an argon atom.
i.e.

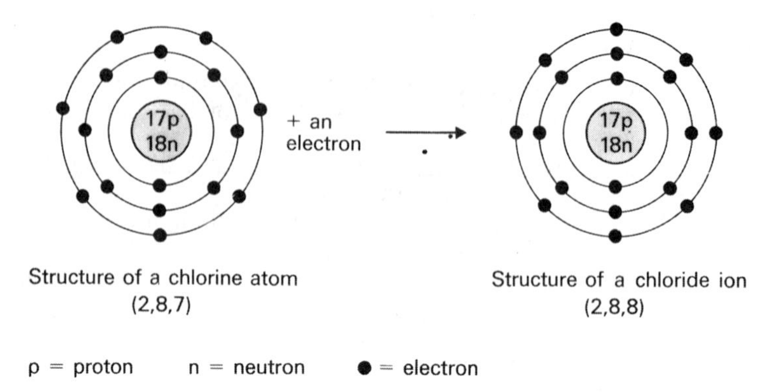

By reasoning similar to that used in the case of sodium you should realize that a chloride ion (Cl^-) has been formed. Again the difference in electronic structure between a chlorine atom and a chloride ion results in the two having completely different chemical properties. The chloride ions (Cl^-) present, for example, in common salt (Na^+Cl^-) are unreactive and quite harmless, whereas chlorine gas (Cl_2) is reactive and poisonous.

Points for Discussion

1. Why is this structure called a chloride ion and not an argon ion?
2. Why is it not neutral?
3. Why is the charge one negative unit?
4. Would you expect its chemical properties to be similar to those of a chlorine atom or an argon atom? Give your reasons.

The Bond Between Sodium and Chlorine

We have seen that a sodium atom needs to *lose* one electron to become stable and that a chlorine atom needs to *gain* one electron to become stable. Thus it would seem reasonable to suppose that when sodium and chlorine atoms react together the chlorine atom would gain an electron from the sodium atom, and this is what we believe does happen.
i.e.

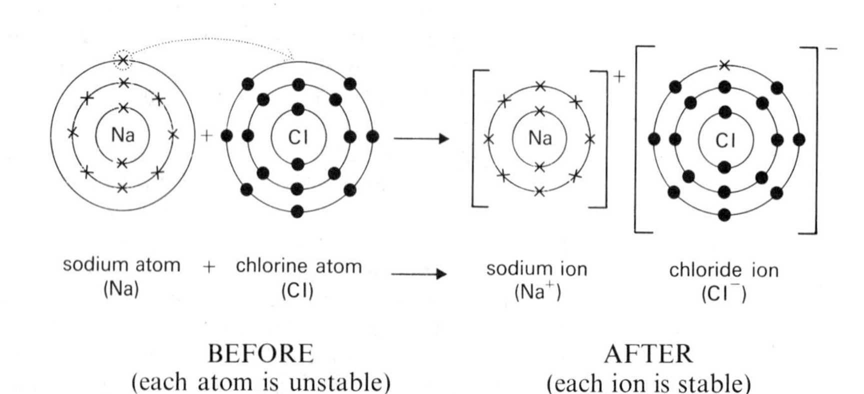

Note: All electrons are identical. Those present in the atoms of sodium and chlorine in the example have been given different symbols in order to show the transfer more clearly.

As the nucleus always remains unchanged during bonding it is not usually shown in the diagrams.

We are saved!

A satisfactory conclusion

The ions formed are oppositely charged and so will be attracted to one another (opposite charges attract). This attractive force between the two ions is called an *ionic bond* which is sometimes also referred to as an *electrovalent bond.*

You may be led into thinking that a sample of sodium chloride was thus made up of millions of discrete (separate) pairs of sodium and chloride ions, but this is not the case (see Figure 7.5). The mutual attraction between the oppositely charged ions produces a giant structure of millions of ions which are arranged in a three-dimensional recurring pattern called a crystal lattice. There are no separate sodium chloride pair-units in the solid state.

The Bonding in Magnesium Oxide

We can now apply the ideas so far learnt to magnesium oxide.

(a) Draw the electronic configuration of a magnesium and an oxygen atom.

(b) What must an atom of magnesium do in order to attain a stable electronic configuration?

(c) What must an atom of oxygen do in order to attain a stable electronic configuration?

(d) Can you thus suggest how atoms of the two elements would combine together to form the compound magnesium oxide? Draw an electronic structure equation as you did for sodium chloride.

In this case you should have found that ions again would be formed, but because two electrons must be lost by each magnesium atom and two electrons gained by each oxygen atom, the ions formed would be Mg^{2+} and O^{2-}. You are now in a position to answer the questions posed at the beginning of this section. Write down the answers to these questions in your notebook.

The Bonding in Magnesium Chloride

Draw the electronic structures of a magnesium atom and a chlorine atom in your notebook. You will notice that in this case the magnesium atom needs to *lose* two electrons and the chlorine atom needs to *gain* only one electron. Thus the chlorine atom can 'accept' one of the electrons the magnesium atom needs to lose and in doing so the chlorine atom attains a noble gas configuration of electrons. This still leaves the magnesium with an electron in its outer shell, which it must lose to become stable. How can this be done? The answer is that the magnesium transfers this remaining outer electron to *another* atom of chlorine, so making this

second atom of chlorine stable and itself attains a stable noble gas configuration in the process.
i.e.

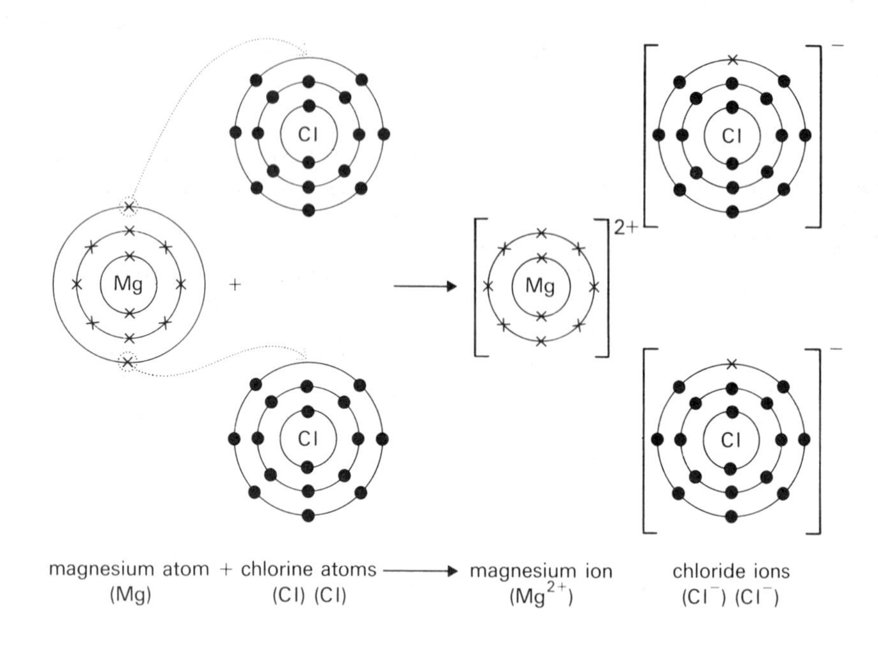

In this case two atoms of chlorine combine with one atom of magnesium to form the compound magnesium chloride ($MgCl_2$). Note that, although there are twice as many chloride ions as magnesium ions, the number of positive and negative charges in the compound is still the same, and the compound as a whole is therefore neutral. This is because a magnesium ion carries two unit positive charges whereas the chloride ion carries one unit negative charge.

Thus we can now explain why atoms combine in the ratios we found by experiment and calculation in the last section.

Further Examples Using your knowledge of ionic bonding, work out the ions produced and hence the formulae of the compounds formed when: (a) potassium combines with oxygen, (b) calcium combines with fluorine, (c) aluminium combines with oxygen.

In general, ionic compounds are formed when atoms of one element (or a radical) need to *gain* electrons and atoms of the other element (or radical) need to *lose* electrons in order to achieve a noble gas electronic configuration. This usually occurs when a non-metal combines with a metal. This is a simple guide only and you will learn later that there are a few exceptions, e.g. aluminium chloride.

Properties of Ionic Compounds

1. Ionic compounds are composed of two or more different kinds of oppositely charged ions.
2. These oppositely charged ions attract one another and form a large three-dimensional lattice, called a giant structure, which is held together by the inter-ionic electrostatic attraction. Thus ionic compounds are usually crystalline solids.

3. Because of the great attraction between the ions, a large amount of energy has to be used to separate them. You will remember from your work on the kinetic theory, that it is not until the 'particles' are separated that a solid can melt and eventually boil. Thus ionic compounds usually have high melting and boiling points. Refer to Table 5.5 (page 139) where melting points are given in relationship to structure. Which chlorides do you think are ionic? Check your answer by reference to Table 5.6 (page 140).
4. Ionic compounds when molten, or in aqueous solution, conduct an electric current (Section 6.1).
5. Ionic compounds are usually soluble in water (Section 3.5), but do not dissolve in organic solvents such as ethanol, benzene or tetrachloromethane.

Covalent Bonding

You will realize that there are a great many substances which do not have the properties of ionic compounds. They are non-crystalline and have low boiling points; many are actually gases at room temperature. How are the atoms of these substances joined together?

At the beginning of the chapter we mentioned molecules, which we said were groups of atoms, and you probably know that gases such as chlorine, oxygen, and nitrogen consist not of single atoms but of pairs of atoms joined together. How can two chlorine atoms, each of which needs to *gain* an electron to achieve the stable configuration of electrons, become bonded together to form a molecule?

Draw the electronic structures of two chlorine atoms. Can you suggest any method by which each atom can achieve the required stable configuration? You will soon find that this is not possible by transferring electrons (i.e. forming an ionic bond). The answer is that the two chlorine atoms *share* electrons. This may be represented as follows:

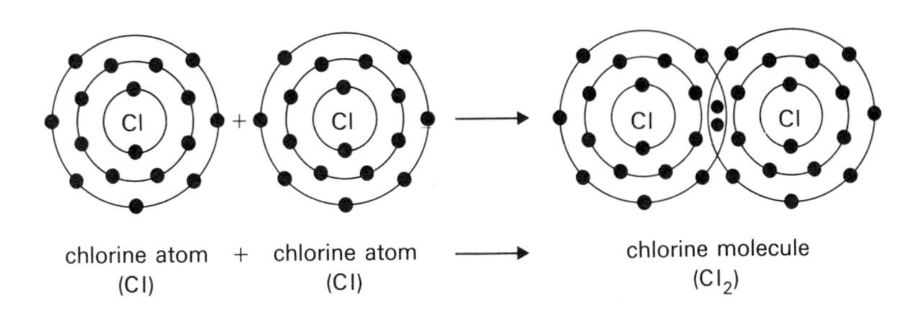

[Only the electrons in the outer shell of an atom are used to form bonds by sharing electrons, and in future diagrams of this type only the outer shell of electrons of an atom will be shown.]

As you will see, each chlorine atom shares one of its electrons with the other one so that a pair of electrons is shared between the two atoms. Each chlorine atom now has eight electrons in its outer shell, six of which belong entirely to that atom, the other two being shared. This shared pair of electrons bonds the two atoms together and the bond is called a *covalent bond.*

When two or more atoms are joined together by means of covalent bonds a molecule is formed. In this case the covalent bond is specifically

formed between two chlorine atoms, thus chlorine will exist as separate molecules (Cl_2). These molecules have only a very weak attraction for each other and so do not form a lattice structure as in compounds where the bonding is ionic. Thus in covalent substances the *intermolecular* forces (i.e. the bonds between the molecules) are weak, but the *intramolecular* forces (i.e. the bonds in the molecule itself) are strong.
i.e.

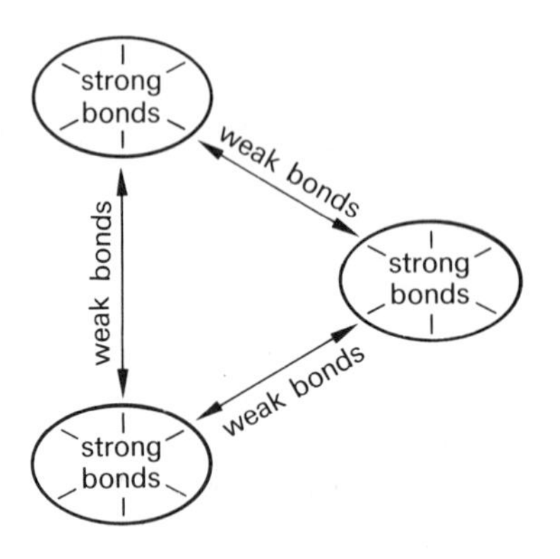

The Bonding in Oxygen and Nitrogen Gases

Draw the electronic configurations of two oxygen atoms. What must each of these atoms do to achieve a stable noble gas structure of electrons? How will they achieve this? Again transfer of electrons would be impossible, but a sharing of electrons could take place. In this case each oxygen atom must share two of its electrons with a neighbouring oxygen atom because both need to gain two electrons. The combination may be represented as follows.

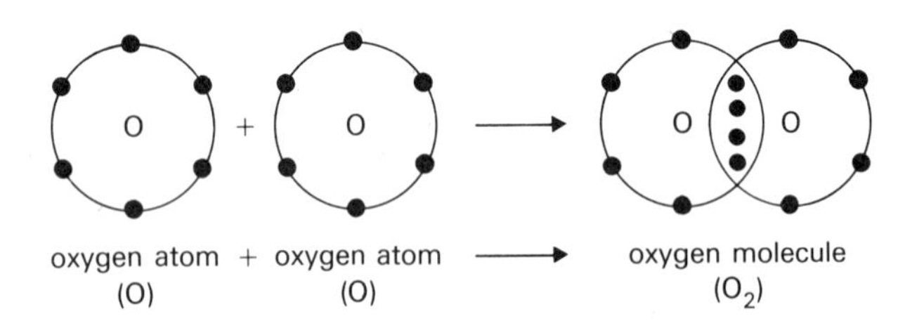

(Remember that only the outer shell of electrons is shown.)

In the oxygen molecule so formed, each atom has eight electrons encompassed by its outer shell, four of which entirely belong to the atom, and four others which it shares. As *one* covalent bond consists of *one pair* of shared electrons, an oxygen molecule contains a *double* covalent bond. It is usual to represent the double covalent bond between two atoms by a double line, e.g. O=O represents the two oxygen atoms bonded together by a double covalent bond. Similarly Cl—Cl represents two chlorine atoms bonded together by a single covalent bond. Triple covalent bonds also exist.

Using these ideas on covalent bonding, see if you can work out how the nitrogen molecule is formed.

Covalent Bonding Between Dissimilar Atoms

So far we have only considered covalent bonding between atoms of the same element, but it is equally possible for atoms of different elements

to form covalent bonds, providing that both need to gain electrons. This occurs when atoms of two different non-metals combine. For example, let us consider hydrogen reacting with chlorine. Both atoms need to gain one electron in order to attain a noble gas configuration and so a single covalent bond is formed.
i.e.

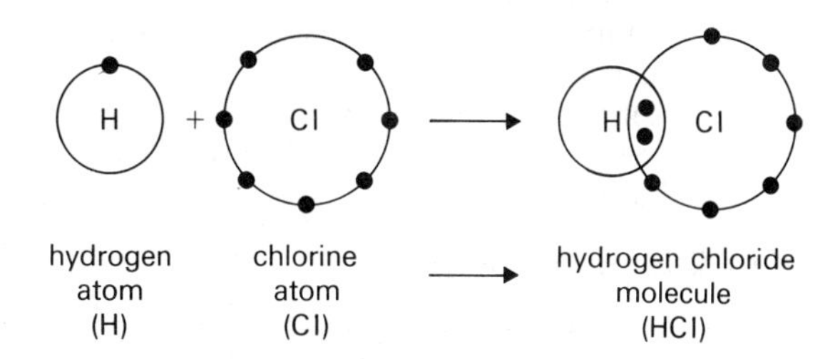

In the case of hydrogen combining with oxygen, we have one atom (hydrogen) needing to gain one electron, but the other atom (oxygen) needs to gain two electrons. By sharing one electron from each atom we would obtain the following structure:

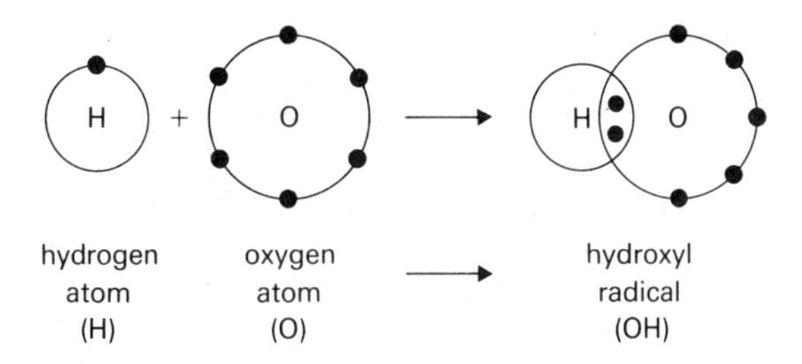

The hydrogen atom would have become stable but the oxygen 'atom' still needs to gain another electron. How may it do this? The answer is that another hydrogen atom must share its electron with the oxygen, and in addition to stabilizing itself, it also stabilizes the oxygen 'atom'.

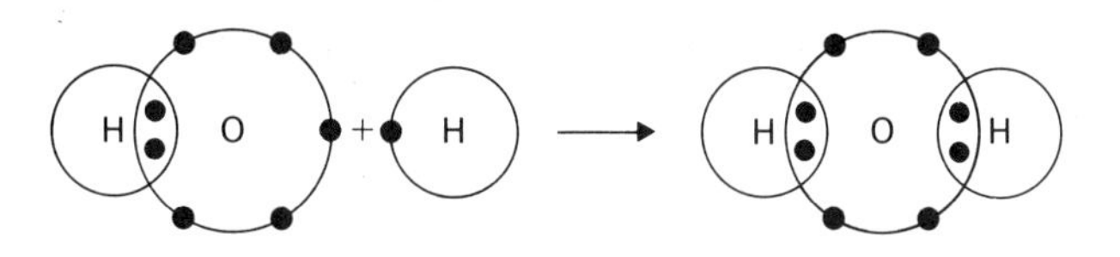

Overall,

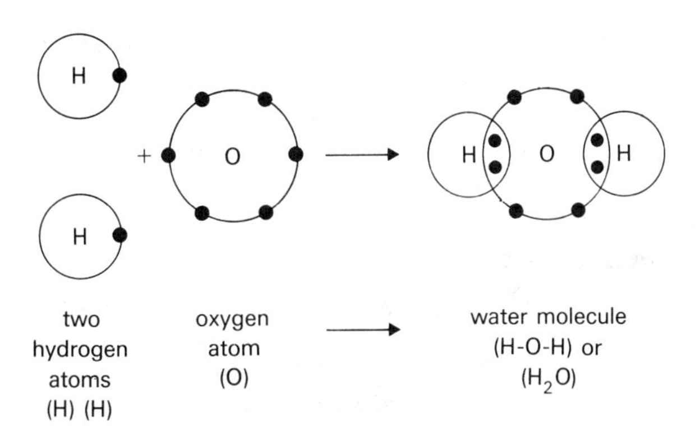

Thus you can now appreciate why the formula of water is H_2O and not HO or H_3O or HO_2 etc. The oxygen atom forms two covalent bonds, one with each hydrogen atom, and is said to show a covalency of two, whereas the hydrogen atom, which forms one covalent bond, is said to show a covalency of one.

Using these principles see if you can work out the covalent molecular structures formed when (a) hydrogen atoms combine with nitrogen atoms to form ammonia, (b) carbon atoms combine with oxygen atoms to form carbon dioxide, and (c) silicon atoms combine with fluorine atoms to form silicon tetrafluoride. In each case draw the electronic structure of the resultant molecule and state if the bonds formed are singly covalent, double covalent, etc.

Properties of Covalent Compounds

1. Covalent compounds consist of two or more different atoms linked together by covalent bonds to form individual molecules.
2. The molecules formed have only a very weak attraction for each other and so can easily be separated, thus covalent compounds are usually gases or liquids with low melting and boiling points.
3. Covalent compounds are often insoluble in water, but dissolve more readily in organic solvents.
4. Covalent compounds do not conduct electricity.
5. The bonds are directional.

Predicting if Elements Will Form Ionic or Covalent Bonds

In general, covalent compounds form when *both* atoms need to *gain* electrons. You should compare this with ionic compounds which form when one atom needs to *gain* electrons and the other to *lose* electrons. Thus atoms of elements which need to lose electrons form ionic bonds, whereas those which need to gain electrons either completely or by sharing can form both ionic and covalent bonds.

As you know, an atom achieves a greater stability by losing or gaining electrons in order to attain the electronic structure of the *nearest* noble gas. Atoms with one, two or three electrons in their outer shell will tend to lose electrons when they form compounds, whereas atoms with five, six or seven electrons in their outer shell will tend to gain electrons. An atom can lose or gain one electron fairly readily, but it is more difficult to lose or gain two electrons and quite difficult to lose or gain three electrons (Section 5.4).

Which of the following elements will gain, and which will lose, electrons on compound formation: oxygen, aluminium, magnesium, phosphorus, sulphur, chlorine, lithium? Consequently, which of these elements can form both ionic and covalent compounds and which can form ionic compounds only?

Atoms which have four electrons in their outer shells would be expected to either gain or lose four electrons. However, for reasons outlined in Section 5.4 it is virtually impossible for any atom to completely gain or lose four electrons, and the only way in which such atoms can achieve the noble gas configuration is by sharing four more electrons, i.e. by forming four covalent bonds. Thus the compounds of carbon and silicon are always covalent.

The Uniqueness of Hydrogen

Of all the elements hydrogen is unique, in that its atom has only one electron. Such an atom can become stable in three ways:

(a) by losing this electron and forming a positive ion, H^+, which cannot, however, exist on its own;

(b) by completely gaining an electron and forming the negative ion, H^-, and achieving the stable electronic configuration of helium in the process;
(c) by sharing an electron as in some of the previous examples of covalent bonding.

See if you can work out the structures of sodium hydride (NaH) and methane (CH_4).

Valency (or Combining Power)

The valency or combining power of an element may be defined as *the number of electrons an atom of that element must lose or gain, either completely or by sharing, in order to attain a noble gas configuration.* Many elements have more than one valency, but for the moment the above definition will suffice.

Thus as a sodium atom must *lose one* electron to attain a noble gas configuration it has a valency of *one*. Similarly a chlorine atom must *gain one* electron to attain a noble gas configuration, therefore it also has a valency of *one*. What would the valencies of atoms of the following elements be: phosphorus, sulphur, calcium, magnesium, nitrogen, oxygen?

Radicals

A *radical* is a group of atoms which usually forms the non-metallic part of a compound and which can remain intact through many different chemical reactions, behaving in many ways like a single atom and always exhibiting a constant valency. For example, the compounds sodium carbonate, Na_2CO_3, calcium carbonate, $CaCO_3$, and magnesium carbonate, $MgCO_3$, all contain the carbonate radical, CO_3. The valency of this radical is always two.

Radicals have no independent existence. You will never find a bottle of 'carbonate' or 'sulphate' on the shelves of a chemistry laboratory, but they can exist as free ions in aqueous solution. Most radicals form the non-metallic part of a compound so their ions are negatively charged, e.g. CO_3^{2-}, SO_4^{2-}. The important exception is the ammonium radical NH_4, which behaves as though it were the metallic part of a compound and forms a positive ion NH_4^+.

Some of the more common radicals are listed in Table 1.7, together with their formulae, valencies, and corresponding ions.

Table 1.7 Some common radicals

Name of radical	*Formula of radical*	*Stable ion formed*	*Valency*
Nitrate	NO_3	NO_3^-	1
Nitrite	NO_2	NO_2^-	1
Sulphate	SO_4	SO_4^{2-}	2
Hydrogen sulphate	HSO_4	HSO_4^-	1
Sulphite	SO_3	SO_3^{2-}	2
Thiosulphate	S_2O_3	$S_2O_3^{2-}$	2
Carbonate	CO_3	CO_3^{2-}	2
Hydrogen carbonate	HCO_3	HCO_3^-	1
Hydroxyl	OH	OH^-	1
Phosphate	PO_4	PO_4^{3-}	3
Cyanide	CN	CN^-	1
Permanganate	MnO_4	MnO_4^-	1
Dichromate	Cr_2O_7	$Cr_2O_7^{2-}$	2
Ammonium	NH_4	NH_4^+	1
Alkyl radicals	R	R^+	1

QUESTIONS CHAPTER 1

Note: the first few questions are of the 'multiple choice' type. Answer these questions by writing the letter before the answer that you think is correct, in your notebook. In no circumstances should you mark the textbook.

1. When pollen grains are suspended in water and viewed through a microscope, they appear to be in a state of constant but erratic motion. This is due to

A convection currents
B small changes in pressure
C small changes in temperature
D the bombardment of the pollen grains by molecules of water
E a chemical reaction between the pollen grains and the water

2. Bromine has an atomic number of thirty-five. How are the electrons of a bromine atom arranged in their energy shells?

A 2, 8, 18, 7
B 2, 10, 8, 15
C 10, 8, 2, 15
D 2, 8, 12, 13
E 2, 8, 8, 17

3. An increase in temperature causes an increase in the pressure of a gas because

A it increases the mass of the molecules
B it causes the molecules to combine together
C it decreases the average velocity of the molecules
D it increases the average velocity of the molecules
E it increases the number of collisions between the molecules

4. Isotopes are atoms of the same element which have

A the same mass number but different atomic numbers
B the same mass number and the same atomic number
C different atomic numbers and different mass numbers
D the same number of neutrons but different numbers of protons
E the same number of protons but different numbers of neutrons

5. In a compound of magnesium (Mg = 24) and nitrogen (N = 14), 36 g of magnesium combine with 14 g of nitrogen. The simplest formula for the compound is

A MgN; B Mg_2N; C Mg_3N; D Mg_2N_2; E Mg_3N_2

6. An atom of chlorine which has seventeen protons and eighteen neutrons in the nucleus is written $^{35}_{17}Cl$. The number of neutrons present in an atom of bromine which is represented as $^{81}_{35}Br$ is

A 46; B 35; C 81; D 116; E 17

7. Sodium atoms and sodium ions

A are chemically identical
B have the same number of electrons
C have the same number of protons
D both react vigorously with water
E react and form compounds with covalent bonds

8. When magnesium combines with chlorine

A each magnesium atom gains two electrons
B each chlorine atom loses one electron
C a covalent bond is formed
D the compound formed contains equal numbers of magnesium and chlorine ions
E the compound formed will be a solid at room temperature

9. An ionic bond is often formed when

A the combining atoms need to lose electrons in order to attain a noble gas configuration
B the combining atoms both need to gain electrons in order to attain a noble gas configuration
C two non-metallic elements react together
D a metallic element combines with a non-metallic element
E two metallic elements react together

10. A molar solution contains

A one mole of a substance dissolved in a litre of water
B one mole of a substance dissolved in 100 cm^3 of water
C one mole of a substance in a litre of solution
D one molecule per litre of solution
E one molecule per litre of water

11. One mole of carbon dioxide

A contains the Avogadro Number of carbon dioxide atoms
B contains the same number of molecules as 1 g of hydrogen
C contains the same number of molecules as 32 g of oxygen
D has a mass of 28 g
E contains 602×10^{23} molecules of carbon dioxide (C = 12, O = 16)

12. 28 g of nitrogen

A contains 6×10^{23} atoms of nitrogen
B contains 6×10^{23} molecules of nitrogen
C contains two moles of nitrogen
D is heavier than 28 g of hydrogen
E contains the same number of atoms of nitrogen as 35.5 g of chlorine (N = 14, Cl = 35.5)

13. A new illuminated sign erected at Heathrow Airport contains 2 g of a rare monatomic gas (X) of atomic weight 20. How many atoms of X are present in this sign?

A 6×10^{22}
B 6×10^{23}
C 6×10^{24}
D 12×10^{23}
E 10×10^{23}

14. An atom has a diameter of about 3×10^{-8} cm and the diameter of its nucleus is about 9×10^{-13} cm. Calculate the ratio of the diameter of the nucleus to the diameter of the atom.

15. You have seen how the results of a number of different experiments help to build up the evidence in favour of the particulate theory of matter. In a similar manner devise four pieces of evidence which might lead you to believe that a load of coal had been delivered at your house today while you were out.

16. Oleic acid molecules are of the order of 10^{-7} cm in diameter (thickness). If 0.01 cm^3 of oleic acid spreads out on the surface of water as far as possible, what would be the approximate area of water covered?

17. Explain, in terms of the kinetic theory, the differing properties of a solid, a liquid, and a gas.

18. How does the kinetic theory explain (a) Boyle's Law, (b) Charles' Law?

19. State the five main points of Dalton's Atomic Theory and say which of these statements had to be modified later.

20. (a) How many moles of sodium carbonate are present in 200 cm^3 of a 2M solution of the salt?

(b) How many moles of sodium chloride are there in 500 cm^3 of 0.1 M sodium chloride?

(Na = 23, C = 12, O = 16, Cl = 35.5)

21. (a) What mass of magnesium sulphate would be required to make a litre of 1M solution of this salt?

(b) What mass of sodium hydroxide would have to be dissolved in 100 cm^3 of water in order to produce a 0.5M solution?

(Mg = 24, S = 32, O = 16, Na = 23, H = 1)

22. (a) What is the mass of a mole of (i) calcium nitrate, (ii) zinc sulphate?

(b) What are the masses of a mole of (i) propanol (C_3H_7OH), (ii) tetrachloromethane (CCl_4), (iii) benzene (C_6H_6)?

(c) How many moles are present in (i) 100 g of calcium, (ii) 13 g of zinc, (iii) 28 g of nitrogen, (iv) 8 g of S?

(C = 12, H = 1, Ca = 40, N = 14, O = 16, S = 32, Zn = 65, Cl = 35.5)

23. (a) 3.5 g of nitrogen combined with 2 g of oxygen to form an oxide of nitrogen. What is the formula of this oxide?

(b) On analysis, a compound was found to contain 2.3 g of sodium and 0.8 g of oxygen. What is the formula of the oxide?

(N = 14, O = 16, Na = 23)

24. What information regarding the atomic structure of an element is given by its mass number and its atomic number? Illustrate your answer by reference to $^{16}_{8}O$, $^{32}_{16}S$, $^{35}_{17}Cl$, $^{37}_{17}Cl$.

25. The elements W, X, Y, and Z have atomic numbers, respectively, of seven, nine, ten, and eleven. Write the formula for the compound you would expect to form between the following pairs of elements and indicate the type of bonding present.

(a) W and X
(b) X and Z
(c) X and X
(d) Y and Y
(e) Z and Z

26. Draw the electronic configurations of the following compounds,

(a) hydrogen sulphide (H_2S)
(b) calcium chloride ($CaCl_2$)
(c) trichloromethane ($CHCl_3$)

27. Draw a diagram to show how you would use two gas jars, one containing a colourless gas of vapour density 14 and the other a green gas of vapour density 35.5, to illustrate gaseous diffusion.

State what you would see after leaving your apparatus intact for two days. (J. M. B.)

28. Two plugs of cotton wool, one soaked in concentrated hydrochloric acid and the other in a concentrated solution of ammonia, are used to seal the ends of a horizontal glass tube. After a time a white ring forms nearer to one end of the tube than the other. Explain the formation and position of the ring in terms of the movement of gaseous molecules. (J. M. B.)

29. (a) For each of the following statements, indicate whether it is true or false and then briefly (one sentence) give the reason:

(i) all the atoms of an element contain the same number of protons;
(ii) the calcium ion has two protons less than the calcium atom;
(iii) the atom of chlorine has one electron in its outer 'shell';
(iv) the atomic number of aluminium being thirteen, there are five electrons in the outer 'shell' of its atom.

(b) Give the electronic structure of an atom of sodium (atomic number eleven). (A. E. B.)

30. What is a covalent bond?
Name and give the structural formulae of two compounds containing chlorine which have only covalent bonds. (J. M. B.)

31. (a) Complete the following sentence in your notebook. The Avogadro Number 6.02×10^{23} is the number of

(b) A packet of sugar contains crystals of average mass 3.42×10^{-3} g. How many sugar molecules will an average crystal contain? (Formula of sugar: $C_{12}H_{22}O_{11}$.) (J. M. B.)

32. Chlorine has an atomic number of seventeen and exists mainly in isotopic forms, A and B, of atomic masses thirty-five and thirty-seven respectively. State the number of

(a) electrons in each atom of A,
(b) protons in each atom of B,
(c) neutrons in each atom of B,
(d) electrons in each ${}^{35}_{17}Cl^{-}$ ion.

Calculate the A:B ratio in ordinary chlorine gas (molecular mass = 71). (J. M. B.)

33. Briefly explain

(a) why a gas exerts pressure on the walls of its container,
(b) how and why this pressure is affected by an increase in temperature. (J. M. B.)

The following questions require rather longer answers.

34. Name the three types of particles occurring in most atoms and give their relative masses and charges.

What do you understand by (a) the mass number, and (b) the atomic number of an element?

Draw a diagram showing the electronic structure of an atom, the nucleus of which carries a positive charge of nine units. What will be the valency of this element?

Explain what is meant by a covalent bond. Illustrate your answer by an example of a compound containing one or more covalent bonds. (A. E. B.)

35. (a) The atomic number and atomic mass of nitrogen are seven and fourteen, respectively. Give a labelled diagram showing the structure of this atom.

(b) Indicate the relative charges and masses of the particles making up the nitrogen atom.

(c) Ammonia is a covalent compound. Give a diagram showing how the atoms in the ammonia molecule are bonded.

(d) Outline how you would prepare and collect a specimen of dry ammonia. (A. E. B.)

36. Fluorine (9); Neon (10); Sodium (11); Magnesium (12).
F Ne Na Mg

The following refer to the above four elements, the atomic numbers of which are shown in brackets:

(a) What *two* facts about the structure of the atom of any *one* of these elements can be deduced from its atomic number?

(b) State the numbers of electrons in successive electron shells of the magnesium atom.

(c) If the symbol for the sodium ion is written Na^{+}, write similar symbols for the ions of fluorine and magnesium.

(d) Write the chemical formula for (i) magnesium fluoride, (ii) sodium fluoride.

(e) Name and explain briefly, with the aid of a diagram, the type of chemical bond linking atoms of fluorine in the molecule F_2.

(f) Caesium and sodium are both alkali metals. In what respect do the structures of atoms of these two elements resemble each other? (W. J. E. C.)

2 The Materials and Methods with which a Chemist works

2.1 HOW CHEMISTS OBTAIN 'PURE COMPOUNDS'

properties of water

If a chemist is to examine the properties of a substance, he must first of all ensure that it is pure. For example, it would be useless to use river water when investigating the properties of water, as some of the properties you may determine may be those of the impurities. The situation is seldom as simple as this and it may not be obvious that a substance is impure, i.e. that it contains two or more components. Sea-water, for instance, may look like pure water, but as you know it contains salt, so other properties in addition to the appearance must be used to find out if the compound is pure.

In this chapter you will learn some of the techniques a chemist uses in order to purify compounds. You will also learn some simple ways of determining the purity of a compound.

Solution, Filtration, and Crystallization

Experiment 2.1

† To Investigate the Effect of Adding (a) Salt and (b) Sand to Water *

Apparatus

Test-tubes, test-tube holder, filter funnel, filter paper, beaker, spatula, filter stand, glass rod, wash-bottle.

Sodium chloride, sand, and deionized or distilled water.

Procedure

(a) Half fill a test-tube with deionized water and add a spatula measure of sodium chloride. Shake the tube.

(b) Repeat (a) but use sand in place of the sodium chloride.

(c) Fold a filter paper and insert it in the funnel. Moisten the filter paper and make sure it fits against the side of the funnel. Pour the salt and water mixture from (a) into the funnel, collecting any liquid passing through, in a test-tube. Use the wash-bottle and glass rod to direct the stream of liquid into the filter funnel as shown in Figure 2.1.

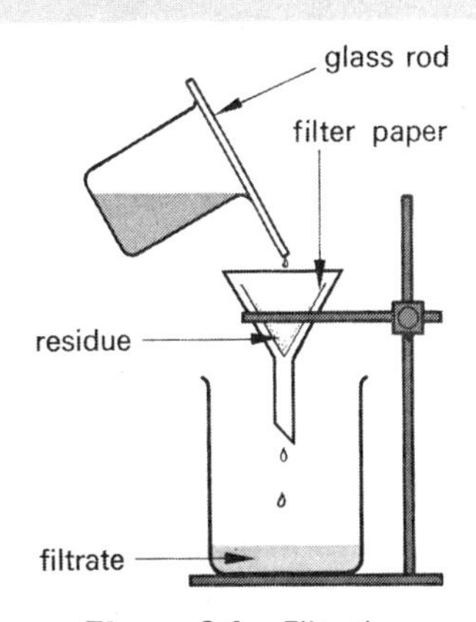

Figure 2.1 Filtration

Points for Discussion

1. If a solid disappears when stirred or shaken with a liquid it dissolves. Which of the two solids dissolves?
2. How can you regain an undissolved solid from water?
3. The liquid which dissolves a substance is called the solvent and the substance dissolving is called the solute. Together they form the solution, i.e. solute + solvent = solution.
4. The solid remaining in the filter paper after filtration is called the *residue*. Which of the two filtrations produced a residue?
5. The liquid passing through the filter paper is called the *filtrate*. Name the filtrate in (a) and (b).
6. Consider a mixture of salt and sand. How would you separate the two constituents? From what you have seen you should be able to conclude that the following procedure would serve the purpose.

Experiment 2.2
† To Separate a Mixture of Salt and Sand *

Apparatus

100 cm^3 beaker, filter paper, filter funnel, filter stand, tripod, gauze, Bunsen burner, asbestos square, stirring rod, conical flask.
A mixture of sand and salt, deionized water.

Procedure

(a) Transfer about 10 g of the mixture of sand and salt into the beaker.

(b) Half fill the beaker with deionized water. Place the beaker on the gauze supported on a tripod, and heat the water until it boils.

(c) Filter the hot solution as in Experiment 2.1, wash out the beaker with deionized water and collect all the filtrate in the conical flask.

(d) Wash the sand which remains in the filter paper with a little deionized water. Add the washings to the filtrate already collected. Keep the filtrate for the next experiment.

Points for Discussion

1. Why was the sand in the filter paper washed with a small volume of water?
2. Why were the washings added to the filtrate?
3. How would you dry the sand?
4. How could you recover the salt from the filtrate?
5. You could centrifuge off the liquid if you wished instead of filtering it. This involves placing the whole mixture in a special tube and spinning it in a centrifuge. This action is comparable to a housewife spinning her washing in a spin dryer. Centrifugal force results in the solid sand collecting in the end of the tube and the liquid may then be poured off.

Experiment 2.3
† To Crystallize Salt from a Solution of Salt in Water *

Apparatus

Evaporating basin, tripod, gauze, Bunsen burner, asbestos square, beaker small enough to allow the evaporating basin to rest on top, microscope slide, stirring rod, hand lens.
Salt solution from Experiment 2.2.

Procedure

(a) Transfer the solution of salt from Experiment 2.2 to the basin.

(b) Place the basin on a gauze supported on a tripod and heat it with the Bunsen burner.

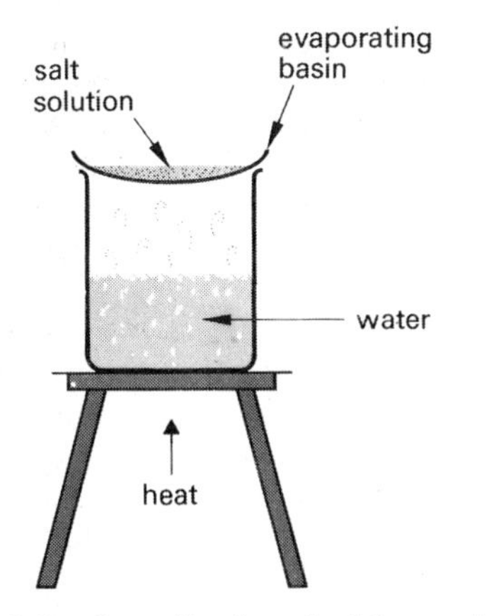

Figure 2.2 Crystallization of salt from solution

(c) When about half the liquid has evaporated, remove the basin and place the beaker half full of water on the gauze. Replace the basin on top of the beaker and heat the beaker as in Figure 2.2. When the water boils the steam given off heats the basin. The beaker of water acts as a *water bath*.

(d) From time to time remove a drop or so of the hot solution by means of a stirring rod and spot it on to a microscope slide. Allow it to cool.

(e) When crystals form on the slide, stop heating the beaker. Remove the evaporating basin and allow it to cool. It may be necessary to evaporate the solution to dryness before any solid is obtained.

(f) If any liquid remains, filter it off so that the solid formed is collected in the filter paper.

Points for Discussion

1. This process is called *crystallization*. The escape of the water vapour is *evaporation*. The crystalline solid remaining in the filter paper may be dried between pieces of filter paper.
2. Examine your crystals carefully with a hand lens and compare them with the pure sample displayed. Are your crystals as white as they could be? If you are not satisfied with your sample you may be able to purify your sample further by using the next technique.

ecolourization

Many chemicals contain traces of coloured materials as impurities, and your salt may do so. These impurities may usually be removed by boiling the substance in solution with a small quantity of activated charcoal, and then filtering off the charcoal from the solution. Activated charcoal has a very large surface area which enables it to *ad*sorb the coloured impurities and retain them, so that the filtrate becomes free of the colour. It is important that the charcoal is added to the solution *before* it is heated, and not added to the boiling solution.

Experiment 2.4
† To Remove the Colour from Brown Sugar *

Apparatus
100 cm³ beaker, tripod, Bunsen burner, asbestos square, stirring rod, spatula, filtering apparatus.
Brown sugar, activated charcoal.

Procedure

(a) Dissolve about 5 g of brown sugar in 50 cm³ of water in a small beaker.

(b) Add three spatula measures of activated charcoal and heat the suspension until it boils. After a few minutes filter the suspension. Note the colour of the filtrate in comparison with the original solution.

Points for Discussion

1. Do you think the method has removed the colour from the brown sugar?
2. Why do you think it is inadvisable to add an excessive quantity of charcoal? For most purposes the mass of the charcoal should be 1–2 per cent of the mass of the crude solid. If this does not prove satisfactory the experiment can be repeated with a fresh quantity of charcoal. Excess of charcoal may adsorb some of the material which is to be purified.

Distillation and Fractional Distillation

So far we have considered the separation of a mixture of solids. What methods can the chemist use if the sample is a solution of (a) a solid in a liquid and (b) two liquids?

In the case of (a), assuming that the solid is non-volatile (i.e. does not boil easily) then, if the solution is evaporated, the solid may be obtained by crystallization as in Experiment 2.3. However this would mean loss of the liquid. Suppose in Experiment 2.3 you wished to recover the solvent. Have you any idea how you would do this? In industry the recovery of the solvent is of vital importance as it may be toxic, flammable, or expensive. As you will see the next experiment provides yet another way of separating mixtures.

Experiment 2.5

† To Obtain Pure Water from Copper(II) Sulphate Solution *

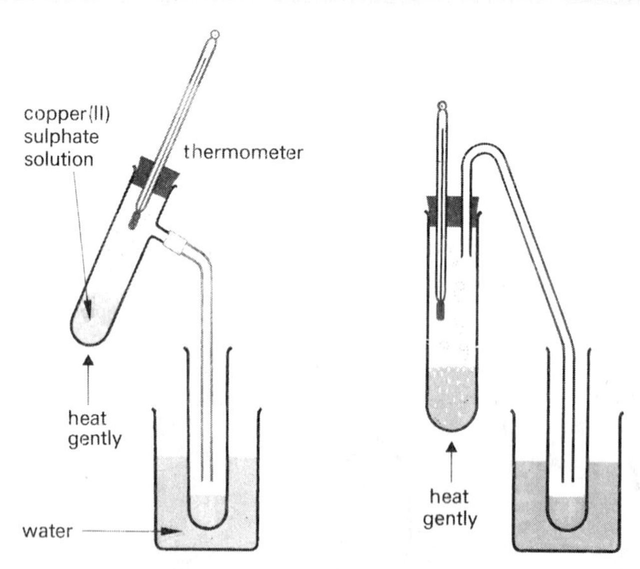

Figure 2.3 Two examples of simple apparatus which can be used to obtain pure water from copper(II) sulphate solution

Apparatus

Filter tube, test-tube, 250 cm³ beaker, rubber tubing, delivery tube, Bunsen burner, asbestos square, stand and clamp, thermometer −10–110 °C.
Copper(II) sulphate solution.

Procedure

(a) Set up the apparatus in one of the ways shown in Figure 2.3.

(b) Pour a few cm³ of copper(II) sulphate solution into the filter tube and reassemble the apparatus.

(c) Heat the solution until it boils and note the steady temperature.

(d) Continue the heating until several cm³ of the condensed liquid have collected.

Points for Discussion

1. Pure water boils at 373 K (100 °C). What was the temperature of the vapour which was later condensed into a liquid in the test-tube? What do you think the liquid was? Your answer could be verified by finding the freezing point and density of this liquid.
2. What would be left in the filter tube eventually?
3. The above operation has separated the water from the dissolved solid which remains in the filter tube. In effect the water has boiled and the vapour passes into the test-tube where it condenses back to a liquid.
4. Repeat Experiment 2.5 using a coloured ink instead of copper(II) sulphate solution.
5. This process of evaporation of a solvent followed by its condensation into a separate vessel is called *distillation*. The liquid collected in this way is called the *distillate*.
6. What was the distillate in 4?
7. How would you test your conclusion?
8. Is the distillate from 4 coloured? From your observations what do you think that ink consists of?

Immiscible and Miscible Liquids

We have so far considered a solid dissolved in a liquid. What procedure do we adopt if our mixture is composed of two liquids? In a mixture of ethanol and water it is impossible to distinguish one liquid from another because they mix in all proportions. Such liquids are said to be *miscible*. Are oil and water miscible? Liquids which do not mix together but form separate layers are said to be *immiscible*. Distillation is not normally used to separate immiscible liquids as there is a more convenient method, illustrated by the next experiment.

Experiment 2.6

† To Separate a Mixture of Methylbenzene (Toluene) and Water ***

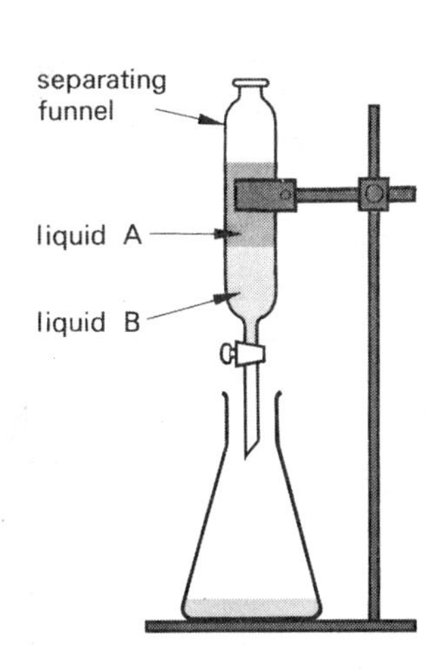

Figure 2.4 Separation of two immiscible liquids

Apparatus

Separating funnel, stand and clamp, conical flask.
Methylbenzene, deionized water.

Procedure

(a) Pour deionized water into the separating funnel until it is about one-third full. Add an equal volume of methylbenzene, and clamp the funnel in the stand as shown in Figure 2.4.

(b) Allow the two liquids to separate. Which is the water layer? How would you find out?

(c) Open the tap of the separating funnel and allow the lower layer to run into the conical flask. Close the tap when you have run as much of the lower layer into the conical flask as you can without allowing any upper layer through.

Points for Discussion

1. In this experiment, was liquid B methylbenzene or water (Figure 2.4)? (The densities of these two liquids are 0.87 g cm^{-3} and 1.0 g cm^{-3} respectively.)
2. In order to ensure that two pure liquids are collected by this method, the last small portion of liquid B and the first small portion of liquid A (i.e. the liquid around the interface) are collected separately and discarded.

Can Distillation Be Used to Separate a Mixture of Miscible Liquids?

Experiment 2.7
†To Separate a Mixture of Ethanol and Water ***

Apparatus
100 cm^3 distillation flask, Liebig condenser, rubber bungs, adapter, conical flask, tripod, sand tray, Bunsen burner, asbestos square, watch-glass, retort stand and clamp.
Ethanol and water mixture (1:2 ratio by volume), anti-bumping granules.

Procedure
(a) Pour a few cm^3 of (i) ethanol, (ii) water, (iii) the mixture, on to separate watch-glasses and try to ignite them.

(b) Pour about 50 cm^3 of the ethanol and water mixture into the distillation flask.

(c) Set up the apparatus as shown in Figure 2.5.

(d) Make sure that the cooling water passes through the condenser, and then heat the sand tray.

(e) Collect the first few cm^3 of distillate in a conical flask and note the temperature at which this has distilled.

(f) Pour a few drops of the distillate from the conical flask on to a watch-glass and try to ignite it.

Points for Discussion
1. Look up the boiling points of water and ethanol in a book of data. From your observations do you consider that the distillate is

A water,
B ethanol,
C a mixture of ethanol and water?

Give two reasons for your answer.

2. Why do you think it is important for water to pass through the condenser in the direction shown and *not* in the reverse direction?

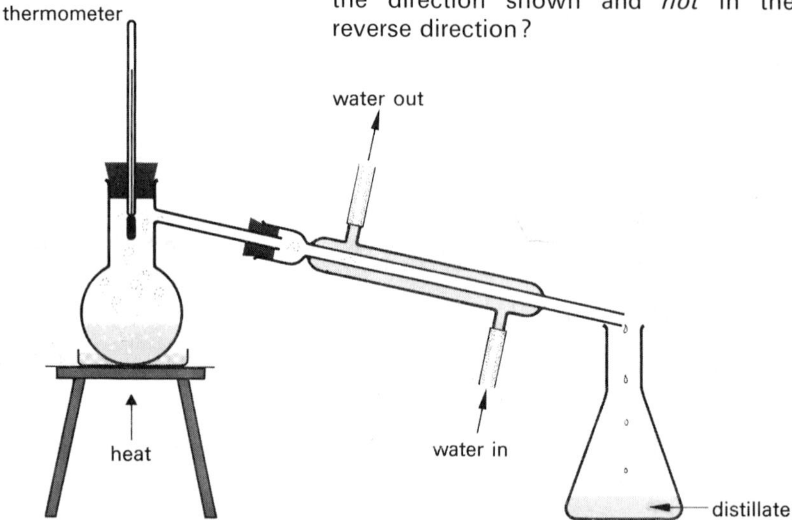

Figure 2.5 Simple distillation

Fractional Distillation

Although the distillate from the previous experiment behaved as though it was pure ethanol, it did in fact contain a small proportion of water. This inability to obtain a *complete* separation of two or more liquids by simple distillation is a problem frequently encountered in the laboratory and industry. A more efficient method of distillation is to use a fractionating column.

Experiment 2.8

†Fractional Distillation of a Mixture of Cyclohexane and Methylbenzene ***

Apparatus
Round bottom flask, fractionating column, 50 cm^3 measuring cylinder, rubber bungs, Liebig condenser, adapter, two conical flasks, tripod, sand tray, Bunsen burner, clamps and stands, thermometer 0–360 °C, glass rings or pieces of glass tubing to pack the column, suitable column lagging if necessary, anti-bumping granules.
Mixture of cyclohexane and methylbenzene (1:1 ratio by volume).

Procedure
(a) Look up the boiling points of the two liquids in a book of data.

(b) Pour about 30 cm^3 of the cyclohexane and methylbenzene mixture into the round bottom flask, and set up the apparatus as shown in Figure 2.6.

(c) Make sure that the cooling water passes through the condenser and heat the sand tray. Allow the mixture to distil very slowly.

(d) When the lowest boiling point component has distilled over, the first distillation should cease. Is the distillate cyclohexane or methylbenzene? If liquid continues to distil over at *all* times how would you know when the *first* liquid had finished distilling?

(e) As the top of the fractionating column is relatively far from the source of heat, it may not be hot enough for the second component to remain as a vapour and distil over. Modify your apparatus and continue the heating so that the second component distils off. Predict the temperature of distillation and check your prediction.

Point for Discussion
The following information will help you to understand what happened in the experiment, and you should also be able to explain why the two liquids are separated more efficiently than they would be by simple distillation.

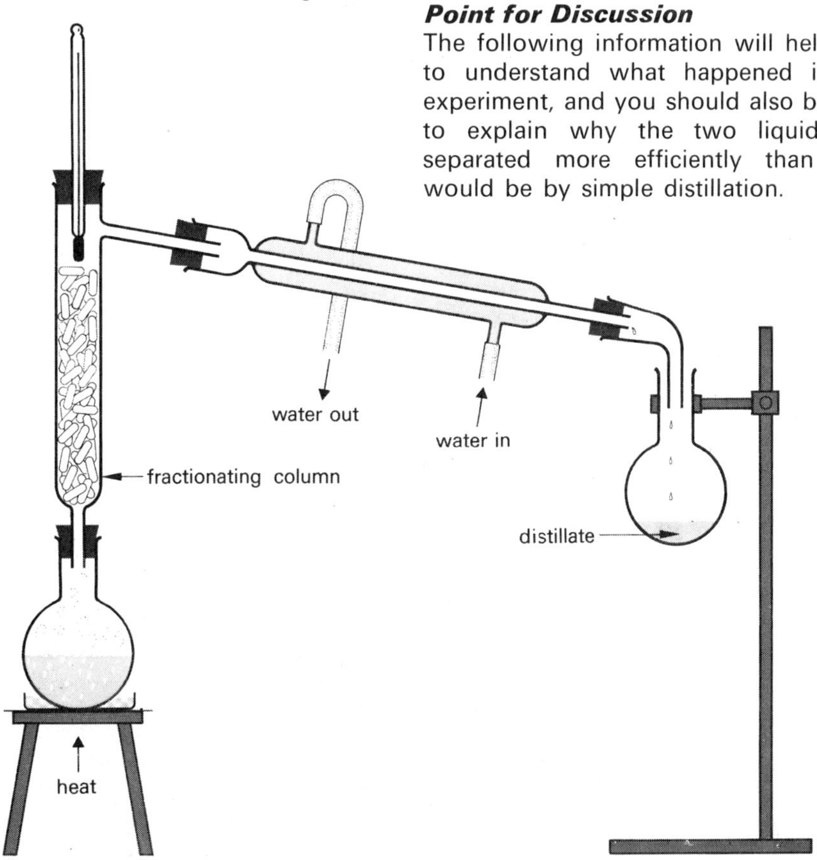

Figure 2.6 Fractional distillation of a mixture of two miscible liquids

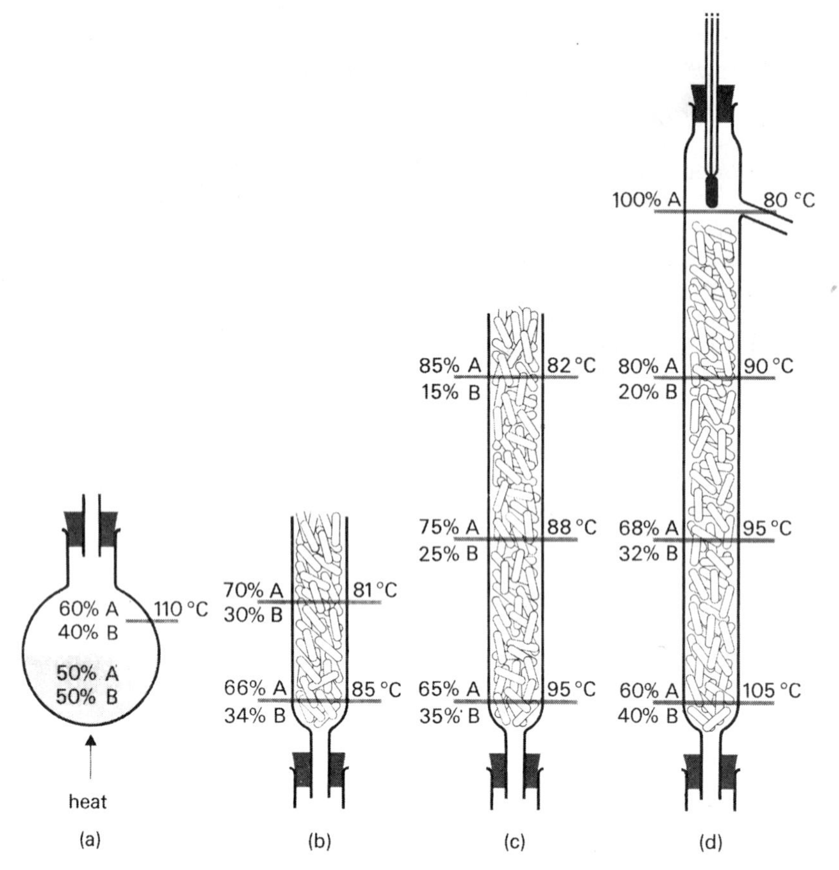

Figure 2.7 Illustration of a fractional distillation (a) initially, (b) after two minutes, (c) after ten minutes, and (d) after twenty minutes

Figure 2.7 illustrates a distillation. Suppose that the flask contains a 50:50 mixture of two miscible liquids, A and B. The boiling point of A is 353 K (80 °C) and the boiling point of B is 413 K (140 °C). (You should be able to decide which of the two liquids used in the experiment corresponds to liquid A.) When the mixture boils the vapour will contain more of the volatile component A than of B, let us say 60 per cent A and 40 per cent B. As soon as the vapour reaches the bottom of the fractionating column, which is cold, it will condense to liquid which has the same composition as the vapour, i.e. 60 per cent A and 40 per cent B. In this first stage we have increased the proportion of A in the mixture in the column from 50 per cent to 60 per cent. This condensed liquid will run back into the flask but in so doing it will meet more hot vapour from the boiling mixture in the flask; this will heat this condensed liquid up again and form additional vapour which rises a little higher in the column. This 'new' vapour now contains more of the more volatile component A than the original 'condensed' liquid, say 70 per cent A and 30 per cent B, and so when it condenses higher up in the column, the 'new' condensed liquid will contain more of the component A, namely 70 per cent. This is the position illustrated in Figure 2.7 (b). As heating continues this progressive vaporization and condensation of this vapour in a higher portion of the column means that two things are happening, (a) that the column is getting hotter with the hottest portion at the bottom, and (b) that there is a gradual increase in the proportion of component A as vapour passes up the column. Figures 2.7 (c) and (d) illustrate the progressive increase in the percentage of component A and the temperature changes within the column. Eventually if the column is efficient, the vapour emerging at the top and entering the condenser will be the pure, more volatile component, namely A. The temperature at the top of the column will still be too low to allow the vapour of B to exist there. Fractional distillation thus progressively increases the proportion of the more volatile component in the vapour.

Points for Discussion

1. What will happen to the proportions of A and B in the boiling liquid in the flask?

2. What is the temperature at the top of the fractionating column at the start of the experiment? When vapour of A reaches the top of the column the temperature will have reached 353 K (80 °C), the boiling point of A. Why is the thermometer inserted to the level of the column outlet rather than into the boiling liquid in the flask?

3. What will happen when all A has been distilled off?

4. Why are the temperatures within the column referred to as a 'temperature gradient'?

5. Figure 2.8 illustrates a Dufton column, which is an efficient fractionating column. Why do you think it is efficient?

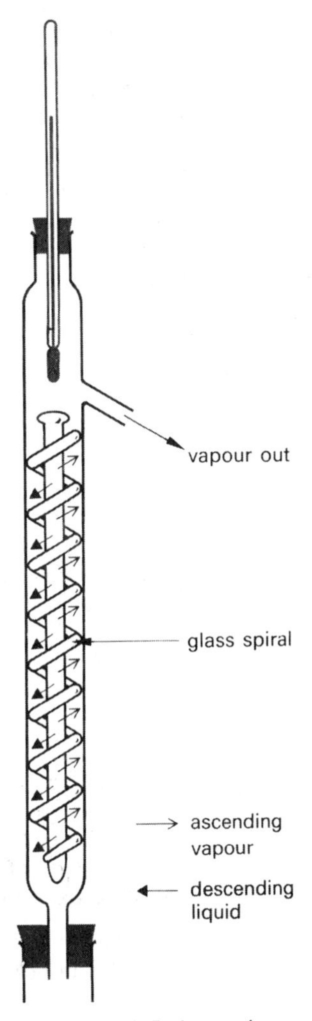

Figure 2.8 A Dufton column

Petroleum (Crude Oil)

Naturally occurring liquids usually consist of a number of different components. Crude oil, for example, contains numerous substances, and it would be extremely difficult to completely separate such a mixture, but it would be possible to divide it into several parts called fractions.

Experiment 2.9

†To Separate the Components of Crude Oil into Several Fractions **

Apparatus

Test-tube (150 × 25 mm) or filter tube, delivery tube, rubber bung, Bunsen burner, asbestos square, test-tubes, tripod, sand tray, thermometer 0–360 °C, glass wool, evaporating basins.
Crude oil.

Procedure

(a) Place a small piece of glass wool in the filter tube, and pour in about 6 cm³ of crude oil (Figure 2.3).

(b) Carefully heat the boiling tube with a small flame and collect the first cm³ of distillate in a test-tube. Record the range of temperature over which this sample is collected.

(c) Collect the next cm³ of distillate in a separate test-tube, and continue collecting subsequent cm³ portions in separate tubes until you have four samples. Note the range of temperature for the collection of each sample.

(d) Note the colour and odour of each sample. Soak separate pieces of glass wool in each of the samples and place them in dry evaporating basins. Ignite each one separately (care!) and record the appearance of the flame.

(e) Record your results in a table as follows:

Fraction number	*Boiling range*	*Colour*	*Odour*	*Appearance of flame*

Points for Discussion

1. What was the particular property of fraction one that caused it to distil before the other fractions?

2. Do you think each fraction is a pure liquid? Explain your answer.

3. The fractions probably correspond to liquids which are used in everyday life such as petrol, paraffin, heavy oil and light oil. Suggest which of these names correspond to each of your fractions.

4. The dark mass remaining in the filter tube is also important. What do you think it is?

5. A large scale version of the experiment is used in the petroleum industry to obtain a larger number of fractions.

6. Each of these fractions is itself a mixture of liquids, some of which may be individually important. For example, petrol contains a number of different hydrocarbons such as pentane and octane. Suggest how the individual components could be obtained from such a fraction.

Fractional Distillation of Petroleum (Crude oil)

Your samples from the distillation of crude oil (Experiment 2.9) boiled over a wide range of temperatures, which indicated that they were not pure. In fact, crude oil contains many hydrocarbons, and separation is a complex process. In industry, separation into main fractions is achieved by the use of a fractionating column, but the design of such a column is more complicated than the glass fractionating column used in the laboratory. The vapours of the various components pass into the

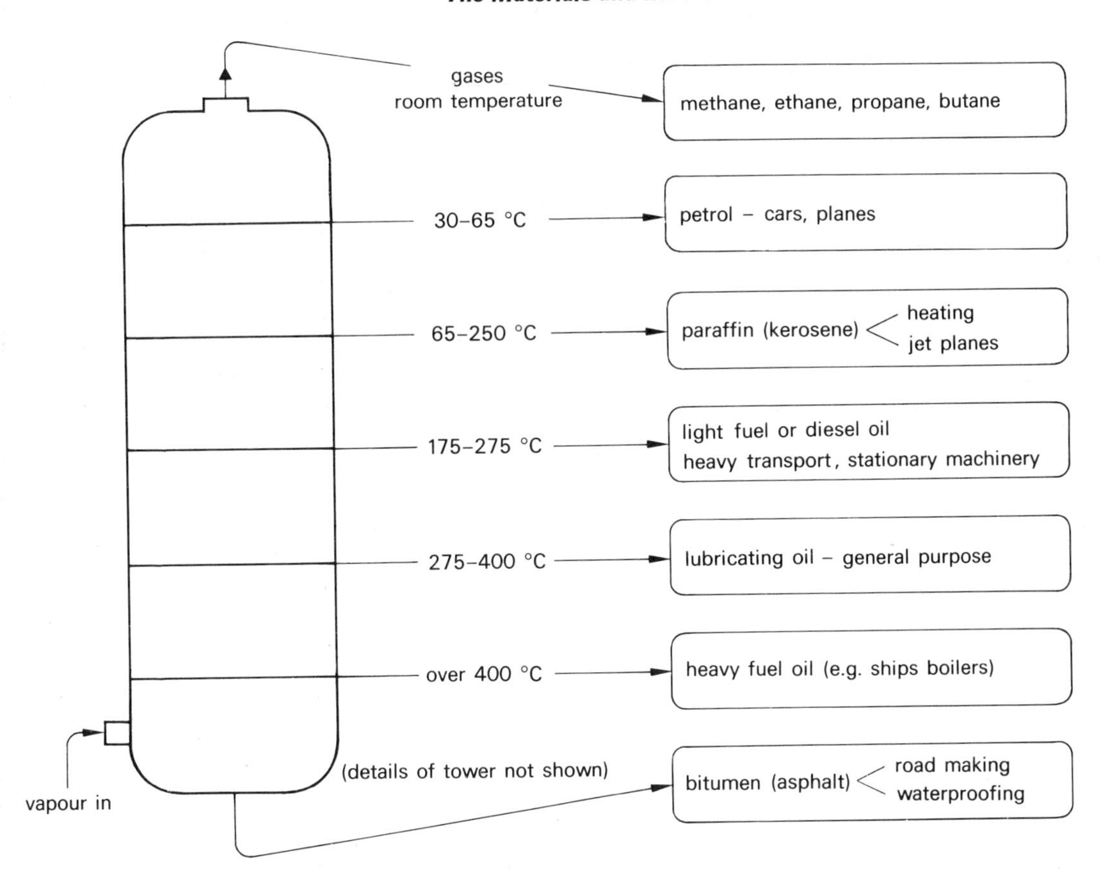

Figure 2.9 Fractional distillation of crude oil

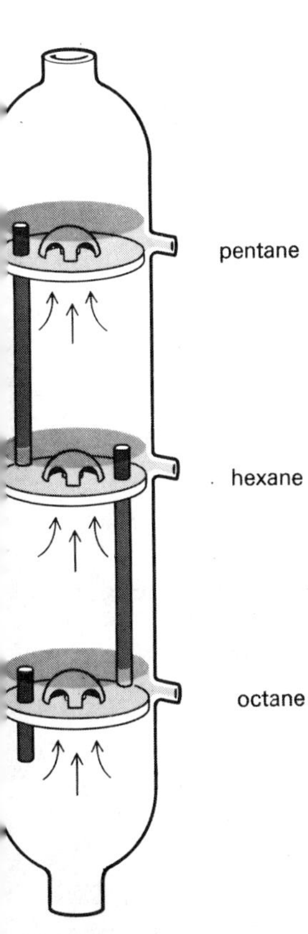

Figure 2.10 Secondary fractional distillation of a fraction from a primary distillation

fractionating tower which contains trays and bubble caps as in Figure 2.10. Each tray is a little cooler than the one below it because it is further away from the source of heat. The percentage of the more volatile components increases up the tower as shown in the earlier example (Figure 2.7).

Figure 2.9 is simplified but it shows that the first separation yields several fractions. It must be realized that this separation is only a first stage and that each of the fractions obtained is itself made up of a number of different compounds. If an individual compound or component is required then the fraction must be fractionally distilled again. Even then the compound produced may not be pure and further fractional distillation might be necessary. The whole process of separation of the components of crude oil is really a multistage process. Fortunately industry does not usually require pure compounds and such fractions as petrol, kerosene, etc., although still a mixture of compounds, are used in that state.

Chromatography

A more recent technique of separation is that of *chromatography*. The name is derived from the Greek 'chromos' which means colour, and arises from the fact that early work was carried out on coloured substances. A simple illustration is to dip a piece of chalk into ink and allow the piece of chalk to stand upright in a small dish of water. The water rises up the chalk and takes the components of the ink with it, so that bands of differing colour appear on the chalk. This phenomenon depends upon two conflicting trends, (a) the adsorption of the substances on to the chalk or paper and (b) the solubility of the substances in the solvent. These will vary from substance to substance so that each material will travel a different distance. We can think of some of the components of the ink as being 'clingers', that is, they tend to cling to the surface of the solid chalk particles. We say that they are *adsorbed*. Some components will cling more tightly than others. As the water ascends the chalk it will also tend to dissolve some of the components, but again some will dissolve more easily than others. The end result is that the components will be spread out along the length of the chalk (or at least as far as the liquid will have travelled) according to their clinging and dissolving tendencies. The ink has been separated into more than one component.

Experiment 2.10

† To Separate the Components of Black Ball Point Ink *

Apparatus

Screw-topped jar, chromatography paper or filter paper.

Solvent, e.g. propanone (acetone) or ethanol.

Procedure

(a) Take a piece of chromatography or filter paper about 1.5 cm wide and slightly less than the height of the jar in length. Fold in two along its length as in Figure 2.11.

(b) Pour the solvent into the jar to a depth of about 0.4 cm.

(c) About 1 cm from one end of the paper mark a small but intense dot in black ball point ink. Stand the paper in the jar with the marked end to the bottom of the jar. Screw on the top of the jar.

(d) Allow the jar to stand undisturbed until the level of the ascending liquid is about 1 cm from the top. Remove the paper and dry it.

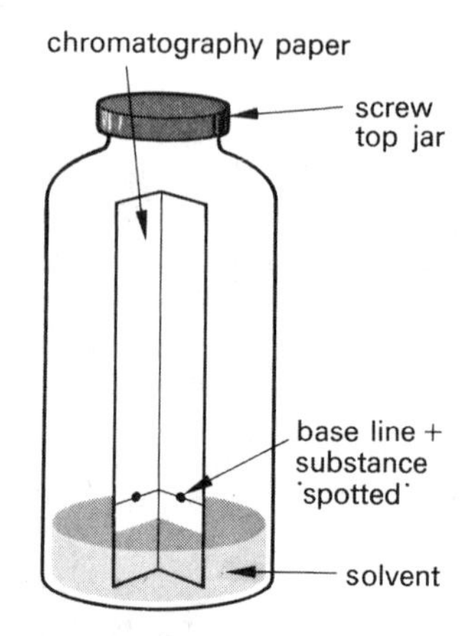

Figure 2.11 Chromatographic separation of the components of ball point ink

Points for Discussion

1. Is black ball point ink a single substance? Using your observations explain your answer.

2. The resulting separation of the components into several parts at varying heights on the paper is called a *chromatogram*.

3. In such an experiment the resulting chromatogram was composed of three coloured spots one above the other. Which of the following statements is correct?

A The lowest spot was the least adsorbed and the highest the most adsorbed.
B The middle spot was the least soluble of the three spots, and the lowest spot the most soluble.

C The lowest spot was the most adsorbed and the highest spot the least adsorbed.

Experiment 2.11

†Is the Green Pigment in Nettles a Single Substance? *

Apparatus

Pestle and mortar, teat pipette with finely drawn jet, Petri dish, filter paper, chromatography paper.
Nettles and propanone (acetone).

Procedure

(a) Place some chopped nettles in the mortar and just cover them with a little propanone. Crush and stir them by means of the pestle.

(b) Lay the filter paper over the top of the Petri dish, and using the teat pipette carefully spot one drop of the extract of nettles on to the centre of the paper as in Figure 2.12. Allow the spot to dry before adding another spot to the same area. Repeat this procedure about four times so that the extract is concentrated in the centre of the paper.

(c) Using the teat pipette drop propanone on to this concentrated extract and allow the solvent (propanone) to spread out as far as possible. Repeat this procedure about four times.

Points for Discussion

1. How many coloured bands can you see on your chromatogram? There may be others which are not separated by this method. The green pigments are chlorophylls, and the yellow are xanthophylls.

2. Do you think the colouring matter in nettles is a pure substance?

3. Which of the coloured bands has been adsorbed the most?

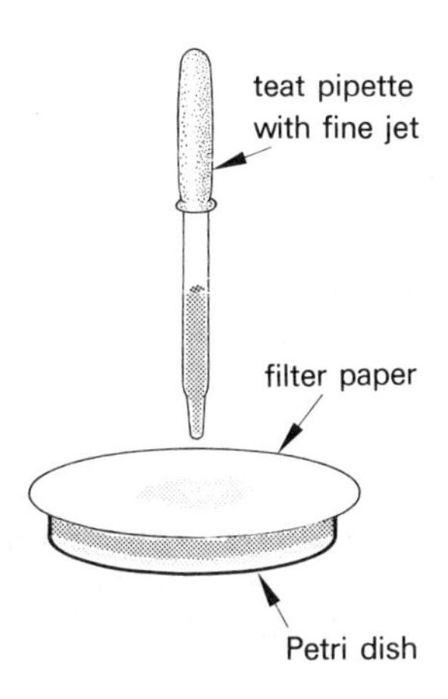

Figure 2.12 Chromatographic separation of the components of the green colouring matter of plants

Chromatographic Separation of Colourless Materials

Although the term chromatography comes from the Greek word for colour, a chemist may wish to know if a mixture contains a colourless component. You may be shown the use of ultra-violet light in this respect, but remember that you must not look at the ultra-violet source; ultra-violet light is extremely dangerous to the eyes.

Summary

You will realize from your work on chromatography that the basic principle involves a stationary phase (on which substances are adsorbed to varying degrees) and a moving phase which causes the substances to move with it. The experiments you have seen have involved a liquid solvent as the moving phase, but the techniques can be extended to use a gas as the moving phase. This is called *gas chromatography*, and is used as an analytical tool in hospitals, industry, research, and agriculture. For example, the chemical analysis and separation of a group of very similar substances, such as in a petroleum fraction, is only made feasible by such processes as gas chromatography.

Sublimation

Finally we turn our attention to a method of separation which is simple in principle but of more limited application.

Experiment 2.12
† To Examine the Effect of Heat on Ammonium Chloride *

Apparatus
Rack of test-tubes, test-tube holder, spatula, Bunsen burner, asbestos square. Ammonium chloride.

Procedure
Place a spatula measure of ammonium chloride in a test-tube and heat it, gently at first and then more strongly for a minute or so. Record your observations. Keep the test-tube and its contents undisturbed until the next experiment.

Points for Discussion
1. What normally happens to a solid when it is heated?
2. Does this solid behave normally when heated? Explain your answer.
3. Your observations can be explained in two ways:

(a) The ammonium chloride changed chemically, i.e. the white solid at the top of the tube was *not* ammonium chloride, but a different chemical formed as the product of heating the ammonium chloride.

(b) The ammonium chloride could have missed out the normal liquid phase and changed directly into a vapour, which in the cooler part of the tube resolidified unchanged, i.e. the white solid at the top of the tube is *still* ammonium chloride which has simply changed state and then reformed.

How can we decide which of these ideas is correct? The next experiment should provide an answer.

Experiment 2.13
A Simple Chemical Analysis **

Principle
A chemist can often identify an unknown substance by analysis. If he suspects that the compound is ammonium chloride, he would test for an ammonium compound and for a chloride. If in Experiment 2.12 hypothesis (a) is correct, the original

compound will give a positive result for both tests but the product will not. If hypothesis (b) is correct the original compound and the product will both give a positive result for each test.

Apparatus
Rack of test-tubes, test-tube holder, test-tube from Experiment 2.12, Bunsen burner, asbestos square, spatula.
Ammonium chloride, silver nitrate solution, sodium hydroxide solution, dilute nitric acid, and red litmus paper, sodium chloride, magnesium chloride, potassium chloride and ammonium sulphate.

Procedure
1. If you are not familiar with the tests for ammonium compounds and chlorides, look up these tests (see pages 79 and 90 respectively) and practice the tests on the ammonium and chloride compounds provided.
(Note that in the test for an *ammonium* compound when an alkali is added and the solution warmed, a reaction occurs which liberates *ammonia* gas. It is this gas and *not* the ammonium compound which is being tested for by the damp red litmus paper. However, the fact that ammonia gas is liberated confirms that the compound which reacted with the alkali was an ammonium compound.)

2. Devise a suitable procedure to confirm whether idea (a) or (b) is correct.

Points for Discussion
1. You should have found that idea (b) was correct. The process is called *sublimation* and the substance formed in the *cooler* part of the tube is called the *sublimate.*

$$\text{Solid phase} \underset{\text{cool}}{\overset{\text{heat}}{\rightleftharpoons}} \text{Vapour phase}$$

2. Other examples of substances which sublime are benzoic acid, anhydrous aluminium chloride, and iron(III) chloride.

Experiment 2.14
†To Separate a Mixture of Sodium Sulphate and Ammonium Chloride *

Apparatus and Procedure
Devise a simple method of separating the mixture. Confirm that your idea has been successful by testing appropriate samples for an ammonium compound, a chloride, and a sulphate (see page 89).

Points for Discussion
1. What were the positive tests given by the sublimate?

2. Do you think that sodium sulphate sublimes?

3. Did you succeed in achieving your separation?

Summary

So far in this chapter we have considered some of the techniques which may be used to separate mixtures. The following table is a brief summary of these methods.

Technique	*What it separates*
Solution and filtration	An insoluble substance from a soluble one
Crystallization	A crystalline solid from its solution
Decolorization	Impurities which discolour the sample
Separating funnel	Immiscible liquids
Distillation (simple)	A liquid from a solution, or two liquids with widely differing boiling points
Fractional distillation	Liquids with boiling points close together and a better separation of other liquid mixtures
Chromatography	Substances which are adsorbed to differing extents on columns or paper, and which differ in their solubilities in particular solvents
Sublimation	Solids, one of which sublimes on heating

2.2 CRITERIA OF PURITY

The methods described so far have been methods for separating mixtures. How can we be sure that the components have been separated efficiently, or in other words, how can we be sure that they are now pure? Any pure substance has its own distinctive physical and chemical properties. Recognition of a substance may be possible by using our sense of smell, sight or taste, although these are not always advisable; it might be unwise to taste or smell the substance. Even when such properties can be used to identify a substance, they do not indicate the *degree* of purity; for this purpose we must adopt methods which are capable of more exact measurements. If we determine the boiling point, freezing point, density, refractive index, etc. for a given substance, and if these values all agree with those obtained from a book of data, it is more than likely that the substance is pure. If a quick indication of the degree of purity is required we usually determine the melting point if the substance is a solid, and its boiling point if it is a liquid. Under normal laboratory conditions these techniques are suitable only for melting points and boiling points up to about 633 K (360 °C).

Experiment 2.15

† To Determine the Melting Point of (a) Pure Naphthalene and (b) Impure Naphthalene (a Mixture of Naphthalene and Camphor) ***

Apparatus

Test-tube (125 × 25 mm), capillary tube, thermometer 0–110 °C, wire stirrer, bung with V notch for the stirrer, clamp and stand, Bunsen burner, asbestos square, white tile, spatula, fuse wire.
Naphthalene and camphor.

Procedure

(a) Place a few crystals of naphthalene on the white tile and crush these to a fine powder by means of the spatula. Keep some of the powder until later.

(b) Seal the end of a capillary tube in a Bunsen flame and allow it to cool. Push the open end of the tube through the powder and up against the flat end of the spatula. Push the naphthalene to the bottom of the tube using the fuse wire. Repeat if necessary until about 0.5 cm length of tube is filled.

(c) Attach the tube to the bulb of the thermometer by means of two small pieces of rubber cut from the end of a piece of small bore rubber tubing (Figure 2.13).

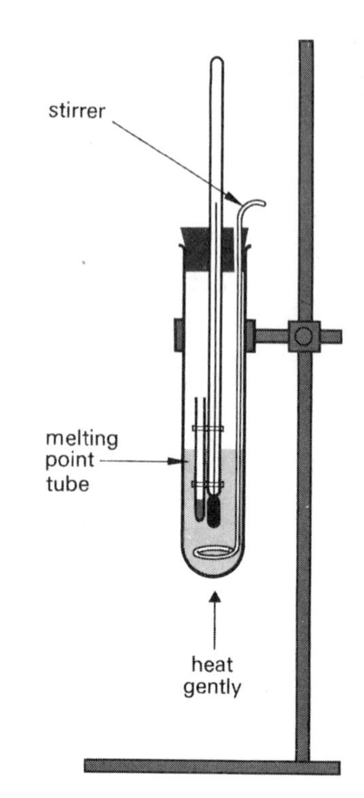

Figure 2.13 Melting point determination

(d) Insert the bung carrying the thermometer and stirrer into the test-tube, which is half filled with water. Clamp the tube upright in the stand. The open end of the capillary tube should not be in the water.

(e) Carefully and *slowly* heat the test-tube with a small flame. Stir the water constantly and note the temperature at which the solid melts.

(f) Prepare a sample of impure naphthalene by mixing the rest of the crushed naphthalene with a little crushed camphor. Repeat the appropriate procedures so as to determine the melting point of the impure naphthalene.

(g) Compare the results obtained by the other groups in the class (e.g. in columns on the blackboard). Find the class average for the m.p. of pure naphthalene. Why would it not be sensible to find the class average for the impure sample?

Points for Discussion

1. What was the purpose of the water in the apparatus?

2. Which of the two, the pure sample or the impure sample, has the sharpest melting point (i.e. the smallest melting range)?

3. Use a book of data to find the melting point of pure naphthalene.

4. Do impurities alter the melting point of a solid? If so, how? Does the class average suggest the original sample of naphthalene is really pure?

5. Can you account for the much wider range of melting point found by the class for the impure naphthalene?

Experiment 2.16
† A More Accurate Method of Determining a Melting Point **

Apparatus

Test-tube, 250 cm^3 beaker, tripod, gauze, Bunsen burner, asbestos square, thermometer 0–110 °C, clamp and stand, stop-clock.
Naphthalene.

Procedure

(a) Put about 2 cm depth of naphthalene into the test-tube and clamp it in a beaker which is half filled with water (Figure 2.14). Place the thermometer in the tube.

(b) Heat the beaker on a gauze supported on a tripod. When the naphthalene melts note the temperature, and continue heating until the temperature of the melt rises another 10 degrees. Use the thermometer to stir the liquid.

(c) Lift the clamp and test-tube free of the water, and wipe the tube.

(d) Stir the molten naphthalene with the thermometer, record the temperature and at the same time start the stop-clock. Record the temperature every half minute. When the naphthalene is beginning to solidify, continue to stir the pasty mass, but when it becomes stiff leave the thermometer embedded in it. Continue to record the temperature every half minute until it has dropped a further 10 degrees.

(e) Remove the thermometer by replacing the tube in the beaker of hot water and remelting the naphthalene.

(f) Draw a graph of your results, bearing in mind that the factor you control (time) is always plotted on the horizontal (x) axis, and the dependent variable (temperature) along the vertical axis (y).

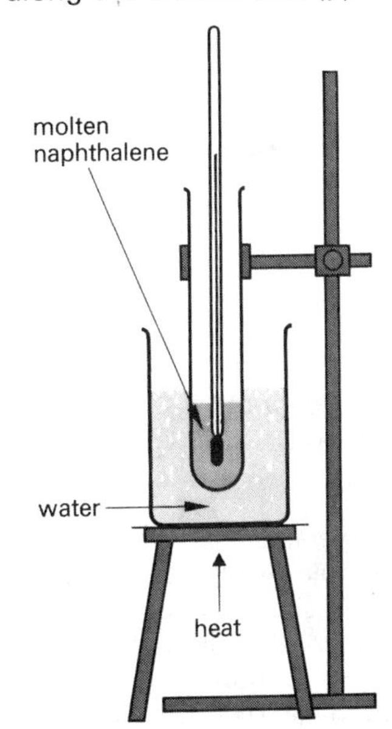

Figure 2.14 Apparatus for determining the cooling curve for naphthalene

Points for Discussion

1. When the naphthalene was removed from the source of heat and allowed to cool, did the temperature *continue* to fall at a steady rate? Your graph should provide an answer.

2. As the molten naphthalene begins to solidify it gives out its latent heat of fusion and this keeps the temperature constant while it is freezing. This constant temperature is the melting point and it is a more accurate value than that obtained in Experiment 2.15.

Summary

Pure solids melt sharply within a narrow range of temperature. Impure solids melt over a wider range of temperature and less sharply, and the depression of the melting point is directly proportional to the amount of impurity present.

We may make use of these facts to confirm our observations regarding the purity of a substance. If the observed melting point is lower than it should be, then the presence of an impurity is indicated. Recrystallization of the substance should furnish a sample which will give a higher melting point. If this value agrees with the correct melting point as obtained from a book of data, this will indicate a pure substance. If not, further recrystallization may be necessary.

In addition, these facts may be utilized in the so-called mixed melting point test for identifying an organic substance. If an unknown solid Z is suspected of being, let us say, benzoic acid (m.p. 394 K (121 °C)), then a quantity of Z is mixed with some benzoic acid (e.g. two parts Z to eight parts benzoic), and the melting point determined. If Z is benzoic acid then the determined melting point should be 394 K (121 °C). If not, then Z will act as an impurity in the benzoic acid and depress its melting point. The determined melting point will now be lower than it should be. In this method it is important to make the proportion of the known substance in the mixture the largest, as otherwise the melting point of the unknown substance Z will be the important factor.

If the melting point of the substance is between about 363 K (90 °C) and 493 K (220 °C) then a water bath is not suitable for melting the sample. In these instances a convenient heating liquid is dibutyl phthalate which can be heated to about 493 K (220 °C) without decomposition. Above this temperature it is probably better to use an electrically heated apparatus.

Boiling Points

Similarly a pure liquid will have a reasonably sharp boiling point, although this may vary over a range of several degrees with variations in atmospheric pressure.

Experiment 2.17

†To Determine the Boiling Point of Ethanol ***

Apparatus

Simple distillation apparatus (see Figure 2.5), small round bottom flask, beaker in which the flask will fit, tripod, gauze, Bunsen burner, asbestos square, thermometer 0–110 °C, conical flask, measuring cylinder (10 cm³).
Ethanol, antibumping granules.

Procedure

(a) Put a few antibumping granules into the flask and pour in about 5–10 cm³ ethanol.

(b) Assemble the apparatus for simple distillation (Figure 2.5).

(c) Immerse the lower end of the flask in a beaker of water resting on a tripod and gauze so as to avoid contact between the flask and the naked flame; ethanol is flammable. Note that the bulb of the thermometer is placed in the neck of the flask level with the outlet from the flask to the condenser.

(d) Heat the beaker of water slowly and when the ethanol begins to distil into the conical flask, note the temperature recorded by the thermometer. This will be the temperature of the vapour, and if the liquid is pure it will also be the temperature at which vapour is in equilibrium with the boiling liquid, i.e. the boiling point of the liquid.

Points for Discussion

1. What was the boiling point of the liquid?

2. Even if the temperature of the water is 20 degrees above the temperature of the liquid, the boiling point will not alter. Why?

3. Why do you think that this method is less accurate than the melting point determination?

Experiment 2.18

† To Determine the Boiling Point of a Mixture of Carbamide (urea) and Ethanol***

Apparatus

As for Experiment 2.17 or as in Figure 2.15(a).

Ethanol, carbamide.

Procedure

(a) Use a mixture of 10 cm^3 ethanol and 2 g carbamide.

(b) Put the bulb of the thermometer into the liquid in the flask and <u>not</u> into the vapour at the outlet of the flask to the condenser.

(c) Determine the boiling point of the mixture.

Points for Discussion

1. Does the addition of the carbamide have any effect upon the boiling point of the ethanol? If so, what effect?

2. How does this effect compare with that of the presence of impurities upon the melting point of a solid?

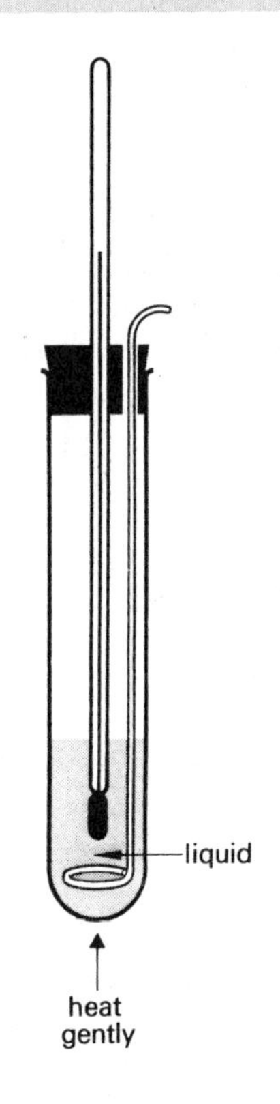

Figure 2.15 The determination of boiling points

Summary

As stated earlier, the boiling point of a pure liquid is reasonably sharp but can vary over a few degrees with variations in atmospheric pressure. This is in contrast to the sharper melting point of a solid. The presence of impurities has the opposite effect to that in solids because they cause a *rise* in the boiling point, and this rise is directly proportional to the quantity of impurity present.

2.3 ELEMENTS AND COMPOUNDS

Suppose we have impure copper(II) sulphate and have tried to purify it by various techniques such as chromatography, filtration, etc. Eventually such methods will fail to separate it further, and the substance will be pure. However, it will still contain more than one kind of atom, i.e. it is a *compound.* A compound may be defined as a substance which is composed of two or more elements which are chemically joined together and which cannot be separated by physical means. Can a compound such as copper(II) sulphate be broken down further? Perhaps more drastic methods, such as heating or the use of electricity, will decompose it.

Experiment 2.19

†The Action of Heat on Copper(II) Sulphate **

Apparatus

Evaporating basin or crucible, tripod, gauze, Bunsen burner, asbestos square, pipe-clay triangle, top pan balance, tongs. Copper(II) sulphate crystals.

Procedure

(a) Find the mass of the crucible (or the evaporating basin).

(b) Measure accurately about 10 g of copper(II) sulphate into the crucible.

(c) Set up the apparatus as shown (Figure 1.20) and heat gently at first and then strongly. When the reaction appears to be complete, stop heating, and allow the crucible and lid to cool. Reweigh.

(d) Reheat the crucible and contents for about three minutes. Cool and reweigh.

(e) Repeat the procedure if necessary until two consecutive weighings agree. Record the masses.

Points for Discussion

1. Was there any difference between the initial and final masses? If so, what conclusion can you draw?

2. Do you think that the effect of heating has broken down the pure compound copper(II) sulphate any further? If so, do you think you have split up the compound as far as possible? Would the effect of electricity be more drastic?

Experiment 2.20

†To Pass an Electric Current through Copper(II) Sulphate Solution *

Apparatus

Source of low voltage d.c. supply, small beaker, two lengths of wire fitted with crocodile clips, carbon electrodes. Copper(II) sulphate solution.

Procedure

(a) Half fill the beaker with copper(II) sulphate solution. Insert carbon electrodes into the solution and connect them by means of the connecting wires to the low voltage supply.

(b) Allow the current to pass for some minutes. Observe any reactions at the electrodes.

Points for Discussion

1. What do you think is formed at the negative electrode?

2. Do you think it is possible to further simplify this product? The following experiment should help you to decide.

Experiment 2.21
To Determine the Effect of Heating Copper in Air *

Apparatus
Test-tube, test-tube holder, Bunsen burner, asbestos square, top pan balance or other suitable balance.
Copper foil.

Procedure
(a) Find the mass of the test-tube. Place a piece of copper foil about 3 cm × 0.5 cm in the tube and find the new mass.

(b) Heat the tube strongly for about two to three minutes.

(c) Cool the tube and find its mass. Reheat it for a minute. Cool and reweigh.

(d) If necessary repeat the procedure until two consecutive weighings agree.

Points for Discussion
1. What was the purpose of procedure (d)?

2. Was the copper broken down further by heating? Explain your answer.

Summary

The copper(II) sulphate (a pure compound) can be broken down into something simpler, i.e. copper. This is not a purification; the copper(II) sulphate is decomposed. The methods used are much more drastic than the physical methods used previously to separate mixtures. However, heat and electricity do not alter the chemical nature of copper, nor simplify it further. No matter how we try, we cannot further simplify copper. Substances such as copper which cannot be split up into anything simpler by chemical means were given the name elements by Boyle. You will know that the number of electrons possessed by an atom is equal to its atomic number. Thus in any given element, all the atoms have the same number of electrons and react chemically in the same way, even though the element may contain isotopes (page 21).

Metals and Non-metals

've the appearance of specimen

There are about ninety elements which occur naturally. Although we cannot simplify elements by chemical processes, we can subdivide them into two main groups, *metals* and *non-metals*. What properties do we look for in deciding how to classify a particular element? Both physical and chemical properties can be used to make a decision, but before referring to some suitable experiments, it is useful to learn some of the terms which are applied to the properties of materials.

A substance is *malleable* if it can be hammered or flattened without returning to its former shape, and does not fracture.

Ductile materials are those which can be drawn into wire.

Brittle materials are those which are fragile and tend to fracture on bending.

Experiment 2.22
† Physical Differences between Metals and Non-metals *

Apparatus
6 V bulb and holder, lengths of wire fitted with crocodile clips at both ends. Low voltage d.c. supply.
Metal sheet or rods of zinc, aluminium, copper, and lead; pieces of carbon, lump silicon.

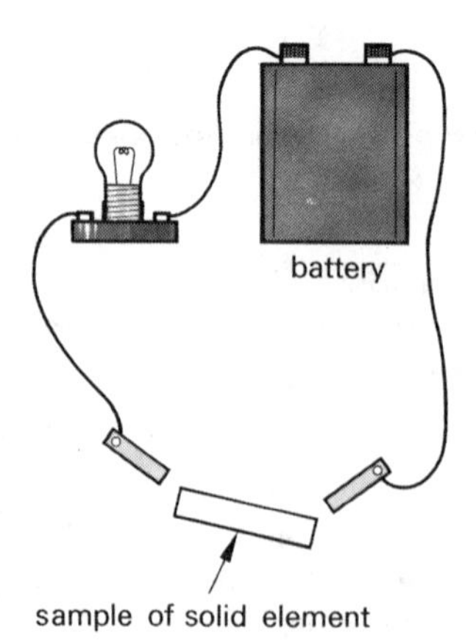

Figure 2.16 The testing of elements for conduction of electricity

Procedure

(a) Observe the appearance of each specimen. Record your results in a table.

(b) Try hammering a little of each specimen if this is allowed.

(c) Set up the incomplete circuit as shown in Figure 2.16.

(d) Take each specimen in turn and connect it into the circuit by means of the crocodile clips. Make a note of your observations.

(e) Hold a little of each specimen in a pair of tongs and heat it in a Bunsen flame. Does the specimen conduct heat? Does it melt?

(f) Find the melting points and boiling points of the elements from a book of data.

Points for Discussion

1. We could add other measurable physical properties such as density and specific heat. From your observations can you now divide the specimens into two classes? Which of these classes contains the metals?

2. Some elements (e.g. carbon) are not easy to classify as metallic or non-metallic using only physical properties. In these cases chemical properties provide powerful additional evidence. For example, some metals react with acids to liberate hydrogen (see Experiment 3.3); non-metals never do.

3. The following observations were made with regard to three elements X, Y and Z. Deduce whether X, Y and Z are metals or non-metals.

X is a black brittle solid which sublimes when heated. It is a poor conductor of electricity.
Y is a hard malleable solid, melting point 1728 K (1455 °C).
Z is a solid, density 7.1 g cm^{-3}, which melts at 692 K (419 °C) and boils at 1180 K (907 °C). It conducts electricity.

2.4 COMPOUNDS AND MIXTURES

In the work in this chapter we have been concerned with the separation of mixtures, and in the course of these experiments we have introduced two other terms to define the nature of the substances involved in a mixture. These terms are *compounds* and *elements*. A mixture may be composed of two or more compounds and/or elements. You will now further investigate the difference between elements, compounds, and mixtures.

Experiment 2.23
† To Investigate the Properties of the Elements Iron and Sulphur **

Apparatus
Rack of test-tubes, test-tube holder, Bunsen burner, asbestos square, watch-glasses, splints, spatula, magnet, hand lens, access to fume cupboard.
Iron filings, powdered sulphur, carbon disulphide, dilute hydrochloric acid.

Procedure
(a) Place samples of the two elements on separate watch-glasses, and examine the appearance of each, using a hand lens.

(b) Heat a little of each specimen in a test-tube.

(c) Place each specimen on a piece of paper and stroke the underside of the paper with the magnet.

(d) Using no more than a quarter of a spatula measure of each specimen in a test-tube, add approximately 1 cm^3 carbon disulphide (in a fume cupboard). Shake well and allow the tubes to stand in a test-tube rack in the fume cupboard until any solid residue settles. Using a teat pipette transfer a little of the liquid from each tube on to separate watch-glasses and allow the liquid to evaporate in the fume cupboard.

(e) To a little of each specimen in separate test-tubes, add about 2 cm^3 of dilute hydrochloric acid. Test any evolved gas for hydrogen as on page 74.

Experiment 2.24
† To Prepare a Compound from Iron and Sulphur **

Apparatus
Tin lid, tripod, Bunsen burner, asbestos square, access to fume cupboard.
Iron filings and sulphur.

Procedure
(a) Measure out about 5 g of iron filings and 5 g of powdered sulphur. Mix the two together and pour half the mixture into the tin lid. Keep the other half.

(b) Place the tin lid on the tripod, in a fume cupboard, and heat the lid.

(c) When reaction occurs remove the burner.

(d) Allow the residual solid to cool and use it in (e).

(e) Carry out the same procedures as in Experiment 2.23 on (i) the mixture which was reserved in (a) and (ii) the solid from (d), but test any evolved gases for hydrogen sulphide also (page 229).

(f) Set out your observations for the elements, the mixture, and the compound, in the form of a table as follows.

	Sulphur	*Iron filings*	*Mixture*	*Compound*
1. Appearance				
2. Effect of heat				
3. Magnet				
4. Carbon disulphide				
5. Action of dilute hydrochloric acid				

Points for Discussion

1. Do the elements still retain their original properties when mixed together?

2. Do the elements retain their original properties after they have been joined together to form a compound? Explain your answer.

3. Are the properties of the compound an average of those of the components or are they quite different? Explain your answer.

4. The compound formed in Experiment 2.24 is iron(II) sulphide (FeS). A mole of iron atoms weighs 56 g and a mole of sulphur atoms 32 g. In what ratio should they be mixed to form the compound? (Refer to Section 1.6.) The actual ratio given was nearer two moles of iron filings to three moles of sulphur. Why was the sulphur in excess?

5. When a rod of iron is heated to red heat, what immediately happens when the burner is removed? How does this compare with what you saw in Experiment 2.24 when the burner was removed? What do you think could be continuing to provide the energy after the burner has been removed?

Summary

What conclusions can we draw from the results? Both iron and sulphur are elements. The mixture and compound contain these elements, yet as we see from the results, their properties are entirely different. In the case of the mixture, iron and sulphur still exist as elements and therefore exhibit their own properties, both physical and chemical. In the case of the compound the properties are entirely different from those of its constituent elements (Figure 2.17). These differences are summarized in Table 2.1.

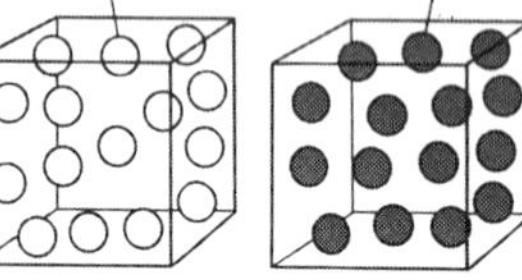

ELEMENTS All atoms of an individual element are alike

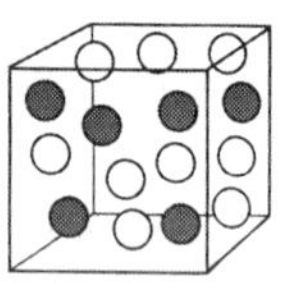

MIXTURE Atoms of the individual elements retain their own identities

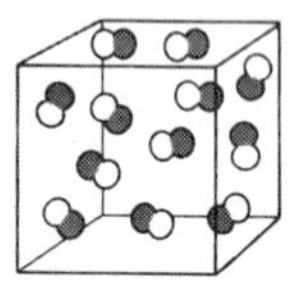

COMPOUND Individual 'units' composed of two atoms in this case

Figure 2.17 An illustration of the differences between elements, mixtures, and compounds

Table 2.1 Properties of compounds and mixtures

Compound	*Mixture*
1. Always homogeneous	May be homogeneous or heterogeneous
2. Always contains the same elements in a fixed and definite mass ratio, i.e. of constant composition	The composition is variable, e.g. 50:1, 1:20, etc.
3. The melting and boiling points are fixed and constant, although the boiling point may vary with atmospheric pressure	Does not melt at a single temperature, and the melting point varies with composition Boiling point is not constant and varies with composition
4. Cannot be decomposed by physical or mechanical means	Separation is possible by physical or mechanical means, e.g. using a magnet, filtering, distilling, etc.
5. Its properties are totally unlike those of its constituent elements	Properties are those of its constituent elements and/or compounds

When a compound is formed from its elements, the elements are *chemically combined* together and no longer retain their individual reactions. The compound formed has a completely new set of reactions. As a further example, consider the two elements sodium and chlorine. Sodium is a soft grey metal which can be easily cut and which reacts very vigorously with water. Chlorine on the other hand is a poisonous gas which reacts to a small extent with water to produce an acidic solution which bleaches damp litmus paper. However, the compound formed from these two elements (sodium chloride, common salt) shows no metallic properties, does not react with water and you 'eat' it every day in or with your food.

2.5 AN INVESTIGATION INTO THE NATURE OF 'LIQUID X'

Each of the techniques introduced in this chapter may not seem important in itself, but collectively they can give a lot of information about a particular substance. The following problem is posed to show how many of these techniques may be used to track down an unknown substance.

Aim

(i) To determine whether the 'unknown' substance X is (a) a pure liquid or (b) a mixture.
(ii) If it is (a) to find out if it is an element or a compound.
(iii) If it is (b) to separate it into its components and identify them.

How do you go about this?

Your first thought will probably be that the task is almost impossible as X could be one of many thousands of different liquids. In scientific investigations into unknown materials the first thing to do is to record as much information about the material as possible, and then to decide which information is useful for your purpose. It is all too easy to approach this type of problem in a disorganized way by making random and unrelated tests. Suppose you want a book on zoology from the library. You do not start at the first shelf you come to and look through the books until you obtain what you want. You would go first to the science section, and then look for the subsection zoology. You would finally reach your goal by looking for the author's name under its correct alphabetical sequence.

Let us now try to solve our problem in a similar logical way. What broad classifications could liquid X fit into? If it is a mixture it could consist of:

A a gas dissolved in a liquid,
B a mixture of two or more miscible liquids,
C a solid dissolved in a liquid.

If it is a pure liquid it could be:

D an element or E a compound.

Is X a Mixture?

1. How would you show the presence of a dissolved gas in a liquid? Hint: lemonade is a liquid containing a dissolved gas. How could you show that this is so?
2. How would you find out if X is a mixture of two or more liquids? Experiment 2.8 may provide an answer.
3. How would you find out if X contains a dissolved solid? Experiment 2.3 may help.
4. Each of these questions can be answered by using a procedure which you have recently seen demonstrated. Suggest how this apparatus can be used to provide an answer to these three questions.

The next step in the investigation may not be necessary depending upon your conclusions so far.

Is X an Element or a Compound?

1. How many elements are there, and how many of these are liquids at room temperature? Could X be one of these?
2. If you are still investigating liquid X at this stage, you will have probably concluded that it is a pure compound. Which characteristic property of a pure liquid helps to identify it? Your investigation should have already provided this necessary information about X, and further use of a book of data will help to simplify the problem.

The Final Stage

You have narrowed the investigation to a few substances. With the help of your teacher it should now be possible to reach a solution to the problem.

QUESTIONS CHAPTER 2

1. Four of the following substances have something in common. Which is the odd one?
A Nickel C Sulphur E Sodium
B Copper D Silver

2. Which of the following groups of liquids is miscible with water?

A Ethanol, acetic acid (ethanoic acid) and propanone (acetone)
B Ethanol, benzene, and propanone
C Ethanol, benzene and xylene
D Ethanol, propanone and xylene
E Acetic acid, benzene and propanone

3. Which of the following statements about salt water is FALSE?

A It boils at a higher temperature than pure water
B It freezes at a higher temperature than pure water
C Its density is greater than that of pure water
D Its appearance is similar to that of pure water

4. You are given a mixture of xylene and a dilute aqueous solution of potassium chloride. Xylene boils at 140 °C. Xylene is immiscible with water. Which of the following methods would you use to separate the mixture as far as possible?

A Filtration D Chromatography
B Use a separating funnel E Sublimation
C Distillation

The following questions require rather longer answers.

5. Write down as many examples as possible in which 'filtering' takes place in daily life.

6. You are camping by a muddy river which is your only source of water. Describe how you would improvize from your camping equipment a method of obtaining (a) clear water for washing and (b) pure water for drinking.

7. Your father has run out of distilled water for his car battery. You have an electric kettle. Using this and any other suitable kitchen equipment, how would you produce some distilled water for him?

8. Although water is called the 'universal solvent' some substances, e.g. sulphur, oil and paint are insoluble in water. Name suitable solvents, one in each case, for these substances.

9. From the following methods of separating substances from each other, (a) filtration, (b) evaporation, (c) sublimation, (d) fractional distillation, (e) paper chromatography, select the most suitable for each of the following processes:

A to separate a mixture of petrol and paraffin
B to separate the various colours in red rose petals
C to obtain salt from sea water

10. A substance A is a colourless solid. The melting point of A is 67 °C. Describe briefly how you would purify a sample of impure A which melts at 64 °C. How could you show that the sample you have prepared is pure A? (Solid A is insoluble in water but dissolves in ethanol.)

11. You are given a colourless solid B. How would you find out whether it was a single substance or a mixture of two or more solids?

12. Name the method of purification you would use in each of the following:

(a) to separate a mixture of red and blue inks
(b) to remove the 'cloudiness' from limewater
(c) to obtain pure water from salt water
(d) to separate a mixture of tetrachloromethane and water

13. State TWO differences between a compound and a mixture. What is meant by the fractional distillation of liquid air? (J. M. B.)

14. The table below gives some details about the properties of three compounds.

Use this information to devise a scheme for obtaining pure dry samples of the compounds from a mixture of the three, and carefully describe the procedure.

Give one property of calcium fluoride in which it differs from calcium chloride.

How could you prepare a dry sample of calcium fluoride from potassium fluoride solution? (J. M. B.)

Compound	*Heat*	*Cold water*	*Hot water*
Naphthalene	Sublimes	Insoluble	Insoluble
Calcium fluoride	No effect	Insoluble	Insoluble
Potassium chloride	No effect	Fairly soluble	More soluble

15. Draw two columns headed 'Metal' and 'Non-metal' respectively. In the first column list *three* physical and *two* chemical properties which are characteristic of metals. In the second column give the corresponding contrasting properties of non-metals which distinguish them from metals. Given a mixture of iron filings and flowers of sulphur, and any other necessary reagents, how would you prepare (a) crystals of rhombic sulphur, (b) hydrogen, (c) hydrogen sulphide? (Diagrams of the apparatus are not required.)
(J. M. B.)

16. Name the products formed when aluminium hydroxide reacts with sodium hydroxide solution.

How could you use this reaction to obtain a dry sample of calcium hydroxide from a mixture of aluminium and calcium hydroxides? Outline only the essential steps in the procedure.
(J. M. B.)

17. Describe and explain the changes observed when each of the following mixtures is heated:

(a) ammonium chloride and sodium chloride
(b) iron and sulphur
(c) calcium hydrogen carbonate (calcium bicarbonate) solution
(d) concentrated sulphuric acid and potassium nitrate
(e) concentrated sulphuric acid and copper
(J. M. B.)

18. (a) It is required to separate a mixture of *three* solid dyes; one red, one yellow, and one blue. The following facts are known about the dyes. The blue and yellow dyes are soluble in cold water, while the red dye is insoluble. When an excess of aluminium oxide is added to a stirred, green aqueous solution of the mixed blue and yellow dyes and the aluminium oxide is filtered off and washed with water, it is found that the solid residue is yellow and the filtrate blue. When the yellow solid is stirred with ethyl alcohol and the mixture filtered, the solid residue is white and the filtrate yellow.

Describe how you would obtain dry samples of the three dyes.

(b) A green dye is known to be a hydrate. Describe how you would determine the percentage of water of crystallization in the hydrate.
(J. M. B.)

19. Describe briefly how you would separate a pure sample of the first-named substance from the impurity in each of the following mixtures:

(a) iron turnings contaminated with oil
(b) sodium chloride crystals contaminated with glass
(c) hydrogen sulphide contaminated with hydrogen chloride
(d) water contaminated with copper(II) sulphate (cupric sulphate)
(e) copper powder contaminated with magnesium powder
(J. M. B.)

3 Some Important Chemicals

By the end of your first few chemistry lessons you begin to realize that chemists have a language of their own. Some of the basic ideas and techniques were introduced earlier, and in this chapter you will learn how to recognize acids, bases, and salts as three important types of substances frequently used in laboratory work.

3.1 ACIDS

Common Laboratory Acids

Most people have heard of acids, and to a beginner the term may conjure up vague impressions of highly dangerous, corrosive, and fuming liquids. Although this often is not the case, acids must always be treated with great care and handled sensibly, even when dilute. You may be able to name some of the common acids used in the laboratory, and you may know some of their formulae. Two acids which are familiar to you in everyday life are hydrochloric acid which is secreted in the stomach as an aid to digestion, and ordinary vinegar which contains ethanoic (acetic) acid. Acids can be obtained as pure substances, although they are normally used as *concentrated* or *dilute* solutions, according to whether they have been dissolved in a small or large volume of water respectively. The names, formulae, and appearances of some common laboratory acids are summarized in Table 3.1.

The first three acids in the table are often referred to as the mineral acids because they were first obtained from minerals. Acids used rather less frequently in the laboratory include citric acid, tartaric acid, and oxalic acid, all of which are white solids when pure. Citric acid occurs in many fruits, especially those of the citrus variety, and lemon juice may contain up to 10 per cent of the acid. Tartaric acid is found in grapes, and small quantities of the very poisonous oxalic acid occur in rhubarb leaves and sorrel. All three solids are soluble in water.

Before you investigate the properties of acids you must remember an important laboratory rule. *Never dilute a concentrated acid by adding water to it.* The heat produced may cause the water to turn to steam and the acid to spray out into the air.

Properties of Acids

Introduction Many compounds are classified as acids and it is important to learn and understand why they are grouped together in this way. Acids have certain characteristic chemical properties which will become familiar to you as you examine some of the more common ones. Once the ideas are understood you may be able to decide whether an unfamiliar material is an acid or not by conducting some simple tests on it.

Table 3.1 Some common laboratory acids

Acid	*Formula*	*Appearance of pure substance*	*Appearance of concentrated solution*	*Appearance of dilute solution*
Sulphuric	H_2SO_4	Dense, fuming, oily liquid	Colourless, oily liquid	Looks like water
Hydrochloric	HCl	A gas called hydrogen chloride	Resembles water, often fumes, sharp acrid smell	Looks like water
Nitric	HNO_3	Dense, fuming, oily liquid	Colourless when pure, often yellow-brown due to impurities, usually kept in brown bottles	Looks like water
Ethanoic (Acetic)	CH_3COOH	Colourless, dense liquid called *glacial* ethanoic acid because it easily freezes to an ice-like solid, smells very strongly of vinegar	Rarely used	Looks like water, smells of vinegar

Taste

You will remember that vinegar contains ethanoic (acetic) acid. Which two of the following terms best describes the characteristic taste of vinegar?

A sweet, B sharp, C sour, D bitter, E mild.

Do you think that the same adjectives can be applied to the taste of lemon juice, which contains citric acid? Most acids in dilute solution have sharp, sour tastes but the tasting of materials in a chemistry laboratory is not a recommended practice. It should be obvious that it would be foolish to taste an unfamiliar substance in an attempt to decide whether or not it is an acid, as many chemicals are poisonous. This property (taste) is thus rarely, if ever, used in deciding what a chemical is.

Colour Changes Involving Acids—Indicators

It has been known for a long time that acids can change the colour of certain materials, particularly those occurring naturally as the colouring matter of plants. You may know for example that red cabbage, which is naturally a bluish-purple, becomes deep red when placed in vinegar, an acid. Many plant extracts can be made to change colour by the addition of an acid and usually the colour change can be reversed by a member of another group of chemicals called *alkalis*. Substances which act in such a way are called *indicators*.

There are many commercial indicators such as litmus (an extract from lichen), phenolphthalein, and methyl orange, but if you would like to prepare an indicator yourself full instructions are given in the next experiment.

Experiment 3.1
† A Home-made Indicator **

Apparatus
Suitable source of coloured plant material such as bluebells, delphiniums, or red roses. Almost any fruit or flower which is coloured will suffice but the yellow varieties are not good. Pestle and mortar, round bottom flask and vertically mounted condenser, tripod, gauze, asbestos square, filter paper and funnel, access to balance, corked container for the indicator.
Ethanol or propanone (acetone).

Procedure
(a) Weigh the plant material you are going to use and place it in the mortar together with the solvent, a mixture of equal volumes of water and propanone or ethanol. You should use about 12 cm^3 of mixed solvent for every gramme of plant material. Grind up the mixture for a minute or two so that the solvent can penetrate through the broken plant tissues.

(b) Transfer the mixture to a round bottom flask and arrange the apparatus as shown in Figure 3.1. Adjust the water flow through the condenser jacket until a fine, steady stream flows and heat the contents of the flask until the solvent is boiling gently.

(c) Boil the mixture for about twenty minutes, allow it to cool, and then filter it into a suitable container so that a clear, coloured solution is obtained, free from solid plant material. If the colour is weak it may be advisable to concentrate the indicator before use by evaporating off some of the solvent.

Points for Discussion
1. Why do you think the water is run in at the *bottom* of the condenser?

2. What is the purpose of the vertically mounted condenser?

3. Your indicator may contain more than one coloured pigment. How could you find out if this is so? If you have been successful in preparing an indicator containing a mixture of colours you will be able to use it as a 'universal indicator' in later experiments.

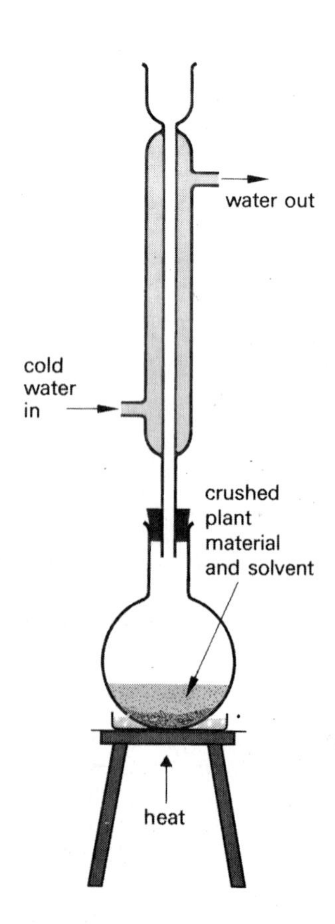

Figure 3.1 Distillation under reflux

Experiment 3.2
† To Investigate how Acids Affect Indicators *

Apparatus
As many home-made indicators as possible, solutions of methyl orange, neutral litmus, and phenolphthalein, rack of test-tubes, white tile or filter paper, teat pipette, spatula, 100 cm^3 beaker. Dilute solutions of hydrochloric, nitric, and sulphuric acids, tartaric or citric acid.

Procedure
(a) Prepare a dilute solution of tartaric or citric acid by dissolving half a spatula measure of the solid in about 10 cm^3 of water in a beaker.

(b) Place about 2 cm^3 of one of the four acid solutions into a test-tube and add two drops of one of the indicators. Allow the solutions to mix and note any colour change. If the home-made indicators are too weak to give a good colour change, place a drop or two on a white tile or piece of filter paper and add one drop of an acid solution. This method can also be used for the commercial indicators.

(c) Repeat (b) using the other acids and indicators until each acid has been tested with each indicator.

(d) Record your results in a table as follows:

Indicator	*Colour in hydrochloric acid*	*Colour in nitric acid*	*Colour in citric acid*	*Colour in sulphuric acid*

Points for Discussion
1. Does each acid produce the same type of colour change with any one particular indicator?
2. Can acids be detected by using indicators?

The Action of Acids on Metals

Another characteristic of dilute acids is the way in which they react with certain metals, although there are exceptions to the rule.

Experiment 3.3
† To Investigate the Behaviour of Dilute Acids with Metals *

Apparatus
Rack of test-tubes, splints, Bunsen burner, asbestos square.
A range of metals such as zinc, iron, magnesium and copper; solutions of dilute hydrochloric, sulphuric, nitric, and tartaric or citric acids.

The Test for Hydrogen Gas
If a gas is liberated in a test-tube reaction it can be tested for hydrogen as follows. Trap some of the gas by holding the thumb tightly over the mouth of the test-tube for a few seconds (unless effervescence is rapid). Remove the thumb and almost at the same time apply a lighted splint to the mouth of the tube. If the evolved gas is hydrogen, a 'squeaky pop' is heard as the mixture of hydrogen and air reacts with a miniature explosion.

The wrong and right way of testing for hydrogen

Procedure
(a) Place a small sample of magnesium metal in a test-tube and cover it with a few cm^3 of a dilute acid. Observe carefully what happens and test any evolved gas for hydrogen.

(b) Repeat (a) using other reagents until each metal has been tested with each acid and sniff *cautiously* by wafting any gases evolved from the reaction between nitric acid and metals towards your nose.

(c) Set out your results in a table as shown. Use the word effervescence where appropriate.

Metal under test	*Reaction with hydrochloric acid*	*Reaction with sulphuric acid*	*Reaction with nitric acid*	*Reaction with tartaric (or citric) acid*

Points for Discussion
1. Which of the acids seems to behave rather differently from the others?

2. Magnesium, zinc, and iron are fairly reactive metals. They react with typical dilute acids such as hydrochloric acid and sulphuric acid in a similar way. Summarize this similar behaviour.

3. Which of the metals behaved differently with a typical dilute acid? Other unreactive metals such as lead do not liberate hydrogen from dilute acids.

4. Metals which are very reactive, such as potassium, sodium, and calcium, react as in 2 but so violently that these reactions should not be performed in the laboratory.

5. Nitric acid reacts differently with metals. Magnesium will liberate hydrogen from the (very) dilute acid. Other metals, even unreactive ones such as copper and lead, react to evolve oxides of nitrogen instead of hydrogen. One of these has the pungent smell and the colour you may have noticed during the experiment.

6. Remember that if a liquid effervesces with a metal to liberate hydrogen gas it does not *prove* that the liquid is an acid, for certain alkalis will react with some metals to liberate hydrogen (Section 3.2). However, heat is normally required to start a reaction between an alkali and a metal, whereas acids often react in the cold.

The Action of Acids on Carbonates and Hydrogen Carbonates

Most metals form compounds called carbonates (containing metal, carbon, and oxygen atoms, e.g. copper carbonate, $CuCO_3$) and hydrogen carbonates (containing metal, carbon, oxygen, and hydrogen atoms, e.g. sodium hydrogen carbonate, $NaHCO_3$). Hydrogen carbonates used to be called bicarbonates. These react with acids in a characteristic way and there are very few exceptions to the general rule.

Experiment 3.4

To Investigate the Reaction of Acids with Carbonates and Hydrogen Carbonates *

Apparatus

Rack of test-tubes, teat pipette, splints, Bunsen burner, asbestos square.

A selection of solid carbonates and hydrogen carbonates such as sodium carbonate, sodium hydrogen carbonate, and copper carbonate; a solution of sodium carbonate, dilute hydrochloric, nitric, sulphuric, and tartaric acids, calcium hydroxide solution.

The Test for Carbon Dioxide Gas

If a gas is evolved in a test-tube reaction it can be tested for carbon dioxide as follows. Trap some of the gas by holding your thumb tightly over the mouth of the test-tube (unless effervescence is rapid), and place a few cm^3 of limewater (calcium hydroxide solution) in another test-tube standing in a rack. Squeeze the bulb of a teat pipette, remove your thumb from the mouth of the reaction tube and insert the teat pipette so that its tip is just above the reacting chemicals, but not close enough to collect acid spray. Release the pressure on the bulb of the pipette, thus filling it with a sample of the gas being evolved. Remove the pipette, place its tip under the limewater in the other tube and bubble the sample of gas through the liquid. If the limewater turns cloudy or milky the gas is carbon dioxide. (The cloudiness is due to a finely divided precipitate of chalk, known chemically as calcium carbonate.) Remove the pipette without releasing the pressure on the bulb.

Procedure

(a) Place a small quantity of a solid carbonate or hydrogen carbonate in a test-tube and cover it with a few cm^3 of a dilute acid. Record your observations and test any evolved gases for hydrogen and carbon dioxide.

(b) Repeat (a) using the other solid samples and the other acids until the range has been covered.

(c) Pour a few cm^3 of sodium carbonate solution into a test-tube and add a few cm^3 of a dilute acid. Test any gases evolved.

Points for Discussion

1. In all the reactions between carbonates or hydrogen carbonates and dilute acids there was a common product. Use this fact to summarize the results of your experiments.

2. Sodium carbonate was tested in the solid state and in solution. Was the result of the tests the same in each case? Does it matter whether you perform a carbonate test on a solid or on a solution?

The Reaction Between Acids and Bases

All acids will neutralize bases but this property will be examined when you have learned about the properties of bases in the next section.

Acids as Electrolytes

Acids in solution are always electrolytes and this property is dealt with in detail in Chapter 6.

Summary

The most reliable tests for an acid in solution are its action on indicators and on carbonates or hydrogen carbonates. The reaction with metals is not always conclusive but may often be used as a confirmatory test with the other two.

The properties of acids, and those of bases, are summarized in table form at the end of Section 3.5 and a few typical equations are included.

3.2 BASES AND ALKALIS

Introduction

You may already be familiar with the terms base and alkali. The 'parent' term is in fact base, for an alkali is a special kind of base. Acids and bases may be regarded as 'chemical opposites' and when they react together each destroys the other's characteristic properties, forming a neutral substance. You will be studying this opposing action in more detail in Section 3.4, but before that you will work with a number of the common laboratory bases and learn some of their characteristic properties.

The Relationship between Bases and Alkalis

A compound which consists of only an element and oxygen is called an oxide, e.g. copper(II) oxide, CuO. Many elements form compounds containing hydrogen and oxygen in which the hydrogen and oxygen atoms are joined together to form OH groups, and such compounds are called hydroxides, e.g. sodium hydroxide, NaOH.

As a general guide most oxides and hydroxides of *metals* are bases, but whereas all the acids you studied in Section 3.1 are soluble in water, many bases are insoluble. Bases which do dissolve in or react with water form solutions which are given the special name alkali. All alkalis are thus automatically bases and have the same properties as bases except that they also dissolve in water. All alkalis are bases but not all bases are alkalis (cf. all buses are vehicles but not all vehicles are buses).

Neutralization?

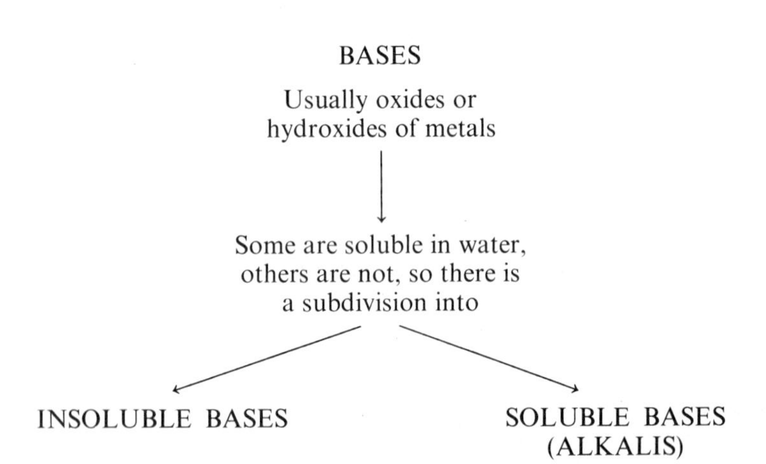

Common Laboratory Alkalis

There are only a few common laboratory alkalis and so it is more profitable at this stage to become familiar with these, rather than to study some of the many insoluble bases, as the alkalis will serve to show all the properties of bases and also one or two additional properties shown only because of their solubility. Study the information given in Table 3.2 and examine as many of the alkalis as possible.

Table 3.2 Laboratory alkalis

Chemical name	*Formula*	*Common name*	*Appearance before dissolving in water*	*Appearance of solution*
Sodium hydroxide	$NaOH$	Caustic soda	White solid as pellets, sticks or flakes	Resembles water
Potassium hydroxide	KOH	Caustic potash	White solid as pellets, sticks or flakes	Resembles water
Calcium hydroxide	$Ca(OH)_2$	Slaked lime (solid) limewater (solution) milk of lime (suspension)	White powder	Resembles water
Ammonia solution	$NH_3(aq)$	'Ammonia solution'	A colourless pungent gas ammonia, NH_3	Both concentrated and dilute solutions resemble water but the concentrated solution has a very powerful smell of ammonia gas—care!

Properties of Alkalis

Experiment 3.5
To Investigate the Effect of Alkalis on Indicators *

Apparatus and Procedure
As in Experiment 3.2 but using dilute solutions of sodium hydroxide. ammonia solution (ammonium hydroxide), and potassium hydroxide instead of the acids. Summarize your findings, together with your earlier ones on acids, in a table as follows:

Indicator	*Colour in acid solution*	*Colour in alkali solution*

Points for Discussion
1. Why do you think insoluble bases do not give colour changes with indicators?

2. An aqueous solution of the gas ammonia behaves in many ways like an alkali, and is called ammonia solution.

3. You should learn the various colour changes of the common laboratory indicators as they provide a simple means of deciding whether a substance is an acid or an alkali. If you are conducting such a test it is always advisable to treat a sample of the water being used with an indicator before testing the unknown solution. This is called a 'blank' test. Why do you think it is necessary?

Experiment 3.6
† The Action of Alkalis on Metals **

Apparatus
Bunsen burner, rack of test-tubes, test-tube holder, splints, asbestos square, teat pipette. Pieces of aluminium foil and magnesium ribbon, dilute solutions of potassium hydroxide, sodium hydroxide, calcium hydroxide, and ammonia solution.

Care must be taken with ammonia fumes

Procedure
(a) Place a few small pieces of aluminium foil in a test-tube and add 2 or 3 cm^3 of dilute sodium hydroxide solution. If there is no sign of reaction warm the mixture carefully. Test any evolved gas for hydrogen.

(b) Repeat (a) but using the other alkalis in turn instead of sodium hydroxide. [Note: care is needed when warming ammonia solution as the unpleasant gas ammonia is always evolved. This occurs whenever ammonia solution is heated and is *not* due to a reaction with the metal.]

(c) Repeat (a) and (b) using magnesium metal instead of aluminium.

Summary

This type of reaction with alkalis and metals is an exception rather than the rule so the conclusions will be summarized for you.

Only two common alkalis (sodium hydroxide and potassium hydroxide), and no insoluble bases, will react with metals to give hydrogen. They never do so until warmed and even then only a few metals (e.g. aluminium and zinc) react in this way. Such metals are said to have an *amphoteric* nature. Compare these conclusions carefully with those for the action of acids on metals. They are summarized at the end of Section 3.5.

Alkalis on the Skin

It is very bad practice to handle any unfamiliar chemical

As an 'experiment' this compares with the tasting of acids. It is very bad practice to handle any chemical unfamiliar to you and this 'test' should never be used as a routine procedure. It is safe to dip a finger into a *dilute* solution of sodium hydroxide or potassium hydroxide. If you then rub your thumb over the solution you will notice the soapy feel of the alkali. (Rinse off with water.) This is characteristic of sodium hydroxide and potassium hydroxide *only*, and you will remember that the same two alkalis are the only ones which liberate hydrogen with amphoteric metals. These two are often called the *caustic alkalis* as in more concentrated solutions they will attack flesh. One of their most important uses is in the manufacture of soap, during which a caustic alkali is heated with an oil or fat. The characteristic soapy feel of these alkalis on the skin is simply due to the fact that the skin always has a thin coating of oil secreted on it and by rubbing the oil with alkali a small quantity of soap is produced. This is not to be recommended as a substitute for commercial soap! You may be allowed to make a sample of soap by using a small scale modification of the industrial process.

Experiment 3.7
The Action of Alkalis on Ammonium Compounds **

Apparatus
Rack of test-tubes, Bunsen burner, asbestos square, test-tube holder, spatula, teat pipette.
Ammonium chloride, solution of sodium hydroxide.

Procedure
Place a spatula measure of ammonium chloride into a test-tube. Add 2 cm^3 of sodium hydroxide solution. Heat the mixture gently until it is almost boiling. *Carefully* sniff the gas evolved in the reaction, but do not do so whilst the mixture is boiling or being heated. In a situation like this it is best to hold the test-tube some distance from your face and then direct some of the evolved gas to your nose by waving a hand over the top of the tube towards your face. Test the evolved gas with moist red litmus paper.

Summary You probably recognized the evolved gas as ammonia. When any insoluble base or alkali is warmed with an ammonium compound the gas ammonia is liberated. This is used as a test for an ammonium compound.

Alkalis as Electrolytes It was stated in the previous section that solutions of acids are electrolytes. Solutions of alkalis also display this property but a fuller consideration of this is left until Chapter 6.

The Action of Bases on Acids As already stated these substances react together and each destroys the properties of the other. These reactions are studied in more detail in Section 3.4.

3.3 STRONG AND WEAK ACIDS AND ALKALIS. THE pH SCALE

Strong and Weak Acids

You have learned of the properties which acids have in common but no attempt has been made to divide acids into smaller groups. Are some acids stronger acids than others or are all acids as acidic as each other? If acids do differ, how can we recognize those which are stronger than others?

Experiment 3.8
† To Compare the Strengths of some Dilute Acids *

Apparatus
Rack of test-tubes, measuring cylinder, glass rod.
2 M solutions of hydrochloric, nitric, and ethanoic (acetic) acids, magnesium ribbon, solutions of methyl orange and litmus.

Procedure
1. (a) Taking great care, pour 5 cm^3 of 2M hydrochloric acid into a clean test-tube standing in a rack. Pour 5 cm^3 of each of the other acids into two more test-tubes in the rack.

(b) Cut off a piece of magnesium ribbon about 5 cm long. Cut off two further pieces of exactly the same length.

(c) Coil each of the lengths of magnesium ribbon into a spiral by winding it around a glass rod.

(d) Drop a coil of magnesium into each of the test-tubes of acid *at the same instant,* making sure that each coil is fully exposed to the acid. Note the extent of each reaction.

2. (a) Place 2 cm^3 of each acid into a separate test-tube in the rack.

(b) Add two drops of litmus solution to each of the acids.

(c) Repeat 2 (a) and (b) using methyl orange solution instead of litmus solution.

Molar Solutions. A Reminder

In order to compare the three acids properly we need to consider an equal number of 'acid units' of each. A molar solution is a chemist's measure of concentration (page 27). You will remember from Section 1.6 that a mole of *any* material contains the same number (the Avogadro Number) of molecules or units. There is, therefore, always an equal number of molecules in equal volumes of different solutions of the same *molarity* (molecular concentration), and so 5 cm^3 of 2M hydrochloric acid contains the same number of dissolved molecules as there are in 5 cm^3 of 2M nitric acid or in 5 cm^3 of a 2M solution of any other solute. Make sure that you fully understand this idea before writing any conclusions about the experiment.

Points for Discussion

1. In procedures 1 (a), (b), (c), and (d) there are five factors which could be varied in the experiment. These are:

(a) the volumes of acid used;
(b) the concentrations of acid used (the number of molecules of acid dissolved in a fixed volume of water);
(c) the masses of metal used;
(d) the surface area of the metal exposed to the acid;
(e) the acidity (strength) of the acid, i.e. one of the acids could be more acidic than the others.

Note: strength or acidity is *not* the same as concentration! These two terms are often confused. Lemon juice and grape juice may be of the same concentration but you can tell by the taste that lemon juice is much stronger, i.e. more acidic, than grape juice.

2. In any experiment it is usual to have only one variable factor so that any differences in the reaction are due to a variation in this factor. Which of the five factors did you keep constant throughout procedures 1 (a), (b), (c), and (d)? If the three acids did not react at the same rate, which factor must be responsible for the difference?

3. The three acids were also compared under identical conditions with two indicators. Did the indicators also show a difference in the strength of the two acids?

Some appear to be stronger than others

Even when equal concentrations of acids are considered, some appear to be stronger (more acidic) than others. This is also true of alkalis. Ordinary indicators such as methyl orange and litmus can only be used to show if a substance is an alkali or an acid but they cannot show how strong or weak they are.

Universal Indicators and the pH Scale

If an indicator is to determine the strength of an acid or alkali it must be capable of showing a variety of colours, each of which corresponds to a certain degree of alkalinity or acidity. Indicators such as methyl orange can only show one colour in acidic solution and one in alkaline solution; finer subdivisions cannot be detected. Special indicators have been produced to show a range of colours corresponding to different degrees of acidity or alkalinity. Such an indicator is called a 'universal indicator'. The plant extract you prepared in an earlier experiment may have been rather like one of the universal indicators but it will almost certainly not show a wide range of colours.

In order to compare acid and alkali strengths in a scientific way a scale of numbers is used, called the pH scale, which ranges approximately from 0 to 14. If a solution has a pH of less than 7 it is an acid. Neutral liquids have a pH of 7 and alkaline solutions have a pH of more than 7. An acid with a pH of 0 or 1 is a very strong acid and strong alkalis have a pH of 13 or 14. There is a complete range of possibilities, with strong acids and strong alkalis the two extremes.

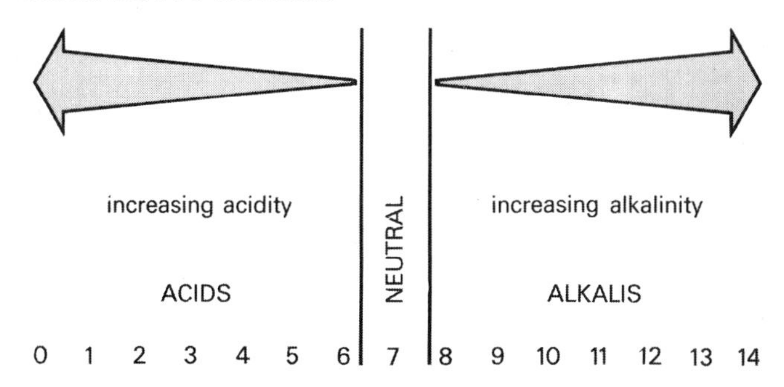

Experiment 3.9

†To Show how a Universal Indicator can Detect the pH of a Solution *

Apparatus

Rack of test-tubes, teat pipette.
A commercial universal indicator solution and a home-made one if available, a range of solutions of known pH, common household materials such as bleach, toothpaste, detergents etc., deionized or distilled water.

Procedure

(a) Wash out each test-tube before making a test. Use tap water first and rinse out twice with a little deionized water.

(b) Pour 2 cm^3 of a liquid of known pH into a clean test-tube, add two drops of universal indicator, shake, and note the colour of the solution.

(c) Repeat (b) using other liquids of known pH and record your results in a table as follows:

pH					
Colour of universal indicator					

(d) Repeat (b) using common laboratory or household materials shaken up with 2 cm^3 of deionized water, and compare the colours of your solutions with the colours of solutions of known pH. Compile a chart or table to illustrate your findings.

Summary Acids and alkalis vary in strength and one way of determining the strength of a particular acid or alkali is to test it with a universal indicator and to note the colour produced. The strength is expressed as a number on the pH scale.

Later in your chemistry course you may use special instruments to determine the pH of a solution. Colorimeters are designed to detect slight colour variations (e.g. in indicators) invisible to the naked eye, and pH meters (Figure 3.2) allow a very accurate measurement of pH to be made directly.

Figure 3.2 Portable pH meter *(Electronic Instruments Ltd.)*

3.4 NEUTRALIZATION

We have already said that acids and bases are 'chemical opposites' which can destroy the properties of each other when mixed together, i.e. they can neutralize each other.

Acids often cause problems in daily life. Sufferers from indigestion usually have excess acid formed in the stomach, farmers may be unable to cultivate certain crops if the soil is acidic, and you probably know that bacteria feeding on food particles trapped between teeth produce acids

Resting between meals. (Is the acid they produce a 'molar' solution?)

which are partly responsible for tooth decay. If a base can destroy the properties of an acid it should be obvious that bases can be used to overcome problems due to the presence of excess acid. The idea can be followed very simply in the following experiment.

Experiment 3.10
Following a Neutralization Process by Using an Acid Drop *

Apparatus
Glass rod, 50 cm^3 beaker.
One or more acid drops, deionized water, universal indicator sodium hydrogen carbonate or a proprietary stomach powder.

Procedure
(a) Crush a piece of an acid drop into a powder, place it in the beaker, add 5 cm^3 of deionized water, and stir. Add two or three drops of universal indicator solution.

(b) Place a second acid drop in the mouth and notice its taste.

(c) Place a little sodium hydrogen carbonate or stomach powder on the *clean* palm of a hand and pick a little up on the tip of your tongue still keeping the acid drop in your mouth. Note if the stomach powder has any effect on the taste of the acid drop. Repeat with more stomach powder if necessary.

Points for Discussion
1. Are acid drops so called because they actually contain an acid?

2. The taste of the sweet is not that of the acid alone as other flavourings are present. Did the mild base (stomach powder or sodium hydrogen carbonate) neutralize the acidity, i.e. remove the acid taste?

3. What happens to the taste if an excess of base is used?

Summary When a base is added to an acid it neutralizes the acid, but if too much base is added the excess can cause problems of its own. In order to neutralize substances effectively no excess of either acid or base must be used. A simple way of neutralizing an acid or base fairly accurately is given in the next experiment. As mentioned earlier, taste is not a recommended way of examining a chemical reaction.

The Use of Indicators to Illustrate Neutralization

The neutralization of an acid drop has shown how a base such as a stomach powder can overcome problems due to excess acidity. No attempt was made to measure the amount of base used as an excess of such a mild base is unimportant and will result in no major discomfort. If a farmer were to add too much base to an acid soil he would obtain a basic soil which may be just as difficult to produce crops from as the original. He would have swung the pendulum from the acid side of the pH scale *through* the neutral point and into the alkaline side. This may or may not be important to a farmer but in chemical reactions it is sometimes essential to be able to mix just the right amounts of acid and base to produce a neutral solution.

When a base is gradually added to an acid the pH of the solution rises (i.e. the acidity falls) and when the solution has a pH of 7 it is said to be neutral. At this point there is neither excess of acid nor excess of dissolved base. If more of a *soluble* base (alkali) is added, the pH continues to rise and the solution becomes alkaline, i.e. contains excess alkali. If the added base is insoluble the solution will first become neutral (because acids can react with bases even if they are insoluble in water) and then stay neutral even if excess base is added, as only dissolved substances can effect the pH of a solution. These changes are summarized in Table 3.3.

Table 3.3 The neutralization cycle

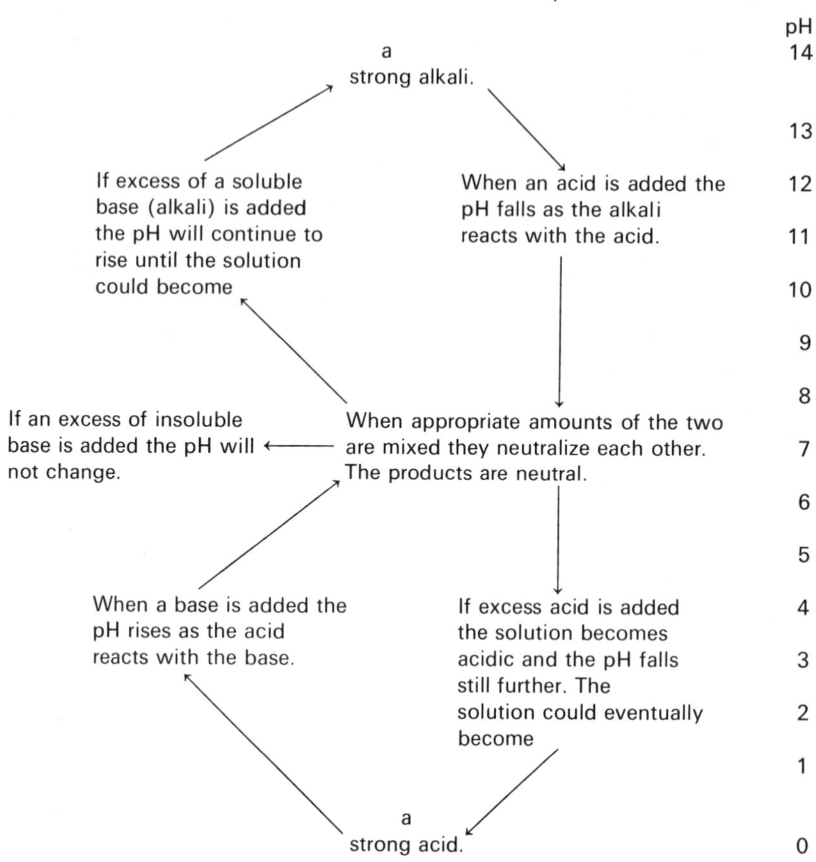

Performing a Neutralization in Order to Determine What is Formed when an Acid is Neutralized by a Base

In the next experiment you will learn how to perform a controlled addition of one liquid to an accurately measured volume of another in such a way that a pure neutral solution is formed. The experiment involves the use of two pieces of equipment which may be unfamiliar to you, namely the *burette* and the *pipette*. You will need to practise using these two pieces of apparatus before you perform the next experiment.

Experiment 3.11

† To use a Pipette and a Burette in Order to Neutralize a Solution of Sodium Hydroxide with Hydrochloric Acid **

Aim
To obtain a neutral solution by mixing sodium hydroxide and hydrochloric acid in the correct proportions. The final solution must be uncontaminated with indicator and free from excess acid or alkali.

Apparatus
25 cm^3 or 10 cm^3 pipette, burette and stand, funnel, white tile or paper, two 250 cm^3 beakers, a conical flask, pipette filler. Phenolphthalein indicator, approximately 0.1M solutions of sodium hydroxide and hydrochloric acid.

Procedure

(a) Label one of the beakers 'acid' and the other 'alkali'. Wash out each beaker with water and then with the appropriate liquid. Pour about 100 cm^3 of each liquid into the corresponding beaker.

(b) Pipette exactly 25.0 cm^3 (or 10.0 cm^3) of the sodium hydroxide solution into a clean conical flask. (Use a pipette filler if possible, but if these are not available take great care not to suck any liquid into your mouth.) A little water in the flask will not affect the result. Add two drops of indicator.

(c) Clean and check the burette, clamp it in position and fill it up with the acid solution. Run a little acid through to fill the tip of the burette. Adjust the level of the meniscus and record the level in your notebook. Place the conical flask on a white tile or piece of white paper below the tip of the burette.

(d) Run the acid into the alkali fairly quickly, shaking the flask all the time, until the colour of the indicator just permanently changes from pink to colourless. Close the tap. Note the new level of the acid in the burette. Record your results as follows:

	Rough titration	*1*	*2*
Final burette reading (cm^3)			
Initial burette reading (cm^3)			
Volume of acid used (cm^3)			

(e) The first result is only approximate as there was no drop-by-drop control near the end of the reaction, so that a slight excess of acid was probably added. Repeat with a further 25.0 cm^3 (or 10.0 cm^3) of alkali after washing out the conical flask. This time run in the acid quickly until you have added about 1 cm^3 less than the volume used in the rough titration. Swirl the contents of the flask and add one drop of acid. Swirl again. Repeat this dropwise addition until the indicator *just* changes colour. Record the readings as before. The volume of acid used in this reaction should represent the volume needed to neutralize accurately the fixed volume of alkali.

(f) As a check on your own technique it is advisable to repeat (e) until two or more titration results agree.

(g) The normal procedure is over, but as you are going to investigate some of the neutral product a little later the whole operation must be repeated again but without the indicator. Use exactly the same amount of acid as in your accurate titration. Keep the neutral solution for the next experiment.

Points for Discussion

1. Why do you think that the conical flask was placed on a white tile or piece of paper?

2. Why is it important to swirl the contents of the flask during the addition of the acid?

3. Why does it not matter if there is water in the flask when you add the 25.0 cm^3 (or 10.0 cm^3) of alkali?

4. Why is a conical flask used and not a beaker?

What is Formed when an Acid Neutralizes a Base?

The neutral solution obtained in the previous experiment has none of the properties of the original acid nor any of the original alkali. You saw no gases being evolved during the reaction, nor did you see any other indication of materials being lost from the conical flask. The atoms present in the original acid and alkali must still be present in the solution but they have 'joined up' in different ways so that new materials have been formed. The acid and alkali were dissolved in water and this is still present. Water is also a product of the reaction, so in order to determine what else, if anything, is present you must evaporate off some of the water.

Experiment 3.12
Is Water the Only Product when an Acid Neutralizes a Base? *

Apparatus
The neutral solution from the previous experiment, Bunsen burner, tripod, gauze, asbestos square, beaker, watch-glass suitable for resting on top of the beaker, hand lens or microscope.

Procedure
(a) Half fill a beaker with water, place it on the tripod and gauze and heat it.

(b) Pour a few cm^3 of the neutral solution onto a clean watch-glass and rest the glass over the beaker as shown in Figure 3.3.

(c) Boil the water in the beaker until the liquid on the watch-glass has evaporated.

(d) Examine a sample under a hand lens or microscope.

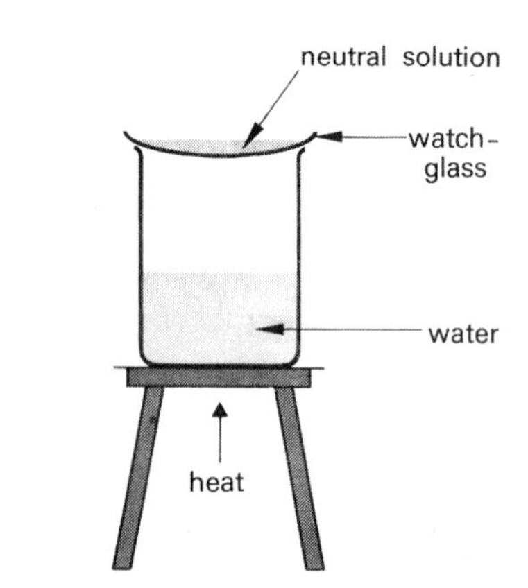

Figure 3.3 A simple evaporation technique

Points for Discussion
1. Was anything present in the neutral solution apart from water?

2. Did you recognize the residue? To confirm your idea ask for a sample of the chemical which you think is the residue and examine it under a microscope or hand lens. Does what you see help to confirm your idea?

3. Was this compound there all the time or was it formed in the reaction between the acid and the alkali?

Summary

When hydrochloric acid and sodium hydroxide neutralize each other, sodium chloride and water are produced. Sodium chloride is common salt and this is in some ways an unfortunate term, as sodium chloride is just one of many chemicals which are collectively called *salts*. Salts are studied in the next section. Any acid and any base can neutralize each other to form a salt and water. There is no other product.

$$\text{Acid} + \text{Base} \rightarrow \text{A salt and water}$$
$$\text{e.g. } HCl(aq) + NaOH(aq) \rightarrow NaCl(aq) + H_2O(l)$$

3.5 SALTS

What are Salts?

In the previous sections you learned of the chemical families called acids and bases and how to recognize them by examining their chemical properties. However, it is quite easy to decide whether an unfamiliar substance is a salt or not if you know its name or formula. It so happens that most of the inorganic chemicals you will use, which are not acids or bases, will in fact be salts.

The parent compound of any salt is an acid. All acids form salts. Sodium chloride is a salt formed by the reaction between hydrochloric acid and sodium hydroxide. Hydrochloric acid always forms salts called chlorides. All metallic chlorides are salts.

If you consider the formulae of hydrochloric acid (HCl, the parent acid) and of sodium chloride (NaCl, the salt formed from the acid) you will see that the salt is formed by replacing the hydrogen in the acid by a metal.

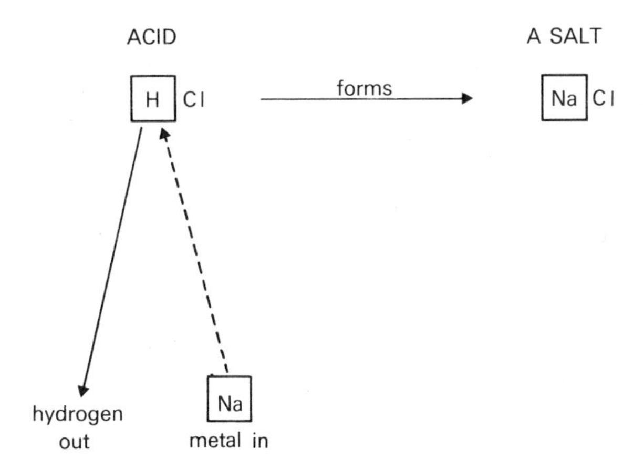

The replacement by the metal was not performed directly for the base sodium hydroxide was used to provide the metal part. Sometimes the reaction can be performed directly by using a metal, but the essential point in preparing a salt, as you will learn in this section, is to use some substance to provide a metal which can substitute for the hydrogen of an acid.

This idea leads us to a definition of a salt, for all common acids contain hydrogen which can be replaced (directly or indirectly) by a metal.

A salt is a substance formed when the hydrogen of an acid is partly or completely replaced by a metal or an ammonium ion.

Hydrochloric acid has only one atom of hydrogen in each molecule and so it can form only one series of salts, the chlorides. Sulphuric acid has two atoms of hydrogen in each molecule and can form two different types of salt. If one of the two hydrogen atoms in a molecule of sulphuric acid is replaced by a metal, the salt formed still contains hydrogen. Such salts are called *acid salts* as some of the hydrogen of the original acid is still present. If all of the hydrogen in the 'molecules' of sulphuric acid is replaced by a metal a *normal salt* is produced. These ideas are extended and summarized in Table 3.4.

Table 3.4 Salts

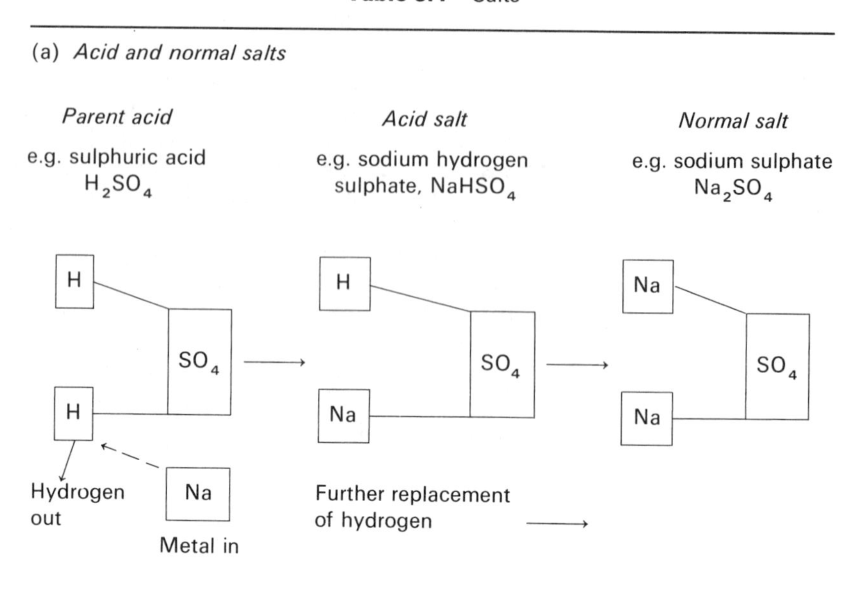

(b) *Salt nomenclature and formulae*

Sulphuric acid, H_2SO_4		*Hydrochloric acid, HCl*	*Nitric acid, HNO_3*	*Carbonic acid, H_2CO_3*	
Part of hydrogen replaced by a metal	All of hydrogen replaced by a metal	Only one series of salts, called chlorides, e.g. NaCl, sodium chloride	Only one series of salts, called nitrates, e.g. $NaNO_3$, sodium nitrate	Part of hydrogen replaced by a metal	All of hydrogen replaced by a metal
Acid salts called hydrogen sulphates, e.g. $NaHSO_4$, sodium hydrogen sulphate	Normal salts called sulphates, e.g. Na_2SO_4, sodium sulphate			Acid salts called hydrogen carbonates, e.g. $NaHCO_3$, sodium hydrogen carbonate	Normal salts called carbonates, e.g. Na_2CO_3, sodium carbonate

Summary

If the name or formula of a substance is one of the following types then it is automatically a salt:

> metallic chlorides, metallic sulphates, metallic hydrogen sulphates, metallic nitrates, metallic carbonates, metallic hydrogen carbonates, and ammonium compounds.

Points for Discussion

1. Name five salts of copper.

2. Name an acid salt of magnesium.

3. Why is sulphur chloride not a salt?

4. What are salts of citric acid called?

5. Salts are important substances in daily life. Epsom salts, washing soda, and limestone are salts. Can you find their chemical names?

6. Make a copy of Table 3.5 in your notebook and fill in the formulae of the salts by using your knowledge of symbols and valencies.

Table 3.5 Salt formulae

Metal radical	*Copper(II)*	*Iron(III)*	*Potassium*	*Magnesium*	*Iron(II)*
Sulphate	$CuSO_4$				
Chloride					
Nitrate					
Hydrogen carbonate					
Carbonate		———			
Hydrogen sulphate					

Chemical Tests for Salts

You have learned how to recognize a salt by its name or formula. Can we decide whether an unfamiliar compound is a salt or not if we do not know its name or formula? In such a case we can try some chemical tests on the compound, but as it could be one of several different types of salt we have to try different tests until one of them proves to be positive. The experiments which follow serve to illustrate some standard tests for some very common types of salt and it is very important that you try these tests for yourself if you are not already familiar with them.

Experiment 3.13
†Chemical Tests for Soluble Sulphates, Soluble Chlorides, Soluble Nitrates and any Carbonate*

Apparatus
Rack of test-tubes, teat pipette.
Solutions of a number of sulphates, chlorides, nitrates, several carbonates (solid or solution), solutions of barium chloride and silver nitrate, calcium hydroxide, dilute hydrochloric and nitric acids.

Procedure
Perform the following tests on the appropriate solutions provided.

(a) The test for a soluble sulphate. If the material to be tested is not already in solution place it in a test-tube, add water to it and shake until dissolved. Add a little dilute hydrochloric acid followed by a solution of barium chloride. A white precipitate (of barium sulphate) confirms that the original substance was a sulphate.

[Note: sulphuric acid gives this test as it is a sulphate.]

This test cannot detect an insoluble sulphate. Why not?

(b) A test for a soluble chloride. If the material to be tested is not already in solution place it in a test-tube, add water to it and shake until dissolved. Add a little dilute nitric acid followed by silver nitrate solution. A *white* precipitate (of silver chloride) confirms that the original substance is a chloride.

[Note: hydrochloric acid also gives this test as it is a chloride.]

(c) A test for a carbonate or a hydrogen carbonate. If you think carefully about a reaction you have seen earlier in this chapter you should be able to devise for yourself a test for a carbonate. You should remember that when an acid is added to a carbonate effervescence occurs and carbon dioxide is liberated. You used the test before as one of the properties of acids, but the test can also be used for detecting a carbonate. The test is also given by hydrogen carbonates. You will learn later how to distinguish between these.

To the suspected carbonate, either in solution or as a solid, add dilute hydrochloric acid. Test any evolved gas for carbon dioxide in the usual way (page 75). If the test is positive the original substance is a carbonate or a hydrogen carbonate.

(d) A test for a nitrate. This can be carried out as indicated on page 240.

Summary

Barium sulphate is precipitated if a sulphate test is positive. The reaction is essentially between sulphate ions (from *any* soluble sulphate) and barium ions from the barium chloride. You will understand why they combine and precipitate and leave the other ions in solution later in this section when double decomposition, lattice energies and hydration energies are discussed.

$$\underset{\text{(From barium chloride solution)}}{Ba^{2+}(aq)} + \underset{\text{(From any soluble sulphate)}}{SO_4^{2-}(aq)} \rightarrow BaSO_4(s)$$

Similarly the precipitation of silver chloride in a positive test for a chloride can be represented as:

$$\underset{\text{(From silver nitrate solution)}}{Ag^{+}(aq)} + \underset{\text{(From any soluble chloride)}}{Cl^{-}(aq)} \rightarrow AgCl(s)$$

These tests are very important. As so often happens in chemistry one type of reaction can have more than one application. For example the test for an acid by using a carbonate can just as well become a test for a carbonate by using an acid, and the same reaction can also be used for the laboratory preparation of the gas carbon dioxide. Three different applications of the same basic principle. Similarly, if these tests are used in conjunction with the tests for acids introduced earlier, you should be able to decide not only whether or not a given substance is an acid, but also which particular acid it is likely to be.

Methods of Preparing Salts

The first preparation is set out in detail so that you can become familiar with the principles involved. Similar principles can be applied in the subsequent experiments.

Experiment 3.14

†To Prepare a Salt by the Action of an Acid on a Fairly Reactive Metal **

Apparatus
250 cm³ beaker, stirring rod, filter paper, tripod, gauze, asbestos mat, Bunsen burner, Petri dish or crystallizing dish, measuring cylinder, spatula, filter funnel, access to fume cupboard.
Iron filings, moderately concentrated sulphuric acid. (Goggles should be worn whenever salt solutions are being evaporated, and whenever acids or alkalis are being heated. This is particularly important here, where the acid may be more concentrated than usual.)

Procedure
(a) Measure out 30 cm³ of the sulphuric acid, pour it into the beaker. Heat it *gently*. Add a spatula measure of iron filings to the acid (preferably in a fume cupboard) and stir. Vigorous frothing will occur.
(b) Continue to add portions of iron filings with stirring until effervescence ceases and there is a slight excess of unreacted metal.
(c) Set up filtration apparatus and filter the liquid into a clean Petri dish or crystallizing dish. You should obtain a clear filtrate. Repeat the filtration if necessary.
(d) The salt has already been made but it is dissolved in water. It remains to isolate and purify the solid salt. If the laboratory is warm, put the dish to one side until crystals form. In general good crystals are obtained only if evaporation is slow but under these conditions crystals usually form readily. If left for any length of time they are likely to go brown.

Note: one way of ensuring slow evaporation for crystallizations is to cover the dish with a piece of filter paper and then pierce a few small holes through the paper. If crystals do not form after a day or two your solution is not sufficiently concentrated and you will have to evaporate off more of the water by boiling before repeating the crystallization procedure. In this particular experiment boiling will cause the salt to change chemically.

(e) Pour off any residual liquid (the 'mother liquor') from the crystals and rinse them several times with a little cold water. Use only a little water with each rinse as some of the crystals will redissolve. Press the crystals gently between sheets of filter paper in order to dry them.
(f) Examine a good crystal under a microscope or hand lens and record its appearance as part of your observations.
(g) Your salt has been made from sulphuric acid and so it should be a sulphate. Perform the sulphate test on a small sample of your product.

Points for Discussion

The ideas and principles behind the various procedures are given in full so that you can apply them to the subsequent preparations.
1. How do you make a record of a preparation like this in your notebook?

The method of reporting work varies from one individual to another. In general the following steps can be applied to any scientific investigation. State the *aim* of the experiment and then the *apparatus* or *materials* used, with a diagram if appropriate. Then, under *method* or *procedure,* outline the various steps you actually take in the experiment but do not yet record anything which happens as a result of these steps. Under *results* or *observations* set out everything that happens during the experiment and finally as a *conclusion* comment on whether the aim of the experiment was achieved, and add any other relevant ideas such as whether it was a good method, a generally useful method and so on. Most conclusions will include at least one equation to summarize the reaction(s) involved.
2. What happened in the chemical reaction?

You should know the answer to this as you have performed a similar reaction when investigating the properties of acids. You noted that most acids react with fairly active metals to liberate hydrogen, which was the gas evolved during the salt preparation. The experiment has also shown that a salt is formed during the reaction. This is another example of the application of one basic idea in more than one way. What else could be prepared by this reaction, apart from a salt?

The salt made in the experiment is iron(II) sulphate. Did your last test prove that it was a sulphate?

$$\text{Metal} + \text{Acid} \rightarrow \text{Salt} + \text{Hydrogen}$$

$$Fe(s) + H_2SO_4(aq) \rightarrow FeSO_4(aq) + H_2(g)$$

3. Are the crystals pure?

You have learned that an acid and a metal may react to produce a salt and hydrogen, and that water is present in which the original acid and the salt formed are dissolved. If you have successfully removed any unchanged acid, any unreacted metal, the water, the hydrogen and the impurities evolved with it, then the crystals are probably reasonably pure.

It should be obvious to you that the hydrogen escaped (with any gaseous impurities) and that you dried the crystals free from water. All of the acid was used up because you added an *excess* of metal. The metal must have consumed all of the acid for otherwise it would have continued to react. The excess metal does not dissolve in, or react with, the water or the salt, and it can be removed by filtration. Do you think that your crystals are reasonably pure?

4. What happens in a crystallization process?

The crystallizing dish or Petri dish allows a large surface area of the liquid to be exposed to the air. The water slowly evaporates and the solution becomes increasingly concentrated until eventually there is not enough water left to dissolve the crystals and they crystallize out from solution.

5. Is this a general method of preparing salts?

Any reactive metal will be attacked by dilute sulphuric acid or hydrochloric acid to form a salt and hydrogen. Each of the metals magnesium, zinc, and iron can combine in turn with both sulphuric and hydrochloric acid to form the corresponding salt. The method is exactly the same except that different chemicals are used. Thus in one reaction you have learned how to prepare at least six salts.

Nitric acid also reacts with metals to form salts, but you will remember that hydrogen is rarely evolved in such cases. Metals which can be used with nitric acid to form salts are magnesium, iron, zinc, copper, and lead but it is essential to conduct these reactions in a fume cupboard as the evolved gases are likely to contain poisonous nitrogen dioxide.

Experiment 3.15
To Prepare a Salt by the Action of an Acid on a Carbonate *

Apparatus

A 250 cm^3 beaker, an evaporating basin, crystallizing dish, stirring rod, Bunsen burner, tripod, gauze, asbestos mat, filtration apparatus, microscope slides, spatula.

Lead carbonate (or any other insoluble carbonate), dilute nitric acid. (Care if lead carbonate is used—lead salts are poisonous.)

Procedure

(a) Pour about 30 cm^3 of the nitric acid into the beaker and add spatula measures of the carbonate, with stirring, until the reaction has stopped and there is a slight excess of carbonate.

(b) Filter off the unchanged carbonate and collect the filtrate in a clean beaker or evaporating basin. Boil the solution in order to concentrate it and occasionally transfer a few drops to a clean microscope slide. If crystals form on the slide the solution is concentrated sufficiently and the Bunsen can be turned off.

(c) Transfer the concentrated solution to a crystallizing dish and allow to crystallize as in the previous experiment.

(d) Collect, wash, and dry the crystals as before.

(e) Test a small sample for either a sulphate, chloride, or nitrate as you think appropriate.

Points for Discussion

1. Express the reaction by means of an equation. Water is formed in the experiment and you should know which gas was evolved as you have studied this reaction before.

2. What is the name of the salt you have made? Did your test on it help to confirm its identity?

3. Do you think your crystals are reasonably pure? If so, explain how you removed each of the impurities that could have been present in the salt.

4. If you were using a carbonate which dissolves in water for this preparation, how could you decide when all the carbonate has been used up, without adding too much acid?

5. From what you know of the reaction between acids and carbonates do you think that this is a general method for the preparation of a salt? State the starting materials you would need to make samples of zinc sulphate and copper chloride by this method.

Experiment 3.16
To Make a Salt by the Action of an Acid on an Insoluble Base *

Apparatus
Copper(II) oxide and the usual equipment for salt preparations.

Aim and Procedure
This is a suitable occasion for you to plan an experiment of your own. The insoluble base is copper(II) oxide. You are required to make pure, dry crystals of copper(II) sulphate using the principles learned so far in this section. When a base reacts with an acid, a salt and water only are formed. Plan each stage before you begin and remember that your crystals should be free from all impurities. Test the product for a sulphate.

Point for Discussion
Write an account of your experiment under the headings aim, apparatus, method, results, and conclusion. Explain the various steps you devised and state whether or not you think this reaction is a general one for the preparation of salts. Give reasons for your answer.

To Prepare a Salt by the Action of an Acid on a Soluble Base

You have already made a salt in this way (*see* Experiments 3.11 and 3.12). This is another general method of making salts but is more limited as there are only a few common alkalis.

Points for Discussion
1. Write equations (in words if necessary) for the reactions between sodium hydroxide and nitric acid, and between potassium hydroxide and hydrochloric acid.

2. As both reactants are soluble, and no gas is evolved in the reaction, it is difficult to 'see' when the reaction is finished. What did you add to enable you to decide this?

Double Decomposition. Lattice Energies and Hydration Energies

Some salts are insoluble in water and cannot be made by any of the methods so far described, each of which has involved the crystallization of a soluble salt from solution. It is a relatively common procedure in chemistry to mix two solutions which then react together to form two new products, one of which immediately separates out from the other, usually as a precipitate. Such a reaction is called a *precipitation* (*double decomposition*). The solid product of the reaction is separated from the liquid by either filtering or centrifuging. Insoluble salts can be made in this way.

At some stage in your chemistry course you will need to learn just what happens in a double decomposition reaction, and at least partly understand why some salts are insoluble in water. We shall briefly explain these factors now, but at this stage you may prefer to go straight on to the next experiment.

Two main factors influence the solubility of an ionic solid in water. The first of these is known as *lattice energy*. As the ions in the solid are held tightly together by ionic bonds, energy is needed to separate them. When an ionic solid dissolves its ions are separated and dispersed by the solvent, and so this process requires energy. Solids which need a great deal of energy in order to separate the ions are said to have a high lattice energy. Solids with relatively low lattice energies *tend* to be soluble, as water (even at room temperature) has sufficient energy to break down the lattice. In such cases the temperature of the water would be expected to fall as it uses some of its own intrinsic energy to separate the ions. However this is not always so as a second factor must be taken into account.

This factor is called *hydration energy* and may bring about the solution of solids which have a relatively high lattice energy without the fall in temperature already mentioned. Separated ions can form new bonds with the water molecules and this stabilizes them again, in the same way that they were stabilized in the solid by ionic bonds. Energy is released when an ion reacts with a water molecule. The process is called hydration and the energy liberated is called the hydration energy. All ions are hydrated in aqueous solution.

It is thus possible for a solid with a high lattice energy to dissolve in water if the ions have a high hydration energy, so that some of the energy needed to break down the lattice is recovered by hydration. Some substances dissolve in water so that the temperature of the water rises. This is because in such cases the hydration energy is greater than the lattice energy. Similarly, if the temperature of the water falls when a solid dissolves it does not necessarily follow that the ions are not hydrated. Both processes could be occurring but the lattice energy is greater than the hydration energy.

When an ionic solid dissolves, its ions are separated and dispersed by the solvent; this requires energy. Separated ions can form new bonds with solvent molecules; this releases energy

If sodium chloride and silver nitrate are added to water both solids dissolve because their lattice energies and hydration energies are suitable. If the two solutions are mixed there will be four different ions moving randomly in the water, i.e.

$$Na^+(aq),\ Cl^-(aq),\ Ag^+(aq),\ \text{and}\ NO_3^-(aq)$$

You know from your work on the kinetic theory that many collisions will take place between the ions, and four possible substances could be formed by such collisions, namely silver chloride, silver nitrate, sodium chloride, and sodium nitrate. The lattice energy of silver chloride and the hydration energies of its ions are such that silver chloride is insoluble, and so whenever silver ions and chloride ions collide they will combine to form solid silver chloride which thus precipitates. The other ions remain in solution as their hydration energies and lattice energies make the other theoretical products soluble. We express such reactions by using ionic equations, in which only those ions which react together chemically are shown.

e.g.

$$Ag^+(aq) + Cl^-(aq) \rightarrow AgCl(s)$$

It is not incorrect to write

$$AgNO_3(aq) + NaCl(aq) \rightarrow AgCl(s) + NaNO_3(aq)$$

but it is slightly confusing as it implies that the sodium and nitrate ions have joined together to form sodium nitrate, whereas in reality the ions continue to move about the solution at random and individually; such ions are referred to as 'spectator ions'.

These ideas lead us to a definition of a double decomposition reaction. 'A double decomposition reaction is one of the type

$$AB+CD \rightarrow AD+CB$$

where AB and CD are soluble and one of the products is either insoluble or a gas.'

The significance of the test for a chloride may now be more obvious to you. Silver chloride is one of the few insoluble metallic chlorides, so that if silver nitrate (which provides Ag^+ ions) is mixed with any soluble chloride (which automatically provides Cl^- ions) silver chloride must be precipitated. The other ions (spectator ions) are really irrelevant and take no part in the reaction.

Refer to the test for a soluble sulphate. Try to explain how the ions react in a similar way to those in the chloride test.

By definition, silver chloride and barium sulphate are salts and so double decomposition is a means of preparing insoluble salts. In the next experiment you will make another insoluble salt as an example of this method.

Experiment 3.17
The Preparation of Lead(II) Chloride by Precipitation (Double Decomposition)*

Apparatus

Teat pipette, stirring rod, measuring cylinder, 50 cm³ or 100 cm³ beaker, centrifuge, centrifuge tubes, and filter paper.

Approximately M solutions of sodium chloride and lead(II) nitrate. (Care—lead salts are poisonous.)

Procedure

(a) Measure 10 cm³ of the sodium chloride solution into a beaker and add 5 cm³ of the lead(II) nitrate solution. Stir.

(b) Transfer a suitable volume of the precipitate, which is in the form of a suspension, into a centrifuge tube and centrifuge.

(c) Discard the liquid, add water to the solid and stir thoroughly. Centrifuge again and discard the water.

(d) Remove the solid from the centrifuge tube and dry between sheets of filter paper.

Points for Discussion

1. Write an ionic equation for the reaction.

2. Could you have used potassium chloride solution instead of sodium chloride? Explain your answer.

3. What was the purpose of procedure (c)?

4. It is often possible to make two salts by one double decomposition reaction. In the experiment you have just conducted you were interested only in obtaining the lead(II) chloride, but two other ions were present in the discarded liquid. Which ions were they and how could you make them join together to form a solid salt?

5. Lead(II) iodide is insoluble in cold water. Which reagents would you need to mix together in order to obtain a precipitate of lead(II) iodide and a solution of potassium nitrate? How would you obtain a separate pure sample of each of the two salts?

Direct Combination

Some salts such as sodium chloride contain only two elements. It is possible to make such salts by reacting the two elements directly. Water is not needed for such processes and so this method is quite different from any of those so far studied.

It is not appropriate at this stage to study any of these preparations in detail. You may be shown one or two demonstrations to emphasize the idea, and you have already made one salt by this method. Can you remember when you heated two elements together to form a single compound which is a salt? (Hint: sulphides are salts.) If you cannot remember see Experiment 2.24.

Similarly you will synthesize sodium chloride (Experiment 11.12) by burning sodium in chlorine. You should refer to this in revision as it is an excellent example of a direct combination producing a salt.

The work on acids, bases and salts is summarized in Table 3.6.

Table 3.6 Acids, bases, and salts—a summary of the main reactions

Reactions		*Gas evolved (laboratory preparation)*	*Salt formed*	*Any other product*
1. *Reactions of acids*				
(a) *Acid + carbonate*	⟶	carbon dioxide	+ a salt	+ water
e.g.				
2HCl(aq) + Na_2CO_3(aq or s)	⟶	CO_2(g)	+ 2NaCl(aq)	+ H_2O(l)
H_2SO_4(aq) + $CuCO_3$(s)	⟶	CO_2(g)	+ $CuSO_4$(aq)	+ H_2O(l)
Exceptions: calcium carbonate + Sulphuric acid (very slow)				
(b) *Acid + metal*	⟶	hydrogen	+ a salt	—
e.g. Zn(s) + 2HCl(aq)	⟶	H_2(g)	+ $ZnCl_2$(aq)	—
Mg(s) + H_2SO_4(aq)	⟶	H_2(g)	+ $MgSO_4$(aq)	—
Exceptions: not given by unreactive metals; usually no hydrogen from nitric acid; very reactive metals are dangerous to use				
(c) *Acid + base*	⟶	—	a salt	+ water
e.g.				
HCl(aq) + NaOH(aq or s)	⟶	—	NaCl(aq)	+ H_2O(l)
$2HNO_3$(aq) + CuO(s)	⟶	—	$Cu(NO_3)_2$(aq)	+ H_2O(l)
Exceptions: none				
2. *Reactions of bases*				
(a) *Base + acid* (see acid + base)				
(b) *Base + ammonium salt*	⟶	ammonia	+ a salt (not usually used as a preparation)	+ water
e.g.				
CuO(s) + $2NH_4Cl$(aq)	⟶	$2NH_3$(g)	+ $CuCl_2$(aq)	+ H_2O(l)
NaOH(aq) + NH_4Cl(aq)	⟶	NH_3(g)	+ NaCl(aq)	+ H_2O(l)
Exceptions: none				
(c) *Caustic alkali + amphoteric metal*	⟶	hydrogen	—	+ complex product
Exceptions: most metals and most alkalis				

Table 3.6 (continued) Acids, bases, and salts—a summary of the main reactions

Reactions		*Gas evolved (laboratory preparation)*	*Salt formed*	*Any other product*
3. *Other reactions to form salts*				
(a) *Element + element* (direct combination)	⟶	———	a salt	———
e.g. $2Na(s) + Cl_2(g)$	⟶	———	$2NaCl(s)$	———
Only given for some metal + non-metal combinations				
(b) *Double decompositions*			often two salts	
e.g. $Pb^{2+}(aq) + 2Cl^-(aq)$	⟶	———	$PbCl_2(s)$	———
Only given by reactions satisfying the definitions				

QUESTIONS CHAPTER 3

1. A solution of sulphuric acid

A is a poor conductor of electricity
B has a pH below 7
C will react with copper to produce hydrogen
D has no reaction with zinc carbonate

2. A liquid turns universal indicator red, attacks some metals to liberate hydrogen, and liberates carbon dioxide when added to hydrogen carbonates. Which of the following is the most reasonable conclusion about the liquid?

A It is an acid salt
B It is a normal salt
C It is a base
D It is an acid
E It is an alkali

3. Which of the following combinations of reactants would NOT be used to produce magnesium chloride?

A Magnesium carbonate and dilute hydrochloric acid
B Magnesium oxide and dilute hydrochloric acid
C Magnesium nitrate and dilute hydrochloric acid
D Magnesium and dilute hydrochloric acid
E Magnesium hydroxide and dilute hydrochloric acid

4. A molar solution of a certain acid has a pH of only 5. Which of the following is the most reasonable explanation for this?

A The acid is too dilute
B The acid is only sparingly soluble in water
C It is not a strong acid
D It reacts with water to produce a high concentration of hydronium ions
E It is a poor electrolyte

5. When sodium hydroxide solution is added to dilute hydrochloric acid in a beaker, which of the following is NOT happening in the beaker?

A The pH of the solution increases
B The hydronium ion concentration falls
C The hydroxide ions neutralize some of the hydronium ions
D The volume of water increases
E The reaction $Na^+(aq) + Cl^-(aq) \rightarrow NaCl(s)$ takes place

6. When potassium chloride solution is mixed with silver nitrate solution,

A a salt cannot be obtained from the mixture
B one salt only can be obtained from the mixture
C two soluble salts can be obtained
D two insoluble salts can be obtained
E one soluble salt and one insoluble salt can be obtained

7. Name the acids found in lemons, milk, and vinegar.

8. Name three mineral acids. Which of these play a part in digestion?

9. What is meant by an indicator? Name two common indicators and give their colours in acid solution.

10. Name two metals which react with dilute hydrochloric acid. What is the name of the gas evolved? Describe how you would test for this gas.

11. Name the gas evolved when a dilute acid reacts with a carbonate or hydrogen carbonate. Describe carefully how you would test for this gas.

12. What do we mean by an *alkali?* Name two common alkalis and say what effect they have on two named indicators.

13. Sodium hydroxide and potassium hydroxide solutions both turn litmus blue and have a soapy feel. What other properties would you expect these substances to have in common?

14. Write the equations in words for the neutralization of an acid by a base. Use symbols to write equations for the following neutralizations: sulphuric acid by potassium hydroxide, and hydrochloric acid by calcium hydroxide.

15. What do we mean by an acid salt? Give the names and formulae of two acid salts.

16. Sea-water is said to contain both chloride and sulphate ions. Describe how you could use a concentrated solution of sea-water to test the truth of this statement. Why must the sea-water be concentrated?

17. Write ionic equations for reactions between solutions of (a) silver nitrate and hydrochloric acid, (b) barium chloride and copper(II) sulphate.

18. Complete and balance the following equations:

$$NaOH(aq) + H_2SO_4(aq) \rightarrow$$
$$CaCO_3(s) + HCl(aq) \rightarrow$$
$$Mg(OH)_2(s) + HNO_3(aq) \rightarrow$$
$$Mg(s) + H_2SO_4(aq) \rightarrow$$
$$Fe(OH)_3(s) + H_2SO_4(aq) \rightarrow$$
$$Cu(OH)_2(s) + HNO_3(aq) \rightarrow$$

19. Identify the following. Explain the reactions taking place.

(a) A green liquid which turns purple when added to sodium hydroxide solution and red when added to dilute hydrochloric acid.
(b) A white precipitate formed by mixing silver nitrate and sodium chloride solutions.
(c) A white precipitate formed by mixing barium chloride solution with dilute sulphuric acid.
(d) A dilute acid which only rarely produces hydrogen when added to fairly reactive metals.

20. Select from the pH values 1, 5.5, 7, 8, and 11 the one you consider most applicable to each of the following solutions: limewater, household soap, hydrochloric acid, lemon juice, sodium chloride. (W. J. E. C.)

21. Identify a colourless liquid which has the properties given below. Explain the significance of *each* test and make a general conclusion about the nature of the liquid. Give equations where appropriate.

(a) When the liquid is warmed with potassium hydroxide solution a gas is evolved which turns moist red litmus paper blue.
(b) When the liquid is added to anhydrous copper(II) sulphate a blue colour results.
(c) When barium chloride solution is added to the liquid, followed by dilute hydrochloric acid, a white precipitate is formed.

22. Use a table of solubilities to answer this question. Samples of two metallic salt solutions, A and B, reacted as described below. For each solution suggest a metal ion that it may contain and explain how you arrived at your answer.

A gave a white precipitate when mixed with an aqueous solution of potassium sulphate but no precipitate was produced when A was mixed with potassium chloride solution.

B gave a white precipitate when mixed with potassium sulphate solution and also when mixed with potassium chloride solution.

23. State in words the changes represented by the following ionic equations:

(a) $H^+ + OH^- \rightarrow H_2O$
(b) $Ag^+(aq) + Cl^-(aq) \rightarrow AgCl\downarrow$
(c) $Cu^{2+}SO_4^{2-}, 5H_2O + aqua \rightarrow Cu^{2+}(aq) + SO_4^{2-}(aq)$
(J. M. B.)

The following questions require rather longer answers.

24. Describe with full practical details how you would (a) use a pipette to measure out exactly 10.0 cm^3 of 0.1 M sodium hydroxide solution, (b) set up and fill a burette with dilute sulphuric acid.

25. Give four general methods of preparing salts. Name the starting materials you would use to make four named salts, each one prepared by a different method, and describe one of the methods in detail.

26. Describe with full experimental details how you would prepare (a) pure dry crystals of magnesium sulphate starting from magnesium oxide, (b) a pure dry specimen of zinc carbonate (insoluble) starting from zinc chloride.

27. Suppose that you have to prepare fairly pure specimens of (a) solid calcium sulphate, and (b) crystalline copper(II) nitrate, starting from marble chips (calcium carbonate) in each case. If you do not know the solubilities of these compounds use a book of data and then suggest how you would proceed to prepare the two samples.

28. The following is an extract from a pupil's notebook. 'Solutions of sulphuric acid and hydrochloric acid are equally concentrated because they each have a pH of 3. They are also as strong as each other for the same reason.'

These statements are inaccurate and show a lack of understanding of the terms 'strength' and 'concentration'. Describe how you would try to help the pupil and explain how you could use routine laboratory apparatus and chemicals to support your arguments.

4 Gases. Air, Oxygen, Hydrogen

4.1 THE PREPARATION AND COLLECTION OF GASES

When you are preparing a gas there are four main factors which you may have to consider. These are (1) the reaction vessel or generator, (2) a purification stage, (3) a drying stage, and (4) the collection of the gas. The two middle stages can sometimes be ignored. This will be decided by the particular gas and the purposes for which it is prepared, but in general you should always attempt to prepare a pure and dry gas unless you are otherwise instructed. If the gas has an unpleasant odour or if it is toxic it must be prepared and collected in a fume cupboard.

The Reaction Vessel

We need some arrangement whereby we can mix the reagents together, and preferably control the mixing, so that the gas is produced at a controlled rate and escapes by a separate route. It may also be necessary to arrange for the apparatus to be heated. Some typical generating devices are illustrated in Figure 4.1 and the choice is governed by convenience, the type of reagent to be employed, and the amount of gas which is required. You will probably be able to suggest generators other than those in Figure 4.1. Which of the arrangements would you *not* use if you wanted a large supply of a gas?

Kipp's generator is a very useful device for producing large quantities of a gas evolved by the reaction between a solid and a liquid. It can be kept ready for action so that as soon as the tap is opened, gas is liberated. Study the diagram carefully and try to explain how Kipp's generator works. If the tap controlling the gas outflow is closed, how does the design of the apparatus prevent a pressure build up and at the same time stop the reaction?

There are two very important points to remember when drawing or constructing a gas generator. These may seem obvious but it is surprising how many pupils forget about them: (1) the gas outlet pipe or tube must always be above the level of the reacting substances; (2) any inlet tube for liquid reagents must be adjusted so as to prevent any gas escaping through the same tube. If the tube has a tap it is a simple matter to close it while the gas is being generated.

Purification

It is difficult to generalize here as any purification process depends upon the degree of purity required and the impurities which may be present. In order to purify a gas completely we have to know the chemistry of the

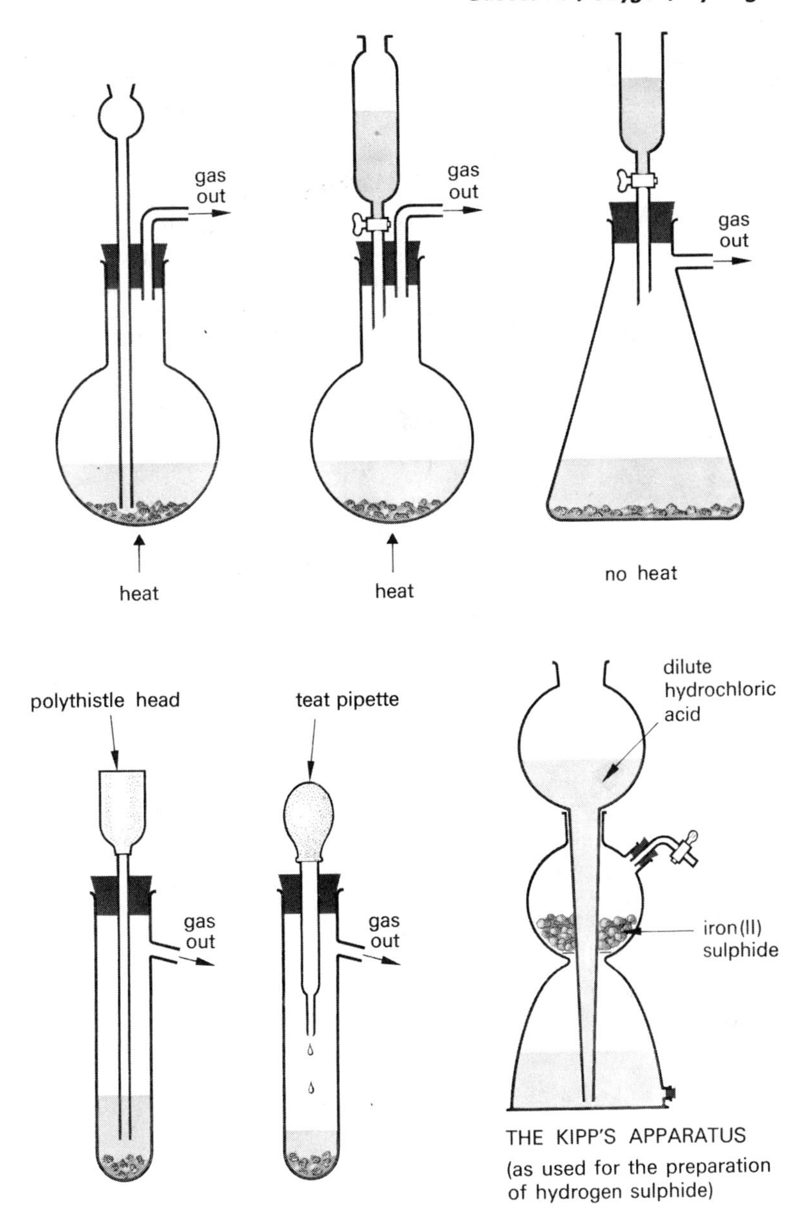

Figure 4.1 Various types of gas generator

reacting substances so that we can decide what side reactions leading to the formation of impurities may also be taking place. At school level we are rarely concerned with the collection of ultra pure gases, and purifying a gas by passing it through water (unless it is very soluble in water) is usually adequate, but in research work and in industry a whole series of purifiers may be used. In 1842 J. B. A. Dumas determined the gravimetric composition of water by burning hydrogen in oxygen. You may like to find out for yourself the apparatus he used to purify the hydrogen. This will give you some idea of the way in which gases could be purified, even as early as 1842.

Drying

If possible you should always dry a gas before collecting it, as water is an impurity, but many gases are more easily collected over water and it would obviously be pointless to dry such gases first! The drying agent normally used is concentrated sulphuric acid. However, some gases, e.g. ammonia and hydrogen sulphide, should not be dried by this method because they react with the acid. In such cases another drying agent has to be used, for example calcium oxide (quicklime) is used to dry ammonia. You should be able to suggest why ammonia cannot be dried by sulphuric acid.

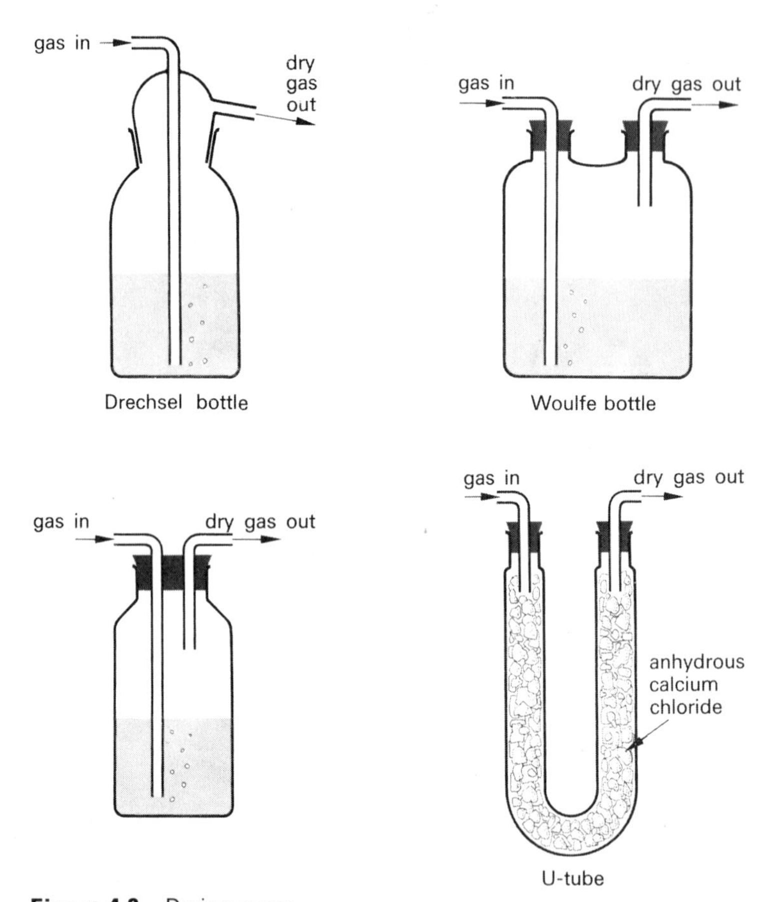

Figure 4.2 Drying gases

A range of typical vessels used for drying gases is shown in Figure 4.2. It should be obvious that if a gas is to be dried by passing it through a liquid the inlet tube must be *below* the liquid and the exit tube *above* the liquid.

Collection

The choice of method needs care and thought, and depends on the properties of each individual gas. There are four main methods of collection, but remember that the gas may have to be collected in a fume cupboard. Whichever method is adopted, it is always essential to allow the first supply of gas to escape without being collected. Why do you think this is necessary?

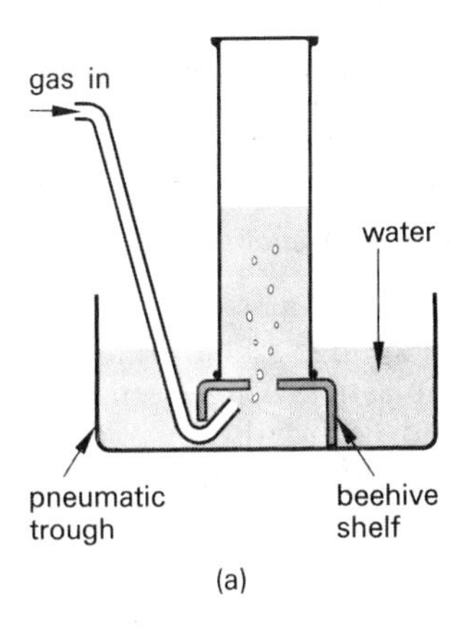

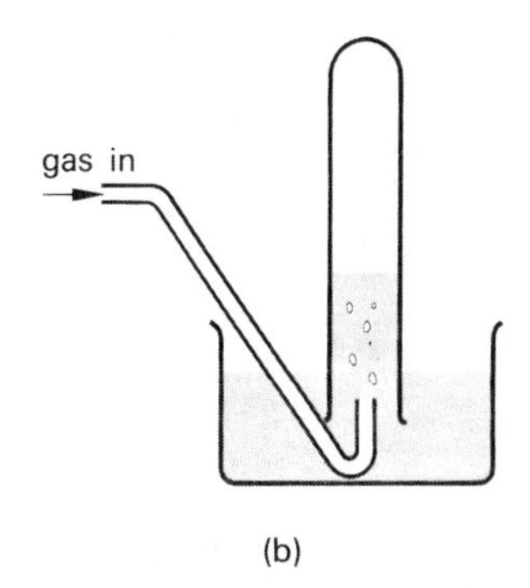

Figure 4.3 The collection of gases which are insoluble in water using (a) a pneumatic trough and (b) a Baco beaker

Over Water

Two typical methods are shown in Figure 4.3. The gas jar or test-tube must be *filled* with water first. This method *cannot* be used if (1) the gas dissolves appreciably in water, (2) the gas reacts with water, or (3) the gas is required dry.

One advantage of this method is that we can always see when a gas jar or test-tube is full of the gas.

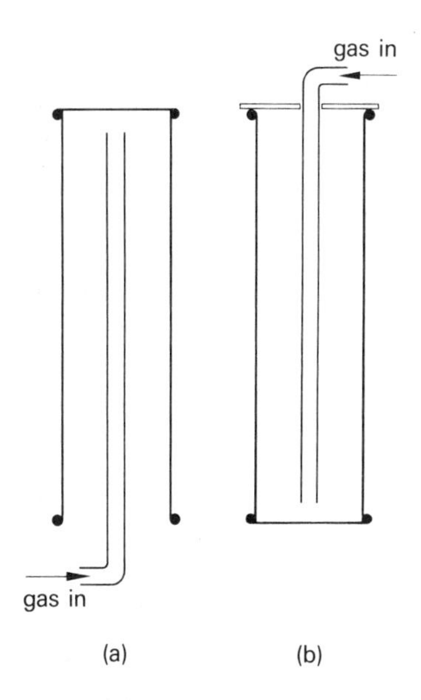

Figure 4.4 The collection of gases which are soluble in water (a) upward delivery and (b) downward delivery

By Downward or Upward Delivery

If a gas is less dense than air it can be collected by an arrangement such as that shown in Figure 4.4 (a). We say that the gas is being collected by *upward delivery*. Conversely, if the gas is denser than air it can be collected by *downward delivery* as shown in Figure 4.4 (b).

These methods are not ideal, for gases diffuse so quickly that it is almost impossible to collect a pure sample of a gas in this way. Also, unless the gas is coloured, it is not easy to decide when the gas jar is 'full'.

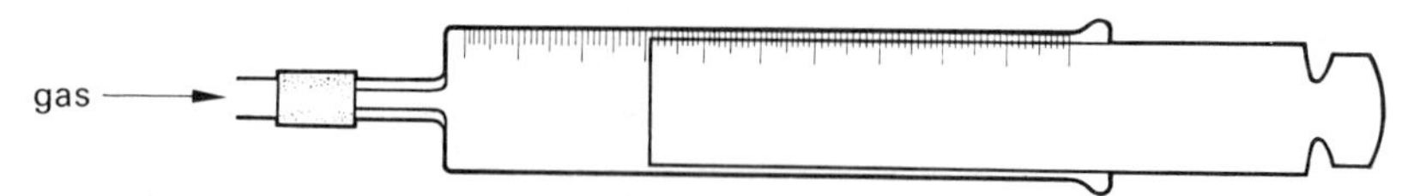

Figure 4.5 Collection of a gas by syringe

In a Syringe

Gases can be collected in a syringe as shown in Figure 4.5. Ground glass syringes are air-tight but plastic syringes tend to leak slightly and the plunger does not always move smoothly.

There are obvious advantages of the method—(1) the sample of gas can easily be transported; (2) it is possible to deliver a known volume of gas into a separate vessel by pushing the plunger to the appropriate graduation mark; (3) if a sample of gas is expelled from the syringe the residual gas remains uncontaminated with air (compare with taking the cover off a gas jar); (4) we know when the syringe is full.

The chief disadvantages are cost (if glass syringes are used) and the fact that if gas is evolved rapidly the syringes are soon filled and the collection may be difficult to control.

Liquefaction and Freezing

If a gas has a suitable boiling point or freezing point it can be collected as a liquid or a solid by placing the receiver in a freezing or cooling mixture. The only gas you are likely to collect as a liquid (up to O level) is dinitrogen tetroxide. A typical arrangement is shown in Figure 4.6.

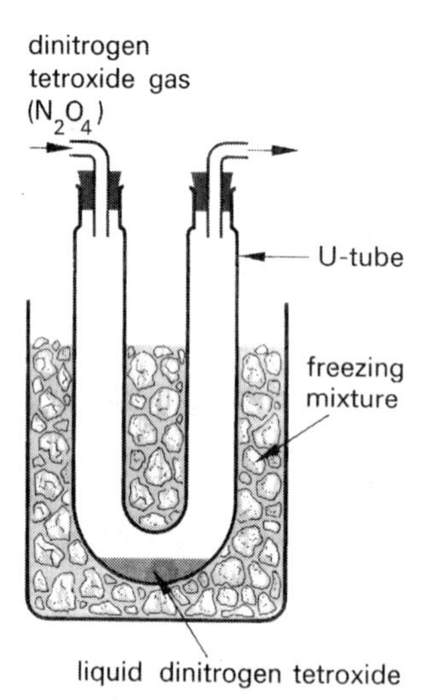

Figure 4.6 The liquefaction of dinitrogen tetroxide gas

Summary

The choice of a collection method depends upon a knowledge of the toxicity, odour, solubility, density, and melting point or boiling point of the particular gas. Study the data in Table 4.1 and answer the following questions.

1. Which gases would you collect over water in an open laboratory?
2. Which gases would you collect over water in a fume cupboard?
3. Which gases could you collect by upward delivery?
4. Which gases can be conveniently collected as liquids?
5. Which gases could you collect by downward delivery?

Name and formula	*Odour*	*Physiological effect*	*Colour*	*Relative density (H = 1)*	*Solubility in water (cm^3 100 cm^{-3})*	*Melting point K (°C)*	*Boiling point K (°C)*
Air	—	—	—	14.4	5.0	—	—
Carbon monoxide, CO	—	Very dangerous	—	14.0	2.5	66 (−207)	83 (−190)
Carbon dioxide, CO_2	Faint, pleasant	Poisonous only at high concentrations	—	22	100	Sublimes at 195 (−78)	
Chlorine, Cl_2	Characteristic choking smell	Poisonous	Pale green	35.5	263	172 (−101)	239 (−34)
Hydrogen, H_2	—	—	—	1.0	2.1	14 (−259)	20 (−253)
Hydrogen chloride, HCl	Characteristic acrid smell	Corrosive, poisonous	—	18.25	46 100	159 (−114)	188 (−85)
Nitrogen, N_2	—	—	—	14.0	1.6	63 (−210)	77 (−196)
Ammonia, NH_3	Characteristic choking smell	Poisonous	—	8.5	80 670	195 (−78)	240 (−33)
Nitrogen monoxide, NO	Unknown	Unknown	—	15.0	5.0	108 (−165)	121 (−152)
Dinitrogen monoxide, N_2O	Faintly sweet, pleasant	Anaesthetic	—	22.0	75	169.3 (−103.7)	185 (−88)
Dinitrogen tetroxide, N_2O_4	Characteristic, unpleasant	Poisonous	Brown (colourless only when *pure*)	46 (varies)	Soluble, decomposes	262 (−11)	294 (21)
Oxygen, O_2	—	—	—	16	3.4	54 (−219)	90 (−183)
Sulphur(IV) oxide, SO_2	Pungent, acrid	Corrosive, poisonous	—	32	4730	200 (−73)	263 (−10)
Sulphur(VI) oxide, SO_3	Sharp, acrid	Corrosive, poisonous	Solid is white	—	Reacts violently	290 (17)	317 (44)
Hydrogen sulphide, H_2S	Rotten eggs	Very poisonous	—	17	260	187 (−86)	213 (−60)

4.2 THE AIR

How Experiments with Copper can be used to Determine the Nature of Air

Experiment 4.1

†What happens when Copper is Heated in Air? **

Apparatus

Crucible, pipe-clay triangle, asbestos mat, Bunsen burner, tongs, tripod, access to balance, scissors.
Copper wire or copper foil.

Procedure

(a) Find the mass of a crucible.

(b) Place about 4 or 5 g of copper foil (preferably cut into small pieces) or copper wire in the crucible and find the mass of the crucible and contents accurately.

(c) Support the crucible on a pipe-clay triangle resting on a tripod and heat it strongly for at least five minutes. Allow to cool and find the mass of the crucible and contents.

(d) Find the mass of the empty crucible after the experiment.

(e) Record all your masses and observations.

Points for Discussion

1. Does the appearance of the crucible contents before and after the experiment help you to decide whether the copper reacted in any way? Explain your answer.

2. As a result of your measurements and observations, which of the following seems to be the most likely conclusion?

A The copper reacted when heated. It decomposed to give off an invisible gas and left a black residue.

B The copper reacted when heated. It combined with some substance to form a new compound.

C The copper did not react in any way when heated.

What Does Copper Combine With When it is Heated?

As there was an increase in mass in the last experiment it should be obvious that when copper is heated in air it combines with something to form a new, black compound. Where could the substance with which it combines have come from? It could hardly be from the flame as this did not touch the metal. It could have come from the air or from the crucible itself. (Refer to the masses of the empty crucible before and after the last experiment.) The next experiment should help to confirm where it came from.

Experiment 4.2

†Heating Copper in the Absence of Air ***

Apparatus

Hard glass test-tube with Bunsen valve to fit, clamp stand, Bunsen burner, asbestos mat.
Pieces of copper foil.

Procedure

(a) Pour 2 cm^3 of water into the test-tube and clamp the tube at an angle.

(b) Place a strip of copper foil in the tube so that it rests about half way down the tube.

(c) Insert the Bunsen valve. The valve will allow a gas under pressure to escape from inside the tube but will not allow any air to enter the tube.

(d) Heat the water so that steam fills the tube and escapes from the valve. Then heat the copper foil strongly and occasionally heat the water so that the tube is always filled with steam.

Points for Discussion

1. Did the copper react to form the black substance when heated this time?

2. How do you think the presence of steam prevented the reaction?

3. What does copper need to combine with in order to make the black substance?

Experiment 4.3

†When Copper is Heated in Air Does it React with all of the Air or Only Part of it? ***

Apparatus

As in Figure 4.7(a) and also delivery tube, test-tube (150 × 25 mm), dish or low form beaker, combustion spoon.
Copper powder or wire, magnesium ribbon.

Procedure

(a) Place some copper in the silica tube and connect the tube to the two syringes. Make sure that there is as little air space as possible in the tube and connectors. One syringe should be full of air (100 cm³ or 50 cm³) and the other empty.

(b) Test the apparatus for leaks by pulling air through from one syringe to the other. The syringes should now read 0 cm³ and 100 cm³ or 50 cm³.

(c) Heat the silica tube containing the copper vigorously and at the same time continually pass air slowly from one syringe to the other through the hot copper. After about three minutes stop heating and cool the silica tube by applying a cold damp cloth. When cold, record the volumes of gas in the syringes.

(d) Repeat procedure (c) until no further change of volume takes place. Record the final volume of the remaining gas.

(e) Disconnect the syringe containing the residual 'air', quickly attach a delivery tube to it, and collect a sample of the gas in a test-tube over water. Do not collect the first few cm³ of gas which are expelled. (Why?) Attach a small length of magnesium ribbon to a combustion spoon, ignite it, and quickly plunge it in the tube of gas. Record your observations.

Points for Discussion

1. Why was the silica tube heated more than once?

2. Why was the tube cooled each time before the volume of gas was noted?

3. Do you think that copper reacts with all of the air or only part of it? If only part of it, what proportion of the air is this active part? Explain your answers.

4. Compare the behaviour of ignited magnesium in (a) ordinary air, and (b) the residual air. Try to explain the difference in behaviour.

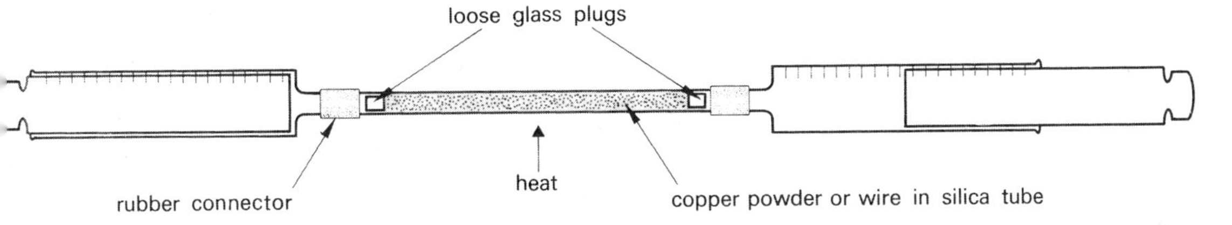

Figure 4.7(a) To determine the percentage of oxygen in the air

The Composition of Air

You have probably realized that the 'active' part of air which was used up by the copper, and is usually needed to make things burn, is oxygen. It occupies about 20 per cent of the air by volume. Is the remaining 80 per cent of the air a single substance or not? We will attempt to answer this latter question by the following experiment.

Experiment 4.4
Two Further Components of Air***

Apparatus
As shown in Figure 4.7(b).

Procedure
Assemble the apparatus as shown in Figure 4.7(b) and turn on the filter pump. This draws a stream of air through the apparatus.

Points for Discussion
1. What was the colour of the silica gel (i) before (ii) after the experiment? What does this change indicate?
2. From your observations of the change in the calcium hydroxide solution, what other gas do you think is present in air?
3. What is the purpose of the calcium chloride tube?

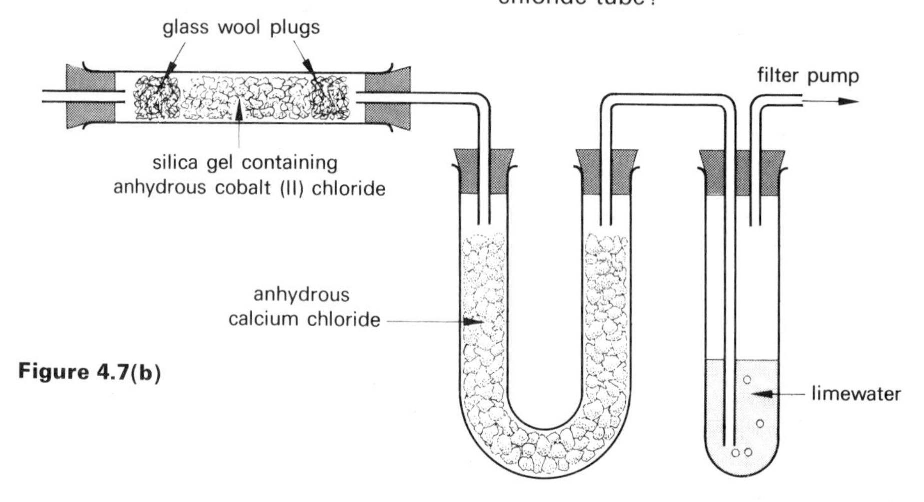

Figure 4.7(b)

Summary

Carbon monoxide from car exhausts has reached dangerously high levels in large cities

Air is a *mixture* of gases and not a compound. You should be able to suggest several reasons for the fact that we know it could not be a compound (Section 2.4). Air contains oxygen, water vapour, and carbon dioxide. When these are removed from the air the residual gas is largely nitrogen, but with more refined techniques it can be shown that small quantities of other gases, such as carbon monoxide and sulphur dioxide, are also present. Concentrations of gases such as sulphur dioxide are likely to be higher near to industrial towns, and carbon monoxide from car exhausts has reached dangerously high levels in large cities during the

Table 4.2 A typical analysis of dry air

Component	*Composition by volume (%)*
Nitrogen	78.08
Oxygen	20.95
Argon	0.93
Carbon dioxide	0.03
Neon	0.002
Other noble gases (helium, krypton, xenon)	0.0006
Methane	0.0001
Hydrogen, and ozone	minute traces

Ordinary air always contains water vapour, the proportion of which varies. Small proportions of solids will be present (e.g. soot, pollen, bacteria) and in industrial areas there may be such gases as hydrogen sulphide and sulphur dioxide. The air above cities and towns is likely to contain detectable quantities of carbon monoxide from the exhaust fumes of vehicles.

rush hour. It is obvious that air does not have a fixed composition; apart from the above variations, the water content varies with temperature and the carbon dioxide level will be higher in, for example, a crowded classroom than it is outside. A 'typical' analysis of air is shown in Table 4.2. Air also contains solid particles, e.g. dust and soot, particularly in industrial areas.

Experiment 4.5
†The Isolation of 'Nitrogen' from the Air ***

Principle and Aim
As nitrogen is very unreactive it is relatively easy to remove some of the more active gases from the air to leave fairly pure nitrogen. Read the following information carefully, refer to Section 4.1 and to Experiment 4.3, and then devise and draw an apparatus for collecting a sample of 'nitrogen' from the air by removing carbon dioxide, water vapour and oxygen. The vessels to be used for each operation, the *order* in which they are to be used, and the way in which the 'nitrogen' is to be collected must be carefully thought out.

Necessary Information
An aspirator is a useful way of obtaining a slow flow of air through an apparatus (Figure 4.8). When water is slowly run into the container, air is forced out through the other tube.

A concentrated aqueous solution of potassium hydroxide absorbs carbon dioxide.

Concentrated sulphuric acid is a good drying agent.

Points for Discussion
1. The 'nitrogen' obtained from the air by this method is not absolutely pure. What do you think it still contains? Could you remove any of these?

2. As you will learn later, pure nitrogen *can* be obtained from the air by the fractional distillation of liquid air.

3. You may be interested to know that an experiment similar to the one you devised led to the discovery of the noble gas argon. In 1892 Lord Rayleigh repeatedly observed that samples of nitrogen prepared chemically always had a slightly different density compared with samples of nitrogen extracted from the air. He reported the respective values 1.2505 g l^{-1} and 1.2572 g l^{-1}. Small though the differences were, they were consistent and Sir William Ramsay thought the discrepancy was due to an impurity in the atmospheric nitrogen. He examined the spectrum of 'nitrogen' obtained from the air and was able to detect the presence of a hitherto unknown element, which he and Rayleigh were later able to isolate in small quantities. They called the gas argon (from the Greek argos, 'idle'), for they were unable to remove it from the air by any chemical means.

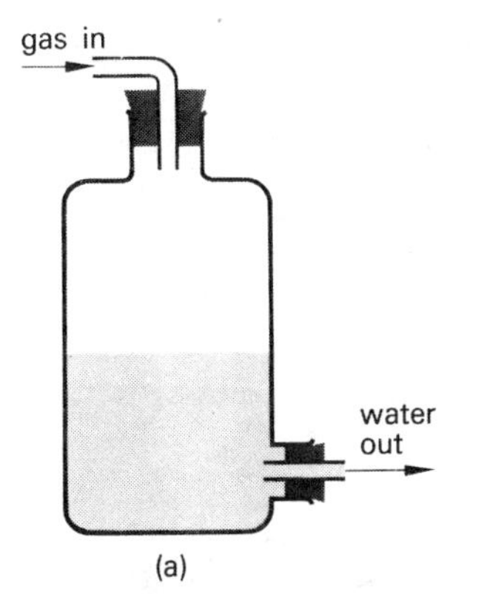

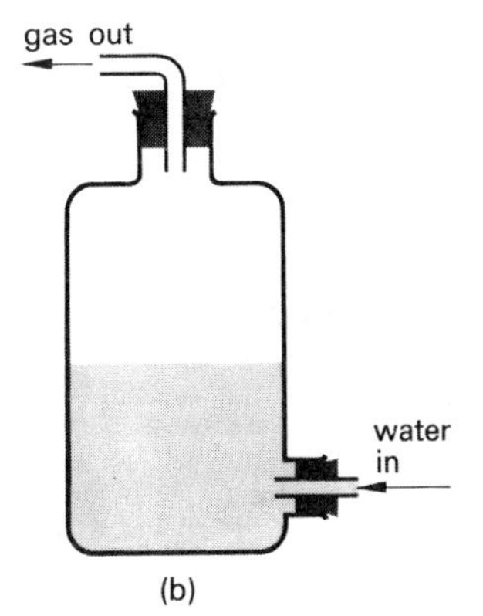

Figure 4.8 The use of an aspirator to control gas flow (a) to draw gas in and (b) to expel

The Solubility of Air in Water. The Effect of Pressure on Boiling Point

Experiment 4.6
†Is Air Soluble in Water? ***

Apparatus
Set up the apparatus as in Figure 4.9. The flask and the delivery tube must be completely filled with water and the end of the delivery tube must be level with the bottom of the bung.

Procedure
Heat the flask until the water boils. Continue the heating until there is no further change. Record any changes which take place.

Points for Discussion
1. How can you account for the gas collected in the tube?

2. Do you consider that air is soluble in water? Have you noticed what happens when a tumbler of water is left standing in a warm room? Can you *explain* what happens?

3. If 100 cm^3 of the air boiled out of water is transferred to a syringe and analysed as in Experiment 4.3, it is found that about 33 per cent of the gas is removed by the copper, compared to about 20 per cent of normal air. Which of the following is the most reasonable conclusion?

A Dissolved air has the same composition as ordinary air.
B Nitrogen is more soluble in water than oxygen, so the proportion of nitrogen in dissolved air is greater.
C Oxygen is more soluble in water than nitrogen, so the proportion of oxygen in dissolved air is greater.

4. Confirm your answer to 3 by reference to Table 4.1.

5. Have you heard of deep sea divers suffering from a painful condition known as the 'bends'? How is it caused?

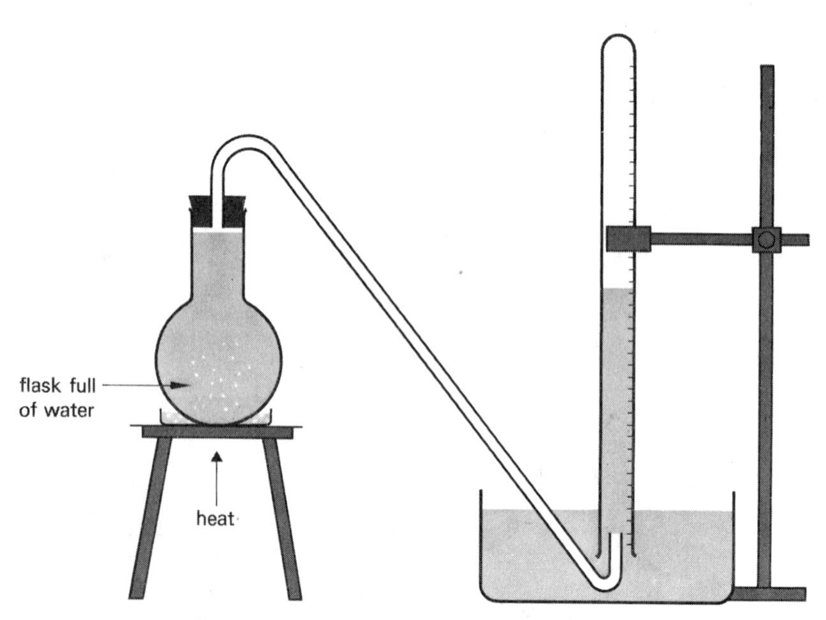

Figure 4.9 To show that air is soluble in water

Experiment 4.7
†The Effect of Air Pressure on Boiling Point ***

Apparatus

Set up the apparatus shown in Figure 4.10. The thermometer should read to 110 °C. Pieces of broken porcelain or antibumping granules should be added to the water in the flask and all connectors should be of pressure tubing. The bungs must fit tightly.

Procedure

(a) Heat the water until it boils. When the temperature of the boiling water is steady, record it, and also note the atmospheric pressure.

(b) Remove the Bunsen flame and connect the pump to the apparatus. Use the pump to reduce the pressure by about 3 cm of mercury.

(c) Again boil the water in the flask and note the steady temperature and the difference in the levels of the mercury in the manometer.

(d) Continue to reduce the pressure in the apparatus in stages of about 3 cm and note the boiling point and pressure difference each time. Four readings at reduced pressure are sufficient.

(e) Plot a graph of 'boiling point of water' against 'pressure'.

Points for Discussion

1. Why do you think that a shield was placed between the flask and the bottle?

2. Why is the large bottle used? Why is the flask not joined directly to the pump?

3. Is the boiling point of water fixed or does it depend upon the atmospheric pressure?

4. When a liquid is heated more of its molecules gain sufficient energy to overcome the intermolecular forces and so escape to become a gas. If the boiling point of a liquid is lowered do you think its molecules are escaping more readily or less readily than before? Try to explain how decreasing the pressure causes this.

5. The solubility of most solids in water increases with a rise in temperature. Do you think this is true of the solubility of gases in water? Give as much evidence as possible for your answer.

6. Why does water boil at about 343 K (70 °C) near the top of Mount Everest?

7. How do you think a pressure cooker increases the rate at which food is cooked?

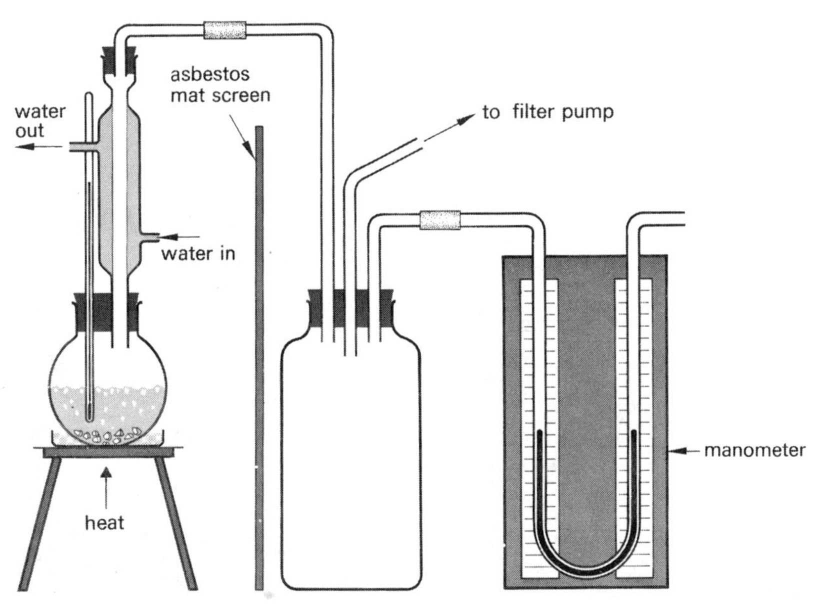

Figure 4.10 The effect of air pressure on the boiling point of water

Combustion, Respiration, Photosynthesis and Rusting

There are important similarities between these apparently unconnected processes.

Combustion

This usually refers to the combination of substances with pure oxygen or oxygen in the air, and is accompanied by a liberation of heat and/or light. However, it is important to remember that the term is also used for similar reactions involving a gas other than oxygen. For example, sodium burns in chlorine to form sodium chloride. This is combustion, even though oxygen is not present.

We have already seen (Experiments 4.2 and 4.3) that magnesium and copper will not undergo combustion unless oxygen is present.

The energy liberated in combustion reactions is of the utmost importance. Fuels are substances used to provide energy, and many fuels are made to liberate such energy during combustion, e.g. coal, petrol, oil, coal gas, and natural gas. Most fuels currently used by man contain both carbon and hydrogen, and the combustion of such fuels leads to the formation of carbon dioxide and water as products. This can easily be demonstrated by burning a candle (a fuel) under an inverted beaker. The flame goes out when most of the oxygen has been used up. A mist of water vapour is observed and this condenses on the inside of the beaker. On testing this 'mist' with universal indicator paper it is found to be slightly acidic, and this acidity is due to some of the carbon dioxide (the other main product) dissolving in the mist.

$$C_xH_y + \text{oxygen} \rightarrow CO_2 + H_2O + \text{energy}$$

Respiration

Aerobic respiration is the process by which an organism obtains its energy from food substances and oxygen. It is a complex multistage process but the overall change can be investigated fairly easily.

Experiment 4.8
The Products of Respiration ***

Apparatus
Arrange as in Figure 4.11. A thermometer and a beaker are also required.

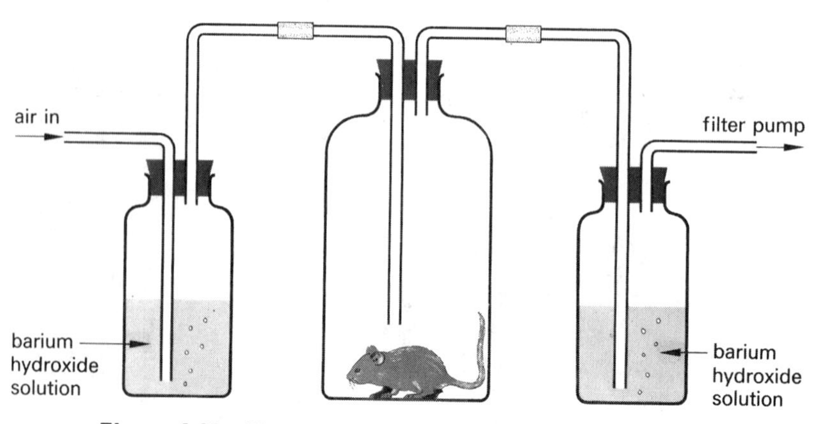

Figure 4.11 The difference between inspired and expired air

Procedure

(a) Place the small animal in the container and ensure a steady flow of air through the apparatus. Barium hydroxide solution is similar to calcium hydroxide solution (i.e. limewater) but is more sensitive as a test for carbon dioxide; the milkiness (due to precipitated barium carbonate) appears more quickly than with limewater. However, limewater is normally preferred as barium compounds are poisonous. Watch for the milky appearance in *both* wash-bottles.

(b) Record the temperature of the air in the room and of expired human air.

(c) Breathe on to a cold glass surface (e.g. a beaker) for a minute or so.

(d) Compare the temperature, water content and carbon dioxide content of ordinary and expired air in a simple table.

Point for Discussion

As a result of your experiments which of the following seems the most reasonable conclusion?

A Ordinary air and expired air are exactly the same.

B Expired air differs from ordinary air by being warmer, containing less carbon dioxide and more water vapour.

C Expired air differs from ordinary air by being warmer, containing more carbon dioxide and more water vapour.

Summary

Many organisms obtain their energy from the reaction between foods (such as sugars) and oxygen from the air. The products are water, carbon dioxide, and energy. Some organisms convert a proportion of the energy into heat, hence the difference in temperature between inspired and expired air. Overall change:

$$\underset{\text{(a sugar)}}{C_6H_{12}O_6(aq)} + 6O_2(g) \rightarrow 6CO_2(g) + 6H_2O(g) + \text{energy}$$

Can you see any similarities between the combustion of fuels containing carbon and hydrogen, and aerobic respiration? When sugar (a fuel) burns in oxygen the energy liberated is dissipated as heat and light (the flame). Aerobic respiration involves the combustion of sugars and other foods so that none of the energy is liberated as light but is in a form which can be used by the organism.

stion or respiration?

Photosynthesis

This is the process by which green plants synthesize carbohydrate foods from their environment. If you are studying biology you will realize that photosynthesis is, like respiration, a multistage chemical reaction, but the overall change can be summarized as:

$$6CO_2(g) + 6H_2O(l) + \underset{\text{(from sunlight)}}{\text{energy}} \xrightarrow{\text{chlorophyll}} C_6H_{12}O_6(aq) + 6O_2(g)$$

You will notice that oxygen is released by the process and carbon dioxide consumed. Are there any similarities between aerobic respiration and photosynthesis? What difference is there? It should be obvious that if respiration was the only process involving oxygen taking place in our environment, the Earth's supply of oxygen would have been used up long ago and the carbon dioxide level in the atmosphere would have built up to a toxic level. Why do you think that this has not happened?

Rusting

This is another environmental change involving oxygen (Section 6.3). Compare photosynthesis, combustion, rusting, and respiration carefully. See what the reactants and products are in each case and look for similarities. These are summarized in Table 4.3.

Table 4.3 A comparison between combustion, respiration, photosynthesis, and rusting

Component	*Combustion*	*Respiration*	*Photosynthesis*	*Rusting*
Oxygen	Often needed, although some combustions proceed in some other gas	Needed for aerobic respiration	Liberated	Needed
Carbon dioxide	Nearly always formed if the 'fuel' contains carbon	Liberated	Needed	Not essential, but the process is speeded up
Water	Always formed if the 'fuel' contains hydrogen	Liberated	Needed	Needed
Energy	Always liberated	Liberated	Needed	Liberated slowly

Other Reactions of Substances in the Air

There are many substances which react chemically with the air, in particular with the oxygen present. Secondary reactions may also occur involving water and/or carbon dioxide. All forms of atmospheric corrosion and phenomena such as efflorescence and deliquescence are reactions involving the air, and these are discussed in detail in appropriate sections of the book.

4.3 OXYGEN AND THE OXIDES

The main divisions of our environment from which we can extract elements are the seas and other waters (the hydrosphere), the Earth's crust (the lithosphere), and the atmosphere. Eighty-eight elements are found in nature, either free or combined, and yet about 99 per cent of the total weight of the hydrosphere and the lithosphere is made up of only ten elements.

Oxygen is by far the most abundant element on Earth. About 49.2 per cent of the lithosphere and hydrosphere consists of combined oxygen (as sulphates, silicates, carbonates, water etc.) and the element is also found free in the atmosphere, about 23 per cent by weight and 21 per cent by volume.

Preparation of Oxygen

Laboratory Preparation

Oxygen was first prepared by Carl Wilhelm Scheele about 1772, and named by Antoine Lavoisier from the Greek 'oxys genon' (acid former) as he thought (erroneously) that the products of combustion with oxygen are always acidic.

Oxygen is conveniently prepared by the catalytic decomposition of hydrogen peroxide solution. A suitable apparatus is shown in Figure 4.12.

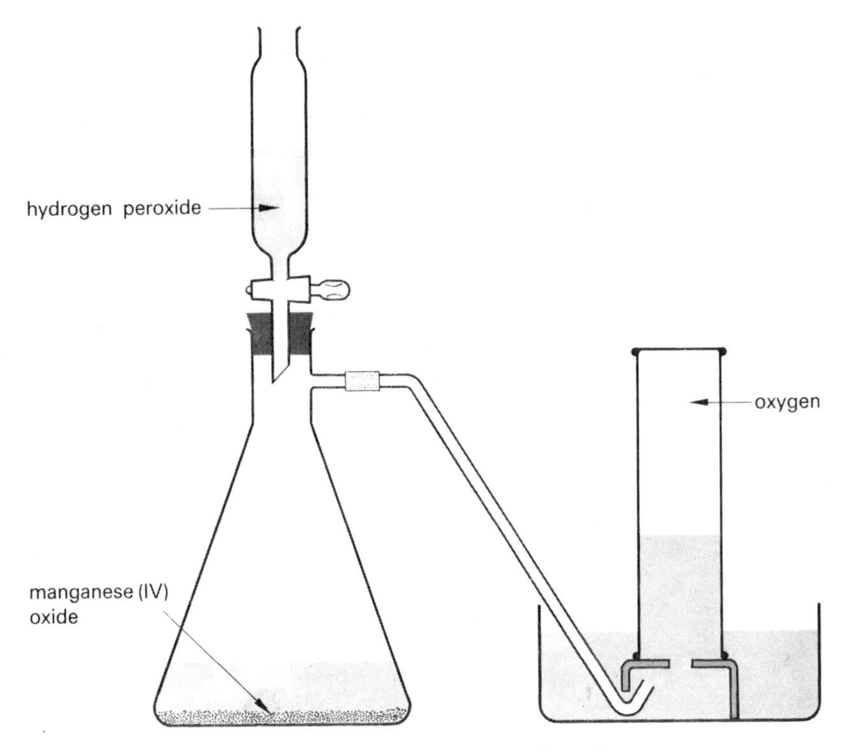

Figure 4.12 The laboratory preparation of oxygen

The catalyst (page 365) is manganese(IV) oxide and the hydrogen peroxide is usually about 'twenty volume' strength.

$$2H_2O_2(aq) \xrightarrow{(MnO_2\ catalyst)} 2H_2O(l) + O_2(g)$$

The oxygen so produced is fairly pure, although if necessary it can be dried with concentrated sulphuric acid.

Industrial Preparation

Oxygen is obtained by the fractional distillation of liquid air (page 394).

Other Methods of Preparation

Oxygen is sometimes prepared by the thermal decomposition (i.e. decomposition induced by heat) of potassium chlorate(V) using manganese(IV) oxide or copper(II) oxide as a catalyst (Experiment 14.6).

$$2KClO_3(s) \rightarrow 2KCl(s) + 3O_2(g)$$

Oxygen is evolved when tri-lead tetroxide (Pb_3O_4) or potassium manganate(VII) (potassium permanganate, $KMnO_4$) is heated, and during electrolysis of many aqueous solutions.

Properties of Oxygen

Physical Properties of Oxygen

Liquid oxygen is pale blue. There is an allotrope of oxygen, the gas trioxygen (ozone), O_3. This is very unstable and need not concern us at this stage but it has commercial uses and is an oxidizing agent.

Study the data on oxygen in Table 4.1. Can you think of any other facts which confirm that oxygen is slightly soluble in water?

General Properties of Oxygen

Oxygen occurs in the same group of the Periodic Table as sulphur, but similarities in their chemistry are not obvious at an introductory level and the two elements will be studied separately.

As oxygen is the second most electronegative (page 146) element it is very reactive. It reacts with most other elements, although it may not appear to do so with some metals because an oxide film is formed which protects the element from further attack. In general, solid elements react more vigorously if their surface area is increased. For example, finely divided lead is *pyrophoric*, i.e. it burns spontaneously when allowed to fall through oxygen. Many of these reactions liberate considerable amounts of energy, which is dissipated as heat, light, or even sound.

As we found in the previous section, respiration, rusting, and combustion are all processes for which oxygen is essential.

Can you think of any other facts which confirm that oxygen is soluble in water?

Chemical Properties of Oxygen

Experiment 4.9

†The Formation of Metallic Oxides **

Apparatus

Oxygen generator and collection apparatus as in Figure 4.12, rack of test-tubes (150 × 25 mm) with bungs to fit, combustion spoon, tongs, asbestos mat and Bunsen burner.

Magnesium ribbon, steel wool, sodium, asbestos paper. The sodium must only be used by the teacher, and even then great care is needed as the test-tubes will be wet. Goggles should be used in this and all similar experiments.

Procedure

(a) Collect three test-tubes of oxygen and seal each one with a tight fitting bung before placing it in the rack.

(b) Note: In all the tests in this and the next experiment the heated substances must be inserted into the boiling tubes while they are in the rack and *not* while held in a test-tube holder. Re-seal each tube immediately after the reaction and keep them for later tests.

(c) Wind a small length of magnesium ribbon around the bottom of the combustion spoon so that a straight piece about 1.5 cm long hangs free. Ignite the ribbon and immediately insert it into a sample of oxygen. (Warning: do not stare directly at the burning magnesium.)

$$2Mg(s) + O_2(g) \rightarrow 2MgO(s)$$

(d) Place some iron wool on the end of the spoon, heat it to red heat and *quickly* plunge it into another sample of oxygen.

$$3Fe(s) + 2O_2(g) \rightarrow Fe_3O_4(s)$$

(e) Place a piece of asbestos paper on the spoon and then add a piece of sodium about the size of a rice grain. Hold the spoon in the flame until the sodium begins to burn and then transfer it to a test-tube of oxygen.

$$4Na(s) + O_2(g) \rightarrow 2Na_2O(s)$$
(Sodium monoxide)

$$2Na(s) + O_2(g) \rightarrow Na_2O_2(s)$$
(Sodium peroxide)

(f) Record all your observations in each case, noting in particular any colour changes, the state of the products (i.e. whether solids, liquids or gases), and the vigour of the reactions.

Experiment 4.10
†The Formation of Non-metallic Oxides **

Apparatus
As in Experiment 4.9 but using sulphur, carbon and red phosphorus instead of the metals. The phosphorus is to be used only by the teacher.

Procedure
(a) Place a small quantity of sulphur on the spoon, ignite it in the Bunsen flame and then plunge it into a sample of oxygen.

$$S(s) + O_2(g) \rightarrow SO_2(g)$$
(Sulphur(IV) oxide)

$$2S(s) + 3O_2(g) \rightarrow 2SO_3(g)$$
(Sulphur(VI) oxide)

(b) Place a small quantity of carbon on the spoon, heat to redness and *quickly* plunge it into a sample of oxygen.

$$C(s) + O_2(g) \rightarrow CO_2(g)$$

(c) With care ignite a little *red* phosphorus on the spoon and place it in a sample of oxygen in the fume cupboard.

$$P_4(s) + 5O_2(g) \rightarrow P_4O_{10}(s)$$

(d) Record all your observations, noting in particular any colour changes, the state of the products, and the vigour of the reactions.

Points for Discussion
1. What *type* of compound is always formed when an element reacts with oxygen?

2. You will notice that two equations are given for two of the reactions. In these cases both reactions do occur but only one is really significant. The main products in the reactions quoted are Na_2O_2 and SO_2.

3. Oxides of metals tend to be basic, and the oxides of non-metals are usually acidic or neutral. How could you decide the nature of the products of the above reactions? Confirm these generalizations by testing your samples.

4. When a substance burns in a gas (particularly oxygen) the process is called combustion. Remember that air or oxygen is not needed for *all* combustion reactions. Do you consider that elements burn more vigorously or less vigorously in air than they do in pure oxygen? Try to explain your answer by reference to the percentage composition of the air.

The Test for Oxygen Gas

Place a *glowing* splint (compare with the test for hydrogen) into a sample of oxygen. Record what happens. This is the usual test for oxygen, but it is useful to remember that there is one other gas (dinitrogen monoxide, sometimes called nitrous oxide, N_2O) which gives the same result, although you are not likely to study this gas until a later stage.

Uses of Oxygen

If a mixture of oxygen and ethyne (acetylene) gases is burned at a jet, temperatures of the order of 3000 K are reached. The oxy-acetylene blowpipe is thus frequently used for welding and cutting metals, supplies of the two gases being obtained from cylinders.

Modern steel-making processes use oxygen to burn out impurities in the molten metal, and liquid oxygen is used in some rocket fuels. Oxygen is used as a respiratory aid in deep sea diving, high altitude flying, climbing, and hospitals.

Oxides

The general preparations of metallic oxides are discussed in Chapter 10, and non-metallic oxides are often prepared by direct combination with oxygen (as in Experiment 4.10). Other methods of preparation are considered when a non-metallic oxide is studied in detail (Chapter 11).

Oxides can be classified as acidic, basic, neutral, amphoteric or peroxides according to their chemical properties, although this classification is by no means hard or fast.

Acidic Oxides

Acidic oxides will react with bases so that they are neutralized and a salt is formed.

These are often non-metallic oxides, although not all non-metallic oxides are acidic. If an acidic oxide is soluble in water its solution behaves as a typical acid and the oxide is said to be the *anhydride* of the acid. For example, sulphur(VI) oxide dissolves in water to form a solution of sulphuric acid. Sulphur(VI) oxide is an acidic oxide and it is the anhydride of sulphuric acid. If an acidic oxide is insoluble in water it may be less obvious that it is acidic, but it will still neutralize bases. Examples of acidic oxides are given in Table 4.4.

Table 4.4 Some typical acidic oxides

Name and formula of oxide	*State at room temperature*	*Action on water*	*Is it an acid anhydride?*
Sulphur(IV) oxide, SO_2	Gas	Some dissolves and some reacts $SO_2(g)+H_2O(l) \rightarrow H_2SO_3(aq)$	Yes, that of sulphurous acid
Sulphur(VI) oxide, SO_3	Solid	Violent reaction $SO_3(s)+H_2O(l) \rightarrow H_2SO_4(aq)$	Yes, that of sulphuric acid
Dinitrogen tetraoxide, N_2O_4	Gas	Reacts to form a solution of *two* acids: $N_2O_4(g)+H_2O(l) \rightarrow HNO_3(aq)+HNO_2(aq)$	This is a *mixed* anhydride, it forms nitric and nitrous acids
Silicon dioxide, SiO_2	Solid	Insoluble	No, but it dissolves in alkali
Carbon dioxide, CO_2	Gas	Some dissolves and some reacts $CO_2(g)+H_2O(l) \rightarrow H_2CO_3(aq)$	Yes, that of the weak, unstable carbonic acid

Basic Oxides

These are oxides which will neutralize acids to form a salt. Bases are often metallic oxides or hydroxides, although not all metallic oxides and hydroxides are bases. Typical basic oxides include CuO, MgO, and FeO

Amphoteric Oxides

Amphoteric oxides will neutralize acids and caustic alkalis to form salts. For example, zinc oxide reacts with acids:

$$ZnO(s)+2HCl(aq) \rightarrow ZnCl_2(aq)+H_2O(l)$$

and with a caustic alkali:

$$ZnO(s)+2NaOH(aq)+H_2O(l) \rightarrow Na_2Zn(OH)_4(aq)$$

Aluminium oxide and lead monoxide are also amphoteric.

Neutral Oxides

Neutral oxides do not react with acids or bases and thus form a class of their own. Typical examples include nitrogen monoxide (NO) and dinitrogen monoxide (N_2O).

Peroxides

Peroxides are oxides which liberate hydrogen peroxide when treated with a dilute acid; they always contain the peroxy link —O—O—. Sodium peroxide (Na_2O_2) is a true peroxide, but dioxides such as PbO_2 are not peroxides as they do not contain the —O—O— linkage nor do they liberate hydrogen peroxide when treated with a dilute acid.

4.4 HYDROGEN

As we stated earlier, about 99 per cent of the total mass of the hydrosphere and the lithosphere is due to only ten elements. Hydrogen is one of the 'top ten' yet it only contributes 0.9 per cent of the mass of the hydrosphere and lithosphere because it is the element with the lightest atoms. It has been calculated, however, that approximately 15.4 per cent of all the atoms in the Earth's crust are hydrogen atoms.

Many people believe that hydrogen atoms are the 'building bricks' of all substances in the universe. It is probable that over 90 per cent of all matter in the known universe is hydrogen, and although the planet Earth has a comparatively small amount of free hydrogen our own sun consists almost entirely of hydrogen and helium. At the enormous temperatures generated in the sun and other stars, thermonuclear reactions take place, resulting in the formation of larger nuclei (i.e. fusion reactions). All other elements could *theoretically* be produced from hydrogen by such reactions. Compounds of hydrogen occur in nature as water, as hydrocarbons such as methane (natural gas) and those present in oil, and as proteins, carbohydrates, and fats of living materials.

The first systematic investigation into the chemistry of hydrogen was by Henry Cavendish in 1766 and the name was given by Antoine Lavoisier from the Greek *hydro genon,* 'water former'.

Preparation of Hydrogen

Laboratory Preparation

Note: It is not always appreciated that mixtures of hydrogen with air or oxygen are very dangerous because of their explosive nature. A naked flame should never be used near a hydrogen generator. Before collecting a large sample of the gas, and before burning it, small samples should be collected in test-tubes and tested with a flame (away from the apparatus) until the gas burns quietly.

You have learned that hydrogen is liberated when a typical dilute acid is added to a fairly reactive metal. Reference to the work in Sections 3.1 and 6.2 should suggest that magnesium, zinc or iron could be used with dilute sulphuric or hydrochloric acids. However, magnesium is rather too reactive, iron is usually very impure and so zinc is normally used. It is best if the zinc is rather impure as hydrogen cannot escape easily from a pure zinc surface because of its high overvoltage (Section 6.3). [Hydrogen is evolved more rapidly if a little copper(II) sulphate solution is added to the generator flask. This creates many tiny zinc/copper cells and hydrogen is discharged very easily from the cathodes (copper) (Section 6.3).] Addition of a copper salt is unnecessary if dilute hydrochloric acid is used in place of sulphuric acid.

Figure 4.13 illustrates the usual laboratory preparation. Check with the data given in Table 4.1 and decide whether you think the method of collection has taken into account the properties of the gas.

$$Zn(s) + H_2SO_4(aq) \rightarrow ZnSO_4(aq) + H_2(g)$$

or, ionically,

$$Zn(s) + 2H_3O^+(aq) \rightarrow Zn^{2+}(aq) + H_2(g) + 2H_2O(l)$$

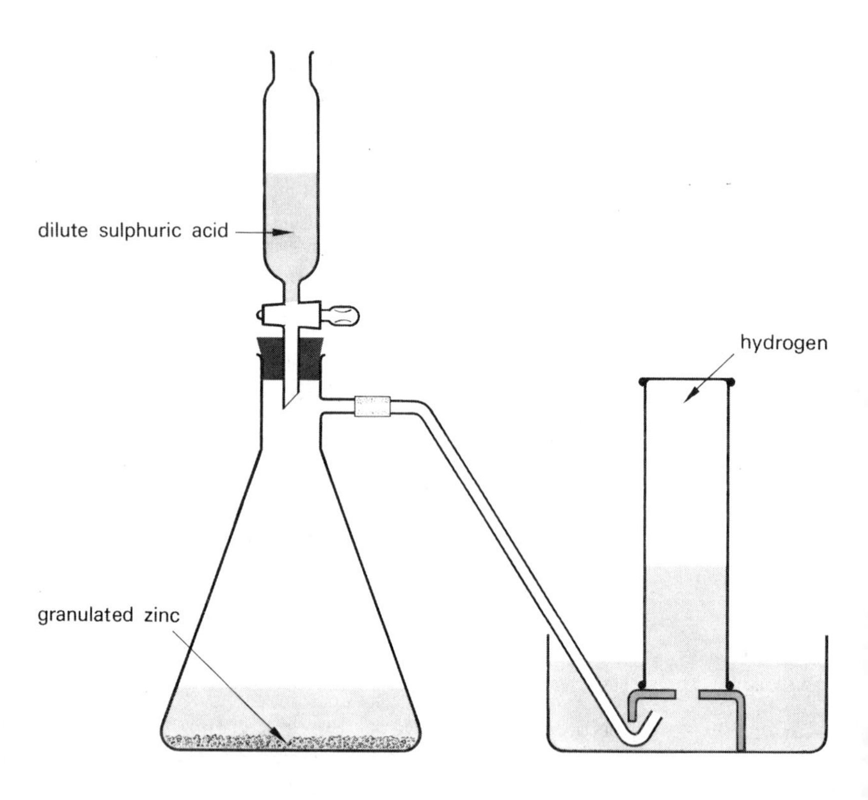

Figure 4.13 The laboratory preparation of hydrogen

Properties of Hydrogen

Physical Properties Study the data on hydrogen given in Table 4.1. In addition, it is worth remembering that hydrogen is the lightest gas and diffuses very rapidly indeed.

How could you confirm by experiment that hydrogen is only slightly soluble in water?

Perform the following tests on samples of hydrogen gas and try to explain which of the physical properties of hydrogen is illustrated by each test, by answering the questions under Points for Discussion.

Experiment 4.11
Some Physical Properties of Hydrogen **

Apparatus
Small scale hydrogen generator, squat dish or low form beaker, delivery tube, test-tubes in rack, bungs, small gas jar and demonstration apparatus as in Figure 4.14.

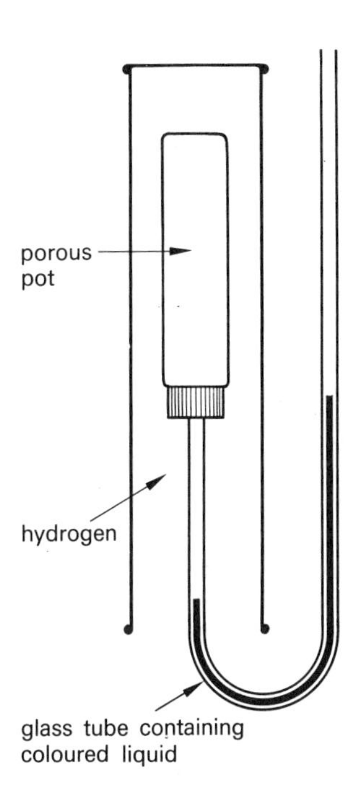

Figure 4.14 To illustrate diffusion

Procedure
(a) Collect two test-tubes of hydrogen over water and seal each with a tight bung or keep each tube inverted under water before use.

(b) Remove the bung from a tube of hydrogen and *immediately* place it below a test-tube of air, mouth to mouth. Wait a few moments and then test each tube with a lighted splint.

(c) Repeat (b) but place the tube of hydrogen on top of a test-tube of air and hold the tubes together for about a minute.

(d) Set up the apparatus shown in Figure 4.14 and collect a sample of hydrogen by upward delivery in a small gas jar. Quickly place the gas jar over the porous pot and wait until no further change is observed in the level of the coloured liquid. Remove the gas jar from the apparatus and record any further changes which take place.

Points for Discussion
1. Do your observations confirm that hydrogen is a colourless, odourless gas?

2. How can you account for your observations during procedure (b)? Does this experiment help to confirm the data about the lightness and diffusion rate of hydrogen? Explain your answer.

3. You should have found from procedure (c) that hydrogen can also move 'downwards'. Is this because hydrogen is a dense gas? How else can you explain what happened? Why did it take the hydrogen longer to move downwards than it did to move upwards? Remember that in Section 1.1 you found that bromine, although a dense gas, could move upwards.

4. Try to account for your observations during procedure (d), and to conclude what you have learnt about hydrogen from this experiment by answering the following questions. What was in the porous pot at the start of the experiment? When hydrogen was in the gas jar outside the porous pot why should it diffuse into the pot? (Remember the definition of diffusion.) Why should air diffuse out of the pot into the jar? If the hydrogen diffused into the pot at the same rate as the air diffused out, there would be no pressure change inside the porous pot and the levels of liquid in the U-tube would remain the same. From what you saw happen to the liquid in the tube, can you say which gas won the diffusion race? Explain your answer. Why did the levels of the liquid return to their original positions even while the gas jar of 'hydrogen' was still there? Try to explain what happened when the gas jar was removed. What can you conclude about the rates of diffusion of hydrogen and air?

General Properties

The two atoms in a hydrogen molecule are joined by a single covalent bond, H—H. When hydrogen reacts this bond must be broken so that the free atoms formed can make new bonds with the other reactant(s). For example, in the reaction

$$2H_2\,(g) + O_2\,(g) \rightarrow 2H_2O\,(g)$$

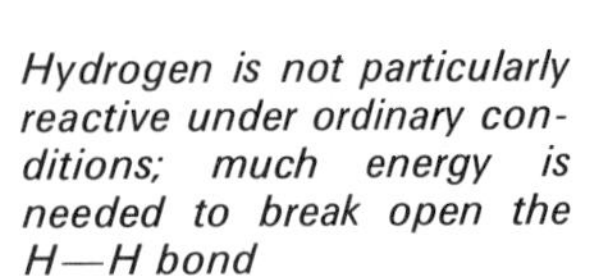

the hydrogen atoms are at first joined together, but in the product they are separate, i.e.

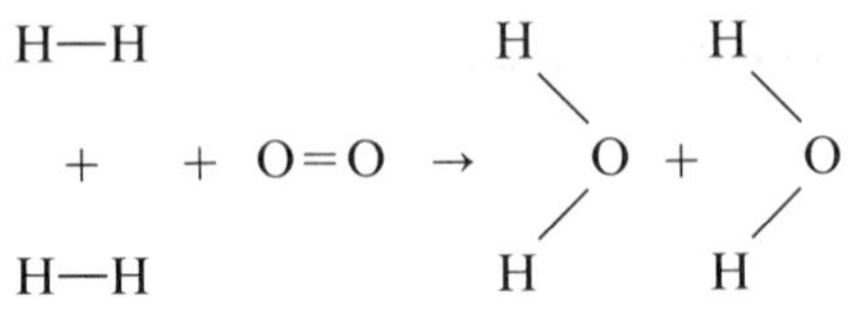

Hydrogen is not particularly reactive under ordinary conditions; much energy is needed to break open the H—H bond

A great deal of energy is needed to break open the H—H bond and so hydrogen is not particularly reactive under ordinary conditions although it forms a highly explosive mixture with air (*see* Atomic Hydrogen, page 124).

Once the bond is broken the hydrogen atoms are exceptional in that they can then do one of three things in order to gain stable electronic configurations (page 38). This is why hydrogen is studied separately; it does not conveniently fit into any one group of the Periodic Table.

Isotopes of Hydrogen

'Ordinary' hydrogen (^1_1H) is sometimes given the special name *protium* to distinguish it from two other isotopes of hydrogen, ^2_1H (deuterium) and ^3_1H (tritium). Draw the structure of an atom of each isotope.

Why do you think there is a more marked difference in physical properties between atoms of ^2_1H and ^1_1H than there is between, say, the two chlorine isotopes $^{37}_{17}Cl$ and $^{35}_{17}Cl$? The difference in boiling point of these two hydrogen isotopes is, for example, 3.3 K.

Atoms of deuterium occur naturally in the proportion of one part in 6400 parts of hydrogen. Thus water, although always assumed to contain hydrogen (^1_1H) and oxygen only, actually contains a small amount of deuterium as well. If water is electrolysed for a long period the residual liquid becomes richer and richer in D_2O (deuterium oxide), as the lighter protium ions are discharged about six times more readily than are deuterium ions. This process is not feasible on a large scale unless cheap electricity (e.g. hydroelectric power) is available, and so large quantities of 'heavy water' are made in Norway. Heavy water is used as a moderator in nuclear reactors. Tritium is radioactive.

Heavy water?

Some physical properties of H_2O and D_2O are as follows:

Some physical properties of H_2O and D_2O

Property	H_2O	D_2O
Boiling point, K (°C)	373 (100)	374.5 (101.5)
Melting point, K (°C)	273 (0)	276.8 (3.8)
Density at 293 K (20 °C) (g cm^{-3})	0.998	1.1

Chemical Properties

Combustion A jet of hydrogen or a gas jar of pure hydrogen burns quietly with a blue flame, when ignited in air or oxygen, to form steam.

$$2H_2(g) + O_2(g) \rightarrow 2H_2O(g)$$

If air or oxygen is first mixed with the hydrogen, an explosion takes place when the flame is applied. Steam is again the product. Mixtures of this type can be very dangerous. You have conducted this kind of reaction on a small scale when testing for hydrogen gas.

Any experiment which proves that steam is the product of the combustion of hydrogen is potentially dangerous, but you may see this as a demonstration. Figure 4.15 illustrates a typical apparatus.

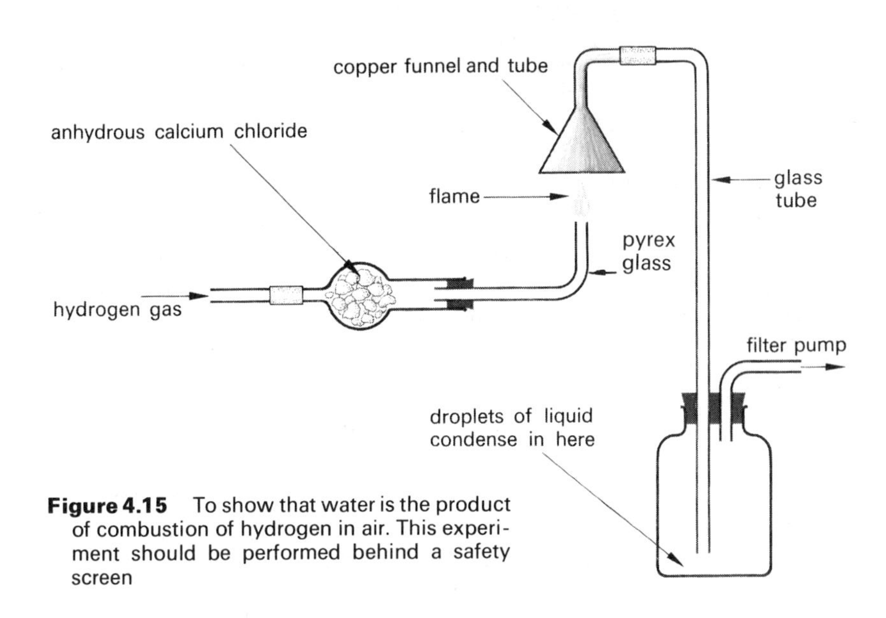

Figure 4.15 To show that water is the product of combustion of hydrogen in air. This experiment should be performed behind a safety screen

Why do you think that a good flow of hydrogen is essential? Why is the gas dried before combustion? How could you prove that the condensed liquid is water?

Hydrogen will not support the combustion of other substances.

Reaction of hydrogen with metals If a stream of hydrogen is passed over hot sodium metal, a white, crystalline solid, sodium hydride, is formed. The compound is ionic and contains the comparatively rare H^- ions.

$$2Na(s) + H_2(g) \rightarrow \underset{(Na^+H^-)}{2NaH(s)}$$

Similar reactions occur with other Group 1 metals and also some metals of Group 2. Why do you think that aluminium does not form an ionic hydride?

Reaction of hydrogen with non-metals Hydrogen reacts under differing conditions with most non-metals to form covalent (and, therefore, usually gaseous) hydrides. All four halogens react with hydrogen but at very different rates (page 145). Typical equations are:

$$F_2(g) + H_2(g) \rightarrow 2HF(g)$$
$$Cl_2(g) + H_2(g) \rightarrow 2HCl(g)$$

The reaction of hydrogen with oxygen has already been mentioned, and its reaction with nitrogen (the Haber process) is considered on page 381. Other non-metallic hydrides such as hydrogen sulphide, H_2S, are not usually prepared directly from hydrogen.

Hydrogen as a reducing agent One of the definitions of reduction (page 204) is a reaction in which hydrogen is added. It thus follows that hydrogen is a reducing agent. Hydrogen also has a strong affinity for oxygen, and when it removes oxygen from compounds it is again, by definition (page 204), acting as a reducing agent.

When a stream of hydrogen gas is passed over a heated metallic oxide, two elements are competing for the oxygen, i.e. the metal and the hydrogen. If the metal 'wins', it keeps its oxygen, remains a metallic oxide, and no reaction with hydrogen takes place. If, on the other hand, the hydrogen succeeds in 'grabbing' the oxygen from the metal a chemical reaction takes place and water (or steam) and the metal are formed. If the metal itself has a strong affinity for oxygen it will not be easy for the hydrogen to remove the oxygen from the metallic oxide.

Metallic oxides can be reduced by hydrogen in an apparatus as shown in Figure 1.21. Typical reactions are:

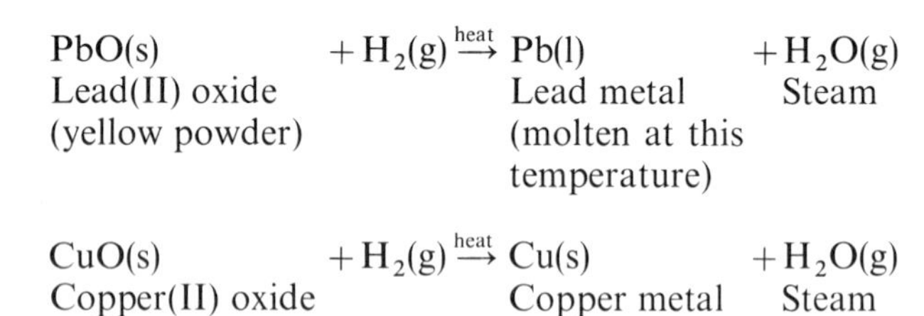

$PbO(s)$	$+ H_2(g) \xrightarrow{heat}$	$Pb(l)$	$+ H_2O(g)$
Lead(II) oxide		Lead metal	Steam
(yellow powder)		(molten at this temperature)	

$CuO(s)$	$+ H_2(g) \xrightarrow{heat}$	$Cu(s)$	$+ H_2O(g)$
Copper(II) oxide		Copper metal	Steam
(black powder)		(red-brown powder)	

In the first reaction the lead(II) oxide has been reduced to lead and the hydrogen has been oxidized to water. What happened in the second reaction?

Hydrogen will *not* reduce the oxides of sodium, calcium, magnesium, aluminium or zinc, and the reaction between hydrogen and tri-iron tetroxide (Fe_3O_4) is reversible.

Addition reactions of hydrogen These are considered in Section 12.1, but the manufacture of margarine, which is an important example of addition of hydrogen, is described at the end of this section.

Atomic hydrogen Hydrogen is not particularly reactive at ordinary temperatures because of the great amount of energy needed to break the H—H covalent bond. If tungsten electrodes are placed in hydrogen gas and an electric arc is struck between them, the hydrogen molecules are dissociated (split up) into free atoms. This atomic hydrogen is much more reactive than the normal molecular hydrogen.

The test for hydrogen *See* Experiment 3.3 for the test for hydrogen.

Uses of Hydrogen

The Manufacture of Ammonia Very large quantities of hydrogen are needed for this process (the Haber process, page 381). Much of the ammonia is then converted into nitric acid, so it could be argued that hydrogen is needed for the manufacture of nitric acid.

The Manufacture of Organic Chemicals Hydrogen is used in the manufacture of many organic chemicals such as methanol and nylon 66.

Welding

A stream of hydrogen atoms from an 'atomic hydrogen torch' is directed on to the surfaces to be welded. The hydrogen atoms recombine to form molecules and in doing so regenerate the enormous amount of energy originally needed to separate them. Temperatures of 3000 K can be reached and the metal surfaces fuse together. A big advantage of this process is that the molten metals are surrounded by an atmosphere of hydrogen so that they cannot become oxidized, as so often happens in other welding processes. Oxidation results in weaker joints.

The Manufacture of Cooking Fats and Margarine

Most animal fats such as mutton fat and pork fat are solids at room temperatures. Most vegetable fats, and a few animal fats, are liquids at room temperatures and are called oils, e.g. olive oil and whale oil. The differences in melting points are largely due to the number of unsaturated bonds (page 310) in the fats; the oils are highly unsaturated.

If the cheap and readily available oils can be made 'less unsaturated' they become solids at room temperatures, and can be easily converted into solid fats for use in cooking or as a butter substitute. The process is termed 'hardening' the oil, and it is achieved by adding hydrogen atoms to the unsaturated molecules. The oil is usually heated and mixed with a finely divided nickel catalyst, and then hydrogen is blown through the mixture under pressure.

QUESTIONS CHAPTER 4

1. A strong, round bottom flask was half filled with water which was then boiled vigorously for five minutes. Heating was discontinued and the flask was closed with a well-fitting rubber stopper. The flask was inverted and cold water was poured on the outside. (a) What would be seen in the flask? (b) Give the reasons for your expected observations. (J. M. B.)

2. When sulphur is burned in a porcelain boat in an atmosphere of oxygen, the mass of the boat and contents decreases. On the other hand, when magnesium is burned in the same way, there is an increase in the mass of the boat and contents. (a) What would you see in the case of sulphur? (b) Name the major product formed when the sulphur is burned in oxygen. (c) What would you see in the case of magnesium? (d) Name the product formed when magnesium is burned in oxygen. (e) Why does one weighing show a decrease and the other an increase in mass? (f) Which product might be a liquid at a temperature of 253 K (−20 °C)? (J. M. B.)

3. A student measured the boiling point of deionized water and found it to be 373.5 K (100.5 °C). The thermometer was accurate and he concluded that the water was impure in some way. Can you think of an alternative explanation? Explain your answer.

4. A liquid Z was found to have a boiling point of 348 K (75 °C) at 760 mm pressure, and of 358 K (85 °C) at 800 mm pressure. At a pressure of 820 mm the boiling point of Z is likely to be

A 359 K (86 °C)
B 361 K (88 °C)
C 363 K (90 °C)
D 365 K (92 °C)
E 367 K (94 °C)

5. Which of the following is an *incorrect* statement about the boiling point of water?

A It rises if the atmospheric pressure increases
B It rises if a solid is dissolved in the water
C It is lower at high altitudes
D It increases when water is heated by a powerful energy source
E It is 373 K at 760 mm pressure

6. Which of the following does *not* reduce the amount of oxygen in the air?

A Aerobic respiration
B Rusting
C Photosynthesis
D The use of petrol as a fuel
E The combustion of natural gas

7. Which of the following statements about hydrogen is untrue?

A It is a neutral gas, almost insoluble in water
B It is a reducing agent

C It will burn in air to form steam
D It diffuses more rapidly than carbon dioxide
E It is prepared by the action of dilute nitric acid on zinc

8. Which of the following statements about oxygen is untrue?

A It is a very electronegative element
B It can be prepared by the catalytic decomposition of hydrogen peroxide
C It forms acidic oxides with most metals
D It has no smell or taste
E It allows most substances to burn more vigorously than they do in air

9. Manganese(IV) oxide (MnO_2) is not a peroxide because

A it catalyses the decomposition of hydrogen peroxide
B it does not form hydrogen peroxide when an acid is added
C it forms chlorine when added to concentrated hydrochloric acid
D it is an oxide of a transition element
E it is insoluble in water

The following questions require rather longer answers

10. Draw a labelled diagram of the apparatus you would use to prepare and collect gas jars of *either* oxygen or hydrogen (not by electrolysis). Describe briefly *three* experiments you have seen that demonstrate chemical or physical properties of the gas that you have chosen above. In any chemical reaction mentioned, name and describe the product(s). Give two important different uses of *each* gas (excluding balloons). (C.)

11. Describe an experiment that you could perform to find, as accurately as you can, the percentage by volume of oxygen in the air. Nitrogen as normally obtained from the air has a slightly greater density than nitrogen prepared from a compound. Give the reason for this greater density.

Give two natural ways by which nitrogen is returned to the soil. Describe what you would observe and say what is formed when (a) solid sodium hydroxide is left in a dish and exposed to the air, (b) copper foil is heated in the air. (C.)

12. Name the products formed when the following react with an excess of oxygen: (a) carbon (b) magnesium (c) hydrogen (d) zinc.

Write equations for the reactions, if any, of these products with (i) dilute hydrochloric acid, (ii) sodium hydroxide solution. If no reaction occurs, write 'no reaction'.

State the type of oxide formed by each of the four elements. (J. M. B.)

13. Give three reasons in each case why (a) air is considered to be a *mixture* of nitrogen and oxygen, (b) water is considered to be a *compound* of hydrogen and oxygen.

Draw a diagram of the apparatus you would use to obtain a sample of the air dissolved in tap water. How would you determine the proportion of oxygen in the air so obtained? How and why would your result differ from the proportion of oxygen in ordinary air? (J. M. B.)

14. Discuss the various methods available for the generation, purification, drying and collection of gases. Point out the advantages and disadvantages of each of the methods you mention.

15. Compare and contrast the processes combustion, respiration and photosynthesis.

5 The Periodic Classification

5.1 HISTORICAL DEVELOPMENT

Introduction

Science is organized knowledge, and for centuries scientists have endeavoured to catalogue and arrange the known facts in some kind of order so that a pattern emerges. The shape of the pattern then helps in the discovery of new facts, and so the work progresses. Now and again a pattern is worked out that is so important that it influences scientific thought and discovery for many years to come.

Up to the middle of the seventeenth century chemical ideas were disordered and chemical classification was non-existent. Once the idea of an element as a substance which could not be broken down into anything simpler had been established by Robert Boyle, scientific workers turned their attention to the discovery and isolation of new elements, and by 1864 sixty-three elements had been isolated and many of their compounds named. The French nobleman, Antoine Lavoisier, attempted to divide the elements into several groups. How would you have tried to solve this problem?

An unusual metal

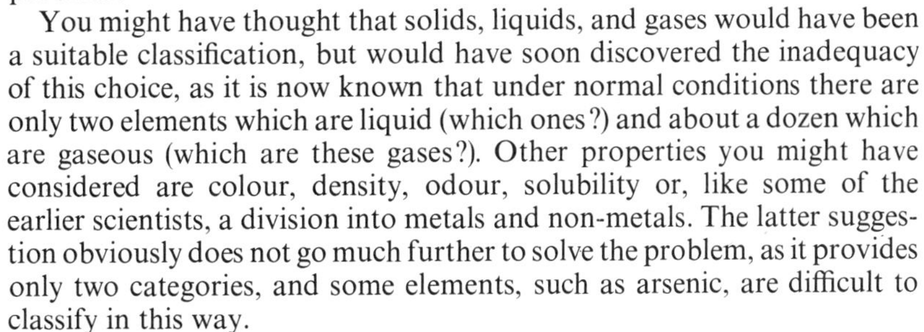

You might have thought that solids, liquids, and gases would have been a suitable classification, but would have soon discovered the inadequacy of this choice, as it is now known that under normal conditions there are only two elements which are liquid (which ones?) and about a dozen which are gaseous (which are these gases?). Other properties you might have considered are colour, density, odour, solubility or, like some of the earlier scientists, a division into metals and non-metals. The latter suggestion obviously does not go much further to solve the problem, as it provides only two categories, and some elements, such as arsenic, are difficult to classify in this way.

For a time the problem seemed insoluble, but as more elements were isolated and their chemical properties examined, similarities between certain elements became apparent. The following experiments will serve to illustrate this point.

Experiment 5.1

†Two Unusual Metals ***

Apparatus

Large trough, sharp knife, tongs, filter paper.

Sodium, potassium, copper, iron metals, and universal indicator.

Procedure

Half fill the trough with water and add a few drops of universal indicator. Remove a piece of sodium from the bottle with the tongs and place it on the filter paper.

Blot to remove the oil. Carefully cut off a piece the size of a split pea. Drop this on the surface of the water. Repeat the above separately with potassium, and suitable pieces of copper and iron.

Points for Discussion

1. How were the metals stored? Why were some of them stored in a special manner?

2. Which metals were soft?

3. Which metals were less dense than water?

4. Which metals react with water?

5. Which metals affect the pH of the water?

6. From your answers, which of the metals are like each other but different from *most* other metals? (i.e. which form a sub-group within the group of metals?)

Experiment 5.2
†Some Coloured Non-metals ***

Apparatus
Rack of test-tubes, teat pipette, spatula, Bunsen burner, asbestos square, test-tube holder, access to fume cupboard. Bromine, carbon, sulphur, chlorine generator (Figure 5.1), universal indicator solution.

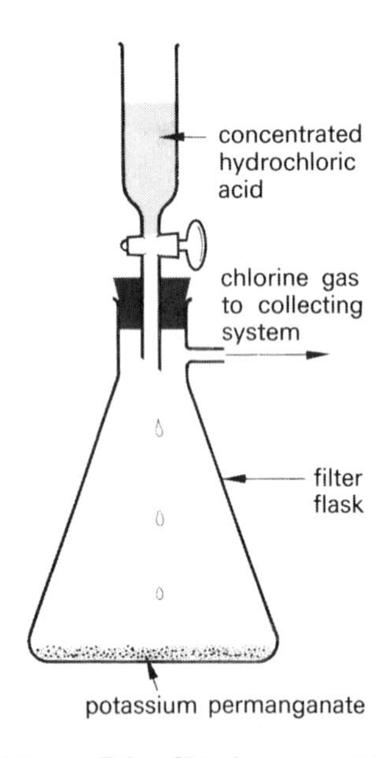

Figure 5.1 Chlorine generator

Procedure
(a) Pass a little chlorine into a test-tube half filled with water containing a few drops of universal indicator solution.

(b) Repeat with three more test-tubes containing water and indicator but using small samples of bromine (one drop from teat pipette), carbon, and sulphur respectively.

(c) Determine whether the two solids can be easily changed into gases by heating small samples of carbon and sulphur in separate dry test tubes. (Care—heat the sulphur slowly in a fume cupboard—it may catch fire.)

Points for Discussion

1. Which elements dissolve in water?

2. Which elements affect the pH of water?

3. Did you notice any odours? If so, of what did they remind you?

4. Which of the non-metals used are more readily converted into gases?

5. Which of the non-metals are like each other, but different from most non-metals?

Summary It should be obvious that we have now found a group of metals which are very different from most metals, and a 'family' of elements within the group of non-metals.

Unfortunately, at the time of Lavoisier it proved difficult to expand this experimental approach any further, but at least chemists were aware that groups of similar elements did exist and they began to look for alternative ways of connecting groups of elements and for an explanation of the similarity in those already known.

The Atomic Theory

Dalton's Atomic Theory did much to stimulate the work of chemists, and during the early part of the nineteenth century the atomic masses of many elements were found. Could these be used in their classification?

This was partly answered by J. F. Döbereiner in 1829. He knew that the metals calcium, strontium and barium had similar chemical properties and that their atomic masses were 40, 88, and 137 respectively. Can you see any relationship between these numbers? As a further clue, if the same pattern holds for all elements, lithium ought to be included in a group with sodium and potassium (found to be similar in Experiment 5.1), and iodine should be included with chlorine and bromine (Experiment 5.2). The mass numbers of these elements are shown in the following table.

	Lithium	*Sodium*	*Potassium*
Mass No.	7	23	39
	Chlorine	*Bromine*	*Iodine*
Mass No.	35	80	127

Döbereiner had noticed that the average of the atomic masses of the first and last elements in each group of three was approximately that of the middle one. He called such a group of chemically similar elements with related atomic masses, *triads*. Unfortunately these groups of three were restricted to only a few elements, but Döbereiner's work was an advance in that it provided a link between the atomic mass of an element and its properties, and stimulated scientists to further work in this direction.

A Döbereiner Triad

The next advance was made by John Newlands, an English scientist, who in 1864 arranged the sixty-three then known elements in order of their atomic masses. He made the remarkable discovery that, to use his own words, 'the eighth element, starting from a given one, is a kind of repetition of the first, like the eighth note in an octave of music'. Study his original table (Table 5.1) in which elements with similar properties form vertical columns.

Table 5.1 The elements as arranged by John Newlands

H	Li	Be	B	C	N	O
F	Na	Mg	Al	Si	P	S
Cl	K	Ca	Cr	Ti	Mn	Fe
Co Ni	Cu	Zn	Y	In	As	Se
Br	Rb	Sr	Ce La	Zr	Di Mo	Ro Ru
Pd	Ag	Cd	Sn	U	Sb	Te
I	Cs	Ba V	Ta	W	Nb	Au
Pt Ir	Os	Hg	Tl	Pb	Bi	Th

You will notice that one of Döbereiner's 'triads' forms part of a vertical column. Döbereiner's 'triads' only apply to a few elements, but Newlands placed *all* the elements then known into a pattern of similar groups. Look at the position of copper in this table. Do you think that it is correctly placed? (Hint: refer to Experiment 5.1.) This is one of the reasons why his work was not fully accepted by his fellow scientists, and after the seventeenth element there were many other discrepancies.

Figure 5.2 Dmitri Ivanovitch Mendeléev—the founder of the modern scheme of element classification

Further Developments

Dmitri Ivanovitch Mendeléev, the son of a Moscow teacher (Figure 5.2), produced in 1869 a table upon which the modern classification of elements is based. He improved upon Newlands' arrangement by leaving gaps for elements which he said had not yet been discovered, and by listing separately some elements which did not appear to fit into any group, i.e. iron, cobalt and nickel (Table 5.2).

Table 5.2 Mendeléev's Table (1871)

Series	*Group 1*	*Group 2*	*Group 3*	*Group 4*	*Group 5*	*Group 6*	*Group 7*	*Group 8*
1	H							
2	Li	Be	B	C	N	O	F	
3	Na	Mg	Al	Si	P	S	Cl	
4	K	Ca		Ti	V	Cr	Mn	Fe Co Ni
5	Cu	Zn			As	Se	Br	
6	Rb	Sr	Y	Zr	Nb	Mo		Ru Ph Pd
7	Ag	Cd	In	Sn	Sb	Te	I	
8	Cs	Ba	Dy	Ce				
9								
10			Er	La	Ta	W		Os Ir Pt
11	Au	Hg	Tl	Pb	Bi			
12				Th		U		

For the first time, elements with similar chemical properties appeared consistently in the same vertical column. You will notice that two of the gaps he left were placed between zinc and arsenic. In the order of the atomic masses then known, arsenic (atomic mass 75) should follow zinc (atomic mass 65) but this would have placed it in a group with dissimilar properties. However, he reasoned that its inclusion in Group 5 was justified by its chemical similarity to phosphorus and antimony, and that the two gaps thus created would be filled by as yet undiscovered elements with atomic masses between those of zinc and arsenic. He even predicted their properties by consideration of the properties of the elements surrounding them. Some of these unknown elements were discovered in his lifetime and the accuracy of his forecasts proved conclusively that his theories were well founded. This was important because, if a theory is to be useful, it must not only explain known facts but enable new predictions to be made from it. One of the elements which Mendeléev predicted would be discovered is now called germanium. It was discovered in 1886, some seventeen years after Mendeléev had published his work, and Table 5.3 shows how remarkably accurate his forecasts were.

Table 5.3 Mendeléev's predictions for germanium

	Mendeléev's predictions (he called the element ekasilicon, symbol Es)	*Actual properties of germanium*
Atomic mass	72.0	72.6
Specific gravity	5.5	5.35
Colour of element	Light grey metal	Dark grey metal
Action with air	Will form an oxide (EsO_2, a white powder) when heated	Forms an oxide (GeO_2, a white powder) when heated
Action on water	Will decompose steam with difficulty	Decomposes steam at red heat
Action on acids	Slight reaction only	Only dissolves in aqua regia
Action on alkalis	No action	Only attacked by fused alkalis
Properties of the oxide	Refractory, specific gravity 4.7, less basic than the oxides of titanium and tin	Refractory, specific gravity 4.70, feebly basic
Properties of the chloride	Formula $EsCl_4$, a liquid at room temperature, boiling at less than 373 K (100 °C)	Formula $GeCl_4$, a liquid at room temperature, boiling point 359 K (86 °C)

This type of classification, in which elements with similar properties recur at regular intervals, is called periodic, and the modern version is referred to as the Periodic Table (see front endpaper). Compare this with Mendeléev's original table. What modifications do you notice? The main differences between this and Mendeléev's original table are:

1. The elements are placed in order of their atomic number instead of their atomic masses. As you will see in the next section, the atomic number is of more significance to the chemist than the atomic mass.
2. An entire new family, the noble gases, has been discovered and placed on the right-hand side of the table.
3. A block of elements has been placed in the centre of the table, and includes those which Mendeléev found it difficult to classify, e.g. iron, cobalt and nickel. These elements are called the transition elements.
4. As new elements have been discovered gaps have been filled, and some man-made elements have been added to the table.

'Bill—it's the halogens'

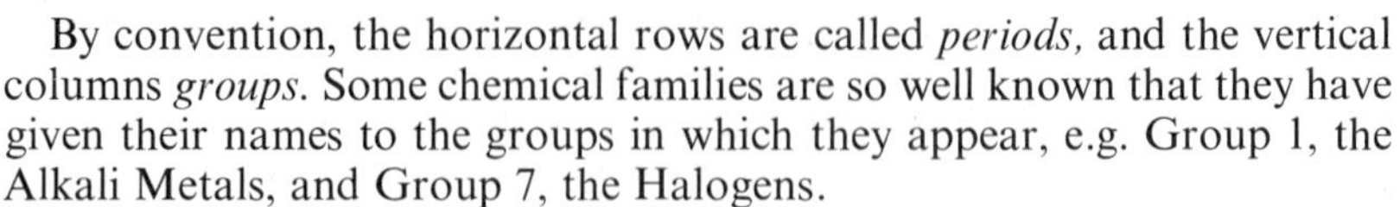

By convention, the horizontal rows are called *periods,* and the vertical columns *groups.* Some chemical families are so well known that they have given their names to the groups in which they appear, e.g. Group 1, the Alkali Metals, and Group 7, the Halogens.

You may well wonder if this concludes the story of the Periodic Table, or if there is a chance for you to add to it. Look at the Periodic Table again, and notice the last seven elements. These have been artificially produced in the United States in recent years, by S. G. Thompson, G. T. Seaborg, G. E. Ghiorso and others, but they are unstable. Can you find the origin of the names of these elements? Russian scientists claim to have prepared element number 104, to be called Kurchatovium, and more recent research suggests the existence of a range of elements of even greater atomic numbers which may be more stable, e.g. elements 110 to 114.

5.2 ELECTRONIC STRUCTURE AND THE PERIODIC TABLE

We have seen that in the modern classification, the elements are arranged in order of their atomic numbers, and not their atomic masses. For a long time scientists thought that the atomic mass of an element was of great fundamental importance, and few imagined that there might be another number characteristic of a particular element and connected with its chemical properties. Mendeléev did suspect, however, that the atomic mass itself was not the real key to the system of classification. Study his original table again, and look particularly at the period starting with silver (Ag). Note the atomic masses of each of the elements in the row. Can you find where Mendeléev had to juggle with two elements in order to make them fit into the correct pattern? If tellurium (atomic mass 127.6) and iodine (atomic mass 126.9) had been placed in order of their atomic masses, iodine would have been found in a group away from, rather than with, the related similar elements fluorine, chlorine, and bromine. A glance at the early table will make this clear. Mendeléev ignored their atomic masses in this case so as to put iodine in the correct gap.

A similar problem arose later when argon was discovered. Look at the modern table. The atomic masses of chlorine, argon, potassium, and calcium are 35.5, 39.9, 39.1, and 40.0 respectively. In which groups would argon and potassium be found if the atomic mass order were to be adopted? Their chemical properties indicate that they would be misplaced in such groups. The noble gas argon should be in the same group as neon and krypton, just as potassium should be in the same group as sodium. Once again, the atomic mass sequence was ignored to make the pieces of the puzzle fit into place.

The use of atomic masses gives no indication as to why the Periodic Table exists at all. You may well wonder why similar elements line up in the same columns; why there are two elements in the first row but eight in the second, and so on. Any other system which could replace atomic masses in explaining the classification of elements should have all the advantages of the old method, but also place tellurium, iodine, potassium, and argon in their correct places without any juggling, and, if possible, explain why the pattern of the Periodic Table emerges. We shall now learn whether the use of atomic numbers in place of atomic masses satisfies these conditions.

Many exciting discoveries were made about atoms in the early years of the twentieth century, and associated with these were explorations into new phenomena such as radioactivity and x-rays. In 1895 W. K. Röntgen had noticed that when a beam of electrons (cathode rays) struck a metal, a highly penetrating radiation was emitted from the metal. He called these radiations x-rays, and we now take their use in routine medical and scientific work for granted. Sir Lawrence and Sir William Bragg, who paved the way for the analysis of complex crystal structures, were just two of the famous scientists working with x-rays after the turn of the century. Another was Henry Moseley.

Moseley, who earlier had studied under Lord Rutherford at the University of Manchester, published his work in 1914 after several years of research at Oxford. He found that the x-radiation produced by a given metal was always the same, and different from that of any other metal. The really important discovery, however, was that the frequency of the x-rays from a given metal could be calculated by using a simple mathematical formula containing a certain whole number. This number, which varied with each element was a fundamental property of the element, and was called the atomic number (Section 1.5). We now know that it represents the number of electrons or protons in the atom. Moseley was able to find many atomic numbers in this way, and although he died at Gallipoli in World War I at the early age of twenty-eight years, he had ensured that elements could be characterized by two numbers, atomic numbers and atomic masses.

Find out the names of the elements having atomic numbers from 1 to 20. Write them down in that order, and make a separate list of the same elements in order of their atomic masses. Compare your two lists.

Atomic Number—An Improvement on Atomic Mass

The use of atomic numbers has produced the same order as the use of atomic masses for the first eighteen elements only. The other two elements, argon and potassium, were two of those which Mendeléev had to juggle into their correct places. The atomic numbers of tellurium and iodine are 52 and 53 respectively, so they also appear in their correct places in the Periodic Table if the atomic number sequence is used, but not if atomic masses are used. We now realize that the use of atomic masses was nearly successful simply because it so happens that the order of atomic masses was almost the same as that of atomic numbers, but atomic mass is not really significant in explaining the chemistry of an element.

We have seen that the use of atomic number is an improvement on the older method. Can it also *explain* the table? The chemical properties of an element depend upon its atomic structure. All chemical reactions involve either a transfer or a rearrangement of the outer electrons of the atoms concerned. The atomic number tells us the number of electrons in an atom, and it seems logical to suppose that this might be in some way connected with the position in the Periodic Table.

From the following data draw the electronic structures of the given elements in (a) Group 1, (b) Group 2, (c) Group 7.

Element	*Group*	*Atomic number*
Li	1	3
Na	1	11
K	1	19
Be	2	4
Mg	2	12
Ca	2	20
F	7	9
Cl	7	17

What do you notice about the number of electrons in the outer energy level of each element in any one group? It is impossible to find any such connection by using atomic masses in place of atomic numbers.

Further Connections Between Atomic Numbers and Chemical Behaviour

It seems reasonable to suppose that if chemical reactions involve electrons in the outer shell, the elements with similar chemical properties will have the same number of electrons in their outer energy levels, and this is exactly what you have just found. Draw the first three periods of the Periodic Table in such a way as to show the electronic structure of each element. Study this carefully. It should be obvious that each row corresponds to the filling up of an energy level with electrons, and when each level is finished a new row is started, so that elements with similar electronic configurations appear in the same group. So far, the use of atomic numbers has 'tied up' the loose ends, and even explained the arrangement of the Periodic Table.

You may still be wondering why there are more elements in the fourth row than in the third, and why there are even more in later periods. This need not concern us very much at this early stage.

There is another point to consider. The early scientists could find no satisfactory place in the table for hydrogen, and it was placed on its own, a sort of rogue element. From a consideration of the knowledge gained about the connection between electronic structure and group number, in which group would you place hydrogen? You would probably place it in Group 1 as it has one electron in its outer shell. Obviously it is not in Group 1, which consists only of metals, yet it does have one 'metallic' property in that it can form a positive ion (H^+) and is thus discharged at the cathode in electrolysis, like the metals. You could also argue that as hydrogen has one electron short of a complete shell, it should be in Group 7 with the Halogens, as they all have one electron short of a complete energy shell. Is hydrogen like the Halogens? You probably know enough already to realize that it is chemically different, and yet, like the Halogens, it can form a negative ion (H^-). Hydrogen is clearly difficult to place in the table. It can either lose an electron to form a positive ion, or gain an electron to form a negative ion. In some cases it also shares an electron pair. Reference to its unique electronic structure makes the behaviour of hydrogen clear. Here again, the use of atomic numbers, from which the configuration is derived, has helped towards an understanding of the problem.

Hydrogen was placed on its own

The ideas developed in this section are very important indeed, and as you learn about the uses of the Periodic Table in the next few sections you should keep referring to your diagrams showing electronic structures, and use them to explain the various observations you make.

5.3 TRENDS ACROSS A TYPICAL PERIOD

Introduction

We have seen that the Periodic Table brought a degree of order into the study of the elements, and we can now take a closer look at some of the ways in which this can help to simplify and correlate the vast amount of apparently unconnected chemical fact. In the next three sections principles should emerge which you may be able to apply in many differing situations throughout the course.

In this section we shall consider the elements and their compounds in a typical period, namely those from sodium to argon.

The Elements Themselves

Physical Properties Examine specimens of each of the elements and study the data given in Table 5.4 on page 136.

Points for Discussion

1. Where are the metals to be found in this typical period?

2. The boiling points and melting points and valency follow a similar pattern from left to right. What is this pattern?

Chemical Properties

Experiment 5.3

†Action of the Elements with Water ***

For purposes of comparison we will study the chemical action of all these elements with the *same* compound, in this case water.

Apparatus

Sodium spoon (if available), small trough, test-tubes, splints, sand paper, rocksil, asbestos paper, delivery tube and bung to fit test-tubes, Bunsen burner, asbestos square safety screen. Sodium, magnesium, aluminium foil, silicon, steam can, tripod, combustion tube.

Procedure—Sodium

(a) Refer to Experiment 5.1.

(b) Carry out the experiment as shown in Figure 5.3 if a sodium spoon is available. Test the collected gas with a lighted splint.

Magnesium

(a) With cold water in the beaker set up the apparatus as shown in Figure 5.4 and leave until a sufficient quantity of gas is available for testing as before.

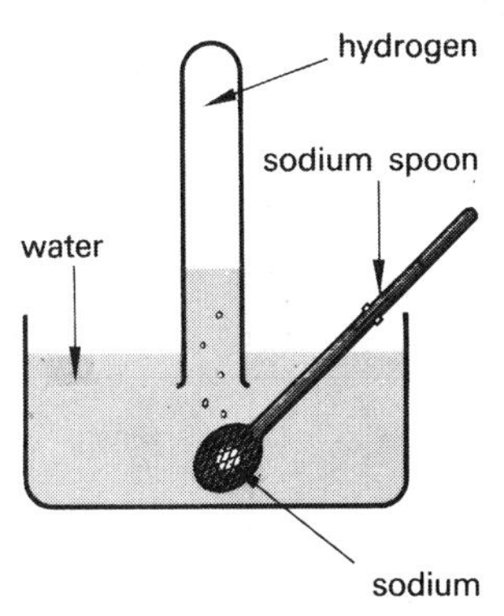

Figure 5.3 The reaction of sodium with water

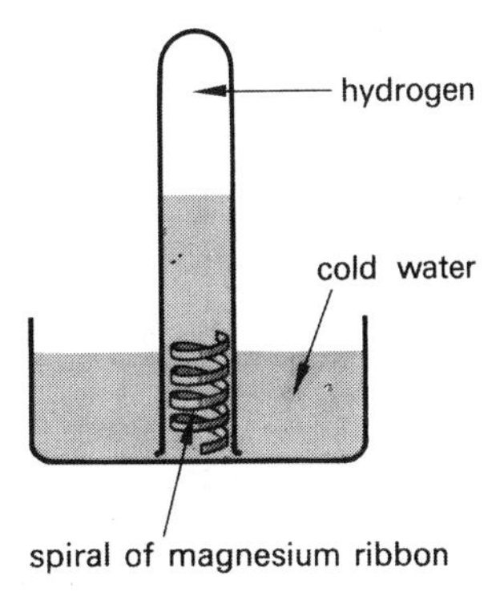

Figure 5.4 The reaction of magnesium with cold water

Table 5.4 Physical properties of some of the elements in Period 3

	Sodium	*Magnesium*	*Aluminium*	*Silicon*	*Phosphorus*	*Sulphur*	*Chlorine*
Appearance	Silvery, stored in oil	Silvery	Silvery	Black solid	Yellow solid, stored under water	Yellow solid	Greenish-yellow gas
Melting point K (°C)	370.8 (97.8)	923 (650)	933 (660)	1683 (1410)	317.2 (44.2)	386 (113)	172 (−101)
Boiling point K (°C)	1163 (890)	1383 (1110)	2743 (2470)	2633 (2360)	553 (280)	717 (444)	238.3 (−34.7)
Density ($g\ cm^{-3}$)	0.97	1.74	2.7	2.33	1.82	2.07	1.56 at 238 K (−35 °C)
Valency	1	2	3	4	3	2	1
Electronic structure	2, 8, 1.	2, 8, 2.	2, 8, 3.	2, 8, 4.	2, 8, 5.	2, 8, 6.	2, 8, 7.

Higher and less stable valencies do exist for P, S and Cl but they are omitted here for purposes of comparison.

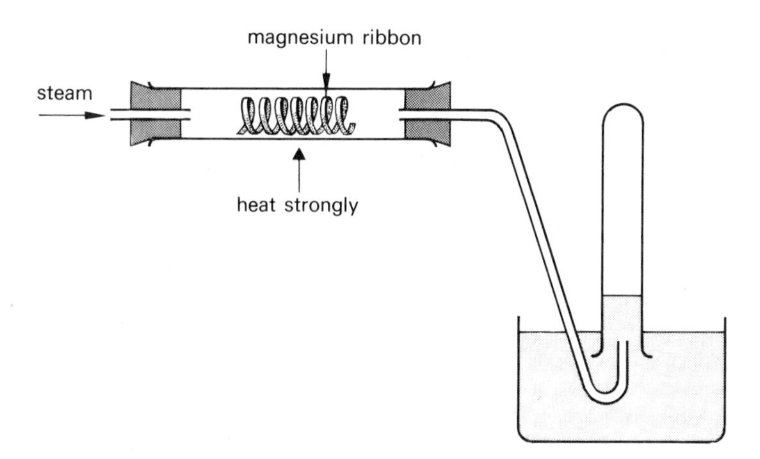

Figure 5.5 The reaction of magnesium with steam

(b) Working behind a safety screen, investigate the action of magnesium on steam as shown in Figure 5.5. Heat the steam can until the water boils and steam passes over the magnesium. Now heat the magnesium strongly so that the steam is passing over the hot metal. Collect and test any evolved gas. Care is needed to prevent 'sucking back' of the water. What precautions must you take to prevent this, and why might it occur?

Aluminium and Silicon
Repeat procedures (a) and (c), first using a small spiral of aluminium foil instead of magnesium, and then a small piece of silicon.

Phosphorus
Under what liquid is it stored? Is it likely to react with cold water? It does not react with steam.

Sulphur
From your knowledge of (a) the oil drop experiment (Experiment 1.7) and (b) the Frasch process for the extraction of sulphur (page 289), do you consider that sulphur reacts with water or steam?

Chlorine
Refer to Experiment 5.2. Does chlorine react with water, and if so how?

Points for Discussion

1. Does the reactivity of the elements with water

A increase from left to right across the period
B decrease from left to right across the period
C decrease and then increase
D increase and then decrease
E show no pattern at all

In general the trend shown in these reactions with water is typical of changes in reactivity of the elements across a period.

2. How is this order of reactivity linked with the number of electrons the elements need to gain or lose to attain a noble gas structure?

3. Would you expect argon to react with water? Explain your answer.

Experiment 5.4
†Chemical Properties of the Oxides ***

Apparatus
Rack of test-tubes, Bunsen burner, asbestos square, test-tube holder and spatula.
Samples of sodium oxide (Na_2O_2), magnesium oxide (MgO), aluminium oxide (Al_2O_3), silicon oxide (SiO_2), phosphorus(V) oxide (P_4O_{10}), sulphur(IV) oxide cylinder (SO_2).
[Note: it is essential that the aluminium oxide be freshly prepared.]

Procedure
(a) Note the appearance of the oxides.

(b) Add a half spatula measure of sodium oxide to a test-tube containing 2.5 cm depth of water and three drops of universal indicator.

(c) Repeat (b) using the other solid oxides.

(d) Bubble a little sulphur(IV) oxide into a test-tube containing 2.5 cm depth of water and three drops of universal indicator.

(e) Place half a spatula measure of aluminium oxide in a test-tube. Add 2.5 cm depth of bench hydrochloric acid. Warm.

(f) Repeat (e) using bench sodium hydroxide solution in place of hydrochloric acid.

Points for Discussion
1. How does the acidity or basicity of the oxides vary across a period? Before you answer read the following additional information. Chlorine oxide, which is explosively unstable, forms an acidic solution with a very low pH value.
When silicon oxide is placed in water, the pH remains unaffected. This does not necessarily mean that it is a neutral oxide. It will, for example, react with a strongly alkaline solution to form a salt, and is thus an acidic oxide. With water its acidity is not apparent because it is insoluble.

2. From your observations in (e) and (f) how would you classify aluminium oxide? Acidic, basic or both?
Such oxides are called *amphoteric* oxides.

3. Do you consider that the oxides in this typical period

A change from strongly basic through amphoteric to strongly acid
B change from acidic through amphoteric to strongly basic
C change from strongly basic through strongly acidic to amphoteric

4. You should be able to link this work on oxides with that on page 118, for you know where metals and non-metals are to be found in a typical period.

Structural Changes Refer to your observations concerning the physical state of the various elements and their oxides, note the appearance of as many of their chlorides as are available, and study the data given in Table 5.5.

Points for Discussion
The physical state of an element is often dependent upon whether its structure is molecular or giant. These two terms are considered in detail in Chapter 7, but at this stage you can regard giant structures as consisting of particles joined together by strong bonds in all directions in such a way that there are no separate units within the structure. Such structures are difficult to 'pull apart' and usually have high melting points and boiling points. Molecular structures consist of separate molecules which are relatively easy to separate and such substances usually have low melting points and low boiling points.
1. On which side of the table (period) do the giant structures generally appear?
2. How can the formulae of the hydrides be explained by the electronic structure of the respective elements?
The structures of some of these elements and their compounds are summarized in Table 5.6.

Table 5.5 Some structural information

	Sodium	*Magnesium*	*Aluminium*	*Silicon*	*Phosphorus*	*Sulphur*	*Chlorine*
Formulae of oxides	Na_2O	MgO	Al_2O_3	SiO_2	P_4O_6 (P_2O_3)	SO_2	
Formulae of chlorides	$NaCl$	$MgCl_2$	Al_2Cl_6 ($AlCl_3$)	$SiCl_4$	PCl_3	S_2Cl_2	—
Boiling points of chlorides K (°C)	1730 (1457)	1680 (1407)	Sublimes at about 453 (180)	333 (60)	347 (74)	408.6 (135.6)	—
Melting points of chlorides K (°C)	1074 (801)	987 (714)	Sublimes at about 453 (180)	266 (−7)	182 (−91)	193 (−80)	—
Formulae of hydrides	NaH	MgH_2	$(AlH_3)_n$	SiH_4	PH_3	H_2S	HCl
Melting points of hydrides K (°C)	Decomposes at 1073 (800)	Decomposes in vacuum at 553 (280)	Decomposes above 423 (150)	88 (−185)	139 (−134)	188 (−85)	158 (−115)
Boiling points of hydrides K (°C)	—	—	Decomposes above 423 (150)	161 (−112)	183 (−90)	212 (−61)	188 (−85)
State of hydride	Solid	Solid	Solid	Gas	Gas	Gas	Gas
Action of water on hydrides	Vigorous with cold water, hydrogen evolved, solution alkaline	Reacts with cold water to liberate hydrogen, solution alkaline	Reacts with cold water to liberate hydrogen, solution neutral	Insoluble	Sparingly soluble	Moderately soluble, solution acidic	Very soluble, solution acidic

Table 5.6 Structures of some elements and their compounds in a typical period

	Sodium	*Magnesium*	*Aluminium*	*Silicon*	*Phosphorus*	*Sulphur*	*Chlorine*	*Argon*
Elements	← Metallic →			← Non-metallic →				
Structure of elements	← Giant metallic lattice →			Giant atomic lattice	← Molecular →			Free atoms
Oxides	← Basic →		Amphoteric	← Acidic →				
Structure of oxides	← Giant ionic lattice →		Giant (ionic-covalent) lattice	Giant atomic lattice	← Molecular →			
Structure of chlorides	← Giant ionic lattice →		← Molecular →					
Structure of hydrides	← Giant ionic lattice →			← Molecular →				

The Special Case of the Transition Elements

The work in this section has been specially chosen to emphasize particular trends across a period, and to make these clear, certain exceptions to the general rules have been ignored. You must not think that the behaviour of natural substances can be always made to fit man-made rules. The patterns which emerge make the work easier to understand and can be applied to other periods, but we must beware of trying to follow them blindly, as exceptions do occur. From Period 4 onwards the eight typical elements in each row are split by a block of elements, the transition elements, which are usually considered separately.

5.4 TRENDS DOWN A METALLIC GROUP

Where do metals predominate in the Periodic Table? Our aim is to find out how the physical and chemical properties of the metals change as the atomic number increases in a typical metallic group (e.g. Group 1—the Alkali Metals) and ascertain if the changes are uniform or haphazard.

Trends in Physical Properties

Table 5.7 lists some physical properties of the Group 1 metals.

Do you observe any trends? These will become more obvious if you plot graphs of some of these properties against atomic number, and draw scale diagrams representing the sizes of the ions and their parent atoms. Why are the atoms larger than their ions?

Trends in Chemical Properties

You will have noticed the gradual change in physical properties of the alkali metals. Is there a similar trend in their chemical properties?

Table 5.7 Some physical properties of the Group 1 metals

	Symbols	*Atomic number*	*Mass number*	*Melting point K (°C)*	*Boiling point K (°C)*	*Radius of atom (nm)*	*Radius of ion (nm)*	*Density (g cm^{-3})*
Lithium	Li	3	7	352 (79)	1590 (1317)	0.133	0.078	0.535
Sodium	Na	11	23	371 (98)	1156 (883)	0.157	0.098	0.971
Potassium	K	19	39	337 (64)	1033 (760)	0.203	0.133	0.862
Rubidium	Rb	37	85	312 (39)	961 (688)	0.216	0.149	1.53
Caesium	Cs	55	133	302 (29)	978 (705)	0.235	0.165	1.90
Francium	Fr	← Unstable radioactive element →						

The Reactions of Lithium, Sodium, and Potassium with Water

In Experiment 5.1 the reactions of sodium and potassium with water were studied. Both metals reacted vigorously to form alkaline solutions. Which of the two metals reacted most vigorously? Using this knowledge predict how vigorous the reaction of lithium with water would be.

Experiment 5.5
† Reaction of Lithium with Water *

Apparatus and Procedure

As for Experiment 5.1 but using lithium metal. Use *very* small pieces, and wear goggles.

Points for Discussion

1. Which gas do you think was evolved during the reaction? How could you test your answer?

2. List the properties which suggest that lithium should be in the same group as sodium and potassium.

3. Was the reaction with lithium

A more vigorous than with sodium or potassium
B less vigorous than with sodium or potassium
C more vigorous than with sodium, but not as vigorous as with potassium
D more vigorous than with potassium, but not as vigorous as with sodium

4. Do you conclude that the metals in Group 1

A gradually become less reactive on descending the group
B gradually become more reactive on descending the group
C display no apparent trend in their reactivity?

5. How would you expect (a) rubidium and (b) caesium to react with water?

Reasons for the Trends in Reactivity

For the Group 1 metals to react similarly they must have something in common. Can you think what this is? (Hint: *see* Section 5.2.)

You should have found that each of their atoms contains one electron in the outer energy level and thus, to form a compound, each must lose this electron in order to attain the stable configuration of the nearest noble gas. To remove an electron from an atom requires energy, and the further away the electron is from the nucleus, the less energy is required to remove it. Look at your diagrams of the electronic structures of lithium, sodium and potassium. Which of the three atoms requires the least amount of energy to remove its outer electron? Which has its outer electron most weakly held by the nucleus? Is this the atom of the most reactive metal?

The factors governing the ease with which an electron can be removed from an atom depend upon the size of the atom. As the atomic size increases:

(a) the outer electron progressively gets further away from the attractive effect of the nucleus,
(b) there are an increasing number of completed electron shells, each of which 'shields' the outer electron from the attractive effect of the nucleus,
(c) the positive charge of the nucleus increases.

The first two effects lead to the outer electron being less strongly attracted to the nucleus, whereas the third effect leads to it being more strongly held. However, this latter effect is not as great as the combination of the other two effects. Thus as the atomic size increases the nucleus has less hold over the outer electron, which can therefore escape more easily. The result of this is that the larger the atomic number of a metal in the group, the more reactive it is. A simpler way of looking at this is to imagine that the outer electron of potassium has fewer orbits to 'jump' into before it escapes from the atom. Similarly, lithium is relatively less reactive because its outer electron has to move through more orbits before it escapes and this needs more energy.

Apply these ideas to the atoms of the Group 2 metals by referring back to their electronic structures. As with all other metals, they need to lose electrons to attain the electronic structure of a noble gas. Will they lose electrons as easily as the atoms of the Group 1 metals? (If not, why not?) Which of the Group 2 metal atoms will lose electrons most easily and will be therefore the most reactive? Will magnesium be more reactive or less reactive than calcium?

Test your answers by performing the following experiment.

Experiment 5.6

†Reaction of Magnesium and Calcium with Water at Room Temperature *

Apparatus and Procedure

Refer to Experiment 5.3 for the reaction of magnesium with water. For the reaction of calcium with water use the procedure outlined in Experiment 5.1.

Points for Discussion

1. Which metal reacted more vigorously with water?

2. Does this agree with the previous conclusion that the lower its position in the metallic group, the more reactive the metal?

5.5 TRENDS DOWN A NON-METALLIC GROUP

In the previous section we found that the elements become more reactive down a metallic group. Does a trend in reactivity also exist in non-metallic groups, and if so, is it similar to that found in metallic groups?

In order to answer this question we shall study a typical non-metallic group, i.e. Group 7, the Halogens.

Trends in Physical Properties

Table 5.8 lists some physical properties of the Group 7 elements. Do you notice any trends? Again these will appear more obvious when you plot graphs of the melting points and boiling points, and draw scale diagrams representing the sizes of the ions and their parent atoms. Why are the ions larger than their parent atoms? Does this relationship hold for the Group 1 metals?

Table 5.8 Some physical properties of the Group 7 elements

Element	*Symbol*	*Atomic number*	*Mass number*	*Melting point K (°C)*	*Boiling point K (°C)*	*Radius of atom (nm)*	*Radius of ion(nm)*
Fluorine	F	9	19	55 (−218)	85 (−188)	0.064	0.136
Chlorine	Cl	17	35	172 (−101)	239 (−34)	0.099	0.181
Bromine	Br	35	80	266 (−7)	332 (59)	0.114	0.195
Iodine	I	53	127	387 (114)	458 (185)	0.133	0.216
Astatine	At	←		Unstable radioactive element			→

Trends in Chemical Properties

Reaction of Chlorine, Bromine, and Iodine with Water

The action of two of these elements with cold water was studied in Experiment 5.2, and you will remember that chlorine and bromine dissolved and reacted to produce acidic solutions.

Experiment 5.7

†The Reaction of Iodine with Water *

Apparatus

Test-tube, Bunsen burner, asbestos square, test-tube holder, spatula.

Iodine, universal indicator paper. Avoid contact between iodine and your skin, and report any spillages immediately.

Procedure

(a) Add a little iodine to water. Stir. Test with indicator paper.

(b) Boil a little iodine with water. Test the resulting solution with universal indicator paper.

Points for Discussion

1. Does iodine appear to react with cold water?

2. Does (b) indicate that a little iodine has reacted with boiling water in a manner similar to that of chlorine and bromine with cold water?

3. From the reactions of the halogens with water, what conclusions can you draw as to which halogen reacts (a) most readily (b) least readily with water?

Experiment 5.8
†Reaction of Chlorine, Bromine, and Iodine with Iron Wool ***

Apparatus

Teat pipette, Bunsen burner, asbestos square, three test-tubes, one with a small hole near the closed end, retort stand and clamp, access to fume cupboard.
Chlorine generator (Figure 5.1), bromine, iodine, iron wool.

Procedure

Working in a fume cupboard, heat the iron wool to red heat, stop heating, and immediately pass chlorine over the iron wool as shown in Figure 5.6(a). For bromine place a few drops of the liquid in the bottom of a test-tube using a teat pipette, as in Figure 5.6(b). (Care.) Clamp the test-tube at the angle shown and insert the iron wool. Heat the iron wool to red heat; this will usually also generate sufficient bromine vapour for reaction to occur.

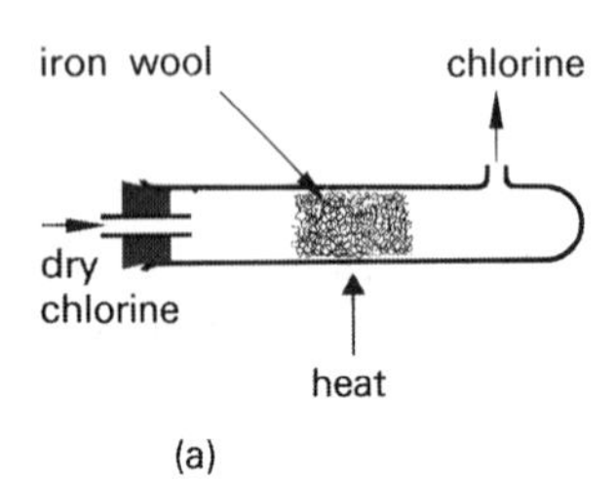

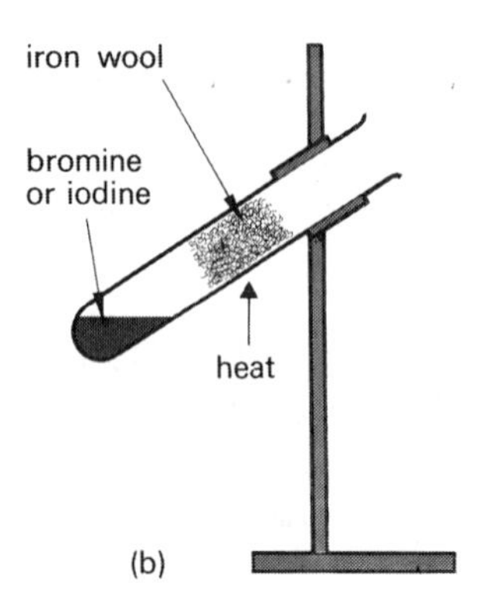

Points for Discussion

1. What did you see when chlorine came into contact with the heated iron wool?

2. Did a similar reaction occur with bromine and heated iron wool?

3. Which of the two elements reacted more vigorously with the iron wool?

4. What compounds were formed in the two reactions?

5. From your answer to 3, what prediction would you make regarding the comparative reactivity of iodine with iron wool? To verify your prediction, repeat the last experiment using iodine instead of bromine. (To vaporize the iodine it will be necessary to heat it as well as the iron wool.)

6. From your experiments do you conclude that the order of reactivity (placing the most reactive first) is:

A iodine, chlorine, bromine
B chlorine, iodine, bromine
C chlorine, bromine, iodine
D bromine, chlorine, iodine

Figure 5.6 The reaction of iron wool and the halogens

Reaction of Chlorine, Bromine, and Iodine with Hydrogen

All these elements react with hydrogen but their reactions are not easily demonstrated in the laboratory, and so the conditions under which the reactions occur will be stated.

Chlorine reacts spontaneously with hydrogen even in diffused sunlight. In bright sunlight the mixture explodes.

Sunlight does not supply sufficient energy to initiate the reaction between bromine and hydrogen. The mixture must be heated to about 473 K (200 °C) before any reaction occurs.

Iodine and hydrogen, even when heated to 773 K (500 °C) and in the presence of a catalyst, do not react completely.

Does this evidence support your conclusion to the previous experiment?

Summary

From the experiments carried out in this section you should have concluded that the elements in this non-metallic group all react similarly and that their reactivity gradually decreases as the atomic number increases, i.e. the order of reactivity is chlorine, bromine, iodine.

Predict how fluorine would react with water, iron wool, and hydrogen.

The Reason the Elements of Group 7 all React Similarly

You will remember from Section 5.4 that all the Group 1 metals reacted in the same way because each had an atomic structure with one electron in the outer energy level, and this electron was always lost when the metal reacted. Can a similar explanation be used to explain why the elements of Group 7 all undergo the same type of reactions?

You will observe that an atom of each Group 7 element has seven electrons in its outer shell and whereas an atom of a Group 1 element undergoes reactions by losing its outer electron so as to achieve the stability of a noble gas configuration, an atom of a Group 7 element does so by gaining an electron. This tendency to gain electrons is a feature of non-metals.

Explanation of the Reactivity Trend

The gradual change in reactivity of the Group 1 elements was attributed to the ease with which the metal atoms *lost* their outer electrons, so it would seem reasonable to expect that the gradual change in reactivity of the Group 7 elements is due to the ease with which their atoms can each *gain* an electron.

In Section 5.4, the three factors governing the attractive force between an electron and the nucleus were discussed, and it was found that as the atomic size increased this attractive force decreased. By referring to the atomic sizes of the atoms of the Group 7 elements, which atom do you think will attract an additional electron most easily? Is this an atom of the most reactive element? Which atom will attract an electron least readily? Is this an atom of the least reactive element? Your answers should have led you to conclude that the reactivity of the Group 7 elements depends upon the ease with which their atoms can gain electrons.

Reactivity Trends in Groups Containing Metals and Non-metals

In this section and the previous one, we have considered trends in reactivity in non-metallic and metallic groups, but all the groups cannot be classified as being entirely metallic or non-metallic. For example the elements of Group 4 consist of both metals and non-metals.

We shall now consider how this situation arises. You will remember that atoms of a non-metallic element gain electrons when they react, in order to attain a stable electronic configuration. The ability of an atom to

Electronegativity is the ability to attract electrons

attract electrons to itself is called its *electronegativity*. The three factors affecting this readiness to accept an electron were discussed in the last section.

Across a period from Group 1 to Group 7 the electronegativity increases (Figure 5.7). This is because while the shielding effect and the size of the atom remain approximately constant, the nuclear charge increases, so increasing the atom's power to attract electrons. However, on descending any group, although the nuclear charge increases more rapidly than on moving across a period, this factor is outweighed by the increased shielding effect and by the greater distances of the outer shells from the nucleus. This means that on descending a group, it becomes more difficult for an atom to attract electrons and thus the electronegativity gradually decreases. These effects are represented in Figure 5.7. This results in the more electronegative elements appearing on the right-hand side of the Periodic Table and the less electronegative elements appearing on the left-hand side (Figure 5.8).

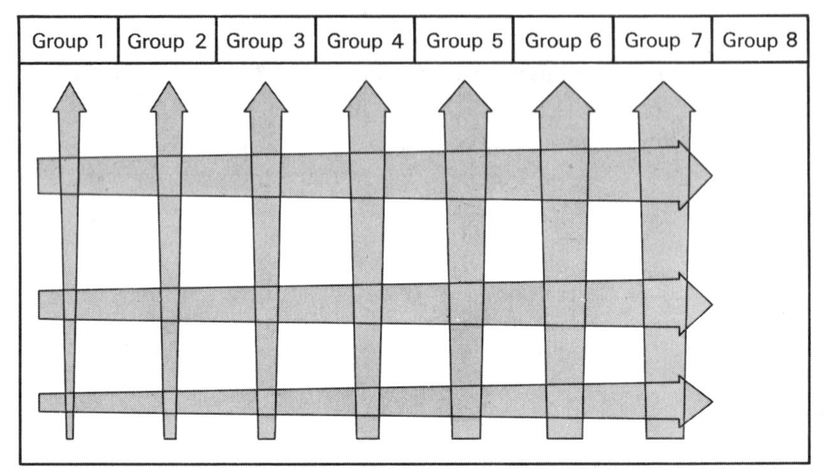

Figure 5.7 Changes in electronegativity in the periodic table. The arrows indicate the direction of increase in electronegativity. The broader the arrow becomes the greater is the electronegativity

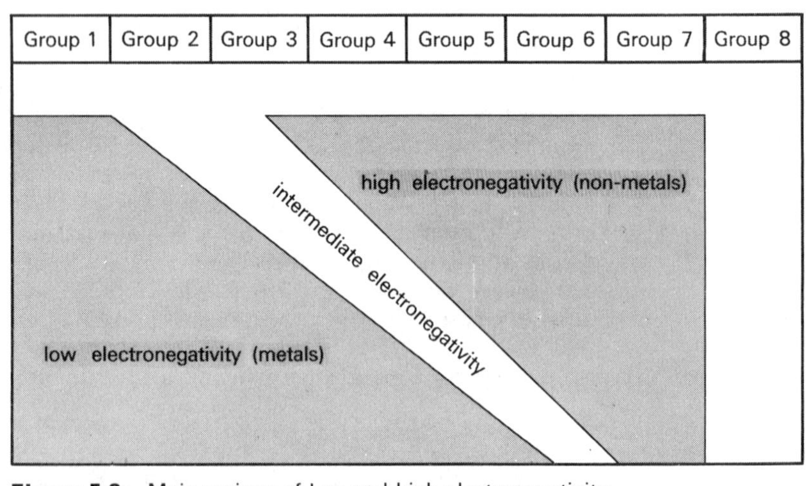

Figure 5.8 Main regions of low and high electronegativity

In general, elements with a high electronegativity are non-metals and those with a low electronegativity are metals and so the groups on the extreme left of the Periodic Table consist entirely of metals and the group on the right-hand side entirely of non-metals. The elements having an intermediate electronegativity usually show both metallic and non-metallic character and so it is possible to have other groups between the two extremes which contain both true metals, true non-metals, and others which are difficult to classify as either.

Reference to Figure 5.8 shows that the Group 4 elements change from high values of electronegativity at the top of the group to low values at the bottom. The properties of these elements listed in Table 5.9 bear out the above generalizations.

Table 5.9 The elements of Group 4

Increasing electronegativity	*Element*	*Metallic or non-metallic*	*Gain or lose electrons*
↑	Carbon	Generally considered non-metallic, but one allotrope, graphite, shows a 'metallic' property, i.e. conduction	Reacts by gaining electrons
	Silicon Germanium	Generally considered non-metallic, but show some metallic properties (e.g. semiconduction)	Usually react by gaining electrons
	Tin	Metallic	Can react by gaining or losing electrons
	Lead	Metallic	Reacts mainly by losing electrons

Figure 5.8 also suggests that other groups show a similar trend. Find out a little more about the elements in Group 5 (Section 11.2) and refer to your work on metals and non-metals (Section 2.3), and see if this suggestion is justified.

This gradual change from non-metallic to metallic character of the elements in the central groups of the Periodic Table means that the general trends in reactivity found in completely metallic, or completely non-metallic groups are not applicable in such cases. Thus these central groups show no general pattern in their reactivity.

Summary

1. The reactivity of the elements in a metallic group *increases* on descending the group. This is because the reactivity depends on the ease with which the metallic atoms lose electrons, and this process becomes easier as the group is descended.
2. The reactivity of the elements in a non-metallic group *decreases* on descending the group. This is because the reactivity depends on the ease with which the non-metal atoms gain electrons, and this process becomes more difficult as the group is descended.
3. In any group the electronegativity decreases going down. This leads to the central groups of the Periodic Table showing a gradual change from non-metallic to metallic character, and so such groups show no general pattern in their reactivity.

QUESTIONS CHAPTER 5

1. In any one group of the Periodic Table, the elements

A become more reactive as their atomic numbers increase
B become less reactive as their atomic numbers increase
C lose electrons more easily as their atomic numbers increase
D lose electrons less readily as their atomic numbers increase
E are all equally reactive

2. Element X is in Group 1 of the Periodic Table. It is likely to be

A a very reactive non-metal
B an element which readily forms X^- ions
C a dense, hard metal with a high melting point
D a light, soft metal with a low melting point
E a dense, soft metal with a high melting point

3. Which of the following statements about the Periodic Table is *not* correct?

A Elements with similar properties occupy the same vertical columns
B There are many more metals than non-metals
C The most reactive metals are in Group 1
D The non-metals are found in the bottom right-hand corner of the table
E The elements are listed in order of atomic number

4. The elements of Group 1, going down the column,

A become less reactive
B have smaller ions
C become less electronegative
D have decreasing atomic numbers
E show an increasing tendency to gain an electron

5. The element with atomic number 10 is likely to have similar properties to the element with atomic number

A 9; B 11; C 16; D 18; E 28

6. The following statements are about elements in the Periodic Table. Which one is *incorrect?*

A Electronegativities in any period increase from left to right
B Electronegativities always increase in a group as the atomic numbers of the elements increase
C Some groups contain both metals and non-metals
D The transition metals show characteristics different from normal metals
E An element low down in Group 1 will react vigorously with an element near the top of Group 7

7. Name (a) three alkali metals, (b) two halogens, (c) two noble gases, (d) one insoluble non-metallic oxide, (e) the most electronegative element, (f) the least electronegative element.

8.

1	2	3	4	5	6	7	0
Li	Be	B	C	N	O	F	Ne
Na	Mg	Al	Si	P	S	Cl	Ar

The above two rows of the Periodic Table may help you to answer the questions below. One of the numbers, symbols or words given at the end of each of the following sentences completes the sentence correctly.

(a) The atomic mass of $^{14}_{7}N$ is
(i) 7, (ii) 14, (iii) 98, (iv) 21, (v) 2.
(b) The number of neutrons in one atom of $^{23}_{11}Na$ is
(i) 0, (ii) 11, (iii) 12, (iv) 23, (v) 253.
(c) A fluoride anion has the same number of electrons as
(i) Be^{2+}, (ii) O, (iii) Cl^-, (iv) Ne.
(d) A sodium cation has a different number of electrons from
(i) O^{2-}, (ii) F^-, (iii) Li^+, (iv) Al^{3+}.
(e) $^{35}_{17}Cl$ and $^{37}_{17}Cl$ are
(i) allotropes, (ii) isomers, (iii) isotopes, (iv) molecules. (J. M. B.)

9. (a) Give the formulae of the chlorides of the following elements and the physical state in which each compound exists at room temperature.

Element	*Formula*	*State*
Carbon		
Hydrogen		
Magnesium		

(b) Give an equation to show what change occurs when hydrogen chloride is added to water. (J. M. B.)

10. This question refers to the elements of the Periodic Table with atomic numbers from 3 to 18. Some of the elements are shown by letters but the letters are not the symbols of the elements.

3 A	4	5	6	7	8 E	9	10 G
11 B	12 C	13	14 D	15	16	17 F	18

(a) Which of the elements lettered A to G,
 (i) is a noble gas
 (ii) is a halogen
 (iii) would react most readily with chlorine
(b) Give
 (i) the formula of the hydride of D
 (ii) the formula of the oxide of C
(c) Indicate whether the bonding in the oxide of C will be ionic or covalent. (J. M. B.)

11. Use the Periodic Table to answer the following.
(a) Is astatine (At) likely to be a solid, a liquid or a gas at room temperature?
(b) Is rubidium (Rb) a metal or a non-metal?
(c) Are the compounds of vanadium (V) likely to be colourless or coloured?
(d) What is the formula of gallium oxide? (The atomic number of gallium (Ga) is 31.) Would you expect this oxide to be basic, acidic, or amphoteric?

12. Suppose that analysis of moon rocks has led to the discovery of three new elements of identical atomic structure to elements already known, except that they have only *one* electron in the first energy shell. The elements have been named Shepardonium, Mitchellonium and Roosonium after the men who brought back the samples. The atomic number of Shepardonium is 10. It is a bright shiny metal with a low density and melting point. Predict two properties in each case for Mitchellonium (atomic number = 13) and Roosonium (atomic number = 16).

13. The following are ionic and atomic radii (in nm) of members of the same group of the Periodic Table.

	Atomic radius	*Ionic radius*
A	0.133	0.078
B	0.157	0.098
C	0.203	0.133
D	0.216	0.149
E	0.235	0.165

(a) Is this a metallic group or a non-metallic group?
(b) Which element would have the lowest atomic number?
(c) Which element would be the most reactive?

14. The following data applies to five elements in the Periodic Table.

Element A valency 1, m.p. 370 K (97 °C), atomic number 11.
Element B valency 4, m.p. 1683 K (1410 °C), electronic structure 2, 8, 4.
Element C valency 1, m.p. 172 K (−101 °C).
Element D valency 1, electronic structure 2, 8, 8, 1.
Element E valency 0, electronic structure 2, 8.
(a) Which elements are in the same group?
(b) Which element is in the same group as carbon?
(c) Name two elements which would react very vigorously together.
(d) Which element has a giant structure?
(e) Which element is most likely to form only covalent bonds?
(f) Which element would resemble helium?
(g) Which element could most easily form a negative ion?

15. (a) Sodium and aluminium have atomic numbers of 11 and 13 respectively. They are separated by one element in the Periodic Table, and have valencies of one and three respectively. Chlorine and potassium are also separated by one element in the Periodic Table (they have atomic numbers of 17 and 19 respectively) and yet they both have a valency of one. Explain the difference.

(b) The halogens, fluorine, chlorine, bromine and iodine, show a gradation in properties. Illustrate this by reference to their ease of combination with hydrogen, and the ease of replacement of one halogen by another. (J. M. B.)

16. The elements sodium, magnesium, aluminium, silicon, phosphorus, sulphur, chlorine and argon form a series in the Periodic Table. For each of *four* of these elements state:

(a) the electronic structure
(b) the formula for its hydride and the reaction, if any, between hydride and water
(c) the formula of an oxide and whether this oxide is acidic, basic, amphoteric or neutral (J. M. B.)

17. Explain clearly why the atomic number of an element is more important to the chemist than its atomic mass or mass number.

18. Explain the three factors which govern the ease with which an electron can be removed from an atom. Do the elements of Period 3, from left to right of the table, become more or less electronegative? Explain your answer in terms of atomic structure.

19. Write short accounts of the work of Döbereiner, Newlands and Mendeléev on the classification of the elements. Quote two improvements on Newlands' arrangement which were made by Mendeléev.

20. Illustrate the gradation in properties of the elements of Period 3 by considering (a) their reaction with water, (b) the type of oxide formed.

6 Electrochemistry

6.1 ELECTROLYSIS

Introduction

You will, by now, be familiar with a number of experiments in which chemical changes are brought about by heat. Heat is a form of energy; another form of energy is electricity. If heat energy can produce a chemical change it is possible that electrical energy can produce a similar result. In this section you will be studying the passage of electricity through different substances and finding out what changes, if any, take place during the process.

Experiment 6.1

†To Find Out Which Substances Will Conduct Electricity *

Note: You carried out a similar experiment (Experiment 2.22) but in that case you dealt only with metallic and non-metallic elements. In this experiment you will be using solid compounds as well as elements. You will be using the same apparatus as in Figure 2.16; look back to page 64 to check the arrangement.

Apparatus

Six volt battery or alternative direct current (d.c.) supply, six volt bulb in holder, connecting wires fitted with crocodile clips.

Small strips or samples of copper and other metals including mercury, samples of wood, rubber, plastic, carbon in the form of graphite, crucibles or small dishes containing solid samples of tartaric acid, sodium hydroxide, paraffin wax, cane sugar, sulphur, naphthalene, etc.

Procedure

Set up the circuit (see Figure 2.16, on page 64) and bridge the gap by allowing the two crocodile clips to touch and so test that the lamp will light. Now place the piece of copper in the gap and touch each end with a clip. Note the effect on the lamp. Repeat with each of the other materials.

List the various substances under two headings, conductors and non-conductors.

Points for Discussion

1. Has this experiment confirmed your earlier findings? Do the compounds behave like metals or non-metals with respect to the electric current?

2. Did you notice any sign of a chemical change when the current passed through the conductors?

3. Why is it that metals conduct electricity whereas few non-metals will do so? A notable exception is carbon, in the form of graphite.

Before you can try to answer point 3, you must be quite sure that you understand the nature of electricity. An electric current is a flow of electrons. An electron, you will remember, is a minute sub-atomic particle carrying unit negative charge. When you know more about the structure and bonding of metals you will understand how it is that they are able to conduct electricity. At this point we will just say that free electrons, produced by chemical action in a battery or by mechanical action in a dynamo, can be handed on from one metal atom or carbon atom to another, rather like buckets of water being passed along a human chain of fire fighters, and that this passage of electrons from atom to atom cannot take place in any other material.

Experiment 6.2

†Does Electricity Have Any Different Effect When Substances Are Molten? **

Apparatus
Six volt battery or similar source of d.c. supply, six volt lamp and holder, connecting wires, carbon rods in wooden holder, crucibles, pipe-clay triangle, tripod, Bunsen burner, tongs, asbestos square.

Paraffin wax; sugar; naphthalene, sulphur (flammable); lithium chloride; mercury; sodium hydroxide (very dangerous); lead iodide, lead, and lead bromide (poisonous—must be used in a fume cupboard). You may be allowed to try one or two for yourself, but your teacher will probably demonstrate most of them.

Procedure
Set up the circuit as shown (Figure 6.1). Half fill a crucible with one of the substances to be tested and, supporting it on a pipe-clay triangle on a tripod, heat gently until the sample just melts. Repeat with the other samples, using a fresh crucible for each and scraping the electrodes clean after each test. Record the effect on the lamp in each case and any sign of chemical activity in the vicinity of the electrodes.

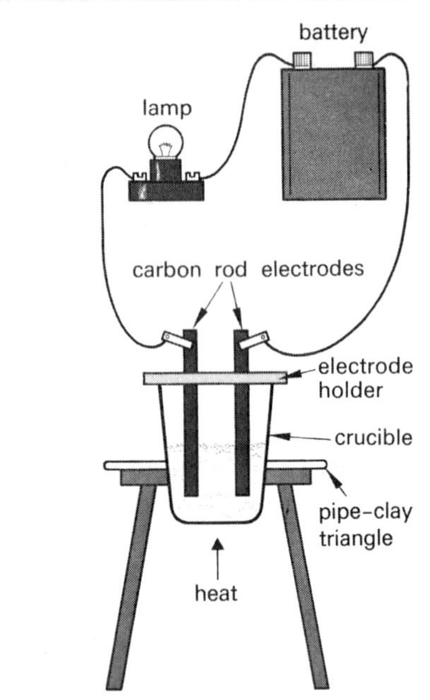

Figure 6.1 Investigating the effect of electricity on molten substances

Record your results in a four column table under the headings shown below.

Substance	*Whether it conducts when solid*	*Whether it conducts when molten*	*Any chemical change?*

Points for Discussion

1. Look at the list of substances which conduct when molten but not when in the solid state. What type of bonding is shown by all these substances?

2. Look at the list of substances which do not conduct even when molten. Some of these are elements and others are compounds. Notice that all these compounds have covalent bonding.

3. You probably noticed bubbles of gas round the electrodes in some cases. What do you think was causing this activity? Remember what we said at the beginning of the section about the possibility of electrical energy producing chemical change.

4. From your results you should be able

to pick out three groups of substances according to how they behave with regard to an electric current: (a) substances which conduct electricity when solid *or* liquid—these are the metals and graphite; (b) substances which do not conduct electricity under any circumstances—these are either non-metallic elements or compounds with only covalent bonding; (c) Substances which do not conduct electricity when solid but do conduct when molten, and in the process undergo some kind of chemical change—these are ionic compounds. It is the latter class of substances that we shall be considering in this section.

5. We have found that, although these ionic compounds are non-conductors in the solid state, they do become conductors when molten. Another way of making them 'liquid' would be to dissolve them in water, instead of melting them. Will these compounds also become conductors in aqueous solution? The next experiment should provide the answer.

Experiment 6.3

†To Determine Whether Some Substances Conduct Electricity When in Aqueous Solution *

Apparatus
Six volt battery or alternative d.c. supply, six volt lamp and holder, two carbon electrodes in holder, 50 cm³ beaker, connecting wire.
Deionized water, aqueous solutions of sodium chloride, lithium chloride, potassium iodide, sodium hydroxide, sulphuric acid, cane sugar.

Procedure
Set up the circuit as shown (Figure 6.2). Half fill the beaker with deionized water, put in the electrodes and note any effect on the lamp. Repeat with the solutions given, washing the beaker out after each test. Also look for any signs of chemical activity around the electrodes. Record your results in a table.

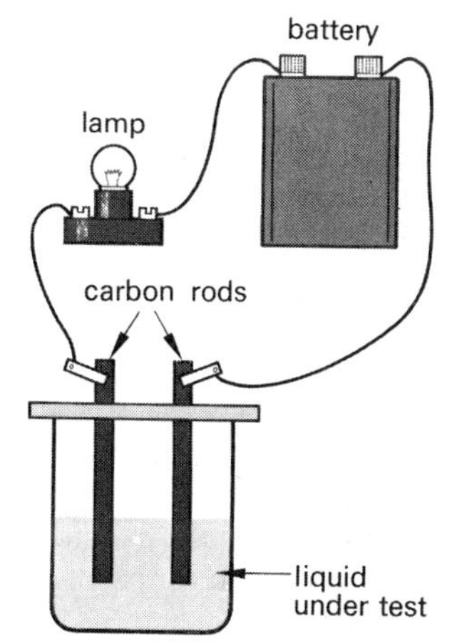

Figure 6.2 To determine whether some substances will conduct electricity in aqueous solution

Points for Discussion

1. Look at your list of substances which when dissolved in water, allowed the current to pass. Are they

A all elements
B all ionic compounds
C all covalent compounds
D some covalent and some ionic compounds?

2. You should have found that the substances which are non-conductors when solid but allow a current to pass when molten or dissolved in water, are all ionic compounds. You will remember that these are compounds which consist of ions of opposite charges, firmly bonded together in a crystal lattice. An ion is formed when an atom either loses or gains one or more electrons. An electric current is a flow of electrons. Obviously this conducting ability of ionic compounds under certain circumstances must have something to do with the fact that they are made up of charged particles. Why then do they conduct only when molten or dissolved? The answer, of course, is that in these cases the ions are no longer held in the rigid giant structure but are free to move. These moving ions must enable the current to pass and somehow in the process chemical changes take place.

3. To try to find out exactly how these ionic compounds can act as conductors of electricity, we will examine the effect of passing a current through a fused salt in more detail.

Experiment 6.4

To Determine What Happens When an Electric Current is Passed Through Fused Lead Bromide ***

Apparatus

Six volt battery or alternative d.c. supply, six volt lamp and holder, hard glass test-tube (150×25 mm), carbon rods in wooden holder, connecting wire, retort stand and clamp, Bunsen burner, asbestos square, access to fume cupboard.
Lead bromide.

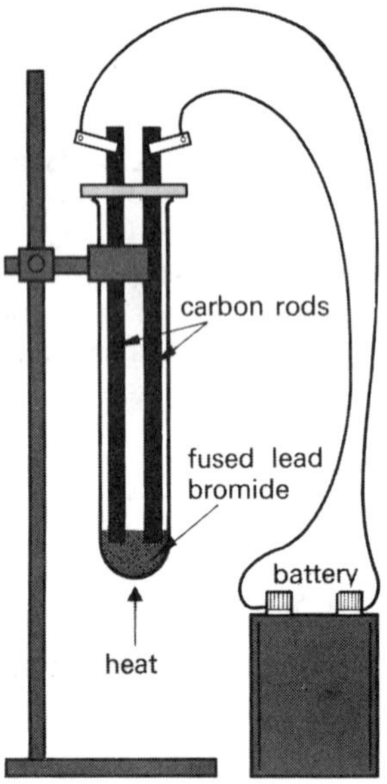

Figure 6.3 The electrolysis of fused lead bromide

Procedure

Set up the circuit as shown (Figure 6.3). Place about 4 cm depth of lead bromide in the tube. Remove the carbon rods and holder and heat the lead bromide gently until it melts. Replace the rods and as soon as the lamp lights, remove it from the circuit. (This lowers the resistance of the circuit.) Continue heating from time to time so that the salt remains molten and allow the current to pass for about five minutes, noting any chemical change that takes place. Remove the rods and allow the tube to cool. Examine the contents of the test-tube; in particular, note what you can see through the bottom of the tube.

[Caution: lead bromide and the vapour produced during the experiment are poisonous. The experiment should be carried out in a fume cupboard.]

Points for Discussion

1. What was the colour of the vapour produced during the experiment? You have seen a similar vapour before. What do you think it was? Did it appear throughout the melt or around an electrode, and if so which one: that connected to the positive, or to the negative pole of the battery?

2. When you examined the contents of the tube after cooling you would probably see through the bottom of the tube that a bead of metal had been formed near to one of the electrodes. What is this metal? Was it near to the electrode connected to the positive or to the negative pole of the battery?

3. The last experiment will have shown you that when an electric current passes through molten lead bromide, a chemical change occurs resulting in the production of a red-brown vapour and a grey bead of metal. The lead bromide has been decomposed by the current into bromine vapour and metallic lead.

Summary

1. Metals will conduct electricity when either solid or liquid. They are not chemically changed during the process and are simply called *conductors*.
2. Non-metallic elements, with few exceptions (notably graphite), do not conduct electricity. Covalent compounds do not conduct in the solid state nor when molten, and they do not usually conduct in solution. It is possible, however, for some covalent compounds to *react* with water to produce a conductive solution.
3. Ionic compounds do not conduct when solid but become 'conductors' when molten or dissolved in water. They are always decomposed during the process. Such substances are called *electrolytes*.

Some Definitions

An *electrolyte* is a compound which, when fused or dissolved in water, conducts an electric current and is decomposed in the process.

Electrolysis is the chemical change which takes place when an electric current passes through a fused or dissolved electrolyte.

An *electrode* is the metal or carbon rod by which the current enters or leaves the electrolyte.

The *anode* (+) is the electrode from which the electrons leave the electrolyte.

The *cathode* (−) is the electrode at which the electrons enter the electrolyte.

What Happens during Electrolysis?

If we now consider the electrolysis of lead bromide we may be able to work out the sequence of events which ends with the production of lead and bromine.

We know that lead bromide is an ionic compound and that the solid form consists of a giant structure of lead and bromide ions. When we heat the solid it fuses (melts) and the strong bonds holding the ions in the crystal lattice are broken (Kinetic Theory, Section 1.3). Molten lead bromide is made up of lead ions, Pb^{2+}, and bromide ions, Br^-, all moving in a random fashion. Putting a pair of electrodes into the melt, however, seems to direct the movement of the ions. The electrodes are connected to a battery or other source of direct current and so electrons are being 'pumped' from the negative pole of the battery to the cathode. Because the electron flow enters the electrolyte at the cathode, it is negatively charged and another way of defining a cathode is to say that it is the electrode by which electrons enter an apparatus from an external circuit. The electrons cannot 'jump' across from the cathode to the anode and yet they must be continually leaving the cathode and arriving at the anode in order to return to the battery and complete the circuit. How are the moving ions connected with this electron flow? Why should the lead ions, Pb^{2+}, and the bromide ions, Br^-, move towards the electrodes? Are they attracted to one electrode only or both?

...y attracted to one ...e or both?

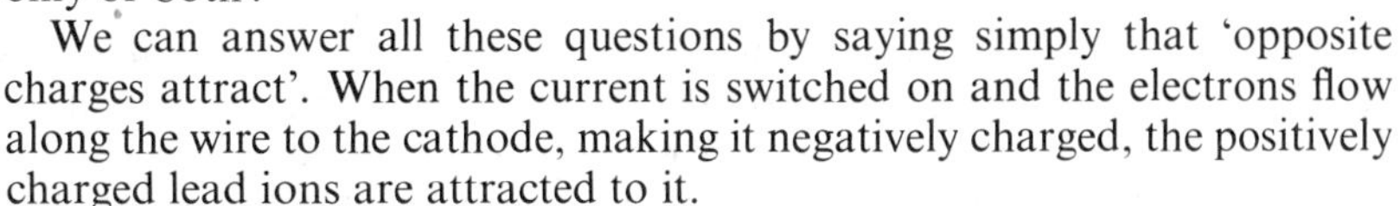

We can answer all these questions by saying simply that 'opposite charges attract'. When the current is switched on and the electrons flow along the wire to the cathode, making it negatively charged, the positively charged lead ions are attracted to it.

At the cathode each lead ion receives two electrons from those flowing from the battery, through the connecting wire and the carbon rod, and so becomes a lead atom:

$$Pb^{2+} + 2e^- \rightarrow Pb(l)$$

The lead atoms bond together to form the giant structure of metallic lead.

The negatively charged bromide ions are attracted to the positively charged anode. As each bromide ion reaches the anode it gives up an electron and becomes once again an atom of bromine:

$$Br^- \rightarrow Br + e^-$$

(Remember that bromide ions are formed when each bromine atom gains an electron, consequently when a bromide ion loses an electron to the anode it must become a bromine atom again.)

The bromine atoms join in pairs to form bromine molecules,

$$2Br \rightarrow Br_2(g)$$

and so bromine vapour is evolved.

The electron removed from the bromine ion is passed on through the carbon electrode and the copper connecting wire to the battery, thus the current is kept flowing.

During electrolysis, therefore, the ions are discharged and this giving of electrons to the ions by the cathode, and taking of electrons from the negatively charged ions by the anode, keeps the current flowing in the circuit. Within the electrolyte, however, the electrons are not handed on like buckets of water in a chain, as they are in metals, but are carried by the ions, which can be compared to one group of individuals taking full buckets of water from the tap to the fire and others taking empty ones back from the fire to the taps (Figure 6.4).

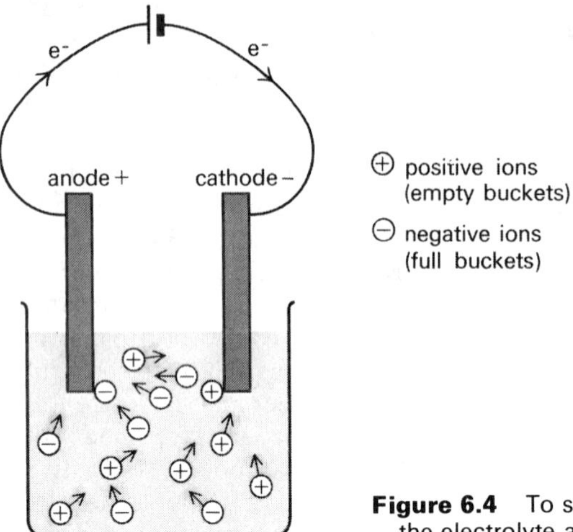

Figure 6.4 To show the electron flow within the electrolyte and in the external circuit

The ions which carry positive charges and which are therefore attracted to the cathode are called *cations*. Metal ions and hydrogen ions are cations. *Anions* are those ions carrying negative charges and are attracted to the anode; these are the ions of non-metals and of radicals such as sulphate, nitrate, and carbonate.

Does Water Play Any Part in Electrolysis?

We have seen that not only do ionic compounds conduct electricity when fused but that they will also pass a current when dissolved in water. As you already know, when an ionic compound is dissolved in water, the ions are jostled out of the crystal lattice by the water molecules which then surround them (page 94). So when an ionic compound is dissolved, the same free ions are produced as when it is fused. Are the same products obtained at the electrodes in both cases? The next experiment should provide the answer to this question.

Experiment 6.5

†To Compare the Results of Electrolysing the Same Salt in the Molten State and Also in Aqueous Solution *

Apparatus

Six volt battery or similar d.c. source, connecting wire, crucible, pipe-clay triangle, tripod, Bunsen burner, carbon electrodes in holder, small glass cell fitted with carbon electrodes, two 75 × 10 mm test-tubes, one 100 × 16 mm test-tube, retort stand and clamp, litmus paper, splints, asbestos square.
Anhydrous lithium chloride.

Procedure

(a) Half fill the crucible with solid lithium chloride and set up the circuit as for Experiment 6.4. Heat the powder until it just melts; insert the electrodes and observe the reaction at each electrode. Note particularly whether there is any evolution of gas at the cathode. Try to recognize the smell of the gas produced at the anode (care!), and test it with moist blue litmus paper.

(b) Half fill the larger test-tube with water, add a spatula measure of lithium chloride and shake the tube until the powder dissolves. Support the cell by the clamp and pour enough solution into it to just cover the electrodes. Set up the circuit as shown (Figure 6.5). Fill the small test-tubes with the lithium chloride solution and invert them over the electrodes. Hold each test-tube in position so that it just covers the top of the electrode and collect any gases evolved. When the tubes are full, test the gas from the anode with moist blue litmus paper and that from the cathode with a lighted splint.

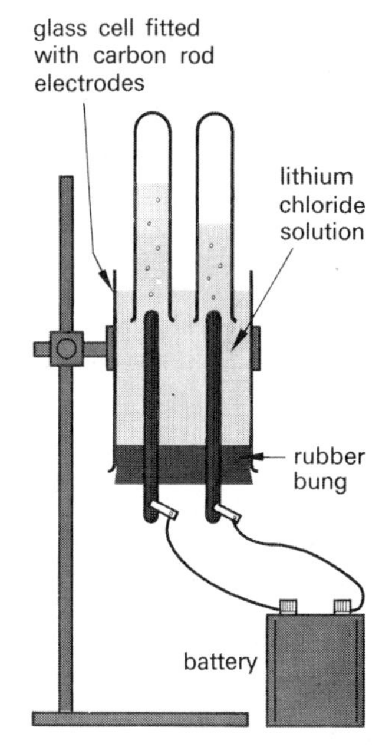

Figure 6.5 The electrolysis of lithium chloride solution

Points for Discussion

1. Did you notice any difference in the products, (a) at the anode, (b) at the cathode, in the two experiments?

2. What do you think are the likely products of the electrolysis of *fused* lithium chloride? (If you are not sure, refer to Experiment 6.4 again.) Did you see any evidence to support your theory? Lithium is a soft silvery metal with a low melting point which forms a white powder when heated in air. Chlorine is a pale greenish-yellow gas with a smell reminiscent of the swimming baths or domestic bleaching liquids.

3. You probably noticed the same smell from the gas evolved at the anode in experiment (b). If your strip of blue litmus paper went pink and was then bleached, you can take it as confirmation, in this case, that the gas contained chlorine.

4. Draw up a table to list and compare the products at the anode and cathode in experiments (a) and (b).

The Electrolysis of a Dissolved Salt Gives a Different Result from the Electrolysis of the Same Fused Salt. Why?

When fused lithium chloride was electrolysed, metallic lithium was produced at the cathode, but when an aqueous solution was electrolysed hydrogen gas was the product at the cathode. Where did the hydrogen come from? An element produced at the cathode is formed by adding electrons to the positive ions of that element, so the hydrogen must have been formed from hydrogen ions and, as there is no hydrogen in lithium chloride, the hydrogen ions can only have come from the water. This is, perhaps, surprising as we have already learnt that water is a covalent compound and in Experiment 6.3 we found that pure water would not conduct electricity. A very small fraction of water molecules (one in 550 million) *does* split up into ions. The ions formed are hydrogen ions (H^+) and hydroxide ions (OH^-). We can show this by an equation

$$H_2O(l) \rightleftharpoons H^+(aq) + OH^-(aq)$$

Notice the sign $\rightleftharpoons$. This means that the reaction is reversible (Section 14.2) and hydrogen and hydroxide ions can combine again to form water. The hydrogen ion is a hydrogen atom that has lost its single electron and so is the same as a proton. It is therefore a very tiny ion which is unstable on its own and immediately joins up with a water molecule to form a hydronium (oxonium) ion (H_3O^+). Therefore the equation $H_2O(l) \rightleftharpoons H^+(aq) + OH^-(aq)$ is an oversimplification and we usually write:

$$H_2O(l) + H_2O(l) \rightleftharpoons H_3O^+(aq) + OH^-(aq)$$

Any sample of pure water therefore contains a very small percentage of hydronium and hydroxide ions. An aqueous solution of an electrolyte contains the positive and negative ions which make up the compound and also hydronium and hydroxide ions from the water (Figure 6.6).

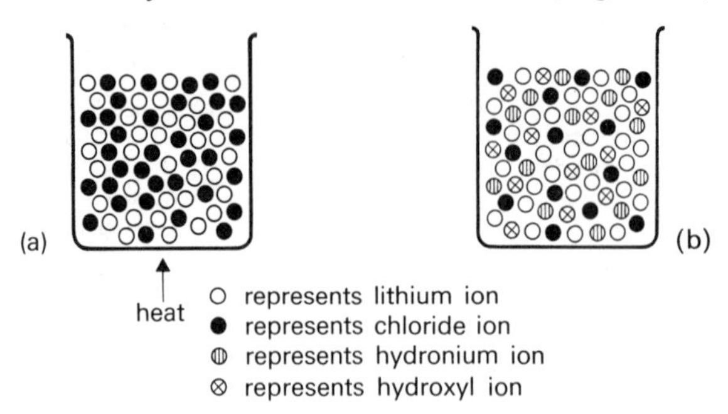

Figure 6.6 The ions present in (a) a molten salt (lithium chloride) and in (b) its aqueous solution

The fact that there are more types of ion in an aqueous solution than there are in a molten salt does not fully explain why, in the case of lithium chloride, lithium was the product at the cathode from the melt, but hydrogen was evolved when the solution was electrolysed. Why was hydrogen discharged in preference to lithium? Why was not a mixture of both elements produced?

When there is more than one ion with the same kind of charge, both types of ion are attracted to the electrode but usually only *one* type can be selected and discharged. This selection depends mainly on the combination of two factors, (a) the ease with which the different ions gain or lose electrons and, (b) the relative concentrations of the different ions.

In the case of the lithium chloride solution, the positive ions present are the lithium ions Li^+ and the hydronium ions H_3O^+. Hydronium ions gain electrons very much more easily than lithium ions, and so although the concentration of hydronium ions is less than that of lithium ions the hydronium ions are *preferentially discharged.* Each ion takes an electron from the cathode. A water molecule and a hydrogen atom result,

$$H_3O^+ + e^- \rightarrow H_2O + H$$

The hydrogen atoms join in pairs to form hydrogen molecules:

$$2H \rightarrow H_2(g)$$

Hydrogen gas is therefore evolved at the cathode. The lithium ions remain unchanged in the solution.

Similarly the two kinds of negative ion, chloride (Cl^-) and hydroxide (OH^-) are attracted to the anode. The chloride ion is in much greater concentration than the hydroxide ion in this case although it loses its electron less easily, so chloride ions are preferentially discharged.

Each ion gives up an electron to the anode:

$$Cl^- \rightarrow Cl + e^-$$

Chlorine atoms bond in pairs to form chlorine molecules:

$$2Cl \rightarrow Cl_2(g)$$

Chlorine gas is evolved at the anode. Most of the hydroxide ions remain unchanged in the solution, although in practice some oxygen may be evolved from them.

Remember:

$Li^+Cl^-(s) \rightarrow Li^+(aq) + Cl^-(aq)$ Many free ions as $\rightarrow$ goes almost to completion.

$2H_2O(l) \rightleftharpoons H_3O^+(aq) + OH^-(aq)$ Few ions as $\rightleftharpoons$ lies well over to the left.

However

$$Li^+ + e^- \text{ not easy}$$

$$H_3O^+ + e^- \text{ very easy}$$

Therefore in spite of low concentration, hydronium ions are discharged.

You might think that as there are comparatively few hydronium ions present in the solution these would soon be used up, but remember the equation which illustrates the decomposition of water molecules into ions:

$$2H_2O(l) \rightleftharpoons H_3O^+(aq) + OH^-(aq)$$

The sign means that the reaction is reversible and that while water molecules break up into ions, the ions are combining to form water molecules. When both these reactions are going on at the same rate, a state of balance is obtained and the total number of molecules split into ions remains constant at about one in 550 million. If, however, a hydronium ion is removed by being discharged at the cathode, the equilibrium is upset and so another water molecule breaks up and provides another hydronium ion to balance it again. Thus the molecules of water in an aqueous solution provide a huge reservoir of hydronium and hydroxide ions. In this case, as only the hydronium ions from the water are discharged, there is a build-up of hydroxide ions in the solution. Also at the anode the chloride ions are discharged and so the solution gradually becomes a solution of lithium hydroxide.

Summary

An anion is an ion carrying one or more negative charges. It is attracted to the anode during electrolysis.

A cation is a positively charged ion which is attracted to the cathode during electrolysis.

Hydrogen and the metals form positive ions.

Non-metals and radicals form negative ions.

The charge on the ion is the number of electrons the atom has to gain or lose to attain a stable noble gas electron configuration. (This is, of course, its valency and if you remember the table of combining powers that you learnt in Section 1.7, you will be able to write the symbols for all the common ions correctly.)

Symbols for some common ions

Hydrogen	H^+	Chloride	Cl^-
Hydronium	H_3O^+	Sulphate	SO_4^{2-}
Sodium	Na^+	Hydroxide	OH^-
Copper(II)	Cu^{2+}		

Write the symbols for the following ions, potassium, magnesium, aluminium, bromide, nitrate, and iodide in your notebook.

The decomposition of molten electrolytes simply produces one of the two components at each electrode. Solutions are more complicated as there are more ions present. When two different ions are attracted to the same electrode, the ion is discharged which gives up or takes electrons the more easily or is present in the greater numbers. Different experimental conditions such as temperature or the material of the electrode may influence the ease with which an ion is discharged, but at this stage all you need to remember is which type of ion is selected in the electrolysis experiments that follow. In the next section you will learn the correct order of readiness for discharge of all the common ions and the reason for this order.

You now have all the information necessary to predict the products at the electrodes in the majority of the experiments that follow. The next experiment is written in some detail so that you will see how such experiments should be set out.

Experiment 6.6
The Electrolysis of an Aqueous Solution of Sodium Chloride *

Apparatus

Six volt battery or similar d.c. supply, glass cell, stand and clamp, carbon electrodes in holder, connecting wire, two 75×10 mm test-tubes, litmus paper, splint. Saturated solution of sodium chloride.

Procedure

Set up the circuit as shown in Figure 6.5 and carry out the electrolysis of sodium chloride solution using the same method as for lithium chloride solution (Experiment 6.5). Collect any gases evolved and test them as before, with damp blue litmus paper and a lighted splint.

Points for Discussion

1. Can you explain why the gases evolved at the electrodes are the same as those produced in the electrolysis of lithium chloride solution?

2. The account given below shows how an electrolysis experiment could be explained using this experiment as an example. You should use the same sub-headings for other examples of electrolysis.

Explanation

Ions present:

Sodium ions, Na^+, chloride ions, Cl^- (from sodium chloride,

$$Na^+Cl^-(s) \rightarrow Na^+(aq) + Cl^-(aq)\ ).$$

Hydronium ions, H_3O^+, hydroxide ions, OH^- (from the water,

$$H_2O(l) + H_2O(l) \rightleftharpoons H_3O^+(aq) + OH^-(aq)\ )$$

At the anode (+ve)

Chloride and hydroxide ions attracted. Chloride ions selected and discharged. Each ion gives up an electron to the anode and becomes a chlorine atom:

$$Cl^-(aq) \rightarrow Cl + e^-$$

The atoms join in pairs to form chlorine molecules:

$$2Cl \rightarrow Cl_2(g)$$

Chlorine gas and a little oxygen are evolved at the anode.

At the cathode (−ve)

Hydronium and sodium ions attracted. Hydronium ions selected and discharged. Each ion takes an electron from the cathode and forms a hydrogen atom and a water molecule:

$$H_3O^+(aq) + e^- \rightarrow H_2O(l) + H$$

The hydrogen atoms bond together in pairs to form hydrogen molecules:

$$2H \rightarrow H_2(g)$$

Hydrogen gas is evolved at the cathode.

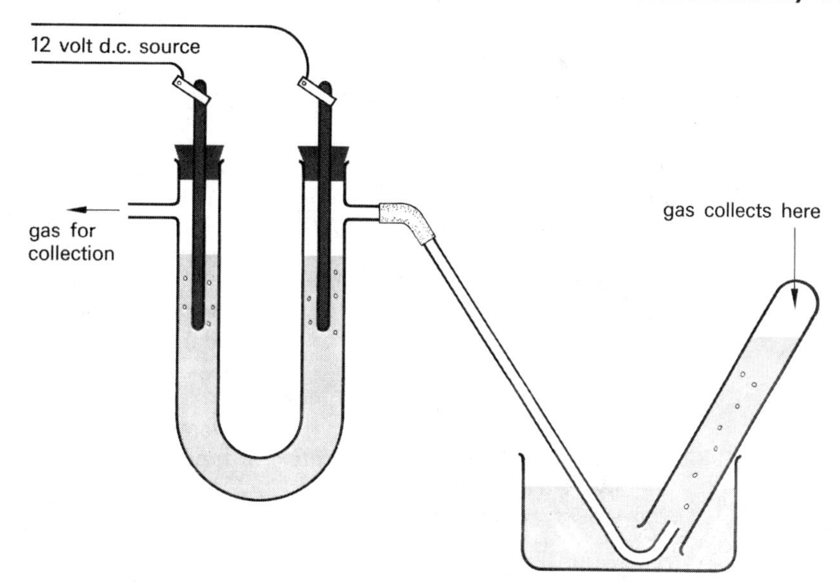

Figure 6.7 Side-arm U-tube used as a voltameter

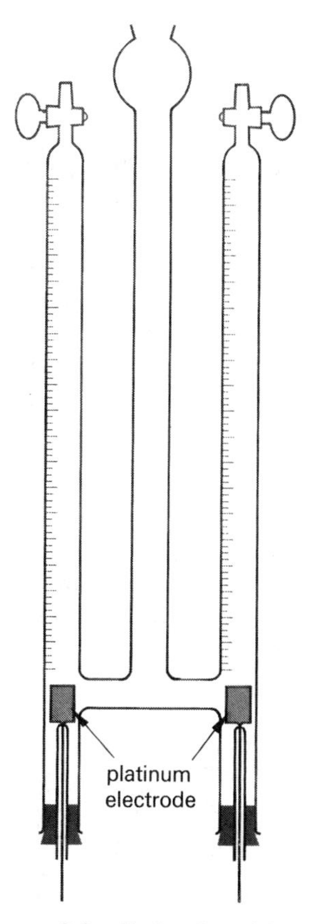

Figure 6.8 Hofmann voltameter

Changes in the solution

Hydroxide ions, left in the solution after their hydronium ion partners have been discharged, accumulate round the cathode as do the sodium ions, which are attracted to the cathode but are not discharged. An increasing concentration of sodium hydroxide is consequently produced in the neighbourhood of the cathode.

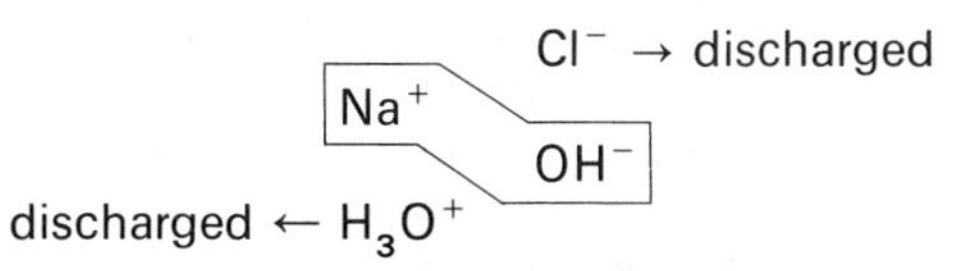

Note: any piece of apparatus in which electrolysis is carried out is called a *voltameter*. The small glass cells that you have already used are voltameters. The side-arm U-tube shown in Figure 6.7 is a convenient voltameter to use for the collection of gases evolved at the electrodes, as is the Hofmann voltameter used in the next experiment (Figure 6.8).

Experiment 6.7
†The Electrolysis of Hydrochloric Acid using a Hofmann Voltameter ***

Apparatus
A Hofmann voltameter with *carbon* electrodes, connecting wire, 12 volt battery or other d.c. source, splint.
Blue litmus paper, solution of 2M hydrochloric acid.

Procedure
Fill the voltameter with the hydrochloric acid, closing the taps when the liquid fills the outside tubes. Connect the voltameter to the battery or low voltage d.c. source. When the side tube above the cathode contains about 20 cm^3 of gas, hold an inverted test-tube over the top, open the tap, and collect the gas in a test-tube. Close the test-tube with the thumb and then test the gas with a lighted splint.

Collect a sample of the gas formed in the tube above the anode in a similar manner and test this gas with a piece of damp blue litmus paper.

Points for Discussion
1. You would notice that some of the bubbles of gas, formed at the anode, disappeared before they reached the top of the tube and so the gas given off at the anode was much slower to collect in the tube than the gas given off at the cathode. Why do you think this happened? If you smell the liquid that is emptied out of the voltameter at the end of the experiment it may help you to answer this question.

2. Write up the experiment in your notebook under the headings as given for Experiment 6.6, including the 'Explanation' and 'Changes in the Solution'.

Experiment 6.8
†The Electrolysis of Copper(II) Sulphate Solution using Platinum or Carbon Electrodes ***

[Note: this experiment can be carried out using either the Hofmann voltameter as in the previous experiment or the small glass electrolysis cell used in Experiment 6.6. The voltameter is filled with a fairly concentrated solution of copper(II) sulphate.]

Procedure
Carry out the electrolysis as for Experiment 6.6 or 6.7. Collect a test-tube of the gas given off at the anode and test it with a glowing splint. Look carefully for any signs of chemical activity around the cathode. When the current has been passing for about five minutes, remove the cathode and examine it.

Points for Discussion
1. Which ions will be attracted to the cathode? Predict which ion will be preferentially discharged. Does this agree with your observation as to what happened at the cathode?

2. Which ions will be attracted to the anode? Which will be preferentially discharged? The correct answer to the last question is that the hydroxide ions are discharged at the anode in preference to the sulphate ions.

Each ion gives an electron to the anode:

$$OH^-(aq) \rightarrow OH + e^-$$

Now the OH radical has no independent existence and reacts immediately with three other hydroxide radicals to form water and oxygen, thus:

$$4OH \rightarrow 2H_2O(l) + O_2(g)$$

The ionic equation for the reaction at the anode when hydroxide ions are discharged is generally written:

$$4OH^-(aq) \rightarrow 2H_2O(l) + O_2(g) + 4e^-$$

3. As the copper ions (point 1) and hydroxide ions are discharged, the concentration of the unchanged sulphate and hydronium ions increases, i.e. the solution gradually changes to sulphuric acid.

discharged ↖ Cu^{2+} + SO_4^{2-}

H_3O^+ + OH^- → discharged

[Note: did you write the equation for the discharge of the copper ions correctly?

$$Cu^{2+}(aq) + 2e^- \rightarrow Cu(s)$$]

Experiment 6.9

†The Electrolysis of Sodium Chloride Solution using a Hofmann Voltameter and an Indicator (a variation of Experiment 6.6) ***

Apparatus

A Hofmann voltameter with *carbon* electrodes, 12 volt battery or other d.c. source, connecting wire, splint.
Solution of sodium chloride coloured purple with litmus solution.

Procedure

Fill the voltameter with the sodium chloride solution and connect it to the battery as in Experiment 6.6. Note any colour changes in the litmus. Collect the gas evolved at the cathode and test it with a lighted splint.

Points for Discussion

1. Describe and explain the colour change that occurred round the anode.

2. Do you think that the presence of the litmus solution made any difference to the products of the electrolysis?

3. Why do the bubbles of gas, evolved at the anode, fail to reach the top of the tube at first?

4. Describe and explain the colour change that occurred round the cathode. (Hint: it is the presence of hydroxide ions that turns litmus blue.)

5. The electrolysis of sodium chloride solution is an important manufacturing process. Can you name three important products of this process?

Experiment 6.10

†The Electrolysis of Dilute Sulphuric Acid (Sometimes called the Electrolysis of Water) ***

Apparatus and Procedure

This experiment can be done using any of the voltameters already mentioned but the volume relationship of the gases is more easily found using a Hofmann voltameter. Carry out the electrolysis as in Experiment 6.9 using, as the electrolyte, water into which a little dilute sulphuric acid has been stirred.

Let the current run until a considerable volume of gas has collected in the tube above the cathode, and then compare the volumes of gas in the two tubes.

Open the tap above the cathode and collect the gas forced out in an inverted test-tube. Keeping the tube still inverted, hold a lighted splint to the mouth of the tube. Repeat with the gas from the other tube but this time use a brightly glowing splint.

Points for Discussion

1. Which ions will be attracted to the anode? Predict which of these will be preferentially discharged.

2. Write up the experiment as you have been shown, explaining how hydrogen gas is produced at the cathode (with ionic equations) and using the information in 1 to explain the changes at the anode. If you are not sure how it is that oxygen gas is evolved, look back at the explanation after Experiment 6.8.

3. As the hydronium and hydroxide ions are discharged the equilibrium $2H_2O(l) \rightleftharpoons H_3O^+(aq) + OH^-(aq)$ is disturbed and more water molecules ionize. These ions are in their turn discharged and so, as the electrolysis continues, water molecules are used up. As the number of sulphate ions and hydronium ions from the sulphuric acid remain unchanged, the concentration of sulphuric acid increases and only the water is decomposed.

The Volume Composition of Water

You will have noted that the volume of hydrogen gas produced from the decomposition of water is approximately twice the volume of oxygen. As the gases were measured at the same temperature and pressure, this experiment can be used to demonstrate the volume composition of water. Allowing for the fact that oxygen is more soluble in water than is hydrogen, the volume composition can be taken as two volumes of hydrogen to one volume of oxygen.

Consider again the ionic equations for the discharge of the two types of ion,

at the anode:

$$4OH^-(aq) - 4e^- \rightarrow O_2(g) + 2H_2O(l)$$

at the cathode:

$$2H_3O^+(aq) + 2e^- \rightarrow H_2(g) + 2H_2O(l)$$

As the same current is flowing through both electrodes the ionic equations can only be compared when they involve the same number of electrons. The ionic equation for the reaction at the cathode becomes:

$$4H_3O^+(aq) + 4e^- \rightarrow 2H_2(g) + 4H_2O(l)$$

Notice that the same quantity of electricity produces twice as many molecules of hydrogen as of oxygen and also twice the volume of hydrogen as oxygen. You will learn more about this relationship between numbers of molecules and volumes of gases later in the course.

Experiment 6.11
The Electrolysis of an Aqueous Solution of Sodium Hydroxide ***

Predict the products at the electrodes, the changes in the solution, and the relative volumes of any gases formed; then carry out the electrolysis of sodium hydroxide using a Hofmann voltameter and platinum electrodes. Write up the experiment in the usual way.

Were your predictions correct?

The Part Played by the Electrodes

So far we have used electrodes made of either carbon or platinum. These are 'inert' electrodes as they do not take part in any chemical changes which occur, but simply act as passers-on of electrons. In the following experiment you will use copper electrodes, which are not 'inert', and you will be able to compare the result with that of Experiment 6.8.

Experiment 6.12
The Electrolysis of Copper(II) Sulphate Solution using Copper Electrodes *

Apparatus

Six volt battery or other d.c. source, connecting wire, 100 cm³ beaker.

Two pieces of clean copper foil 2 cm × 10 cm, copper(II) sulphate solution.

Procedure

Set up the circuit as shown (Figure 6.9). If you bend the copper foil over the side of the beaker it will remain in place quite easily. Let the current pass for about five minutes. Watch for any changes at the electrodes, e.g. gases given off. Remove the electrodes and examine them.

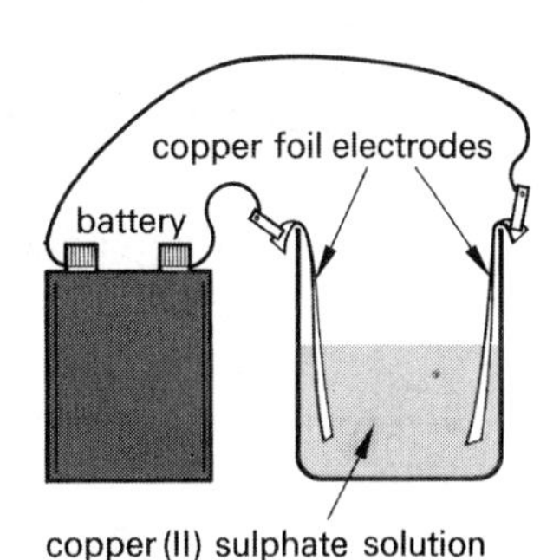

Figure 6.9 The electrolysis of copper(II) sulphate solution using copper electrodes

Points for Discussion

1. Is there any change in the copper foil which was used as the cathode? What do you think has happened? Remember that metallic elements, produced by chemical action, often appear as fine powders and not as shiny metals.

2. Is there any visible change in the appearance of the anode?

3. Was there any difference in the activity round the anode, between this experiment and Experiment 6.8?

Explanation of the Changes That Occur

You will probably have identified the red-brown powder on the cathode in Experiment 6.12 as copper and will have noted that this result was the same as when you used a platinum cathode. The result at the anode was quite different in the second experiment where you should have seen no sign of a gas being evolved. Although hydroxide and sulphate ions were attracted neither of these were discharged. What *did* happen at the anode? Electrons must have been given up by something or the current would have stopped. Where did these electrons come from? If you were to repeat the experiment and run the current for about half an hour you would find no lessening in the intensity of the blue colour of the copper(II) sulphate solution, which is due to the blue hydrated copper(II) ion ($Cu^{2+}(aq)$), although copper ions were continually being discharged at the cathode. Copper ions must also have been *entering* the solution and at the same time electrons were being given up at the anode. Could the copper atoms *comprising the anode* have given up electrons and entered the solution as ions? i.e.

$$Cu(s) \rightarrow Cu^{2+}(aq) + 2e^-$$

At the cathode other copper ions were being discharged:

$$Cu^{2+}(aq) + 2e^- \rightarrow Cu(s)$$

Look carefully at the two ionic equations. What can you say about the amount of charge concerned with the solution of one atom of copper from the anode and the deposition of one atom of copper on the cathode? What do you think is the relationship between the weight of copper dissolved off the anode and the weight of copper deposited on the cathode in your last experiment?

The next experiment should show whether your answer to the last question was correct.

Experiment 6.13
†To Determine the Change in Weight of the Electrodes During the Electrolysis of Copper(II) Sulphate Solution *

Apparatus

Six volt battery or alternative d.c. supply, rheostat, ammeter, balance, connecting wire, 100 cm^3 beaker.

Two pieces of copper foil or gauze about 6 cm × 3 cm, copper(II) sulphate solution, deionized water, ethanol, propanone (acetone).

Procedure

Weigh separately the two pieces of clean, dry copper foil or gauze and record the weights. Set up the circuit as for the previous experiment but include a rheostat and an ammeter. Use the rheostat to keep the current steady at about 150 milliamps. Pass the current for about forty minutes, then remove the electrodes, carefully rinse by dipping them in deionized water, then in ethanol and finally in propanone. When completely dry, weigh the electrodes separately and record the masses.

Points for Discussion

1. What change occurred in the mass of (a) the anode and (b) the cathode? Do these results help to verify your hypothesis?

2. How does the mass lost by the anode compare with the mass gained by the cathode?

3. Does this further confirm the theory? Explain your answer.

4. Write up the experiment and the answers to the questions in your notebook as before.

Uses of Electrolysis

Electroplating

The previous experiment shows how a metal can be plated on to the cathode. This is the principle behind electroplating. Objects can be plated with a very thin, even coating of metal to make them appear more decorative, as when gold or silver is plated on to nickel, or to protect them from rust and corrosion, as when chromium is plated on to iron (Figure 6.10).

Figure 6.10 Bumpers emerging from an electroplating bath *(Wilmot Breeden)*

The object to be plated is made the cathode and the anode is usually made of the metal which is to coat the cathode. The electrolyte must be a solution of a salt of this same metal. The cathode is made of conducting material and so is usually a metal but in some cases it can be a non-conductor covered with a thin layer of powdered graphite. Real leaves can be treated in this way. They are then copper plated and thus make attractive brooches.

You may like to try some plating for yourself but make sure that the object is free from grease and rust and use a small current and a low concentration of electrolyte.

Extraction of Metals

A number of very active metals are extracted from their compounds by electrolysis. Sodium, calcium, and magnesium are obtained by the electrolysis of the fused chlorides, and aluminium from fused aluminium oxide.

Why do you think sodium and potassium cannot be extracted by the electrolysis of aqueous solutions of their salts?

Purification of Copper

As very small quantities of impurities cut down the electrical conductivity of copper to a large degree, it is important that copper, which is to be used for conducting purposes, should be extremely pure.

Slabs of impure copper are made the anodes and thin sheets of pure copper are the cathodes of a cell in which copper(II) sulphate solution is the electrolyte. On passing the current, pure copper dissolves off the impure anode and is eventually deposited in a pure state on the cathode. The impurities, including silver, form a sludge below the anode, called 'anode mud'.

Manufacture of other Elements and Compounds

Gases such as hydrogen, oxygen and chlorine are produced commercially by the large-scale electrolysis of a variety of solutions and melts.

Further details of these and other important electrolytic processes are given in Section 14.3 and will help to emphasize further the importance of electricity as a source of energy and as a very useful tool for bringing about chemical changes.

Oxidation and Reduction in Electrolysis

Oxidation (page 205) can be defined as the removal of electrons from a substance.

Reduction can be defined as the addition of electrons to a substance.

Neither reaction can occur without the other and the complete process, i.e. oxidation of one substance with consequent reduction of the other, is known as a *redox* reaction.

The reactions which occur at the electrodes during electrolysis are reactions in which electrons are removed or added and so they are redox reactions. For example, in the electrolysis of fused lead bromide (Experiment 6.4), the bromide ions give up electrons at the anode and are oxidized to bromine atoms, the lead ions gain electrons from the cathode and are reduced to lead atoms. You should be able to apply these terms to all the examples of electrolysis studied in this section.

6.2 THE ELECTROCHEMICAL SERIES

Introduction

You have already met several methods of classifying substances in chemistry; elements, mixtures, and compounds and the Periodic Classification are two examples. In this section you will be learning of another method of classification, known as the *electrochemical series*.

Experiment 6.14
†A Reaction Between a Metal and a Solution of a Salt *

Apparatus

Boiling tube, filter funnel, filter stand, filter paper, spatula, stirring rod.
Approximately 1.5 g powdered zinc and approximately 10 cm^3 0.5 M copper(II) sulphate solution.

Procedure

Stir the powdered zinc into the copper(II) sulphate solution a little at a time until the blue colour of the solution disappears. Filter the mixture and examine the residue.

Points for Discussion

1. What do you think the red-brown powder consists of?

2. Why did the copper(II) sulphate solution lose its colour?

3. What happened to the powdered zinc?

4. You probably thought this rather a trifling experiment but its significance will be more apparent after you have done Experiment 6.15.

Experiment 6.15

Can a Chemical Reaction Produce an Electric Current? *

Apparatus

100 cm^3 beaker, connecting wire, Avometer or 0–3 voltmeter.
Strips of copper and zinc foil (2 cm × 8 cm), dilute sulphuric acid.

Procedure

Set up the circuit as shown (Figure 6.11). Note the reading on the Avometer and look for any signs of chemical activity at the surfaces of the metals.

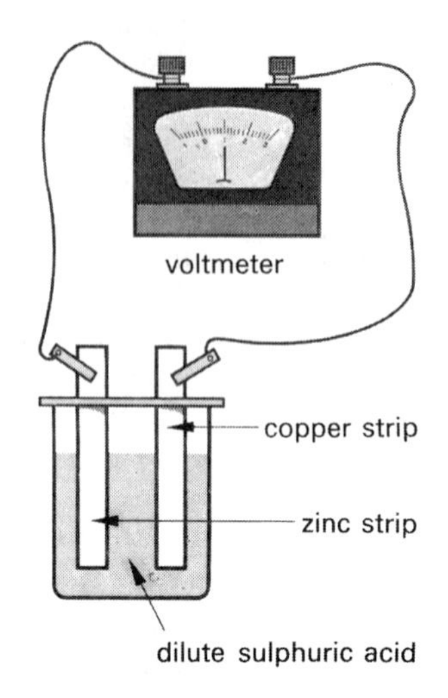

Figure 6.11 A chemical reaction produces an electric current

Points for Discussion

1. When an electric current flows there must be a source of electrons. This is normally a battery or a low voltage supply. In this experiment you had no battery or low voltage supply. Did a current flow? If so, what kind of process must have taken place?

2. The gas given off at the copper strip is hydrogen. Remember that hydrogen is discharged at the cathode so the copper strip must be a cathode, and by definition must be receiving electrons from an outside circuit. Where do you think these electrons are coming from?

What Have We Learned About the Reaction Between Copper and Zinc?

In Experiment 6.14 you probably decided that the red-brown powder was metallic copper in a finely divided state, and as the blue colour disappeared from the solution the hydrated copper ions (which are blue) were changing to copper atoms:

$$Cu^{2+}(aq) + 2e^- \rightarrow Cu(s)$$

As the powdered zinc disappeared it must have 'dissolved', i.e. changed to zinc ions:

$$Zn(s) \rightarrow Zn^{2+}(aq) + 2e^-$$

Let us now consider the net result of these two reactions:

$$Zn(s) - 2e^- \rightarrow Zn^{2+}(aq)$$
$$Cu^{2+}(aq) + 2e^- \rightarrow Cu(s)$$

If we add these two equations we find that we can write the complete reaction as:

$$Zn(s) + Cu^{2+}(aq) \rightarrow Zn^{2+}(aq) + Cu(s)$$

Electrons have been lost by the zinc atoms and gained by the copper ions.

In Experiment 6.15 (which used the same two metals) an electric current passed along the external wire. As hydrogen gas was evolved at the copper strip, the reaction

$$2H_3O^+(aq)+2e^- \rightarrow H_2(g)+2H_2O(l)$$

must have occurred, i.e. electrons must have been arriving at the copper strip to discharge the hydronium ions. These electrons must have travelled along the wire from the zinc strip, in which zinc atoms gave up electrons and entered the solution as zinc ions:

$$Zn(s)-2e^- \rightarrow Zn^{2+}(aq)$$

Thus the two reactions are similar. In both cases the zinc atoms lost electrons; in the first case these electrons were used to discharge copper ions and in the second they gave a negative charge to the copper, which then attracted and discharged the hydronium ions.

It would seem, therefore, that copper atoms will transfer electrons less easily than zinc atoms; copper might be described as a kind of 'sink' for electrons, readily accepting them from other metals which are willing to lose them. We say that copper is more electronegative than zinc, i.e. the copper has a greater tendency to hang on to its valency electrons than has the zinc atom (Section 5.5).

Experiment 6.16 (a)

†To Place the Metals in Order of Their Tendency to Transfer Electrons ***

Principle

In the next series of experiments you will try to put some of the common metals in order of their ability to lose electrons and then compare this order with the order of their chemical activity.

When we connected the copper and zinc strips by an external circuit after placing them in the dilute acid, we knew that the copper was more electronegative than the zinc and we could get some measure of this difference by the reading of the voltmeter. If we connected magnesium and copper strips in a similar manner and obtained a larger voltage, then we would know that magnesium had a bigger tendency to lose electrons than zinc had, i.e. magnesium was less electronegative than zinc.

$$Mg(s) \longrightarrow Mg^{2+}(aq)+2e^-$$

$$Zn(s) \longrightarrow Zn^{2+}(aq)+2e^-$$

In this case we have used copper as our standard or zero and compared other metals to it, but this has been quite an arbitrary choice. In the following experiment we will measure the electronegativity of the metals relative to carbon, which being a non-metal will have a tendency to gain electrons rather than lose them and so will have a higher electronegativity than any of the metals.

Apparatus

100 cm³ beaker, Avometer or 0–3 voltmeter, connecting wire, carbon electrode. Strips of magnesium, zinc, aluminium, iron, lead, copper, dilute sulphuric acid, mercury(II) chloride solution.

Procedure

Connect the carbon electrode to the positive terminal of the Avometer and stand the electrode in the beaker which has been half filled with dilute sulphuric acid. Connect the negative terminal of the Avometer to the magnesium strip. Dip the strip in the acid and record the highest voltage registered. Repeat with the other strips of metal. Record your results in a table, putting the metal which gave the highest voltage first. Note particularly the voltage reading for the aluminium strip and repeat the experiment with this strip after dipping it in mercury(II) chloride solution.

Experiment 6.16 (b)
The Particular Case of Sodium***

Apparatus
As for the previous experiment but replace the beaker of acid with a folded strip of filter paper soaked in the acid and supported on a microscope slide. Tongs, knife, filter paper.
Sodium.

Procedure
Place the carbon electrode under the filter paper and touch the top of the paper above the carbon momentarily, with a small piece of sodium, held in the crocodile clip at the end of the connecting wire (Figure 6.12). Record the highest reading on the Avometer and add this reading to the table of results made out for the previous experiment.

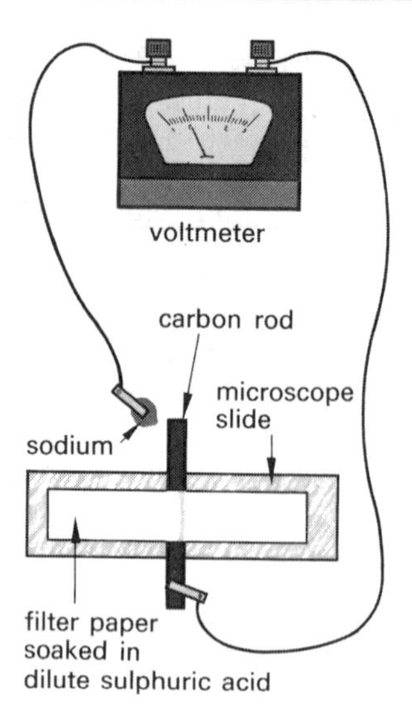

Figure 6.12 Comparing the electronegativity of sodium and carbon

Points for Discussion
1. Which metals in your list do you consider will have the greatest tendency to lose electrons, i.e. which will have the lowest electronegativity, those at the top of the table or those at the bottom?

2. Why was sodium not used in the same way as the other metals?

3. Why do you think aluminium gives a different result when amalgamated with mercury? This question will be considered later in the section as you probably do not have sufficient information to answer it at present.

Electrode Potential

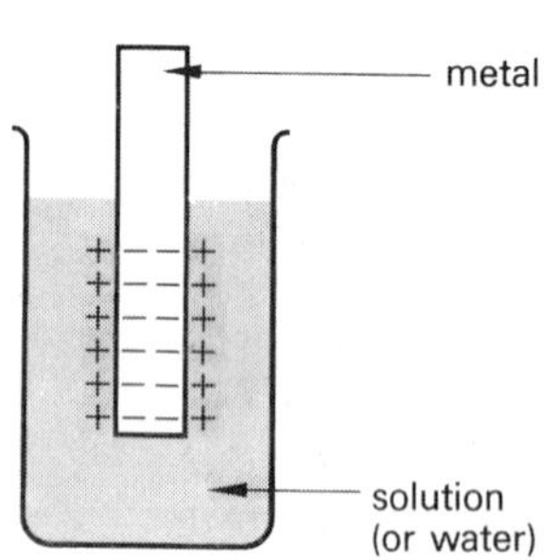

Figure 6.13 The 'double layer' of charge between a metal and a solution

When a metal is placed in water or in an ionic solution, it tends to start to 'dissolve', i.e. the surface layer of atoms change to ions which enter the solution, leaving the valency electrons behind on the metal. This residual negative charge on the metal now attracts the positive ions in the solution and a 'double layer' of opposite charges is formed (Figure 6.13). Thus a potential difference is set up between the negative charge on the metal and the positively charged ions that surround it. The greater the tendency of the metal atoms to lose electrons, the greater this potential difference will be, although the concentration of the ions in the solution will also affect the value of the electromotive force (e.m.f.) produced. For purposes of comparison molar solutions are used, and the potential difference between a metal and a molar solution of its ions is known as its *electrode potential*.

The value of the electrode potential of a single element cannot be measured directly since any completed circuit must contain two electrodes. In practice we measure the *difference* between the electrode potential of an element and a standard hydrogen electrode. This is known as a *standard electrode potential* of the element.

The values for the standard electrode potentials for the more important metals are given in Table 6.1. Notice that the metals above hydrogen have negative values for electrode potential while those below hydrogen have

Table 6.1 Standard electrode potentials of the more common metals

Element	*Standard electrode potential (V)*
Sodium	−2.71
Magnesium	−2.36
Aluminium	−1.67
Zinc	−0.76
Iron	−0.44
Lead	−0.12
Hydrogen	0.00
Copper	+0.34
Mercury	+0.78
Silver	+0.79
Gold	+1.50

positive values. This convention is used because an element such as magnesium, which ionizes readily, will be left with a higher concentration of residual electrons than hydrogen under the same conditions. Conversely an element such as copper, which ionizes less readily, may have a lower concentration of residual electrons than hydrogen and is positive with respect to hydrogen.

Points for Discussion

1. Compare the positions of the metals in the table of standard electrode potentials with their positions in the Periodic Table. Can you see any pattern emerging? Remember that the position of an element in a period, as well as its position in a vertical group, is connected with its chemical activity.

2. Can you explain, in terms of electronic structure, why magnesium should have a lower electrode potential than sodium?

3. Where, in a table of standard electrode potentials, would you expect to find (a) potassium, (b) carbon?

The Relative Chemical Activity of the Metals

We have now seen that it is possible to place the metals, hydrogen, and carbon in order of their electrode potentials and that this order is governed by the relative ability of the atoms to lose electrons. We know also that chemical reactions often involve the gain or loss of electrons, so it will be reasonable to suppose that there will be some relationship between the order of the electrode potentials and the order of the chemical activities of the elements.

In the following experiments you will compare the activities of the more common metals in several different chemical reactions.

Experiment 6.17

†To Compare the Ability of the Metals to Displace Each Other from Solutions *

Apparatus

Test-tube rack, six test-tubes (100 × 16 mm).

Clean strips of copper, aluminium, zinc, and magnesium, clean iron nails, separate solutions of copper(II), lead(II), silver, mercury(I), and magnesium nitrates, and a solution of iron(II) ammonium sulphate. (Remember that mercury salts are very poisonous.)

Procedure

(a) Before beginning the experiments prepare a table for the results. Put the names of the solutions at the head of seven vertical columns and the names of the metals at the left of five horizontal rows.

(b) Pour the solution of copper(II) nitrate into a test-tube to a depth of about 2 cm and repeat with the other solutions in the other test-tubes. Place a strip of copper in each test-tube. Record your results in the table, noting any colour change in the solutions and the formation of any precipitate or deposit on the copper.

(c) Repeat the experiment with strips of the other metals and fresh samples of the solutions of the metallic salts.

(d) From your table draw up a 'replacement series' for the metals you have used such that any metal in the list will replace all those below it from solutions of their salts.

Points for Discussion

1. How does your series compare with the one in which the metals are placed in order of their electrode potentials?

2. Consider what happens when the zinc strip is placed in the copper(II) nitrate solution. A reaction will take place only if this produces a more stable situation. The zinc metal and the copper ions can remain unchanged or, if zinc is a better donor of electrons than copper, the two will be more stable as zinc ions and copper atoms and this can be achieved by an exchange of electrons. As we have seen zinc *is* the better donor of electrons and so the exchange takes place.

The reaction can be written:

$$Zn(s) \rightarrow Zn^{2+}(aq) + 2e^-$$

$$Cu^{2+}(aq) + 2e^- \rightarrow Cu(s)$$

Adding the equations:

$$Zn(s) + Cu^{2+}(aq) \rightarrow Zn^{2+}(aq) + Cu(s)$$

Write ionic equations in a similar manner for (a) the reaction between magnesium and iron(II) ions, (b) the reaction between lead and mercury(I) nitrate solution, and (c) the reaction between zinc and lead nitrate solution.

3. These reactions are all examples of redox (page 208) reactions. Say which substance is oxidized and which is reduced in each case, giving reasons for your answers.

4. Why do the anions not appear in the ionic equations?

Reaction of the Metals with Air

This is complicated by the fact that the metals can react with other substances in the air in addition to oxygen. Even so, the order of reactivity is comparable to the order of electrode potential and of displacement, because once again it is the metals that most readily lose electrons that react most easily with oxygen. Thus sodium oxidizes instantly on exposure to air and for this reason has to be stored in a hydrocarbon oil. Iron rusts rapidly in moist air but this is a reaction in which both oxygen and water take part (Section 6.3). Copper only reacts with atmospheric oxygen on heating, as does mercury, and this latter reaction is reversible. Silver does not react with the oxygen in the air although traces of sulphur compounds in the atmosphere may cause it to blacken because of the formation of silver sulphide. Gold is not affected by the atmosphere under any conditions.

If magnesium, aluminium, and zinc occupy the same relative positions in this 'action with air' series as they have done in the series of electrode potentials, then we should expect magnesium and aluminium to be rapidly attacked by atmospheric oxygen and zinc to oxidize quite quickly. Examine pieces of these metals that have been exposed to the atmosphere for some time. Scrape the surface of each metal with a knife or rub it with abrasive paper. What difference do you notice? The metals *do* oxidize, but the oxide forms a very thin protective layer on the metal surface and this prevents any further action by the atmosphere.

The fact that aluminium is not attacked by steam (page 137) whereas the less reactive zinc is, shows that the oxide layer on aluminium is more effective in masking the metal than is the oxide layer on zinc. This would also account for the comparatively low value obtained for the electrode potential of aluminium in Experiment 6.16. How did the value for its electrode potential alter when the oxide layer was removed? Dip the piece of aluminium you have just examined in a little mercury(II) chloride solution and leave it on the bench for a short time. The white feathery growths that you see are hydrated aluminium oxide, formed by the action of atmospheric oxygen and water vapour on the newly exposed surface of the aluminium. This test reveals the true, but usually hidden, reactivity of the metal.

Reactions of the Metals with Water or Steam

Revise the experiments that you carried out in Section 5.3 on the action of water on sodium and magnesium and the action of steam on heated magnesium and aluminium. From their positions in the table of electrode potentials and what you already know of the chemical activities of these metals, try to predict how zinc, iron, lead, and copper will react with water or steam. Check your predictions by carrying out the following experiments.

Experiment 6.18

†To Determine the Reactions, if any, when Zinc, Iron, Lead, and Copper are Heated in Steam ***

Apparatus

Clamp and stand, hard glass test-tube (125 × 16 mm) fitted with a bung and delivery tube, small trough or squat beaker, Bunsen burner, asbestos square, six test-tubes in rack, spatula, teat pipette, damp glass wool.

Zinc powder, iron filings, lead shot, copper turnings.

Procedure

(a) Pack the glass wool loosely into the bottom of the test-tube to a depth of about 3 cm and using the teat pipette saturate the glass wool with water. Clamp the test-tube horizontally and introduce a spatula measure of zinc powder into the tube so that it covers an area of approximately 1 cm^2 in the middle of the tube. Set up the remainder of the apparatus as shown (Figure 6.14).

(b) Heat the zinc powder in the tube gently at first and then much more strongly. Move the flame backwards and forwards along the tube so that the water in the rocksil boils and the steam passes over the hot metal. Collect two or three test-tubes of any gas evolved.

Note: as soon as you have collected sufficient gas, and *before* removing the Bunsen, lift up the stand and the apparatus attached to it. This will bring the end of the delivery tube out of the water and the stand can then be put down further along the bench. What would probably happen if you did not take this precaution?

(c) Test any gas collected in the test-tubes with a lighted taper, and examine the residue in the hard glass test-tube carefully.

(d) Repeat the experiment using in turn, the iron filings, the lead shot, and the copper turnings.

Write up the experiment in the usual way.

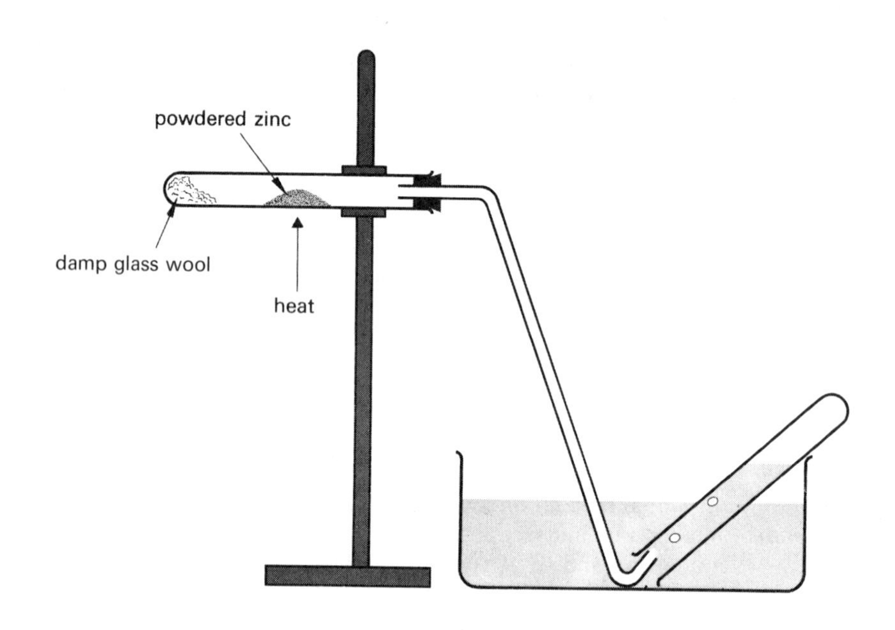

Figure 6.14 Heating zinc in steam

Points for Discussion

1. Did the metals react as you had predicted? Perhaps you expected to see a reaction with lead. Look again at the table of electrode potentials. Lead is only a little less electronegative than hydrogen so we would not expect a great deal of reaction between the two elements which are competing for the oxygen in steam. Lead does react slightly with superheated steam at high temperatures but no significant reaction takes place at the temperature you would use in the experiment.

2. The white powder remaining after the reaction with zinc is zinc oxide. You may have noticed that it was yellow when hot. Write the equation for the reaction between zinc and steam.

3. The black powder formed in the reaction with iron is tri-iron tetroxide (Fe_3O_4). The equation for this reaction is:

$$3Fe(s) + 4H_2O(g) \rightleftharpoons Fe_3O_4(s) + 4H_2(g)$$

The sign $\rightleftharpoons$ means that the reaction is reversible. Note that this is a slightly unusual reaction. Neither iron(II) oxide nor iron(III) oxide is formed although iron usually has a valency of two or three. The formation of tri-iron tetroxide in this case indicates that it is the most stable of the three oxides.

4. List the metals tested in order of their reactivity with water or steam. Note that aluminium is again 'out of order.' Explain why this is so.

Reactions of the Metals with Acids

Look at the position of hydrogen in the list of electrode potentials. It would seem reasonable to suppose that hydrogen will occupy a similar position in the displacement series that you drew up at the end of Experiment 6.17. Acids are solutions containing hydrogen (hydronium) ions and in the same way that a metal high in the list, such as zinc, will replace the ions of a metal lower in the list, such as copper, then any metal above hydrogen in the series should displace hydrogen ions from solution, i.e. hydrogen gas from acids. Using this information you should now be able to predict the reactions, if any, of the metals with hydrochloric and sulphuric acids. Check your predictions with the results of the following experiments.

Experiment 6.19
To Determine the Reactions of the Metals with Dilute Hydrochloric and Sulphuric Acids *

Apparatus
Rack of test-tubes.
Strips of clean magnesium, aluminium, zinc, iron (nails), lead, and copper foil, hydrochloric and sulphuric acids (dilute).

Procedure
Pour hydrochloric acid into each test-tube, six in all, to a depth of about 6 cm and add a different strip of metal to each tube. Note and compare the rates at which the hydrogen is evolved. Repeat the experiment using sulphuric acid instead of hydrochloric acid. Record your results in a table.

Points for Discussion

1. Did you predict the results of the experiment correctly? You should be able to explain why there was very little reaction with aluminium and none with lead.

2. Why do we not use sodium in this experiment?

3. We do not use the other common bench acid, nitric acid, as it is an oxidizing agent as well as an acid and a simple displacement reaction occurs only in the case of magnesium and the very dilute acid.

The Electrochemical Series

You have now done sufficient experimental work with the more common metals to see that the order of their electrode potentials is the same as the order of their chemical activity. (Calcium is a notable exception here.) This is true also for all the other lesser known metals. This arrangement of the metals, and hydrogen, with regard to both electrical character and chemical behaviour is known appropriately as the *electrochemical series*. It is an extremely useful concept in that it gives an easily memorized overall pattern of the comparative reactivity of the metals and their ions.

History is Chemistry in Reverse

It is usual to write the electrochemical series starting with sodium as the most active and least electronegative of the common metals, and ending with gold as the least active and most electronegative. If we reverse this list we find that the metals are now in the order in which they have been discovered and used by mankind. The reason for this is that the more chemically active the metal, the stronger are the bonds that it forms with other elements in nature, and the more difficult it is to extract it from its compounds. Gold and silver are so unreactive that they are found 'native', that is uncombined, and were known and used by the earliest civilizations. Men of the Bronze Age were able to prepare a crude form of the alloy by

Table 6.2 The electrochemical series

Element	*Electrode potential*	*Effect of air (oxygen)*	*Action with water*	*Action with dilute hydrochloric acid*	*Method of extraction*
Sodium	−2.71		Hydrogen evolved in the cold		
Magnesium	−2.38				
Aluminium	−1.67				
Zinc	−0.76				Roasting to the oxide, then reduction with carbon or carbon monoxide
Iron	−0.44				
Lead	−0.12				
Hydrogen	0.0				
Copper	+0.34				
Mercury	+0.78				
Silver	+0.79	No action			
Gold	+1.50				

roasting ores that contained both copper and tin. The Iron Age came later because the iron could only be smelted when some type of draught or air current was used in the wood or charcoal furnace. By the time of the Roman occupation of Britain, gold, silver, mercury, bronze, iron, lead, and brass (a copper–zinc alloy) were known and used, but it was not until many hundreds of years later, when new sources of energy became available, that the last group of metals, the active ones at the top of the electrochemical series, were discovered. Sodium, for example, had never been seen until 1807, when Humphrey Davy carried out his now famous experiment in which he electrolysed fused sodium hydroxide.

Extraction of the Metals

Where gold and silver are found native, no *chemical* reaction is needed for their extraction; only crushing and washing are necessary to obtain the metals in a comparatively pure state. An impure form of copper is made by roasting the ore in a stream of oxygen. The metal is then purified by electrolysis (Section 6.1).

Lead, iron, and zinc are obtained by reducing the heated oxides with carbon or carbon monoxide.

Aluminium, magnesium, and sodium compounds need so much energy for their decomposition that this can only be brought about by electrolysis. Aluminium is obtained by the electrolysis of the fused oxide and magnesium and sodium by electrolysing their fused chlorides.

Thus the method of extracting a metal is related to its position in the electrochemical series, the energy needed becoming less and the ease of extraction becoming greater the lower the position of the metal in the series.

Summary of the Electrochemical Series

This is conveniently expressed in the form of a table. Copy the table (Table 6.2) into your notebook and fill in the blanks from your own observations and results.

6.3 VOLTAIC CELLS AND RUSTING

The Simple Cell

You have seen (Experiment 6.15) that if a strip of copper and a strip of zinc are placed in dilute sulphuric acid and the strips are connected by an external wire, an electric current flows along the wire and its presence can be detected by a sensitive voltmeter. At the same time hydrogen gas is evolved at the copper strip and the zinc strip dissolves. You have learnt also that this chain of events occurs because zinc is less electronegative than copper and loses its valency electrons more easily. These electrons pass along the external wire from the zinc to the copper, giving the copper a negative charge so that it attracts the positively charged hydronium ions from the acid solution. These are discharged and hydrogen gas is evolved. Thus an electric current is produced as a result of chemical action.

An arrangement such as this, where the energy liberated from a chemical reaction is obtained in the form of an electric current, is known as a *voltaic cell*. In the reverse process (electrolysis) electrical energy is used to bring about chemical changes in an electrolytic cell.

The simplest type of voltaic cell consists of two different metals in contact with an electrolyte and the zinc–copper–sulphuric acid combination is a typical example of a simple cell (Figure 6.11).

Such a cell is, however, of little practical use because as soon as the circuit is completed, the current begins to drop rapidly. You probably noticed this happening during Experiment 6.15. The main reason for this falling off of the current is the production of hydrogen gas at the copper surface. The processes involved in the evolution of hydrogen gas at a metal electrode are not fully understood, but it is known that the hydrogen *atoms* first discharged form a strongly *ad*sorbed layer on the surface of the copper, slowing down the action and needing extra energy to remove them. This extra energy, which has to be taken from the total electrical energy of the cell, is known as *overvoltage*.

Contributory factors to the falling off of current in a simple cell are the changes in concentration in the vicinity of the electrodes—an increase in zinc ions in the neighbourhood of the zinc strip and a decrease in hydronium ions near the copper strip. Various devices have been used in other types of voltaic cell to overcome the defects of the simple cell. The Daniell cell and other cells are considered in the *Supplementary Text*.

Experiment 6.20

†To Find the Conditions Needed for Iron to Rust *

Apparatus

Four dry test-tubes (two with rubber bungs), labels, test-tube rack, test-tube holder, Bunsen burner, asbestos square. Iron nails, calcium chloride (anhydrous), glass wool, paraffin wax.

Procedure

Label the test-tubes A, B, C, and D.

Tube A: place a nail in the tube and leave uncovered.

Tube B: wet the inside of the tube and leave the nail uncovered.

Tube C: place the nail in the tube, put in a plug of glass wool and then add anhydrous calcium chloride on top of the plug to a depth of about 2 cm.

Tube D: place the nail in the tube, add sufficient water to cover the nail and heat until the water has boiled for several minutes. Cover the surface of the water with melted paraffin wax and close the tube with a rubber bung.

Leave the tubes in a rack undisturbed for a week, then remove the nails and examine them. If there is no change in the nail in tube A, leave it for a further week.

Points for Discussion

1. Moisture was excluded from one of the test-tubes. Which was this? What prevented moisture reaching the nail?

2. Why was the water boiled in test-tube D? What was the function of the paraffin wax?

3. What conditions do you consider are necessary for rusting to take place?

4. Rusting is brought about by the action of voltaic cells. What do you think the electrolyte is? What could be the anodic and cathodic areas?

Rust Prevention

All methods of preventing rusting involve the exclusion of both air and water. Painting, covering with oil or grease, chromium or nickel plating, enamelling, tinning, and galvanizing are among the best known methods adopted. The use of metal coatings is effective until the metal film is broken. The extent of the corrosion which then occurs depends on the metal that has been used.

Tin Plating

Tin plating is used for canning food and is made by immersing sheet steel in molten tin. As long as the coating is undamaged, no rusting takes place, but once the protective film is punctured, rusting occurs more quickly than it would if there were no tin present. Tin, being more electronegative (i.e. below iron in the electrochemical series) becomes cathodic with respect to iron. Iron, which is the anode of the cell, rapidly dissolves and rusting occurs.

Galvanizing

Here the iron or steel is covered with a thin layer of zinc. The zinc itself is protected by a natural layer of oxide, and so forms an effective barrier against rusting. If the surface is damaged so that the iron is exposed to the atmosphere, a cell is set up as in the other examples, but in this case, as zinc is higher than iron in the electrochemical series, it becomes anodic with respect to iron and thus the zinc dissolves but rusting of the iron does not take place. The effectiveness of this method of protection is increased by the fact that the zinc ions react with hydroxyl ions from the water (cf. the similar reaction with iron(III) ions) to precipitate zinc hydroxide, which blocks up any cracks in the zinc surface thus stopping the corrosion.

Sacrificial Anodes In the two examples of rust prevention that we have considered in detail, tin plating and galvanizing, the metal forming the anode of the cell corrodes away while that forming the cathode is protected. This is used in some cases to protect iron and steel from corrosion where more usual methods of rust protection are not possible. For example, a ship's hull will sometimes have blocks of zinc or magnesium bolted to it. As magnesium (or zinc) is higher in the electrochemical series than iron, it becomes anodic with respect to iron, and dissolves as cells are created. The iron is thus prevented from rusting by the 'sacrifice' of the magnesium. The iron hull remains unchanged and the magnesium can be easily replaced. Iron piers and underground steel pipe lines can be treated similarly. The process is known as 'cathodic protection' and the magnesium forms a 'sacrificial anode'.

Points for Discussion

1. Explain why corrosion often takes place in water pipes where a steel pipe is connected to a copper pipe.

2. Ornamental gates are often made of wrought iron which is a relatively pure form of iron. Why is rusting not such a problem in this case as it would be if the gates were made of steel?

3. Instead of using a sacrificial anode, a modern method of preventing rusting of large steel structures is to pump electrons into the steel by means of a low voltage direct current. Explain how this will prevent corrosion.

4. The chemistry of rusting is considered in *Further Topics*.

6.4 FARADAY'S LAWS

Michael Faraday (1791–1867) was a brilliant scientist. He invented the dynamo and did a great deal of pioneer work on the chemical effects of electricity. To him we owe the terms anode, cathode, and ion (Greek: ana, 'up'; kata, 'down'; odos, 'path'; ion, 'a traveller'). His most important contribution in the field of electrolysis was his discovery of the relationship between the amount of electricity used and the quantity of chemical change that it produced.

In the experiments that you will do in this section, you will be repeating some of Faraday's work and you may be able to formulate, from your own results, the laws of electrolysis which he first worked out in 1833.

Some Electrical Units

Before you can understand Faraday's work and the results he achieved, you must know the meaning of several of the units used. You are probably already familiar with the ampere as the unit of current strength and the volt which is a measure of potential difference. Quantity of electricity is measured in *coulombs* and this is found by multiplying the current in amperes by the time in seconds that the current flows. Thus if a current of two amperes flows for five minutes, the quantity of electricity used is $(2 \times 5 \times 60)$ coulombs, i.e. 600 coulombs.

Calculate the quantity of electricity produced by

(a) a current of one-half an ampere flowing for ten minutes;
(b) a current of ten amperes flowing for two hours.

Finding the Relationship between the Quantity of Electricity used and the Mass of the Product

A solution of copper(II) sulphate is electrolysed using copper electrodes. The copper cathode is weighed before and after the electrolysis. The quantity of electricity is found from the current strength and the time. The experiment is repeated many times using different quantities of electricity. The experiment is too lengthy and exacting to be carried out in a school laboratory.

The following is a typical set of results:

Current (A)	*Time (s)*	*Quantity of electricity (C)*	*Mass of copper (g)*
0.5	600	300	0.099
1.0	600	600	0.198
1.5	600	900	0.297
2.0	600	1200	0.396
2.5	600	1500	0.495
2.0	900	1800	0.594
2.0	1000	2000	0.660
2.0	1200	2400	0.792

Draw a graph of these results, plotting quantity of electricity along the x axis and mass of product along the y axis.

Points for Discussion

1. What can you conclude from the shape of your graph about the relationship between the quantity of electricity used and the mass of a substance produced during electrolysis?

2. Faraday carried out many similar experiments in which he measured the masses of the products of electrolysis and the quantity of electricity used. From his results Faraday formulated his *First Law of Electrolysis:* 'The mass of a substance dissolved off, or produced at, an electrode during electrolysis, is proportional to the quantity of electricity which passes through the electrolyte.'

The Relationship Between the Masses of Different Substances Released by the Same Quantity of Electricity

The discovery of this relationship was another of Faraday's brilliant pieces of work and is embodied in *Faraday's Second Law*: 'The masses of different elements released by the same quantity of electricity, form simple whole number ratios when divided by their atomic masses'.

We can illustrate this more clearly by giving you some results of experiments with a number of different voltameters. For example, when the same current is passed for the same length of time through a number of different voltameters, the masses of the elements released are as follows (Figure 6.15):

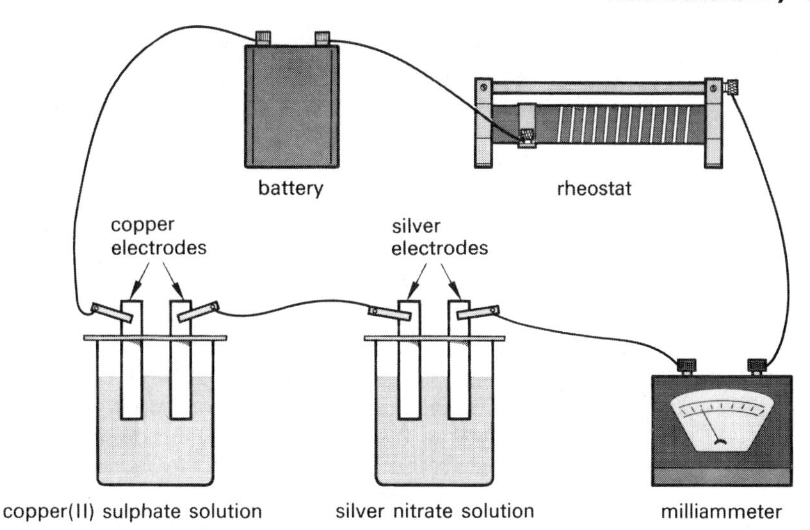

Figure 6.15 Comparing the masses of silver and copper produced by the same quantity of electricity

Silver	*Magnesium*	*Hydrogen*	*Lead*	*Aluminium*
0.535 g	0.060 g	0.005 g	0.515 g	0.045 g

Dividing the mass of each metal by its atomic mass:

$\frac{0.535}{108}$	$\frac{0.060}{24}$	$\frac{0.005}{1}$	$\frac{0.515}{207}$	$\frac{0.045}{27}$
= 0.005	= 0.0025	= 0.005	= 0.0025	= 0.0017

An interesting pattern has now emerged from this division, for the number obtained in the case of silver is the same as for hydrogen, twice that obtained for magnesium and lead, and three times that obtained for aluminium.

Faraday's Laws and the Electronic Structure of the Atom

The *smallest* quantity of electricity that will release one mole of atoms (discharge one gramme-ion) of any element is 96 500 coulombs. This quantity of electricity is termed the *faraday*. One faraday will discharge one mole of silver ions, one mole of sodium ions, one mole of hydrogen ions, but two faradays are needed to discharge one mole of magnesium or calcium ions, and three faradays are needed for one mole of aluminium ions.

As a faraday will discharge a mole of silver ions and as each electron discharges one silver ion it follows that a mole of silver ions will be discharged by a mole of electrons. A faraday can thus be considered as a *mole of electrons*.

Each silver ion carries one positive charge (Ag^+) and one faraday will discharge a mole of silver ions. As two faradays are needed to discharge *a* mole of magnesium ions (i.e. the same number of ions), each magnesium ion must need two electrons for its discharge and so the magnesium ion can be written Mg^{2+}. How would you write an aluminium ion if three faradays are needed to deposit a mole of aluminium atoms?

These facts, derived from Faraday's Laws, can be used to calculate the masses of substances taking part in, and produced by, electrochemical reactions.

Example 6.1

The same current is passed through two voltameters containing silver nitrate solution and dilute sulphuric acid respectively. If 0.108 g of silver is deposited on the silver cathode what mass of hydrogen gas will be liberated?

There are 0.108/108 moles of atoms of silver in 0.108 g of silver.
One mole of atoms of silver needs one faraday
0.108/108 moles of atoms of silver need $1 \times 0.108/108$ faradays
One faraday releases one mole of atoms of hydrogen
$\therefore$ 0.108/108 faradays will release 0.108/108 moles of atoms of hydrogen

= 0.001 moles of atoms of hydrogen

= 0.001 g of hydrogen

Example 6.2

Calculate the mass of magnesium that will be produced during the electrolysis of fused magnesium chloride, if a current of 1.93 A is passed for sixteen minutes forty seconds (1000 seconds).

Quantity of electricity used = (1.93×1000) coulombs
= 1930 coulombs
= 1930/96 500 faradays
= 0.02 faradays

As the magnesium ion is Mg^{2+}, two faradays will be needed for one mole of atoms.

Two faradays produce 24 g of magnesium
0.02 faradays produce *0.24 g of magnesium*

Points for Discussion

1. How many faradays would be required to release a mole of hydrogen molecules from a solution containing hydronium ions? (Remember that the hydrogen molecule is H_2.)

2. In the manufacture of aluminium (Section 14.3) a current of 30000 A was passed through the cell for 24 hours. What mass of aluminium was produced? Give your answer in kilogrammes and take the faraday to be 96 000 coulombs. On the basis of your figures can you suggest why many aluminium works are sited in mountainous country?

3. In an electrolysis experiment the same current was passed for the same time through two cells in series. In the first voltameter 0.081 g of copper were produced and in the second, 0.162 g of copper were formed. If the symbol for the copper ion in the first case is Cu^{2+} what is the symbol for the ion of copper in the second case?

4. A current of 0.36 A produces 0.23 g of lead in ten minutes. Calculate (a) the quantity of electricity used, (b) the number of faradays needed to produce one mole of lead atoms.

Scientific Method

Earlier in the book (Chapter 1), you learnt about the structure of the atom, and much of the information you were given we asked you to take on trust, using it as a working hypothesis by which the chemical behaviour of various substances could be explained. Now, at the present stage of your knowledge and experience, some of the facts which you had to accept blindly can be explained and understood. This work on electrolysis is an excellent example of how the use of experimental facts provides evidence for earlier assumptions and in this way knowledge is expanded and increased.

QUESTIONS CHAPTER 6

1. An electric current was passed through an unknown solution. The gases which were evolved were collected and tested. The gas from the anode bleached damp litmus paper and the gas from the cathode burned with a squeaky pop. The solution was probably that of

A sodium hydroxide
B hydrochloric acid
C nitric acid
D sulphuric acid
E copper(II) sulphate

2. Which of the following statements is the best definition of a cathode?

A It is the negatively charged electrode.
B It is the electrode at which electrons enter the electrolyte.
C It is the positively charged electrode.
D It is the electrode at which hydrogen is evolved.
E It is the electrode at which oxygen is evolved.

3. An electric current is passed through a solution of copper(II) sulphate, using platinum electrodes. The substance liberated at the anode is

A copper　C oxygen　E sulphate
B sulphur　D hydrogen

4. An electric current of 0.1 A is passed through an electrolyte for 1.25 hours. The quantity of electricity used (in coulombs) is

A 0.125　C 125　E 4500
B 0.1　D 450

5. Say whether the following are true or false.

(a) Copper will replace zinc from a solution of zinc nitrate.
(b) Iron will replace copper from a solution of copper(II) sulphate.
(c) Aluminium will replace magnesium from a solution of magnesium chloride.
(d) Iron will liberate hydrogen from dilute hydrochloric acid.
(e) Copper will liberate hydrogen from dilute sulphuric acid.

6. Explain why an electric current passes along a wire joining a zinc rod to a copper rod with both metals dipping into a solution of sulphuric acid.

7. Describe how you would set up a cell to produce electricity from a chemical reaction and explain the working of the cell.

8. The following is a list of standard electrode potentials:

Magnesium	−2.36
Zinc	−0.76
Copper	+0.34
Silver	+0.79

(a) Which two metals, if used together in a cell, would produce the largest e.m.f.?
(b) Explain the meaning of the positive and negative signs.
(c) Which is (i) the most electronegative, (ii) the most reactive, metal on the list?

9. How many faradays are needed to produce:

(a) 2.70 g of aluminium
(b) 6.0 g of magnesium
(c) 10 g of hydrogen
(d) 71 g of chlorine

10. Write ionic equations for the following:

(a) the reaction between zinc and copper(II) sulphate solution;
(b) the reaction between magnesium and dilute hydrochloric acid;
(c) the reaction at the cathode when copper is deposited;
(d) the reaction at the anode when chlorine ions are discharged.

From these reactions, select two substances which have been oxidized and two which have been reduced, giving reasons for your answers.

Give one example of a 'redox reaction' and explain what it means.

11. Explain clearly why, when copper(II) sulphate solution is electrolysed using platinum electrodes, oxygen is the product at the anode, but a different result is obtained when copper electrodes are used.

How is electrolysis used for the purification of crude copper? Describe two other commercial uses for electrolysis.

12. Mercury and bromine are both liquids at room temperature. Explain why only one conducts an electric current. (A. E. B.)

13. Explain why metals in the solid state readily conduct electricity but a salt, such as sodium chloride, needs to be melted or dissolved in water in order to do so, whilst sulphur is non-conducting under all conditions. (W. J. E. C.)

14. Describe two experiments (one in each case) which you could use to demonstrate that zinc is above copper but below magnesium in the electrochemical series.

15. Copy and complete the following paragraph by adding *one* word in each space.

Solid sodium chloride will not conduct electricity. This is because the   in its crystals are not free to move. When the crystals are   strongly, they   and a current will then flow when a voltage is applied. If carbon electrodes are used,   is released at the cathode and   at the anode. The process is called

(J. M. B.)

The following questions require rather longer answers.

16. Describe in detail, three experiments you could carry out to place lead, magnesium and zinc in their correct order in the electrochemical series.

Explain why the tests you choose might not place aluminium correctly in the series.

17. The following is a list of symbols of some of the elements in order of an 'activity series': K, Mg, Al, Zn, Fe, H, Cu, Ag.

(a) Which of these elements will not displace hydrogen from a dilute acid?
(b) Which of these elements has the most stable hydroxide?
(c) A piece of zinc is placed in iron(II) sulphate solution and a piece of iron is placed in zinc sulphate solution. In which solution would there be a reaction and why? Give the equation for the reaction.
(d) From these elements name (i) a metal which reacts with cold water, (ii) a different metal which reacts with hot water but only very slowly with cold, (iii) any other metals which will react when heated in a current of steam.
(e) Name any metals in the list whose heated oxides can be reduced by hydrogen. For one of these metals give an equation for the reaction.
(f) If mixtures of aluminium oxide and iron, and of iron oxide and aluminium, are heated, in which mixture is there a reaction and why? Give the equation for the reaction.

Outline any one experiment by which you could prepare dry crystals of zinc sulphate from a different zinc compound. (A. E. B.)

18. Draw a diagram of an apparatus suitable for the electrolysis of copper(II) sulphate solution using platinum electrodes and for the collection of the products. Give the names and polarities of the electrodes, the names of the products, and the equations for the electrode reactions.

After passing the current for, say ten minutes, what would be the effect of reversing the current and passing it in the opposite direction for about twenty minutes?

In the example given above, electricity is used to bring about a chemical change. Describe a way in which a chemical change can be used to release electrical energy. Carefully indicate the reactions which occur and where they take place.

(J. M. B.)

19. (a) Design and describe a quantitative experiment to find out whether a new alloy is oxidized by the atmosphere.

What results would you expect to observe if the new alloy were (i) easily oxidized, (ii) rust resistant?

(b) If the alloy were attacked, what further experiments would you set up to discover which parts of the air caused the reaction?

(c) How is air prevented from attacking iron (i) on the blades of a lawn mower; (ii) on a dust bin, (iii) on cutlery? (J. M. B.)

20. Two voltameters were connected in series with a battery. The first voltameter had platinum electrodes in dilute sulphuric acid and the second had copper electrodes in copper(II) sulphate solution. After one hour the cathode of the second voltameter had increased its mass by 0.3175 g. Give ionic equations for the reaction at

(a) the cathode of the second voltameter,
(b) the anode of the first voltameter.

Calculate

(c) the mass of the gas evolved at the anode of the first voltameter,
(d) the volume occupied by this gas at s.t.p.

(J. M. B.)

21. (a) Give a detailed description of the preparation of a small sample of lead by electrolysis of a fused compound.

(b) Give an outline of the *industrial* preparation of a metal by electrolysis.

(c) A current of 0.5 A, flowing for six minutes twenty-six seconds, through two cells in series, was found to deposit 0.216 g of silver on the cathode of the first cell and 0.059 g of nickel on the cathode of the second. The atomic masses of silver and nickel are 108 and 59 respectively. Calculate

(i) the quantity of electricity passed through the two cells,
(ii) the quantity of electricity needed to deposit one gramme-atom of silver,
(iii) the quantity of electricity needed to deposit one gramme-atom of nickel.

Comment on the results of your calculations in (ii) and (iii). (C.)

7 Three-dimensional Chemistry

7.1 SOME BASIC IDEAS

Introduction

The Solid, Liquid and Gas States

All substances are built up from atoms or ions, and atoms are often joined together to form molecules, but we have not yet considered how these atoms, ions or molecules are arranged. Are they packed together in a random manner or in an orderly pattern? Does their arrangement affect the physical or chemical properties of a substance? Does an understanding of structure facilitate the synthesis of materials to be used for specific tasks? In this chapter we will try to show that an appreciation of structural principles is essential in helping us to understand the behaviour of both natural and synthetic materials.

Structure has already been discussed in a very simple way in Section 1.3 where the solid, liquid, and gas states were considered in terms of the kinetic theory. The particles in a solid or liquid are closer together than those in gases, and this idea helps us to explain why solids and liquids have different physical properties from those of gases, why liquefaction of gases can be induced by cooling and/or increased pressure, why solids are virtually incompressible, and so on. As the particles in a gas are relatively far apart and move at high speeds, gases do not really have a 'structure' or packing arrangement. This explains why most gases have very similar physical properties, e.g. high rates of diffusion, no shape of their own, always fill their container, and are poor conductors of heat and electricity. To a more limited extent this is true of liquids. The closer packing of the particles in solids explains why solids are different from liquids and gases, but does not explain why many solids are so different from each other. Steel and rubber are both solids, but steel is tough and strong, whereas rubber is soft and elastic. Diamond and graphite are two different solid forms of the *same* element, carbon, and thus consist of carbon atoms only. Why, then, have they such different properties (see Table 7.2)? It seems reasonable to suggest that the properties of any material depend to some extent upon the way in which its atoms, ions or molecules are packed together. The particles in solids are very close together and only *vibrate* about their mean positions. The particles always maintain these same positions in the structure, and varied packing arrangements are theoretically possible. This could account for the wide range of physical properties exhibited by solids. Can you think of two *gases* which show such a marked difference in physical properties as is shown by the two solids steel and rubber? You are probably unable to do so, and the reason could be that, while all gases are alike in having widely spaced rapidly moving particles in no fixed pattern, different solids consist of closely packed particles arranged in a wide variety of different patterns.

Giant (Macro) and Molecular Structures

In Chapter 5 structural ideas were taken a stage further than the solid–liquid–gas concept and solids were divided into giant (macro) and molecular structures. We stated that giant structures tend to have higher melting points, boiling points, heats of fusion, and heats of vaporization than those of molecular structures, and that in general molecular structures of elements (and their compounds) are to be found on the right-hand side of the Periodic Table. This helps to explain further why solids differ in their physical properties but it does not go far enough. We have not yet explained why diamond and graphite are so different, for not only do they each consist of the same kind of atom, but they are also both giant structures. We will continue to develop these ideas in the remainder of this chapter, but so far we have made a hypothesis that differences in physical behaviour could result from varying packing arrangements, and we should first attempt to verify our hypothesis.

Are Particles in Solids Packed Together in Different Ways?

Points for Discussion

1. What do you know about the number of particles in a mole of *any* substance?

2. If equal numbers of particles are considered each time, which would you expect to take up the greater volume, (i) particles with small atomic masses, or (ii) those with higher atomic masses?

3. Would you expect the volume of a fixed number of particles to

A steadily increase as the masses of the individual particles increase,

B steadily decrease as the masses of the individual particles increase,

C remain fixed as the masses of the individual particles increase?

4. Plot a graph of atomic mass (x axis) against the volume of a mole of atoms (y axis) using the data given for some solid elements in Table 7.1. Study the graph and look again at questions 1, 2 and 3. Do you think that elements have their particles packed together in different ways? Look in particular at your graph where two different forms of the same element occur, i.e. sulphur, carbon, and phosphorus.

Table 7.1 The atomic mass and atomic volume of some solid elements

Element	*Atomic mass*	*Volume of a mole of atoms at room temperature (cm^3)*
Lithium	7	13.0
Beryllium	9	5.0
Boron	11	4.3
Carbon (graphite)	12	5.4
Carbon (diamond)	12	3.4
Sodium	23	23.8
Magnesium	24	14.0
Aluminium	27	10.0
Silicon	28	11.5
White phosphorus	31	17.0
Red phosphorus	31	14.0
Rhombic sulphur	32	15.5
Monoclinic sulphur	32	16.2
Potassium	39	45
Calcium	40	26
Scandium	45	14.7
Chromium	52	7.2
Manganese	55	7.4
Iron	56	7.1
Cobalt	59	6.6
Copper	63	7.0

In considering the two forms of any of these elements we are comparing the two volumes occupied by the *same number* of the *same kind* of atom. The volumes are so different that the two forms of the same element must have their particles arranged in dissimilar ways. If compounds are considered instead of elements the same differences apply, so it must be true that the particles in solids are built up in varying ways. However, this is still not a complete answer. These variations could be due to the fact that the particles in the various solids are packed in a random manner but are simply closer together in some materials than in others. Alternatively, it could be that the particles are always arranged in *ordered* patterns which vary from one substance to another. Indeed variations could arise from some combination of these two ideas. The next thing to decide is whether or not atoms, ions, and molecules are arranged in orderly patterns.

Are Particles in Solids Arranged in Symmetrical Patterns?

In a school laboratory we cannot hope to 'see' how solids are constructed as the individual particles are so small, but some useful ideas emerge from a consideration of crystals. You can probably recognize whether or not a given solid is crystalline, but can you explain clearly which factors enable you to make such a decision?

Experiment 7.1
†Watching Crystals Grow **

Apparatus

Overhead projector with slide attachment (or slide projector), Petri dishes or microscope slides, screen, teat pipettes. A rack of test-tubes containing hot concentrated solutions of appropriate reagents such that when a few drops of each hot solution are placed on a Petri dish or microscope slide, crystals form readily but not immediately. If enough microscopes or hand lenses are available they can be used instead of a projector and the work can be done individually. Wires and crocodile clips, low voltage (6 V) supply, scissors.

Two test-tubes half filled with approximately 0.1M lead ethanoate (acetate) and silver nitrate solutions respectively, zinc foil, copper foil. Two pieces of tin foil or granulated tin, dilute tin(II) chloride solution.

Procedure

(a) Place a few drops of each hot solution on to Petri dishes or microscope slides and examine the growth of the crystals by projection on to a screen, or under a microscope or hand lens.

(b) Carefully place a clean strip of zinc metal in the lead ethanoate solution and leave the test-tube undisturbed until suitable crystals have formed. Repeat using the silver nitrate solution and a strip of copper foil.

(c) Pour some tin(II) chloride solution into a Petri dish. Using the two pieces of tin foil as anode and cathode, connect them to a low voltage supply. Dip the electrodes into the solution and keep them well apart.

Points for Discussion

1. Crystals formed in (a) because a hot solvent normally dissolves more solid solute than a cold solvent, so when a hot concentrated solution is cooled, solid crystallizes out. Name the crystals formed

Figure 7.1 A crystal of sodium chloride showing its regular shape. The front edge of the crystal is about 1 cm long

Figure 7.2 Quartz crystals

in (b) and (c). How did these crystals form? A crystal is a solid form which has plane surfaces separated by constant angles (Figures 7.1 and 7.2).

2. You have seen a variety of crystals growing from solutions. It seems as though layer after layer of particles, each invisible under the most powerful light microscope, have been plated over a simple geometrical framework so that the same shape is maintained and the crystals eventually become visible to the naked eye. Imagine the same process in reverse so that the layers would be peeled off one by one. When would this process have to stop? What do you think was the basic structure which started the growth and shape of the crystals at the very beginning? One of the early crystallographers, Abbé Haüy, once pondered over a problem of this kind. A friend showed Haüy (then a botanist) a particularly good specimen of a calcite crystal, and he accidentally dropped it. One of the larger crystals broke off and Haüy was allowed to keep it. He was puzzled by the fact that the broken surface was perfectly flat, and he broke the crystal further to see whether other similar faces could be formed.

He found three directions along which he could break the crystal to obtain a bright flat surface. Easy planes of parting in a crystal are called *cleavage planes* and you may be allowed to cleave some crystals. Haüy thought that he could continue to break each tiny crystal until he would eventually obtain the basic unit from which the whole crystal was constructed. You know enough chemistry to realize that he could never do this, for the basic units are too small, but at least he had the idea of a crystal as an array of simple repeating units and it is said that the excitement caused by this simple episode resulted in Haüy deciding to become a crystallographer rather than continuing as a botanist.

3. Do the particles in solids form orderly or random packing arrangements?

Packing in Solids. A Summary so far

For a crystal to grow from a solution it seems reasonable to assume that a few particles of a substance in solution are linked in a three-dimensional structure and that successive layers of similar particles are formed round this base unit; eventually the crystal grows large enough to become visible. In other words, external order reflects internal order; the shape of the crystal is the same as that of the simplest unit at its centre. As a large range of solids form crystals [from metals to organic compounds such as ethanamide (acetamide)], it would appear that many solids have strictly ordered internal structures.

Thus solids have their particles packed together in different, but always orderly, ways. Some of these arrangements are very complicated but modern methods of structure determination, such as x-ray diffraction, coupled with the use of computers to interpret experimental data, have confirmed the ideas of crystal growth and revealed the secrets of their internal packing. Unfortunately, such methods cannot be used in a school laboratory.

The secrets of the internal packing are revealed

We can now go on to consider various kinds of structure in more detail and we will attempt to account for some of the varying physical properties of solids by showing that their structures at the atomic level are mainly responsible for the bulk properties of the substance. Only relatively simple arrangements will be considered, but an understanding of structural principles has helped to explain how DNA and other complicated biological materials function, why plastics have such useful and interesting properties, and has led to the use of graphite in re-entry cones for space modules, and initiated the manufacture of carbon fibre with its tremendous potential. These are only a few examples, and scientists are always developing new materials. Structural arrangements in substances can be broadly classified as shown in Figure 7.3.

The concepts introduced in this chapter are easier to understand if appropriate models are available for you to examine. All the particles in a giant structure are joined together by strong bonds into a large three-dimensional network in which no free, individual units exist. Molecular structures contain individual, relatively separate units called molecules. The atoms within a molecule are very strongly bonded together but each molecule is joined to those around it by comparatively weak forces (page 36). In a giant structure millions of atoms or ions are arranged in a large three-dimensional pattern. Although a giant structure does not contain separate units, it is built up in an ordered manner having a regular repetition of atoms or ions. Thus for theoretical purposes we can *consider* it as being made up of small units, having exactly the same characteristics as the large structure, but being millions of times smaller. Remember, though, that these repetitive 'units' do not exist by themselves.

By contrast, the outward appearance of a molecular structure may bear no resemblance to the shape of its individual molecules, which are relatively free and may have a different shape of their own. These points will be better understood when a giant structure and a molecular structure have been studied.

Experiment 7.2
To Determine Whether a Given Solid has a Molecular or a Giant Structure *

Apparatus

Three dry test-tubes in a rack, Bunsen burner, asbestos square, tongs, spatula, hand lens or low power microscope. Supplies of graphite, iodine, and potassium chloride.

Procedure

(a) Examine a small sample of each of the solids under a hand lens or microscope.

(b) Place a crystal of iodine in a dry test-tube and, holding it with the tongs, heat gently in a Bunsen flame.

(c) Repeat (b) using separate small quantities of (i) graphite and (ii) potassium chloride instead of iodine.

Points for Discussion

1. It should be obvious from (a) that all three samples are crystalline. One has a molecular structure, one a giant ionic structure, and the other a giant atomic structure.

2. Which one has the molecular structure?

3. Why did it behave so differently on heating?

4. Potassium chloride is an electrolyte. Which of the three solids has a giant atomic structure? Explain your answer.

The Differences Between Giant and Molecular Structures

It seems reasonable to suppose that as giant structures have strong bonds throughout, a great deal of heat energy will be needed to 'unglue' the lattice, i.e. to pull it apart. Consequently, giant structures have high melting points, high boiling points, and high heats of vaporization and fusion. Molecular structures, having relatively weak intermolecular forces, are more easily 'unglued' and have lower melting points, boiling

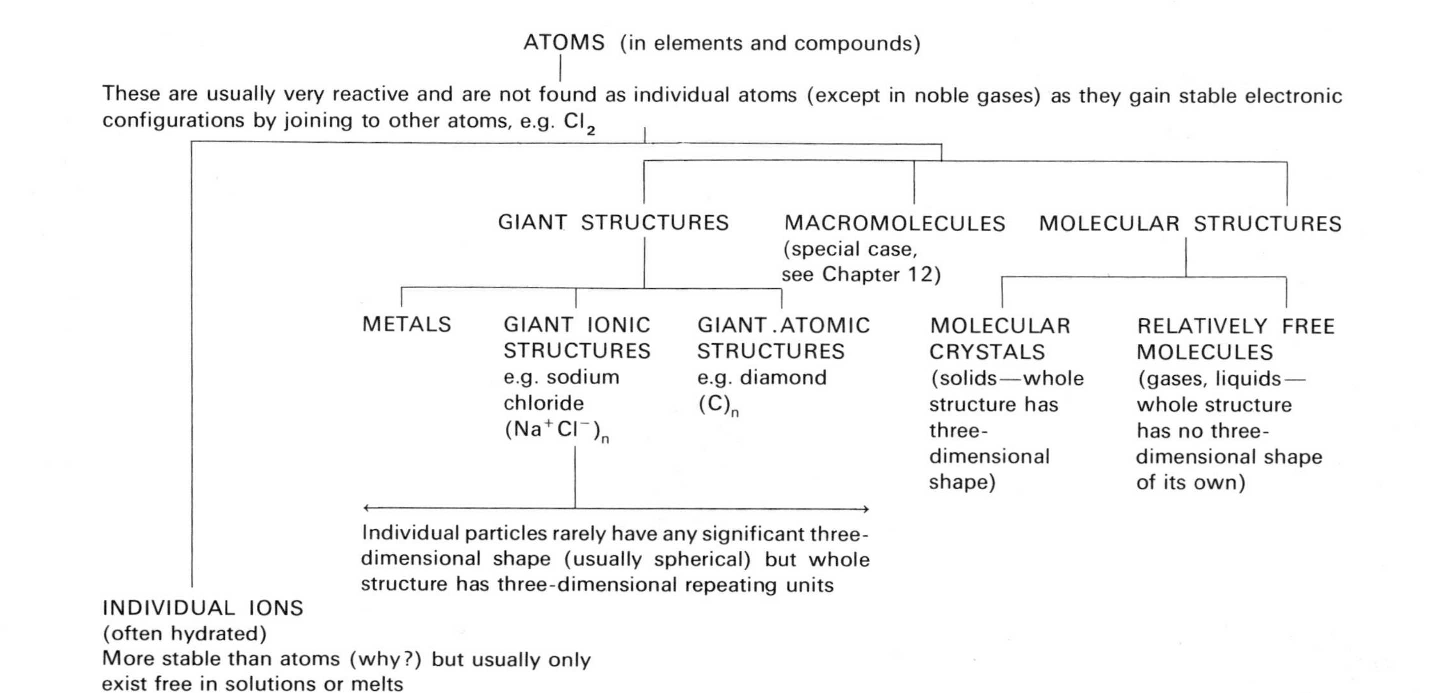

Figure 7.3 Structural divisions of matter

Always consult as much data as possible!

points and heats of vaporization and fusion. This has been shown by Experiment 7.2 and you should remember that giant structures are generally found on the left-hand side of a period and molecular structures on the right-hand side. However, it is important to realize that a substance with a low melting or boiling point does not *necessarily* have a molecular structure; it could be an exception to the general rule. You should always consult as much data as possible before making any conclusion about the structure of an unfamiliar material. In order to be absolutely sure you would have to use techniques such as x-ray diffraction which, as we have pointed out, are not practicable in a normal school laboratory. It is nevertheless true that substances which are gases or liquids at room temperature and pressure are *usually* molecular; their molecules have sufficient kinetic energy under these conditions to overcome the intermolecular forces and become 'unglued'.

7.2 SOME GIANT STRUCTURES. ALLOTROPY

Metals have giant structures. A study of metals provides a fairly simple introduction to structure as a whole, for all the particles in a pure metal are of the same size and type and pack closely together. Their structure is considered in more detail in the *Supplementary Text*.

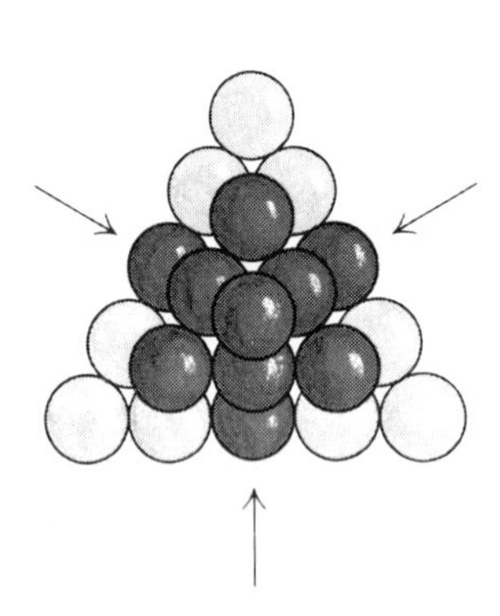

Figure 7.4 Face centred cubic symmetry in metals. The shaded spheres form three sides of a cube. The other faces of the cube are not visible in the diagram. Each face consists of five spheres, four at the corners and one in the centre

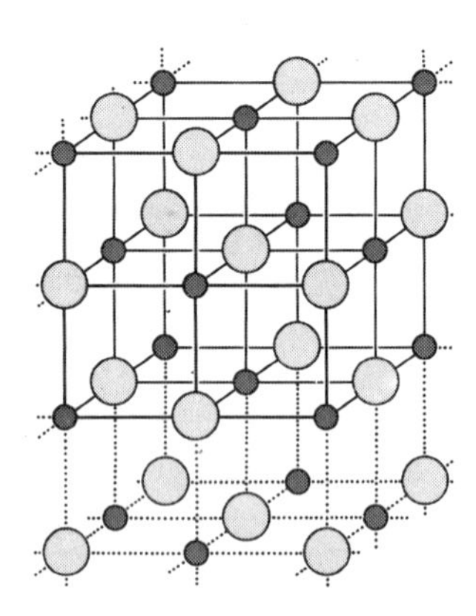

Figure 7.5 The structure of sodium chloride, NaCl. The larger spheres represent the chloride ions. The solid lines indicate the unit cell

Giant ionic structures such as those of sodium chloride and caesium chloride are also considered in *Further Topics*. Figure 7.4 illustrates a typical metallic structure and Figure 7.5 that of sodium chloride.

Giant Atomic Lattices. Allotropy (Polymorphism)

Giant atomic lattices consist of atoms, not ions, and so the bonding is covalent. There are two main types, one consisting only of atoms of a single element, the other having two or more different atoms in the structure, e.g. silicon dioxide, SiO_2. Only the first type will be considered in this book.

Diamond and Graphite

Diamond and graphite are both forms of the same element. Consider the data in Table 7.2. You will notice that their properties differ so markedly that you could well imagine them to be two different elements, and indeed for centuries scientists made that very mistake. In fact both diamond and

Table 7.2 Some properties of diamond and graphite

Property	*Diamond*	*Graphite*
Inter-nuclear distance (nm)	0.154	0.142
Appearance	Colourless, transparent, crystals	Black, shiny, solid
Hardness	Hardest natural substance known	Very soft, flakes easily
Density (g cm^{-3})	3.5	2.2
Electrical conductivity	Non-conductor	Conducts in the direction parallel to the hexagonal planes
Atomic volume at room temperature (cm^3)	3.4	5.4

graphite are forms of the element carbon. This could be readily demonstrated by burning small samples of each separately in an excess of oxygen and collecting the products. Carbon dioxide is the only product in each case, and no solid residue is left if the materials are pure.

When an element can exist in more than one form in the same physical state it is said to exhibit *allotropy* and the different forms are called *allotropes*. Thus diamond and graphite are allotropes of carbon. Allotropes always have different physical properties (which can usually be related to their structures), but normally they have similar chemical properties, although there are exceptions.

If you refer once again to the volumes occupied by the same number of carbon atoms in diamond and graphite (Table 7.1), it is obvious that the atoms in these two solids must be arranged in different ways. These structures have been examined in considerable detail, and an appreciation of the differences has helped to make possible the artificial conversion of graphite into diamond. You may wish to find out for yourself how this can be accomplished.

Examine some crystals of graphite using a hand lens or a microscope. Can you see any sign of external symmetry? Rub a few flakes between your fingers and then examine your skin. Is the graphite breaking up? How do you think that the atoms could be arranged so as to account for the appearance of the crystals, their slipperyness, and the fact that they flake so easily?

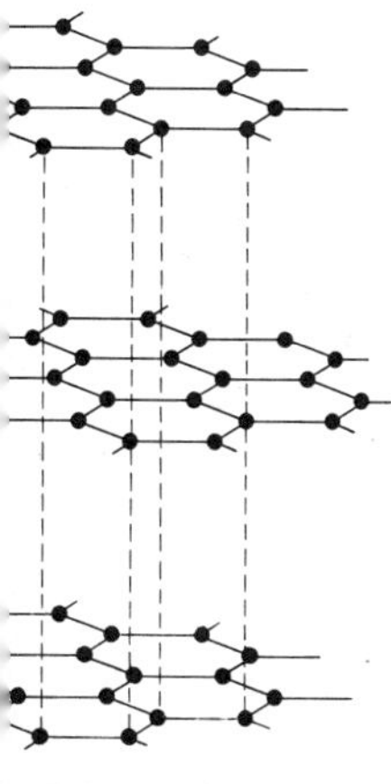

7.6 The structure graphite. The dotted s represent weak ds between the planes

X-ray analysis has shown that the atoms in graphite are strongly bonded together in hexagonal plates which are arranged in layers, but the bonds between the layers are rather weak (Figure 7.6). Each carbon atom in graphite forms only three *strong* bonds. The layers can thus slide over each other and this explains the slippery feel of graphite. Similarly, the flat, plate-like layers explain the appearance of the crystals and the fact that the layers easily flake off. These properties are utilized in pencil leads (how?), and by suspending graphite in oil to act as a lubricating agent. The properties and uses of graphite are neatly explained by its structure (see also Figure 7.7).

Figure 7.7 Mica showing its flat-sheet crystal structure

As a contrast, each carbon atom in diamond forms four strong bonds and no weak ones, each atom being joined to four others arranged tetrahedrally around it (see Figure 7.8). Make sure that you understand this by referring to a model or by building one, as in Figure 7.8. There are no weak bonds in diamond, and there are no flat layers which could slide over each other (cf. graphite). The whole structure is a strong, rigid mass of atoms, and it can be readily appreciated why diamond is so different from graphite. It is the hardest natural substance and is thus used to make cutting instruments and drilling equipment, as well as for jewellery. It is interesting to note that silicon carbide is almost as hard as diamond and has the same tetrahedral array of atoms, although it contains two different kinds of atom.

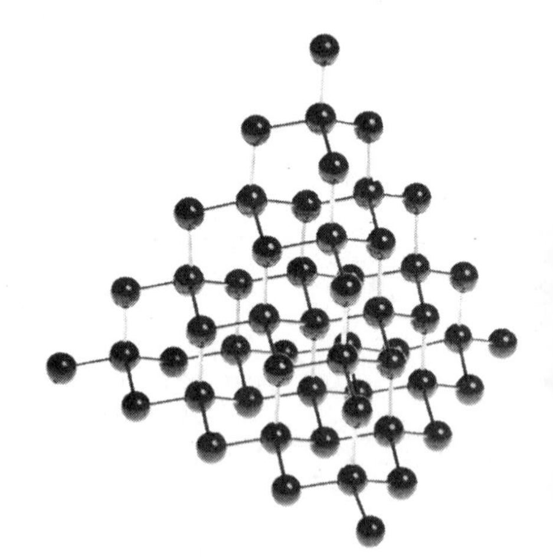

Figure 7.8 The structure of diamond. Each atom is bonded to those around it by four strong bonds

7.3 MOLECULAR CRYSTALS

Sulphur

One of the most familiar substances which forms molecular crystals is sulphur, and this is all the more interesting because, like carbon, it also exhibits allotropy. Rhombic (α) sulphur and monoclinic (β) sulphur are crystalline allotropes which melt at 385 K (112 °C) and 392 K (119 °C) respectively. These low values make a marked contrast to the high melting points of giant structures, and clearly indicate the molecular packing in both allotropes.

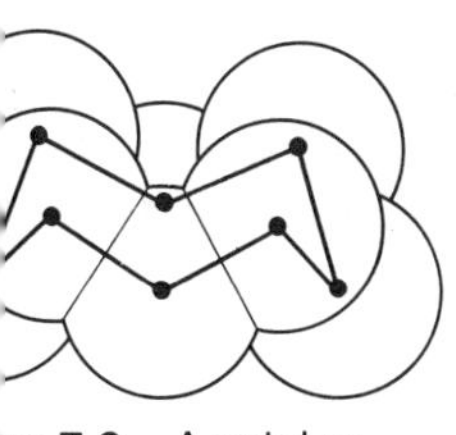

re 7.9 A sulphur
olecule, S_8

It is known that individual sulphur molecules consist of eight sulphur atoms arranged in a 'puckered' ring (Figure 7.9). You may be able to examine a model of a sulphur molecule or to construct one from polystyrene spheres. These molecules can be packed together in different ways to form crystals visible to the naked eye. The shape of these crystals bears no resemblance to that of the individual molecules (puckered rings). You will make samples of two crystalline allotropes and one other form of solid sulphur in this section.

Experiment 7.3
†To Prepare Rhombic (α) Sulphur **

Apparatus
Watch-glass, filter paper, elastic band, hand lens or microscope, teat pipette. A saturated solution of powdered roll sulphur in carbon disulphide (kept in a fume cupboard).

Procedure
Using a teat pipette place a few drops of the solution of sulphur in carbon disulphide on to a watch-glass (in the fume cupboard), and then cover the watch-glass with a piece of filter paper, using an elastic band to keep it in place. Punch a few small holes through the paper so that the solvent can evaporate, and leave in the fume cupboard until the solvent has evaporated. Examine the crystals of α sulphur under a microscope or hand lens. Draw the shape of a typical crystal.

Experiment 7.4
†To Prepare Monoclinic (β) Sulphur **

Apparatus
Filter paper, hand lens or microscope, test-tube and holder, asbestos square, Bunsen burner, paper clip, tongs, spatula. Powdered roll sulphur.

Procedure
(a) Read *all* the instructions before you begin. Place powdered roll sulphur in a test-tube to a depth of about 5 cm. You are going to melt this sulphur and pour it into a container to resolidify. The container should be prepared before heating the sulphur, and a convenient way of doing this is to fold together two thicknesses of filter paper as for filtering, and then clip them together with a paper clip so that a cone is formed.

(b) Heat the sulphur using a *low* flame, preferably in a fume cupboard, and rotate the tube constantly until all the sulphur has just melted. Hold the paper cone over the asbestos square by means of a pair of tongs, and pour the liquid sulphur into it, quickly, but carefully. Place the cone on the asbestos square and leave until a thin solid crust forms over the liquid.

(c) Remove the paper clip, split open the cone, and allow any residual liquid to drain away. (Remember that the liquid is hot.) Crystals of monoclinic sulphur will be seen growing from the crust. Examine them with a hand lens or microscope, and draw the shape of a typical crystal.

Points for Discussion
1. Remember that both allotropes contain the same units (S_8 molecules), but the molecules are arranged in different ways. The allotropes of carbon and

sulphur differ in that diamond and graphite have giant atomic structures, whereas the sulphur allotropes have molecular structures, and in addition the two sulphur allotropes are readily interconvertible but those of carbon are not. Above 369 K (96 °C) (the *transition temperature*) monoclinic sulphur is stable, and below the transition temperature rhombic sulphur is the stable form. It is very difficult indeed to convert diamond into graphite, or *vice versa,* and both forms may be considered to be stable at room temperature, although diamond is changing extremely slowly to graphite under these conditions.

2. The densities of rhombic (α) sulphur and monoclinic (β) sulphur are 2.07 g cm^{-3} and 1.96 g cm^{-3} respectively. In which of the two allotropes are the S_8 molecules most closely packed? (The detailed structure of monoclinic sulphur was not known until 1965.)

3. Unfortunately there are no obvious uses or properties of these two allotropes which reflect their respective structures as do the uses and properties of diamond and graphite.

Experiment 7.5
†Liquid Sulphur and Plastic Sulphur **

Sulphur is unusual in that it forms a third solid structure, and also exhibits allotropy in the liquid state. This experiment provides information about both the other solid form (plastic sulphur) and the liquid allotropes, so record all your observations carefully. These will include colour changes.

Apparatus
Test-tube, test-tube holder, Bunsen burner, 250 cm^3 beaker, asbestos square, spatula.
Powdered roll sulphur.

Procedure
(a) Place powdered roll sulphur in a test-tube to a depth of about 5 cm. Rotate the tube gently in a low Bunsen flame (preferably in a fume cupboard) so that the sulphur slowly melts.

(b) Continue to heat the liquid gently, and move the tube almost into a horizontal position fairly frequently to see how easily the liquid flows.

(c) Bring the liquid to the boil (do not let it catch fire) and then quickly pour it into about 200 cm^3 of water in a beaker.

(d) Wait a few seconds and then use your hands to pick up the product, and gently pull at it to determine its tensile strength, flexibility, etc.

(e) Repeat the flexibility test later in the lesson.

Points for Discussion
1. When a liquid does not flow very easily it is said to be *viscous*, and if it flows freely it is described as being *mobile*. Your observations on liquid sulphur should make reference to these terms.

2. As the tube contained only sulphur throughout the experiment, and there were several colour and viscosity changes, it should be obvious to you that there are different kinds of liquid sulphur, i.e. allotropes of liquid sulphur.

3. Solid sulphur, just below its melting point, consists of S_8 molecules. What will sulphur consist of just after melting? Do you think that there will be any major change, or will the molecules just be further apart? Do the molecules present in liquid sulphur just above the melting point make it viscous?

4. As the temperature is increased the S_8 molecules gain more and more kinetic energy so that they collide more frequently, and the atoms *within* the molecules vibrate more rapidly. You also noticed that the viscosity increases. What do you think might have happened to the S_8 rings to account for this? Not only do the rings open out to form chains of atoms, but also individual chains join together to form very long chains, which have enough kinetic energy (at this temperature) to twist and twine around each other to produce a meshwork of jumbled chains (Figure 7.10). These cannot flow

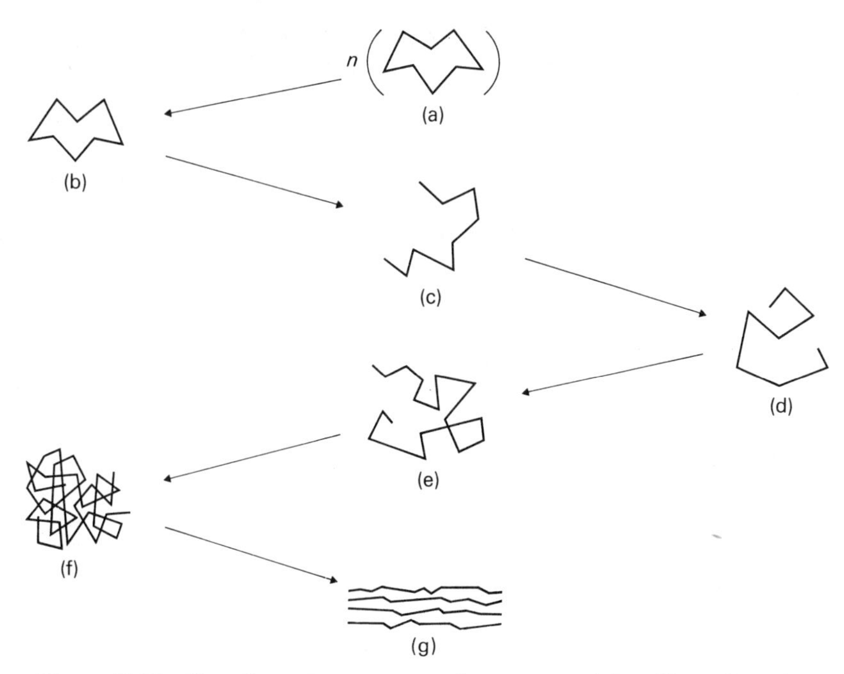

Figure 7.10 The effect of temperature changes on sulphur. Many S_8 molecules forming a solid allotrope (a), become individual S_8 molecules (b) at the melting point. The molecules open up (c and d) and join together to form long chains (e), which become entangled in each other so that the liquid becomes viscous (f). Long free chains of sulphur atoms are found in boiling sulphur and plastic sulphur (g)

over each other very easily and the liquid is viscous. At the same time there is a colour change. Recent evidence suggests that chains of up to 10^6 sulphur atoms can be found in these circumstances.

5. At still higher temperatures the liquid becomes mobile again. Can you suggest a reason for this?

6. When sulphur is heated, progressive changes separate the S_8 molecules, open up the rings, and cause long chains of sulphur atoms to be formed. If these are cooled slowly you might expect to find the reverse process occurring so that eventually S_8 molecules would reform, and they would be arranged to form firstly monoclinic sulphur (as you found in Experiment 7.4), and finally rhombic sulphur, as this is the stable packing arrangement at room temperature.

When you made plastic sulphur, however, you cooled the boiling sulphur very quickly, so quickly in fact that the reverse processes did not have time to take place. What do you think is the structure of plastic sulphur? Does this structure help to explain the elasticity and flexibility of plastic sulphur? Is there any resemblance between the structure of plastics and plastic sulphur (see Section 12.3)?

7. Plastic sulphur, when freshly formed, consists of long chains of sulphur atoms (momentarily 'frozen' liquid sulphur), instead of the S_8 molecules normally found in solid sulphur. Why do you think that plastic sulphur loses its elasticity on standing? Which allotrope of sulphur will it eventually form?

Summary

Sulphur has three solid allotropes, although only two of these are crystalline and exist for any length of time. Each of the crystalline varieties consists of S_8 molecules but they are packed in different ways. The molecules are not held together very strongly so that the melting points are fairly low. When the molecules separate and form liquid sulphur further changes take place so that allotropes of liquid sulphur are formed. At first the molecules stay as S_8 rings but these rings open up at a higher temperature and join together to form chains which intertwine and cause the liquid to become viscous. At still higher temperatures these long tangled chains simplify and the liquid becomes more mobile (Figure 7.10). Apart from the viscosity changes, the relative concentrations of the allotropes of liquid sulphur can be detected by the colour changes. If boiling sulphur at 717 K (444 °C) is suddenly cooled there is not enough time to reverse the sequence and reform the S_8 rings; the plastic sulphur so formed is thus a third form of solid sulphur. On standing, the chains of sulphur atoms in plastic sulphur reform into S_8 rings, and eventually into rhombic sulphur. These changes are summarized in Figure 7.11.

Figure 7.11 The relationships between the various forms of sulphur. N.B. sulphur, if heated rapidly, may melt at 385 K (112 °C) because the transition into monoclinic sulphur has not had time to take place

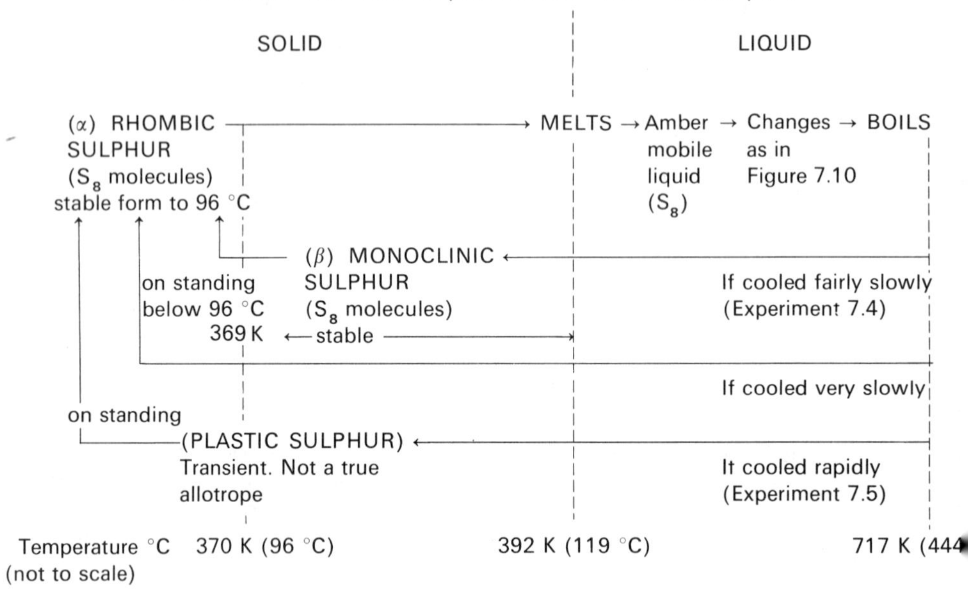

There are other elements which form molecular solids, of course, and many of them also exhibit allotropy. One very common example is phosphorus, and you may already have seen the allotropes 'red' and 'white' phosphorus. You may like to compile a table of data comparing the physical and chemical properties of the allotropes of phosphorus.

Giant Molecules (Macromolecules)

These are dealt with separately in Section 12.3. Briefly, they consist of enormous molecules each containing thousands of atoms, and they include such materials as plastics, fibres (natural and synthetic), carbohydrates, proteins, and fats. One of the simplest proteins consists of molecules each having the composition $C_{662}H_{1020}N_{193}O_{201}S_4$, which gives some indication of the molecular sizes involved.

7.4 THE SHAPES OF INDIVIDUAL MOLECULES

Introduction

So far we have considered the shapes of whole structures, or of crystals and the repeating units within them, but *individual* molecules have shapes of their own which often affect their physical and/or chemical properties.

Molecules contain atoms held together by covalent bonds, each of which consists of one or more pairs of electrons shared between two adjacent atoms. Consider a very simple molecule such as that of gaseous hydrogen chloride, HCl, where there is a single covalent bond between the two atoms. This can be represented as:

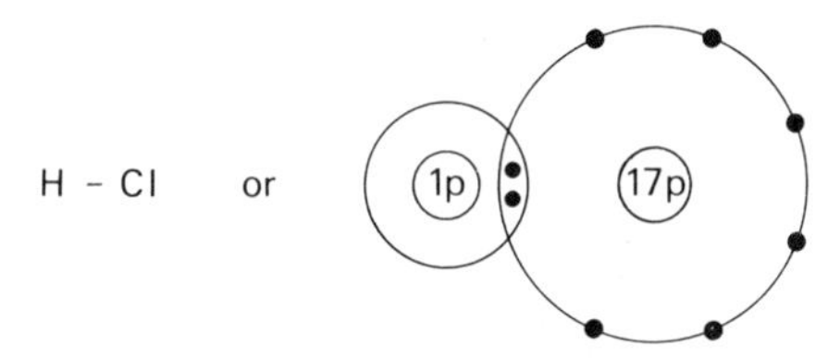

The line between the H and Cl atoms represents a bond and is not part of the shape of the molecule. It should be obvious that a molecule of hydrogen chloride is linear in shape, i.e. the centres of the atoms in the molecule fall on a straight line. It is impossible to arrange a molecule consisting of only two atoms to form any other 'shape'.

When molecules consist of three or more atoms it is possible for them to have a variety of shapes. For example, both ammonia (NH_3) and boron trifluoride (BF_3) molecules have four atoms, but one has a pyramidal shape and the other is triangular. If you go on to study chemistry at a more advanced level you will find that an appreciation of the three-dimensional shapes of molecules can help you to understand the mechanisms of some chemical reactions. For most introductory courses you need only know the shapes of a few simple molecules, and not *why* they are so shaped. A few typical examples follow. However, the prediction of molecular shapes from theoretical principles is a useful and interesting exercise and so we have also explained why these simple molecules have their characteristic shapes, and shown how such shapes can be predicted from simple theoretical ideas in *Further Topics*.

Some Simple Molecular Shapes

The Methane Molecule, CH_4

A methane molecule consists of one carbon atom joined to four hydrogen atoms by single covalent bonds. It is often represented as

but this is *not* the shape of the molecule. It is useful revision to draw the five atoms with their outer electrons arranged to show how they are shared.

The shape of a methane molecule is tetrahedral, i.e. the four hydrogen atoms are arranged tetrahedrally around the central carbon atom. Each H—C—H angle is $109\frac{1}{2}$ degrees. This is illustrated in Figure 7.12, but a model will explain the situation more clearly.

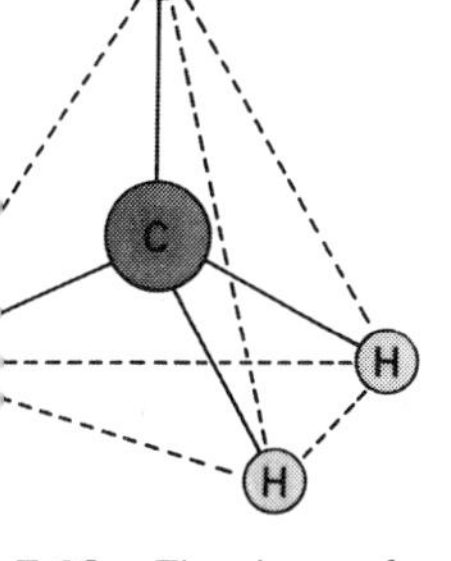

7.12 The shape of a ane molecule

The Ethene (Ethylene) Molecule, C_2H_4

The carbon atoms are joined together by a double covalent bond, and each carbon atom also has two hydrogen atoms attached by single covalent bonds. It is a useful exercise to draw an electron diagram for the molecule. The molecule is often represented as in Figure 7.13, and in this particular case this also indicates the shape. The ethene molecule is flat (i.e. all the atoms are in the same plane), with each H—C—H angle 120 degrees.

The Carbon Dioxide Molecule, CO_2

Draw an electron diagram for the molecule which can be shown as O═C═O. Again this also indicates the shape. The molecule is flat and linear (i.e. all the atoms are in the same plane and in a straight line).

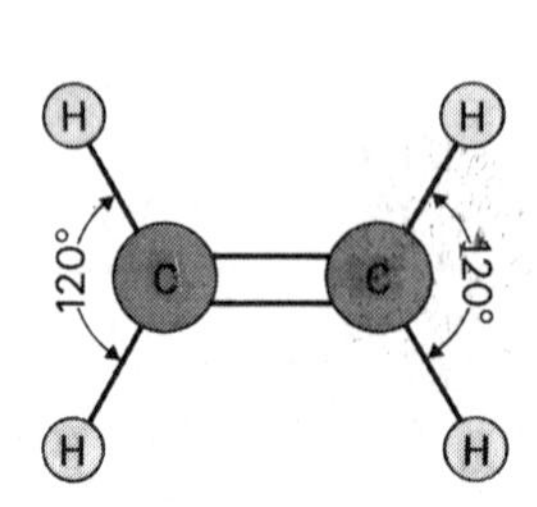

Figure 7.13 The shape of a molecule of ethene (ethylene)

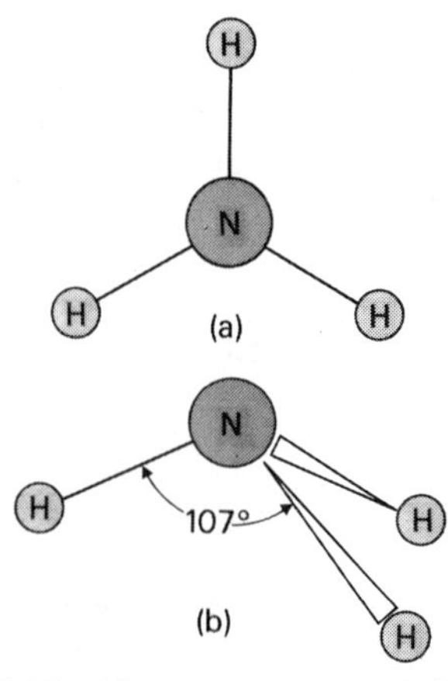

Figure 7.14 The arrangement of the atoms in a molecule of ammonia. The arrangement (a) does not show the shape of the molecule. The pyramidal shape of the molecule is illustrated in (b)

The Ammonia Molecule, NH_3

Draw an electron diagram for the molecule. Figure 7.14 (a) does not represent the shape, for this molecule is three-dimensional. The molecule is pyramidal, with the nitrogen atom at the top of the pyramid. The H—N—H angles are 107 degrees (Figure 7.14 (b)).

The Water Molecule, H_2O

Draw an electron diagram for this molecule. H—O—H does not represent the shape; water molecules are non-linear, i.e. 'bent'. The molecule is more accurately represented by

```
   O
  / \
 H   H
```

and the H—O—H angle is $104\frac{1}{2}$ degrees.

7.5 ISOMERISM

Introduction

Three words frequently confused by young students of chemistry are isotopes (page 21), allotropes (explained earlier in this chapter), and *isomers*. The latter are usually, but not always, confined to the world of organic chemistry, or, in other words, to the compounds of carbon. You may already know that carbon atoms in compounds can be joined together in long chains, which can be 'straight' or branched, and also in rings. Many organic molecules contain both rings and chains of carbon atoms. Some of these structures are composed of thousands of atoms, and indeed there seems no limit to the number of possible permutations. It automatically follows that there are millions of different organic compounds, and some of the very simple types are shown in Figure 7.15.

In each of the examples the normal valencies of carbon (4) and hydrogen (1) have been maintained, but it is often possible to arrange a given group of atoms in more than one way while still maintaining the normal valencies of the constituent atoms. Thus the group of atoms C_2H_6O can be arranged to form two independent molecules:

```
(a)     H  H              (b)    H       H
        |  |                     |       |
     H—C—C—O—H                H—C—O—C—H
        |  |                     |       |
        H  H                     H       H
```

Note that in each case the normal valencies of carbon, oxygen, and hydrogen are preserved, but two different structures are possible; (a) is ethanol (the alcohol found in alcoholic drinks), and (b) is methoxymethane (dimethyl ether). These two molecules are isomers. When two or more structures exist which have the same molecular formula (same type and number of atoms) but different structural formulae (different arrangements of the atoms) the phenomenon is termed isomerism and the individual forms are isomers.

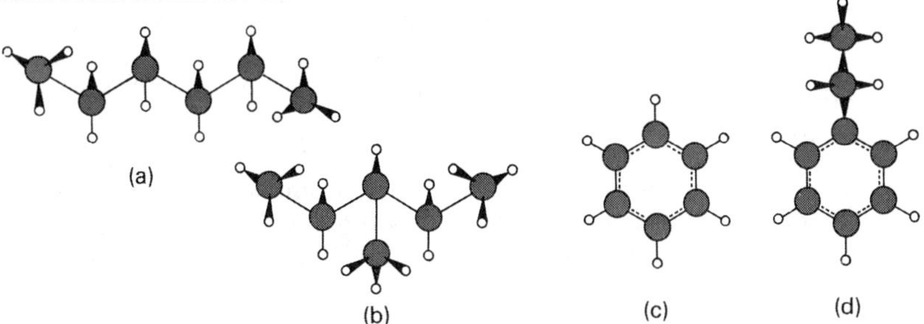

Figure 7.15 The shapes of some organic molecules. (a) shows a chain of carbon atoms in hexane, C_6H_{14}, (b) shows a five-carbon chain with a branch, (c) shows a ring of carbon atoms in benzene, C_6H_6, and (d) shows a ring of carbon atoms with a branch

It may not surprise you to learn that thousands of organic molecules have isomers. We have already said that one of the simplest proteins has the molecular formula $C_{662}H_{1020}N_{193}O_{201}S_4$, and it would indeed be an involved homework to calculate the number of isomers possible.

The different structural arrangements of a given group of atoms each have their own three-dimensional shape. We will now look at a few simple examples of isomerism, restricting our discussion to the case of structural isomerism, although there are other kinds.

Some Simple Examples of Isomerism

Hydrocarbons

Hydrocarbons contain hydrogen and carbon only and they are studied in Section 12.1. One of the simplest hydrocarbons is propane, C_3H_8, which has the structure:

```
    H   H   H
    |   |   |
H—C—C—C—H
    |   |   |
    H   H   H
```

This formula does not indicate the shape of the molecule as it is only a two-dimensional representation. No matter how you try, you will not be able to rearrange these atoms in more than the one structure shown while maintaining their normal valencies.

The next hydrocarbon in the series is butane, C_4H_{10}. This time there is more than one way of arranging this group of atoms, and butane has two isomers:

```
    H   H   H   H                    H   H   H
    |   |   |   |                    |   |   |
H—C—C—C—C—H      and      H—C—C—C—H
    |   |   |   |                    |   |   |
    H   H   H   H                    H   C   H
                                        / | \
                                       H  |  H
                                          H
```

The next member of the series is pentane, C_5H_{12}, which has three structural isomers. Try to draw these for yourself.

These different structural arrangements result in varying physical properties such as melting point, boiling point, and viscosities (if liquid), although the isomeric hydrocarbons are usually very similar chemically.

Ethanol and Methoxymethane

In some isomers the variations are due to the presence of entirely different groups within the molecules, and not just to different chain arrangements. In the example quoted earlier, methoxymethane ($CH_3—O—CH_3$) and ethanol (C_2H_5OH) are structural isomers. They differ chemically as well as physically because they are completely different substances, ethanol containing an —OH group while methoxymethane does not. If suitable models are available many of these points will be understood more easily. Remember that the two isomers of butane only differ physically, as they contain the same functional groups. Ethanol and methoxymethane differ physically and, because they contain different groups, they also differ chemically.

Summary

The third dimension is a relatively new aspect of chemistry to achieve importance. Rapid advances in recent years, particularly in organic chemistry and biochemistry, have clearly indicated that many reactions cannot take place unless the appropriate reacting molecules have the necessary three-dimensional shapes so that they can fit into each other like a lock and key. You may learn later just how important these ideas are in understanding how enzymes work, how proteins are synthesized by living organisms, and how the genetic code is thought to function.

You should now understand some of the ways in which matter is constructed from the basic building blocks, atoms, ions, and molecules. The solid, liquid, and gas states have been compared and the solid state examined in greater detail. Many differing structures have been discussed and some of these have been related to properties and to uses of everyday importance. We have seen that individual molecules, as well as whole structures, have a three-dimensional shape, and that this can often be predicted by using theoretical concepts. Finally, we have learned that individual molecules can sometimes have more than one arrangement of their atoms, each with its own characteristic shape and properties. For a final summation of these ideas have another look at Figure 7.3.

QUESTIONS CHAPTER 7

1. In which substances would you expect to find free atoms at room temperature?

2. Give three ways in which solids differ markedly from gases or liquids, and explain the differences.

3. Explain clearly, in your own words, the differences between giant structures and molecular structures. Give examples to illustrate your answer.

4. Explain the significance of a graph which shows how the gramme-atomic volumes of solid elements vary with their atomic numbers. What form would you expect the graph to take if you did not know about packing variations? Why do the gramme-atomic volumes of (a) diamond and graphite, (b) red and white phosphorus, and (c) α and β sulphur, prove emphatically that atoms can be arranged in different ways?

5. Explain, with examples, what is meant by isomerism.

6. A solid melts at 1020 °C and is an electrolyte when molten. It has

A a molecular structure
B a giant atomic structure
C a metallic structure
D a giant ionic structure
E some other structure

Element	Atomic mass	Melting point K (°C)	Electrical conductivity of solid element	Reaction with oxygen
V	31	863 (590)	Non-conductor	Burns readily
W	40	1124 (851)	Good	Burns readily
X	20	24 (−249)	Non-conductor	No reaction
Y	207	600 (327)	Good	Oxidizes slowly
Z	31	317 (44)	Non-conductor	Burns readily

7. Different solid forms of the same element are called

A isobars   C allotropes   E monotropes
B isotopes   D isomers

8. Liquid sulphur is viscous at certain temperatures. Which of the following is the best explanation for this?

A The sulphur atoms are free and move rapidly to and fro.
B The S_8 molecules expand and restrict the freedom of movement.
C The S_8 molecules open up to form short chains.
D The S_8 molecules open up and join together to form long chains.
E The S_8 molecules are non-streamlined so that smooth flow is restricted.

9. Which of the following pairs are allotropes?

A Carbon (graphite) and sulphur (monoclinic)
B Carbon (diamond) and sulphur (monoclinic)
C Sodium chloride and caesium chloride
D Phosphorus (white) and phosphorus (red)
E ^{31}P and ^{30}P

Questions 10 to 14 are concerned with the substances listed below.

A Sulphur   C Diamond   E Graphite
B Hydrogen   D Sodium

Choose from this list the substance which you think best fits the following descriptions.

10. At room temperature most of the volume occupied by this element is space.

11. This substance has a giant structure but some bonds are weaker than others.

12. This substance has a molecular structure, each molecule containing eight atoms.

13. This substance has a giant structure but it has a relatively low melting point and is quite soft.

14. All bond angles and bond lengths in this substance are identical. It does not conduct electricity and has a high melting point.

15. The data in the table at the top of the page refer to five *elements* lettered V, W, X, Y, Z:

(a) Which pair of elements are metals?
(b) Which pair are allotropic forms of the same element?
(c) Which element could be an inert gas?

(W. J. E. B.)

The following questions require rather longer answers.

16. Suppose that you are trying to convince a fellow pupil that the particles in solids are arranged in regular patterns. Describe the ideas you would use in your argument, and give as many examples as possible.

17. Compare the properties of diamond and graphite and explain how the differences are related to their structures.

18. Describe in detail how you would prepare samples of the two solid crystalline allotropes of sulphur.

19. Describe and explain the changes which take place when sulphur is slowly heated from room temperature to 717 K (444 °C), and then rapidly cooled.

20. Changes from one liquid allotrope of sulphur to another are very rapid but it takes much longer for both 'plastic' and β sulphur to form the α variety. Explain these facts.

21. How many elements can you find which exhibit allotropy? List these together with their melting points and boiling points and suggest the type of structure each may have.

22. Describe the shapes of the following molecules: CH_4, H_2O, BF_3, NH_3, $SiCl_4$, C_2H_4, CO_2, C_2H_6, $BeCl_2$.

8 Oxidation and Reduction. Oxidation Number

Chemists have used the words oxidation and reduction for many years, although the applications of the terms have changed considerably. Oxidation was originally defined as a process in which an element or compound gains oxygen or loses hydrogen, and reduction as the opposite. The definition in its narrowest sense is still used where a simple transfer of oxygen or hydrogen takes place, but the idea has been extended so as to have a much wider significance and to include most chemical reactions; consequently, more comprehensive definitions of oxidation and reduction have been introduced.

Most chemical reactions can be considered as 'battles for electrons' rather than 'battles for oxygen or hydrogen', and the modern definitions of oxidation and reduction thus involve electron transfer; we use one of the most fundamental ideas in chemistry, i.e. that atoms and molecules exchange or share electrons in order to gain more stable structures. This we have discussed in terms of bonding (page 30), electrode potentials (page 170), electronegativity (page 169), and atomic structure, and it should be no surprise to learn that oxidation and reduction are closely related to these ideas.

Definitions

Oxidation and Reduction in Terms of Hydrogen and Oxygen

A substance is said to be oxidized if it gains oxygen or loses hydrogen. Such a reaction is called an oxidation.

$$S(s) + O_2(g) \rightarrow SO_2(g)$$

The sulphur is oxidized to sulphur(IV) oxide by gaining oxygen.

$$H_2S(g) + Cl_2(g) \rightarrow 2HCl(g) + S(s)$$

The hydrogen sulphide is oxidized to sulphur by losing hydrogen.

A substance is said to be reduced if it gains hydrogen or loses oxygen. Such a reaction is called a reduction.

$$N_2(g) + 3H_2(g) \rightarrow 2NH_3(g)$$

The nitrogen has been reduced to ammonia by gaining hydrogen.

$$CuO(s) + H_2(g) \rightarrow Cu(s) + H_2O(g)$$

The copper(II) oxide is reduced to metallic copper by losing oxygen.

If a reaction does not involve a transfer of oxygen or hydrogen we obviously cannot apply these definitions, and instead we must use one of the following alternatives, which are applicable to all oxidation–reduction reactions.

Oxidation and Reduction in Terms of Electron Transfer

Electron transfers are most easily recognized by using ionic equations.

A substance is said to be oxidized if it *loses* one or more electrons.

$$Fe(s) + 2H_3O^+(aq) \rightarrow Fe^{2+}(aq) + 2H_2O(l) + H_2(g)$$

Metallic iron has been oxidized to iron(II) ions by loss of two electrons per atom.

$$Fe^{2+}(aq) - e^- \rightarrow Fe^{3+}(aq)$$

The iron(II) ions have been oxidized to iron(III) ions by losing one electron per ion.

A substance is said to be rèduced if it gains one or more electrons.

$$Fe^{3+}(aq) + e^- \rightarrow Fe^{2+}(aq)$$

The iron(III) ions have been reduced to iron(II) ions by gaining one electron per ion.

Points for Discussion

Study the following equations carefully and decide which substances are oxidized and which are reduced, giving reasons for your answers.

1. $Cu(s) - 2e^- \rightarrow Cu^{2+}(aq)$
2. $Cl_2(g) + H_2(g) \rightarrow 2HCl(g)$
3. $PbO(s) + C(s) \rightarrow Pb(s) + CO(g)$
4. $3Fe(s) + 4H_2O(g) \rightarrow Fe_3O_4(s) + 4H_2(g)$
5. $Zn(s) + Cu^{2+}(aq) \rightarrow Cu(s) + Zn^{2+}(aq)$
6. $PbO(s) + H_2(g) \rightarrow Pb(s) + H_2O(g)$
7. $SO_2(g) + 2Mg(s) \rightarrow 2MgO(s) + S(s)$
8. $2Ag^+(aq) + Mg(s) \rightarrow Mg^{2+}(aq) + 2Ag(s)$
9. $2Ca(s) + O_2(g) \rightarrow 2CaO(s)$
10. $Fe_2O_3(s) + 3CO(g) \rightarrow 2Fe(s) + 3CO_2(g)$

Oxidation and Reduction in Terms of Oxidation Number

You will have seen, from the examples given, how the concept of oxidation and reduction as a transfer of electrons has considerably extended the number of reactions which can be included under this heading. However, this concept also has its limitations as there are cases where it is very difficult to determine, from the equation for the reaction, exactly how and where electrons have been transferred. This difficulty is overcome by using a third concept of oxidation and reduction, i.e. oxidation numbers. Indeed, you may prefer to make this the basis of your interpretation of these reactions, as it is applicable to all.

Oxidation numbers have no real chemical justification apart from their great convenience. They are allotted to elements, whether free or combined, according to some very simple rules. You can regard the oxidation number as being the charge an atom of the element *would* have if it existed as an ion in the compound in question, even if it is in fact covalently bound. This will become clear to you when you have learnt the rules which follow and considered a few examples.

Rules for using oxidation numbers

(a) Atoms of all elements in the elementary state are given the oxidation number zero, e.g. $Cu(s) = 0$, $H_2(g) = 0$, and $Br_2(l) = 0$.

(b) When a non-metal is combined with a metal it has a negative oxidation number (because it would form negative ions if free to do so) which is numerically equal to its electrovalency, e.g. Cl $= -1$ and O $= -2$. *Exceptions* H $= +1$ except in metal hydrides, when it is -1.
O $= -1$ (not -2) in peroxides.

(c) Metals have positive oxidation numbers in their compounds because they would form positive ions if free to do so. If you know with certainty the electrovalency exerted by a metal in a particular situation then its oxidation number will have the same value.

(d) The sum of all the oxidation numbers of the atoms present in a formula of a compound is zero, e.g. $CuSO_4$ (oxidation number of Cu)+(oxidation number of S)+(4 × oxidation number of O) = zero.

(e) The oxidation number of a simple ion is the same as its charge, e.g. $K^+ = +1$, $Cu^{2+} = +2$, and $Cl^- = -1$.

(f) The oxidation number of an ion containing more than one kind of atom is also equal to the charge on the ion, as is the sum of the oxidation numbers of the atoms within the ion, e.g. $NO_3^- = -1$ and (oxidation number of N)+(3 × oxidation number of O) $= -1$; $SO_4^{2-} = -2$ and (oxidation number of S)+(4 × oxidation number of O) $= -2$.

Using oxidation numbers The sum of the oxidation numbers in magnesium oxide, MgO, is zero because magnesium oxide is a compound (rule (d)). The oxidation number of oxygen is -2 (rule (b)), therefore the oxidation number of magnesium in magnesium oxide is $+2$.

$$MgO = \text{zero} \qquad +2-2 = \text{zero}$$

Is this the same as the electrovalency of magnesium in this compound, rule (c)?

The sum of the oxidation numbers in iron(III) chloride, $FeCl_3$, is zero because iron(III) chloride is a compound (rule (d)). The oxidation number of chlorine is -1 (rule (b)). There are three chloride ions in the formula and therefore the total oxidation number contribution from chlorine is -3. The oxidation number of iron in iron(III) chloride is thus $+3$.

$$FeCl_3 = \text{zero} \qquad +3+(3 \times -1) = \text{zero}$$

Is this the same as the electrovalency of iron(III), rule (c)?

In the ammonium ion, NH_4^+, the sum of the oxidation numbers must be the same as the charge on the ion, $+1$ (rule (f)). As the oxidation number of hydrogen is $+1$ (rule (b)) the four hydrogen atoms make a total oxidation number contribution of $+4$. Thus (oxidation number of nitrogen)+(+4) $= +1$, from which the oxidation number of nitrogen is -3 in the ammonium ion. Is this the same as the electrovalency of nitrogen, rule (b)?

$$(NH_4)^+ = +1 \qquad -3+4(+1) = +1$$

In sulphuric acid, H_2SO_4, the oxidation numbers will add up to zero (why?). As each hydrogen atom has an oxidation number of $+1$ and each oxygen atom is -2, their total oxidation number contributions are $(2 \times +1)$ and (4×-2), i.e. -6. The oxidation number of sulphur is thus $+6$ in sulphuric acid. You will notice that in this case the oxidation number of sulphur is not equal to its electrovalency, nor is it negative, because the non-metal sulphur is not combined with a metal in this compound. This clearly shows the limitations of the term 'valency'.

Points for Discussion

1. Calculate the oxidation numbers of each of the elements in the following substances:

(a) H_2S
(b) $CuSO_4$
(c) HNO_3
(d) CH_4
(e) SO_2
(f) H_2SO_3
(g) C_2H_6

2. Calculate the oxidation numbers of each of the elements in the following substances (you will see just how easy the concept is, even when applied to unfamiliar or apparently complicated formulae):

(a) $POCl_3$
(b) $KMnO_4$
(c) K_2PtCl_6
(d) $Na_2Cr_2O_7$.

3. Consider the reactions

$$S(s)+O_2(g) \rightarrow SO_2(g)$$

and

$$H_2S(g)+Cl_2(g) \rightarrow 2HCl(g)+S(s)$$

which were given earlier as examples of oxidation. Decide which of the elements are oxidized in these reactions and then find how their oxidation numbers change in the course of the reaction. When an element is oxidized does its oxidation number increase or decrease?

4. Consider the reactions

$$N_2(g)+3H_2(g) \rightarrow 2NH_3(g)$$

and

$$CuO(s)+H_2(g) \rightarrow Cu(s)+H_2O(g)$$

which were given earlier as examples of reduction. When an element is reduced does its oxidation number increase or decrease?

5. You should be able to use your answers to 3 and 4 to suggest definitions of oxidation and reduction based on oxidation numbers.

6. An element is said to be *oxidized* if its oxidation number increases, i.e. becomes more positive. In the process the element is said to move from a lower to a higher oxidation state. An element is said to be *reduced* if its oxidation number decreases, i.e. becomes less positive. In the process the element is said to move from a higher to a lower oxidation state.

7. Remember that the definitions given in 6 can be used for all oxidation–reduction reactions. Consider the reaction

$$C_2H_4(g)+H_2O(l)+[O] \rightarrow C_2H_4(OH)_2(aq).$$

Both hydrogen and oxygen have been added to ethene (C_2H_4). Nevertheless, the carbon in ethene has been oxidized, as you should be able to show using the oxidation number concept. Could you have reached this conclusion by using either of the other definitions of oxidation and reduction?

8. Working by oxidation numbers only, decide whether the element underlined in each of the following examples has increased or decreased its oxidation number during the reaction, and hence whether it has been oxidized or reduced.

(a) $Cl_2(g)+2K\underline{Br}(aq) \rightarrow 2KCl(aq)+Br_2(aq)$
(b) $\underline{Fe}(s)+CuSO_4(aq) \rightarrow FeSO_4(aq)+Cu(s)$
(c) $2Na(s)+\underline{Cl_2}(g) \rightarrow 2NaCl(s)$
(d) $Cu(s)+\underline{S}(s) \rightarrow CuS(s)$
(e) $\underline{C}(s)+H_2O(g) \rightarrow CO(g)+H_2(g)$
(f) $\underline{C}H_4(g)+Cl_2(g) \rightarrow CH_3Cl(g)+HCl(g)$
(g) $\underline{H_2}(g)+Cl_2(g) \rightarrow 2HCl(g)$
(h) $\underline{Cu^{2+}}(aq)+2e^- \rightarrow Cu(s)$
(i) $\underline{S}O_2(aq)+Cl_2(g)+2H_2O(l) \rightarrow H_2SO_4(aq)+2HCl(aq)$
(j) $4HCl(aq)+\underline{Mn}O_2(s) \rightarrow MnCl_2(aq)+2H_2O(l)+Cl_2(g)$

9. Repeat 8 but this time use either (or both if applicable) of the electron transfer concept or the oxygen/hydrogen concept. Compare your answers with those from 8. Which method do you prefer?

10. Oxidation and reduction in terms of oxidation number are summarized in Table 8.1.

Table 8.1 An oxidation–reduction scale using oxidation number

		Oxidation number or oxidation state		Some oxidation states of common elements: Manganese	Chromium	Nitrogen	Carbon
↑ Oxidation	Reduction ↓	+8	8+				
		+7	7+	Mn_2O_7, MnO_4^-			
		+6	6+	MnO_4^{2-}	CrO_3, $Cr_2O_7^{2-}$		
		+5	5+			HNO_3	
		+4	4+	MnO_2		NO_2, N_2O_4	CO_2
		+3	3+	Mn_2O_3	Cr_2O_3, Cr^{3+}	HNO_2	
		+2	2+	MnO, Mn^{2+}	CrO, Cr^{2+}	NO	CO
		+1	1+			N_2O	
		0	0	Mn	Cr	N_2	C
		−1	1−				C_2H_2
		−2	2−			N_2H_4	C_2H_4
		−3	3−			NH_3	C_2H_6
		−4	4−				CH_4

Oxidizing Agents and Reducing Agents

Redox Reactions

It should be obvious that we cannot have a transfer of oxygen, hydrogen, or electrons without there being both a 'loser' and a 'gainer'. In other words, if one substance in a reaction is oxidized (by gaining oxygen, losing hydrogen, or losing one or more electrons), then another substance must at the same time be reduced (by losing oxygen, gaining hydrogen, or gaining one or more electrons). We cannot have oxidation without reduction (and *vice versa*), and reactions of this type are often termed *redox* (*red*uction–*ox*idation) reactions.

The compound or element which brings about an oxidation is called an *oxidizing agent*. It must follow that an oxidizing agent is itself reduced in a reaction, i.e. in giving oxygen to a substance (or in removing hydrogen or electrons), it is itself reduced by the opposite effect.

Similarly, *reducing agents* bring about reductions, but in doing so they are oxidized. Think about this carefully because the terms oxidation, reduction, oxidizing agent, and reducing agent are frequently confused by young chemists. Most reactants and products in chemical reactions can be described by these terms, e.g.

$$CuO(s) + H_2(g) \rightarrow Cu(s) + H_2O(g)$$

Hydrogen reduces hot copper(II) oxide (by removing oxygen) to metallic copper and is itself oxidized to steam (by gaining oxygen). In this reaction hydrogen is a reducing agent and copper(II) oxide is an oxidizing agent.

Describe, in similar terms, all the other reactions given previously as examples of oxidation and reduction.

Oxidizing Agents and Reducing Agents are Relative Terms

You must not imagine that a certain substance always acts as an oxidizing agent or a reducing agent. These are relative terms. Suppose that a chemical reaction takes place between two substances which both tend to gain electrons. This means that both substances are usually oxidizing agents, i.e. they will tend to gain electrons in reactions and in doing so remove them from something else. However, when two such substances react *together*, the one which *most easily* gains electrons will take them from the other, which is thus forced to act as a reducing agent in this reaction. An oxidizing agent can therefore be made to act as a reducing agent by an oxidizing agent more powerful than itself, and the converse is true of reducing agents. You should therefore use these terms with care, and to be absolutely certain that you are using the correct term, always state the reaction concerned. For example, it is better to say 'hydrogen is a reducing agent for copper(II) oxide' than 'hydrogen is a reducing agent'.

Powerful oxidizing agents (see Table 8.2) and powerful reducing agents (see Table 8.3) usually behave as such, but weaker oxidizing and reducing agents can act as *both* oxidizers and reducers according to the type of substance with which they react.

e.g.

$$2SO_2(g) + O_2(g) \rightarrow 2SO_3(g)$$

Sulphur dioxide is a *reducing* agent in this reaction; it is oxidized to sulphur trioxide.

$$SO_2(g) + 2H_2S(g) \rightarrow 2H_2O(l) + 3S(s)$$

Sulphur dioxide is an *oxidizing* agent in this reaction; it is reduced to sulphur.

Examples of Oxidizing and Reducing Agents

This chapter is not intended to be a comprehensive description of all cases of oxidation and reduction encountered in school laboratories. You must learn to apply these ideas to examples as you meet them throughout the course, and remember that oxidizer and reducer are relative terms.

Laboratory Oxidizing Agents

A range of typical laboratory oxidizing agents is shown in Table 8.2.

Table 8.2 Some common oxidizing agents

Reagent	*Formula*	*External signs of its action*
Acidified (H_2SO_4) potassium manganate(VII) solution	$KMnO_4$	Colour change—purple to colourless
Acidified (H_2SO_4) potassium chromate(VI) solution	$K_2Cr_2O_7$	Colour change—orange to green
Oxygen	O_2	Depends on the other reagent
Chlorine	Cl_2	Depends on the other reagent
Concentrated sulphuric acid (sometimes needs to be hot)	H_2SO_4	Gas evolved, sulphur(IV) oxide
Concentrated nitric acid (sometimes needs to be hot)	HNO_3	Gas evolved, brown nitrogen dioxide
Acidified hydrogen peroxide solution	H_2O_2	Depends on the other reagent

Laboratory Reducing Agents

Some common reducing agents are listed in Table 8.3.

Table 8.3 Some common reducing agents

Reagent	*Formula*	*External signs of its action*
Hydrogen	H_2	Depends on the other reagent
Hydrogen sulphide	H_2S	Deposit of sulphur formed
Sodium and ethanol	—	Depends on the other reagent
Moist sulphur(IV) oxide or its aqueous solution	SO_2 or H_2SO_3	Depends on the other reagent
Neutral or (more usually) alkaline hydrogen peroxide solution	H_2O_2	Oxygen evolved
Metals dissolving in dilute acids	—	Depends on the other reagent
Carbon	C	Depends on the other reagent
Carbon monoxide	CO	Depends on the other reagent
Metals	—	Depends on the other reagent

Summary

The terms oxidation and reduction can be applied to most chemical reactions. For reactions involving a transfer of oxygen or hydrogen it is convenient to use the 'older' definitions. For reactions where electron transfer is clear use the more modern definition. When in doubt (or all the time if you prefer) use the oxidation number concept.

The most powerful oxidizing agents are those which most readily gain electrons. The most powerful reducing agents have a strong tendency to lose electrons.

'Oxidizing agent' and 'reducing agent' are relative terms. You should be familiar with the common laboratory oxidizing and reducing agents (Tables 8.2 and 8.3) and know the kind of reactions they can be used for, what is formed, why they are said to act in this way, and any colour changes associated with them.

QUESTIONS CHAPTER 8

1. What type of reaction is represented by the following equation?

$$Fe^{2+} \rightarrow Fe^{3+} + e^-$$

Name *two* reagents which could be used separately to bring about this reaction. What reagent would you use to test that this reaction had taken place? What would you expect to observe during this test? (J. M. B.)

2. What would you observe if (a) a solution of hydrogen peroxide was added to manganese dioxide (manganese (IV) oxide), (b) an excess of hydrogen peroxide solution was added to (i) a solution of potassium permanganate acidified with dilute sulphuric acid, (ii) lead sulphide? (J. M. B.)

3. Give one example in each case of sulphur dioxide acting as (a) an oxidizing agent, (b) a reducing agent, (c) an acid anhydride. (J. M. B.)

4. (a) Name an oxidizing agent, and write an equation for a reaction in which it is used for oxidation. (b) Name a reducing agent, and write an equation for a reaction in which it is used for reduction. (J. M. B.)

5. Name the product of each of the following reactions, and in each case state whether the reactant named in italics undergoes oxidation or reduction: (a) *hydrogen* burning in oxygen, (b) *chlorine* reacting with iron(II) chloride, (c) *hydrogen* reacting with nitrogen in the presence of a catalyst. (J. M. B.)

6. Calculate the oxidation numbers of each of the elements in the following substances: (a) MnO_2; (b) Cr_2O_3; (c) $(NH_4)_2SO_4$; (d) $KClO_3$; (e) SF_6; (f) P_4O_{10}; (g) N_2O_4.

7. Make an oxidation number table (as in Table 8.1) for a range of compounds of sulphur and chlorine.

8. Which of the following statements describes oxidation?

A Addition of hydrogen to a compound
B A gain of one or more electrons
C An increase in valency (oxidation state) of a metal
D A decrease in the number of negatively charged ions present in the formula of a compound
E Removal of oxygen

9. In the reaction between acidified potassium dichromate and iron(II) ions, which may be shown by the equation

$$Cr_2O_7^{2-}(aq) + 6Fe^{2+}(aq) + 14H^+(aq) \rightarrow 2Cr^{3+}(aq) + 7H_2O(l) + 6Fe^{3+}(aq)$$

A there is both oxidation and reduction
B the colour of the solution changes from green to yellow
C the iron(II) ions are reduced
D the dichromate ions are reduced
E hydrogen ions are reduced

10. The reaction between sulphur dioxide and hydrogen sulphide is as follows

$$2H_2S(g) + SO_2(g) \rightarrow 2H_2O(l) + 3S(s)$$

In this reaction

A there is no oxidation or reduction because both substances are reducing agents
B the most powerful reducing agent of the two will oxidize the other
C hydrogen sulphide is reduced to sulphur
D sulphur dioxide is reduced to sulphur
E both hydrogen sulphide and sulphur dioxide are reduced to sulphur

The following questions require rather longer answers.

11. State what you would see, and explain the reactions which occur, when an excess of hydrogen peroxide reacts with (a) a solution of potassium iodide in dilute hydrochloric acid, (b) a solution of iron(II) sulphate in dilute sulphuric acid, (c) lead(II) sulphide.

Describe what you would see and state whether hydrogen peroxide is acting as an acid, oxidizing agent, or reducing agent in each of the following: (d) the reaction with silver(I) oxide,

$$Ag_2O(s) + H_2O_2(l) \rightarrow 2Ag(s) + H_2O(l) + O_2(g)$$

(e) the reaction with barium hydroxide solution,

$$Ba(OH)_2(aq) + H_2O_2(aq) \rightarrow BaO_2(s) + 2H_2O(l)$$

(J. M. B.)

12. Define the terms oxidation and reduction.

$$Cl_2(g) + 2H_2O(l) + SO_2(g) \rightarrow 2HCl(aq) + H_2SO_4(aq)$$

$$Cl_2(g) + H_2S(g) \rightarrow 2HCl(g) + S(s)$$

$$SO_2(g) + 2H_2S(g) \rightarrow 2H_2O(l) + 3S(s)$$

From the above reaction equations, deduce which of the three gases chlorine, sulphur dioxide, and hydrogen sulphide, is (a) the strongest oxidizing agent, (b) the strongest reducing agent. State clearly how you arrive at your answers.

Describe how you would carry out *two* reactions, the one using nitric acid and the other sulphuric acid, to demonstrate that each of these acids has oxidizing properties. In each case, explain why you classify the reaction as oxidation, and indicate how you would recognize the oxidized product of the reaction. (W. J. E. C.)

13. Describe how you would bring about the following conversions.

(a) $S^{2-}(aq) \rightarrow S(s)$
(b) $Cu^{2+}(aq) \rightarrow Cu(s)$
(c) $CuO(s) \rightarrow Cu(s)$
(d) $PbS(s) \rightarrow PbSO_4(s)$
(e) $MnO_4^-(aq) \rightarrow Mn^{2+}(aq)$
(f) $Fe^{2+}(aq) \rightarrow Fe^{3+}(aq)$
(g) $2I^-(aq) \rightarrow I_2(aq)$

14. Why is it that oxidation and reduction reactions occur together? Select *two* reducing agents and for *each* describe a different reaction illustrating the above statement.

Name an oxidizing gas (not oxygen nor ozone), an oxidizing liquid (not a solution) and an oxidizing solid, and in *each* case give a different substance that can be oxidized by them, writing an appropriate equation. (C.)

15. Describe briefly and write equations for the reactions by which you could convert (a) oxygen gas into a compound containing oxide ions, (b) hydrogen gas into a solution containing hydrogen ions, (c) a solution containing calcium ions into an insoluble compound of calcium, (d) metallic iron into a solution containing Fe^{3+} ions, (e) sulphur into a solution containing sulphate ions, (f) carbon dioxide into a solution containing carbonate ions. (C.)

9 Water. Soaps and Detergents. Hydrogen Peroxide

9.1 Water, the Most Important Chemical

Our lives depend very much upon an adequate supply of pure water and yet it is ironical that some years we are faced with a shortage in the summer months, in spite of the fact that water is an abundant component of our environment. Oceans, lakes, and rivers are obvious sources of water, and it is also found in the atmosphere, and chemically combined in rocks and minerals.

The purest form of natural water is rain water, although this accumulates dissolved gases, particularly carbon dioxide (why?), in its fall through the atmosphere, and may in industrial areas contain solid particles such as soot. Rain water finds its way into rivers and lakes, and as it flows through or over the earth it dissolves more substances and often receives sewage and industrial effluent, so that by the time rivers reach the sea their waters are far from pure. The sea is, in effect, the sink for all our water, and it should come as no surprise to learn that sea-water contains at least 3.6 per cent by mass of dissolved and suspended solids of which about two-thirds is sodium chloride. The other main salts present are magnesium chloride, magnesium sulphate, potassium chloride, and calcium sulphate. Much of the waste of modern civilization is poured directly into the sea, and this includes radioactive waste and unwanted chemicals, some of a highly undesirable nature. At the moment, of course, such waste is diluted to a negligible level although there is growing concern that marine life is already being affected.

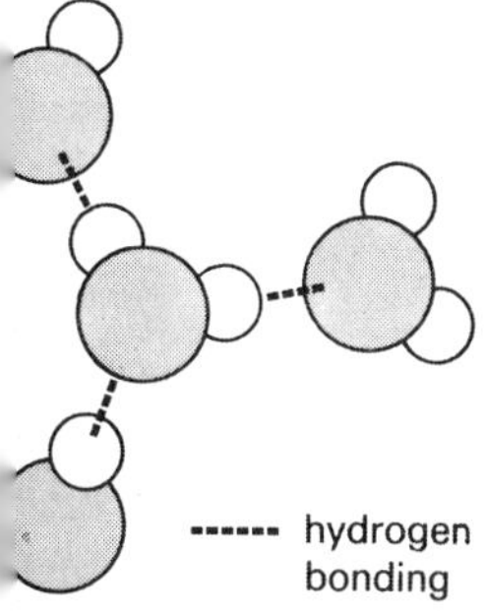

ure 9.1 Hydrogen onding of water molecules into tetrahedral ructures

If you live within reach of the sea you may be able to perform a simple investigation into the solid content of a sample of sea-water. The water which evaporates from the sea and forms rain clouds (to begin the cycle over again) is, on the other hand, very nearly pure. Which laboratory process resembles this natural purification process?

Properties of Water

Physical Properties Water, the liquid form of the compound hydrogen oxide, is a colourless liquid of boiling point 373 K (100 °C) (only at 760 mm mercury pressure) and freezing point 273 K (0 °C). It has much higher boiling and freezing points than other hydrides of Group 6 (e.g. hydrogen sulphide) because of hydrogen bonding. This is also the cause of its solid form (ice) being less dense than the liquid form (the maximum density of water is at 277 K (4 °C)), and accounts for the comparatively high values for its specific heat and heat of vaporization (Figure 9.1).

Water also expands when it freezes, and this is another unusual property. What unfortunate consequence sometimes results from this fact? From your knowledge of structure and the kinetic theory try to explain why you would expect a liquid to *contract* when it freezes and why you would expect the maximum density to be at the freezing point.

Water is a very weak electrolyte; so weak that it may be considered to be a non-electrolyte for all practical purposes. There *is* a slight dissociation into ions however (one molecule in 550 million) and although this is insignificant on its own, it may play an important part in the electrolysis of dissolved electrolytes (Section 6.1). Water also provides a valuable heat exchange medium. Can you think of some examples? How does perspiration cool the body?

Water as a Solvent

We are also drinking a minute amount of dissolved glass

Water has often been described as the 'universal solvent'. It does, indeed, dissolve most of the chemicals familiar to us, even if only to a small extent. For example, each time we drink water from a tumbler we are also drinking a *minute* amount of dissolved glass. This and other similar effects are insignificant, but the fact that water dissolves so many substances can be vitally important to mankind, both beneficially and otherwise. In the former category is the fact that many important chemical reactions take place in aqueous solution. We tend to take this for granted and yet most of the chemistry experiments you conduct take place in water. Similarly the complex chemistry of life which takes place in each of the cells in our bodies does so in aqueous solution; indeed about two-thirds by weight of the human body consists of water. Again, the waste chemicals which our bodies produce are removed from the body by being dissolved in water. On the debit side water also dissolves many substances, of mineral and industrial origin, which we would prefer to have remained insoluble; the task of purifying water for domestic use is made the more difficult by the fact that it is such a good solvent.

Chemical Properties

Water is such a stable and familiar substance that we tend to forget its importance as a chemical. For example, many chemical reactions between gases appear to need water as a catalyst. Water molecules make bonds with a variety of substances. You should be aware of the heat evolved when water reacts with anhydrous copper(II) sulphate as an example of this kind of reaction (page 367). Metal ions do not exist free in aqueous solution; they are usually combined with water. Throughout the course you have met examples of chemical changes directly involving water. How many of these can you remember? Make a list of as many as possible and then look through the work you have already done to see how many you have missed. You should refer to:

(a) the reaction of water or steam on metals (Section 5.3);
(b) the reaction of water or steam on non-metals (Section 5.3);
(c) the reaction of water with sodium and calcium oxides (Chapter 10);
(d) the reaction of water with common non-metallic oxides (Chapter 11);
(e) the electrolysis of aqueous solutions (Section 6.1);
(f) heats of solution (Section 13.2).

Hydrolysis Water reacts chemically with some compounds to form at least *two* products. Such a reaction is called a hydrolysis, and both metallic and non-metallic chlorides provide useful examples.

Anhydrous aluminium chloride is usually prepared by the action of dry chlorine on the metal. The hydrated salt is easily made by the usual methods, but it cannot be converted into the anhydrous salt by evaporating the water, because anhydrous aluminium chloride reacts with (is hydrolysed by) the water:

$$AlCl_3(s) + 3H_2O(l) \rightarrow Al(OH)_3(s) + 3HCl(g)$$

Iron(III) chloride behaves similarly and so moisture must be excluded when either of these anhydrous salts are prepared.

$$FeCl_3(s) + 3H_2O(l) \rightarrow Fe(OH)_3(s) + 3HCl(g)$$

You will notice that hydrogen chloride is produced in both reactions so that solutions of these salts tend to be acidic. Similarly, many non-metallic chlorides are also hydrolysed to form acidic solutions.

Tests for Water

The *presence* of water can be detected by the use of anhydrous copper(II) sulphate (which turns from white to blue) or anhydrous cobalt(II) chloride ($CoCl_2$) which turns from blue to pink. Remember that *any* aqueous solution will react positively to these tests and they do not indicate the presence of pure water.

How would you decide conclusively whether a liquid is pure water?

Water of Crystallization

…ence of water can …ted

Crystals of various salts often contain water molecules which are chemically combined with the salt, and such water is called water of crystallization. It is generally an essential part of the structure of the crystals and gives them their characteristic colour. In cases where this water is driven off, the crystalline shape and colour are lost. Crystals which contain water of crystallization are said to be *hydrated,* and if the water is driven off the substance becomes *anhydrous.* Thus in blue copper(II) sulphate pentahydrate crystals, $CuSO_4, 5H_2O$, each copper(II) sulphate unit has five molecules of water bonded to it. If the water molecules are driven off leaving anhydrous copper(II) sulphate, the characteristic crystalline shape disappears, as does the blue colour.

Other common examples of hydrated salts include:

$Na_2CO_3, 10H_2O$	sodium carbonate decahydrate
$FeSO_4, 7H_2O$	iron(II) sulphate heptahydrate
$CaCl_2, 6H_2O$	calcium chloride hexahydrate

Note that it is possible for a salt to form several hydrates, e.g. $Na_2CO_3, 10H_2O$ and Na_2CO_3, H_2O.

Efflorescent, Deliquescent and Hygroscopic Substances

Experiment 9.1

† To Illustrate Deliquescent, Efflorescent, and Hygroscopic Substances **

Apparatus

Spatula, three watch-glasses, balance sensitive to 0.01 g.

Anhydrous copper(II) sulphate, sodium hydroxide flakes or pellets, clear crystals of sodium carbonate decahydrate.

Procedure

(a) Place two spatula measures of anhydrous copper(II) sulphate on a watch-glass and weigh the glass and contents. Leave in the open laboratory for twenty-four hours.

(b) Repeat (a) using one spatula measure of (i) sodium hydroxide (CARE) and (ii) crystals of sodium carbonate decahydrate.

(c) Reweigh all three samples after the specified time and record all your results and observations.

Points for Discussion

1. When some compounds are exposed to the air they extract water vapour from it and then proceed to *dissolve* in the water. This process is called *deliquescence*. Which of the three substances used in the experiment is deliquescent? Explain your reasoning.

Try to find other examples of deliquescence by examining bottles of chemicals which have been in use for some time. You may find that some deliquescent chemicals have formed solutions in their bottles.

2. Substances which lose water of crystallization to the air are said to *effloresce*. Which of the chemicals used is efflorescent? Explain your answer.

3. Fresh crystals of sodium carbonate decahydrate should be clear. They have the formula Na_2CO_3, $10H_2O$. Crystals which have been exposed to the air become covered with a white *powder* of formula Na_2CO_3, H_2O. Do you think that the 'missing' water of crystallization is necessary for the crystalline structure of sodium carbonate? Explain your answer.

4. *Hygroscopic* substances absorb moisture from the air but do not dissolve in it. Anhydrous copper(II) sulphate and many other anhydrous salts are good examples of hygroscopic substances.

9.2 HARDNESS OF WATER. SOAPS AND DETERGENTS

Introduction

As we mentioned earlier, rivers, lakes, and reservoirs accumulate much dissolved and suspended matter. In order to obtain water suitable for drinking we usually remove the suspended solids by filtration and then sterilize the water so that it is free from harmful bacteria. Chlorine, at a concentration of one part per million (one ppm), is a common sterilizing agent.

The dissolved substances remaining are quite harmless, and some may be beneficial in drinking water, but these dissolved ions can be inconvenient if the water is to be used for other purposes, as the next few experiments will show.

Experiment 9.2
Do Dissolved Ions React With Soap? *

Apparatus

Burette, 10 cm^3 pipette, funnel, conical flask.

Standard soap solution, deionized water, 0.005M solutions of sodium chloride, potassium chloride, magnesium chloride, and calcium chloride.

Procedure

(a) Pipette 10 cm^3 of deionized water into the conical flask. Pour the soap solution into the burette. Add 0.5 cm^3 of soap solution to the water and shake vigorously. Continue adding the soap solution 0.5 cm^3 at a time, and shake after each addition, until a lather is formed which remains for half a minute. Record the total volume of soap solution used.

(b) Wash out the flask and repeat (a) with 10 cm^3 of (i) sodium chloride solution, (ii) potassium chloride solution, (iii) calcium chloride solution, and (iv) magnesium chloride solution instead of the deionized water.

Points for Discussion

1. If water does not readily form a lather with soap it is said to be hard. Compare the volumes of soap solution needed to produce a lather with the various solutions and with pure water. Do some dissolved ions cause water to become hard? Which ions are these? Explain your answer.

2. Explain how the experiment shows that chloride ions are not responsible for hardness. Other negative ions also have no effect.

3. Did you notice the formation of a precipitate in any of the reactions? If so which ones?

How Do Magnesium and Calcium Ions Accumulate in Water?

Salts of calcium and magnesium are widely distributed in nature, particularly those of calcium. Any soluble salts are readily transferred to streams and rivers and these include the relatively abundant sulphates and, to a much less significant extent, the chlorides of calcium and magnesium. The sulphate of magnesium is found to a small extent in some parts of Britain but calcium sulphate, as gypsum or anhydrite, occurs more frequently. Although calcium sulphate is only slightly soluble in water (1:500) this is enough to render the water hard.

Insoluble compounds of calcium and magnesium can, however, be converted into soluble forms by chemical action with water containing carbon dioxide dissolved from the air. Calcium carbonate occurs abundantly in Britain as chalk or limestone, and although it is insoluble the following experiment shows how it can be converted into a soluble compound.

Experiment 9.3

† The Action of Water Containing Carbon Dioxide on Calcium Carbonate *

Apparatus

Test-tube (150 × 25 mm) in rack, spatula, glass rod.
Carbon dioxide generator, finely divided calcium carbonate.

Procedure

Pour about 20 cm^3 of water into the test-tube and add a very small measure of calcium carbonate. Stir to obtain a suspension. Pass carbon dioxide through the suspension until no further change takes place. Record your observations.

Point for Discussion

You saw how the insoluble calcium carbonate gradually disappeared as it reacted with carbon dioxide and water. You have seen this reaction before when carbon dioxide and calcium hydroxide solution form a precipitate of calcium carbonate, but excess carbon dioxide causes the milkiness to disappear, as in the above reaction. Calcium hydrogen carbonate is produced by the reaction and dissolves immediately as it is soluble.

$$CaCO_3(s) + CO_2(aq) + H_2O(l) \rightleftharpoons Ca(HCO_3)_2(aq)$$

$CaCO_3(s)$	$CO_2(aq)$	$H_2O(l)$	$Ca(HCO_3)_2(aq)$
From chalk or limestone *insoluble*	From the air	Rain	*soluble*

This reaction occurs on a large scale in chalk or limestone districts.

Thus natural waters readily accumulate dissolved calcium compounds and, to a lesser degree, magnesium compounds. The amount and, as you will learn later, the type of hardness, depend upon the nature of the rocks over which the water flows.

Stalactites and Stalagmites

The reaction of water, containing dissolved carbon dioxide, on calcium carbonate has other important consequences. Caves are frequently found in limestone areas and as water trickles down through the earth to the roof of a cave, a chemical reaction produces a dilute solution of calcium hydrogen carbonate as described.

As drops of the solution hang from the roof of the cave, water evaporates and the reaction is reversed, resulting in the formation of a deposit of calcium carbonate on the roof.

$$Ca(HCO_3)_2(aq) \rightleftharpoons CO_2(g) + H_2O(g) + CaCO_3(s)$$

During the course of time this deposit builds up and grows down from the cave so that *stalactites* are produced (Figure 9.2). Can you suggest how *stalagmites* are formed? Why is each one usually opposite a stalactite?

Figure 9.2 Stalactites and stalagmites in limestone caves

The Action of Soap With Hard Water

The composition and preparation of soap A detergent is simply something which is capable of cleaning, and soap is one form of detergent. Most soaps are the sodium or potassium salts of organic (carboxylic) acids such as stearic acid, $CH_3(CH_2)_{16}COOH$. Thus a typical soap could be: $CH_3(CH_2)_{16}COONa$, sodium stearate.

Soaps have been in use for a very long time and the basic manufacturing process is to hydrolyse natural oils or fats, which are compounds (esters) derived from organic acids, with sodium hydroxide or potassium hydroxide solution. In the process the fats are broken down to liberate the free carboxylic acids, which then react with the alkali in the usual way to form salts, which are soaps. A typical oil or fat may be considered to contain molecules of the *type:*

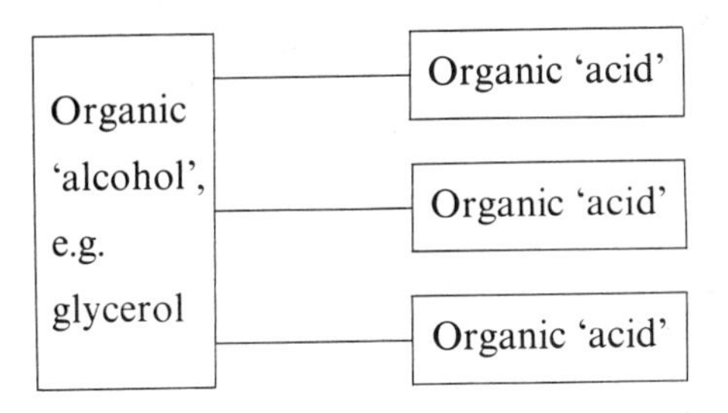

These react with sodium hydroxide to form:

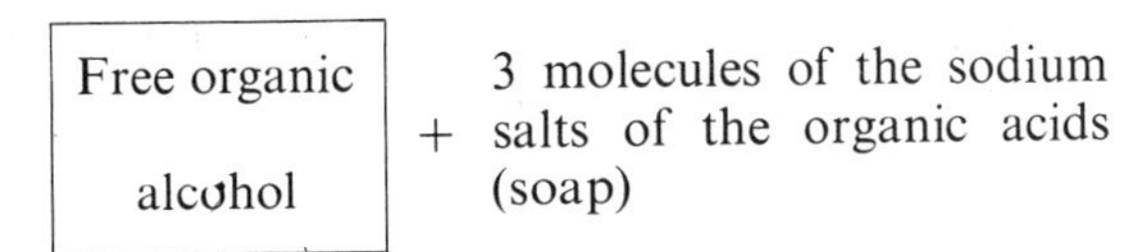

The principle is illustrated in the next experiment.

Experiment 9.4

†A Simple Laboratory Preparation of a Soap **

Apparatus

Test-tubes, beaker, tripod, gauze, Bunsen burner, asbestos square, filter funnel, filter paper, glass rod, spatula, 10 cm^3 measuring cylinder.

Castor oil, approximately 5M sodium hydroxide solution, sodium chloride, deionized water, perfume and colouring matter (if desired). Goggles should be worn when using the alkali.

Procedure

(a) Mix approximately 2 cm^3 of castor oil (an ester derived from organic acids and alcohols) and about 10 cm^3 of the sodium hydroxide solution in the beaker. *Warning:* the alkaline solution is much more concentrated than that which you normally use, so take great care not to spill any. Report any accident immediately.

(b) Slowly bring the contents of the beaker to the boil, stirring constantly. After the mixture has boiled gently for five minutes add about 10 cm^3 of deionized water and five or six spatula measures of sodium chloride. Stir constantly and boil the mixture for a further two or three minutes.

(c) Add the perfume and colouring matter as the mixture cools, and when cool break up the solid pieces with a spatula. Filter or pour off the liquid and wash the solid two or three times with a *little* deionized water.

(d) When the solid is dry test its detergent qualities by scraping a little into a test-tube and shaking with a little deionized water. Are you convinced that you have made a soap? *Note:* the sample is not sufficiently pure to be applied to the skin.

The soap sample is not sufficiently pure to be applied to the skin

What happens when a soap dissolves in hard water? Soaps readily dissolve in water, the acid anions and sodium or potassium cations separating and spreading out:

$$\text{NaStearate(s)} \rightarrow \text{Na}^+\text{(aq)} + \text{Stearate}^-\text{(aq)}$$
(Solid soap) (Soap solution)

When soap is added to hard water there are likely to be Stearate^-, Na^+, Ca^{2+}, and/or Mg^{2+} ions present. The calcium and magnesium salts of stearic acid are insoluble in water. What is likely to happen in this 'soup' of ions? (See double decomposition, Section 3.5.) Did you *see* anything when you mixed soap solution with calcium and magnesium salts in Experiment 9.2? Can you now explain what you saw?

You will learn later that the essential cleansing factor in soap is the acid anion, i.e. the stearate ion in our example. If this acid anion is not available to react with the grease and dirt then the soap cannot function as a detergent.

Points for Discussion

1. What happens to the acid anions first released when soap is placed in hard water? Are they left as free acid anions? If not, why not?

2. Does this explain why the soap will not at first form a lather with hard water?

3. Why does a lather form after further addition of soap?

4. Why is it better to use soap in a fixed volume of hard water, say in a sink or bowl, rather than in running hard water?

Disadvantages of hard water The fact that soap is wasted in precipitating all the calcium and magnesium ions from a sample of hard water before it can function properly is important, but there are other disadvantages too. The precipitate (scum) of calcium and/or magnesium stearate is difficult to remove from fabrics which are being washed in hard water. The calcium and magnesium ions also interfere with dyeing processes.

You have learned that when water containing dissolved calcium hydrogen carbonate (a form of hard water, remember) evaporates, as on the roof of a cave, a deposit of calcium carbonate is formed. This also happens when such solutions are boiled. You may have seen a simple illustration of this at home if you live in a hard water area. The inside of a kettle becomes covered with a deposit of chalk (i.e. 'furred'). This is merely inconvenient, but on a large scale it can be disastrous. If untreated hard water is used in industrial heating systems the pipes carrying the water gradually accumulate deposits of chalk and may eventually become blocked. If the plant has to close down for the pipes to be cleaned a great deal of money is lost. You can imagine how such problems could affect steam engines and laundries.

Water required for laboratory purposes must also be freed from calcium and magnesium ions before it is used.

It is obviously important to remove calcium and magnesium ions from water for a variety of reasons. The process is called *softening* and water which contains only a negligible proportion of calcium and magnesium ions is said to be soft.

Permanently Hard and Temporarily Hard Water

Before we study the methods available for softening hard water we must understand the difference between water which is said to be temporarily hard and that which is said to be permanently hard.

Experiment 9.5

†To Distinguish Between Temporarily Hard Water and Permanently Hard Water *

Apparatus
10 cm^3 pipette, conical flask, burette, Bunsen burner, asbestos mat, three beakers.
Soap solution, calcium sulphate solution (saturated solution diluted by water in the ratio 1:2), calcium hydrogen carbonate solution (made as in Experiment 9.3 but diluted by an equal volume of water), a mixture of the calcium sulphate and calcium hydrogen carbonate solutions (1:1).

Procedure
(a) Arrange for bulk samples of the calcium salt solutions to be boiled for ten minutes and then allow them to cool. This can be shared between groups.

(b) Pipette 10 cm^3 of the original (unboiled) solution of calcium sulphate into a conical flask. Place the soap solution in the burette. Run the soap solution slowly into the conical flask, with vigorous shaking, until a lather is formed which lasts for half a minute. Record the volume of soap solution used.

(c) Repeat (b) using 10 cm^3 of each of the other two unboiled solutions and then with 10 cm^3 of each of the boiled samples.

Points for Discussion

1. One of the calcium salt solutions is principally responsible for temporary hardness, one for permanent hardness, and the other is of course a mixture of the two. Use your results to explain which solution you think is called permanently hard and which is called temporarily hard, explaining your reasoning carefully.

2. Why would it be incorrect to measure out 10 cm^3, boil it, and then titrate it with soap solution?

3. Which of the following would be a suitable definition of temporary hard water?

A Water which never readily forms a a lather with soap
B Water which is softened by boiling
C Water which is unaffected by boiling

4. Samples of water were taken from three towns, P, Q, and L. 10 cm^3 portions of each were titrated with soap solution in the usual way. This was repeated with 10 cm^3 of boiled water from each town. 10 cm^3 of deionized water was used as a reference test. The results were as follows:

Type of water	*Volume of soap solution needed to produce a lather with 10 cm^3 of untreated water (cm^3)*	*Volume of soap solution needed to produce a lather with 10 cm^3 of boiled water (cm^3)*
Deionized	1	1
P	10	10
Q	15	10
L	20	5

What can you conclude about the water from each of the towns? Over which kind of rocks do you think the water flowed in the catchment area for each town? Explain your answers.

5. Why is sea-water not suitable for washing in? Is it hard?

Methods of Softening Water

The basic principle is to remove the calcium or magnesium *ions* from the water. Some processes remove all significant traces of calcium or magnesium, but more frequently the soluble calcium or magnesium compounds are converted into insoluble compounds. Remember that only *dissolved* calcium or magnesium *ions* can react with stearate ions; any insoluble form of calcium or magnesium will not produce hard water. All kinds of hard water can be softened, so the expression 'permanently hard' is rather unfortunate.

The type of process used depends upon the degree and type of hardness, and economic factors.

Boiling As you have learned, boiling can soften temporarily hard water and partially soften a mixture of temporarily and permanently hard water. The reaction is a 'speeded up' version of what happens in stalactite formation and identical to that which produces 'furred' kettles:

$$Ca(HCO_3)_2(aq) \rightleftharpoons CaCO_3(s) + CO_2(g) + H_2O(g)$$

A furred kettle?

The water is rendered soft because the soluble calcium ions (from the calcium hydrogen carbonate) are converted into insoluble calcium carbonate, which is precipitated. The precipitate can be inconvenient but the main objection to the method is the cost of fuel. Remember also that permanent hardness cannot be removed by this method as the sulphates and the chlorides of magnesium and calcium are not affected by boiling their solutions.

The addition of calcium hydroxide (slaked lime) This is called Clark's process. The calcium hydroxide reacts with the calcium hydrogen carbonate as follows:

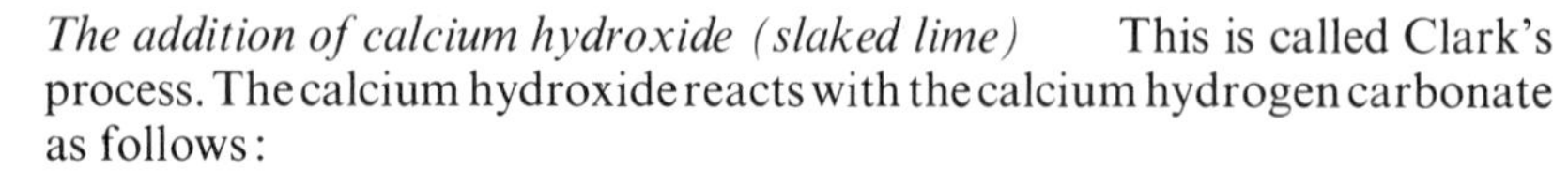

$$Ca(OH)_2(aq) + Ca(HCO_3)_2(aq) \rightarrow 2CaCO_3(s) + 2H_2O(l)$$

Thus the end result is the same as that when temporarily hard water is boiled; the calcium ions are converted into an insoluble form.

Clark's method cannot be used to soften permanently hard water; in fact it would make it harder, for the concentration of Ca^{2+} would be increased. This is because the added calcium ions do not react. When the method is used to soften temporarily hard water only a calculated amount of calcium hydroxide must be added; too much makes the water hard again. The method is only employed at the water works and not in the home.

The addition of sodium carbonate (washing soda) The two methods so far mentioned have shown that if calcium ions can be converted into insoluble calcium carbonate, hard water is rendered soft. In the next experiment you will add sodium carbonate to separate solutions of temporarily hard water and permanently hard water. Predict what should happen chemically in each case (with ionic equations) and explain your prediction. Also predict whether sodium carbonate should soften either or both of the samples, again explaining your prediction, and then perform the experiment to test your ideas. Comment on your results, the feasibility and economics of the process, and its usefulness.

Experiment 9.6
† The Action of Sodium Carbonate on Hard Water *

Apparatus
10 cm^3 pipette, spatula, conical flask, burette.
Calcium hydrogen carbonate solution and calcium sulphate solution (as in Experiment 9.5), deionized water, solid sodium carbonate, soap solution.

Procedure
You know the aim of the experiment. Devise a suitable procedure. Report on the experiment as indicated and if necessary modify your predictions.

Point for Discussion

For economic reasons the last two methods are often combined because calcium hydroxide is cheaper than sodium carbonate. Enough calcium hydroxide is added to remove the temporary hardness and then sodium carbonate is added to remove the permanent hardness. Sodium carbonate could soften both types but the cost would be greater.

Ion exchange This method differs from the previous ones in that Ca^{2+} and Mg^{2+} ions are *removed* from the water and not converted into an insoluble form. An ion exchange resin usually consists of an inert 'backbone' material (e.g. a plastic such as polystyrene) on to which ionic groups are weakly attached. These may be cations such as H_3O^+ or Na^+ (in a cation exchange resin) or anions such as OH^- (in an anion exchange resin).

If a solution containing calcium ions (e.g. hard water) is passed down a glass column packed with a cation exchange resin, the calcium ions are preferentially adsorbed on to the resin and are exchanged with the cations already attached to the resin. If the ions on the resin are sodium or potassium ions the water is made soft because such ions do not react with soap to form insoluble stearates (Experiment 9.2).

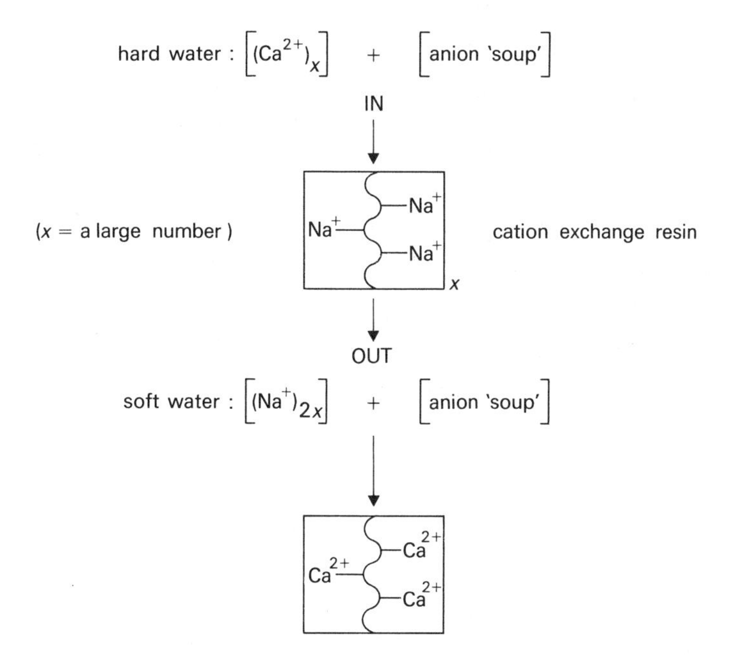

Many ion exchange resins are available for specific purposes (e.g. Permutit). Although this is not strictly connected with hardness of water, the deionized water used in many laboratories is produced by passing tap water through both a cation and an anion exchanger. H_3O^+ and OH^- ions are exchanged for the ions present in the water and they combine together to form more water:

$$H_3O^+(aq) + OH^-(aq) \rightarrow 2H_2O(l)$$

The water emerging from the exchanger is then said to be 'deionized'.

There are many natural ion exchange systems as well as the synthetic ones in general use. Alumino-silicates and other clay-like materials cause ion exchange processes to occur in the soil; such reactions are vital for plant life.

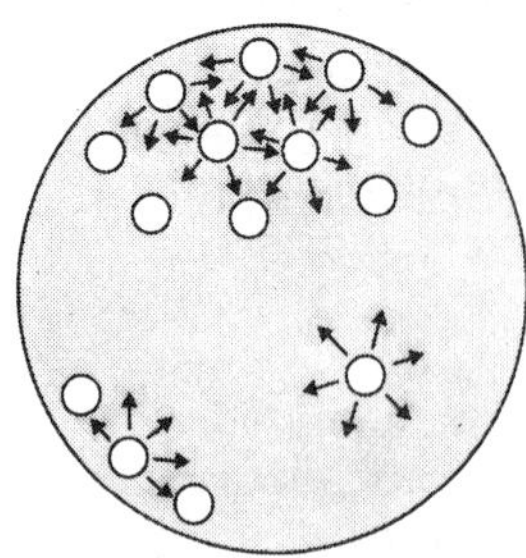

Figure 9.3 Intermolecular forces in a drop of water, the intermolecular forces are shown for a few molecules only and the intermolecular distances are exaggerated

After a period of time an ion exchange resin becomes exhausted when all its original ions have been exchanged for others. Regeneration is a simple matter. A *concentrated* solution containing the original ion (e.g. Na^+) is passed through the resin and the process is reversed so that the replenished resin can be used again.

Phosphate treatment This method is quite different from the others. Complex polyphosphates such as Calgon (highly polymerized sodium phosphate) are added to hard water. The calcium and magnesium ions are not converted into insoluble compounds, nor are they removed completely from the system, but they are *effectively* removed by being 'locked up' as part of a complex phosphate structure. The method has the additional advantage of preventing corrosion of pipes carrying the treated water. Calgon is a component of most washing up liquids.

Soap and Detergents

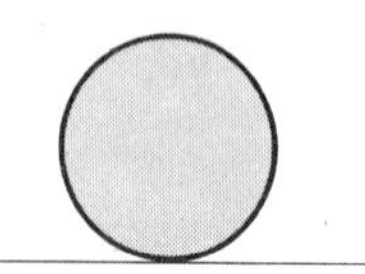

Droplet of water on a surface. The surface is not 'wetted' efficiently

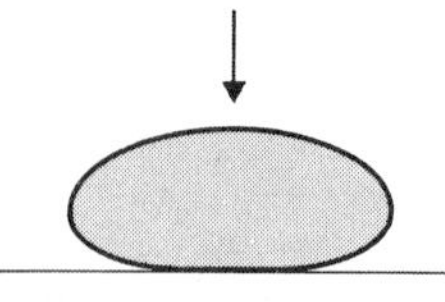

The droplet spreads when a detergent is added. The water becomes a better wetting agent

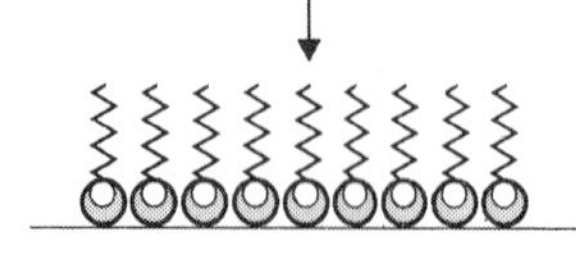

Molecules of water in the droplet are dispersed as the surface tension is lowered by the presence of detergent ions

Figure 9.4 (a)

We all use soaps and other detergents, but it is probably not readily understood what exact functions they perform or what the difference is between a soapy detergent and a soapless detergent.

You may be surprised to learn that water alone is not an adequate wetting agent. As an example, pour some water into a test-tube and then pour it out again. Examine the inside surface of the test-tube, and you will almost certainly find that it is not evenly wet; there are dry patches from which the water has 'shrunk'.

The reason for this is the *surface tension* of the water. The molecules near the centre of a drop of water are attracted equally to the surrounding molecules by intermolecular forces, but molecules on the surface of the drop can only be attracted inwards (Figure 9.3). This results in a net inward pull on the outer molecules in a particular drop and it tends to shrink away from adjacent droplets. This is exactly what happened to the drops of water left on the inside of the test-tube.

If you add a little detergent to some water in a test-tube and then pour it out again you will find that the inside surface is more completely wetted than it was before. This is because a detergent lowers the surface tension of water by weakening the intermolecular forces so that the molecules can spread out and thoroughly wet a surface. This is one of the main functions of a detergent (Figure 9.4 (a) and (b)).

Figure 9.4 (b) Detergents make water a better wetting agent

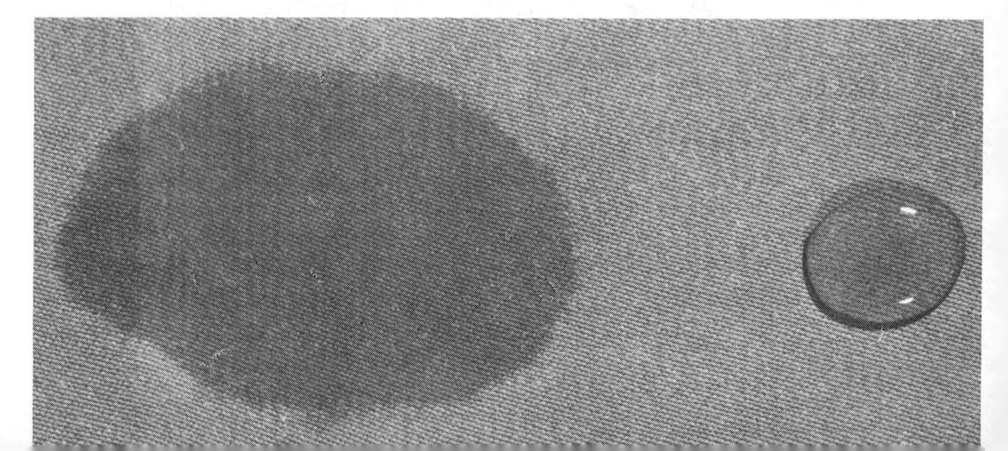

Figure 9.5 Surface tension of water *(Unilever)*

Experiment 9.7

To Illustrate the Effect a Detergent has on the Surface Tension of Water*

Apparatus
Beaker, teat pipette.
Soap solution, a soapless detergent, powdered carbon, hard water, deionized water.

Procedure
(a) Half fill the beaker with deionized water and sprinkle the surface with powdered carbon. Carefully apply one drop of soap solution to the middle of the floating powder. Record your observations.

(b) Thoroughly wash out the beaker and then repeat (a) using a drop of the soapless detergent instead of the soap solution.

(c) Thoroughly wash out the beaker and then repeat (a) and (b) using the hard water instead of the deionized water.

Points for Discussion
1. Use your knowledge of surface tension to explain why carbon floats on deionized water.

2. Explain what happened when a detergent was added to the deionized water. How did the detergents cause this effect?

3. Which type of detergent best lowers the surface tension of hard water?

4. How would you float a needle or a coin on the surface of water as in Figure 9.5?

How Does a Detergent Lower the Surface Tension of Water?

All detergents have the same basic structure. Their molecules have a hydrocarbon chain which is covalently bonded and will not attract water molecules (the chain is said to be *hydrophobic* or water hating), and a small ionic group which readily attracts, and dissolves in, water molecules (a *hydrophilic* part). In the case of ordinary soap, which may contain a large proportion of sodium stearate, the hydrophobic part is the long hydrocarbon chain of $-CH_2-$ units (the 'tail') and the hydrophilic part is the ionic $-COO^-$ group (the 'head') at the end. Soapless detergents have similar structures although the ionic group is then usually sulphate or sulphonate as explained later.

A typical soapy detergent will contain molecules of the type:

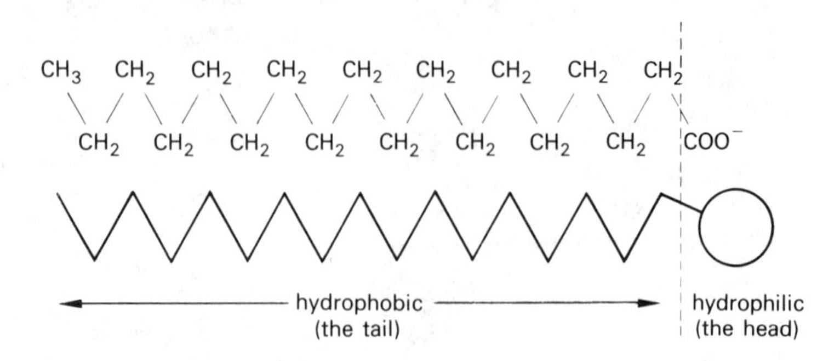

When a detergent is added to water the hydrophilic groups attract water molecules but the hydrophobic groups repel them. The effect is to reduce the intermolecular forces, and consequently the surface tension, so that the molecules of water are spread more evenly.

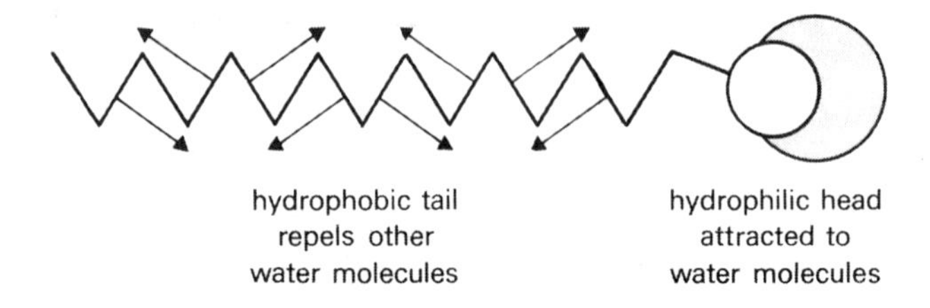

Removal of Dirt

This is another important function of detergents and is fairly easy to understand. It again depends upon the structure of the detergent molecules. Grease is usually of organic origin and always covalently bonded. The hydrophobic tails of the detergent molecules readily 'dissolve' in the grease molecules as they too are covalent. The heads of the detergent molecules, as usual, dissolve in water molecules so that slight agitation dislodges the grease and it is carried away with the water (Figure 9.6). Dirt is removed at the same time because it usually adheres to the grease.

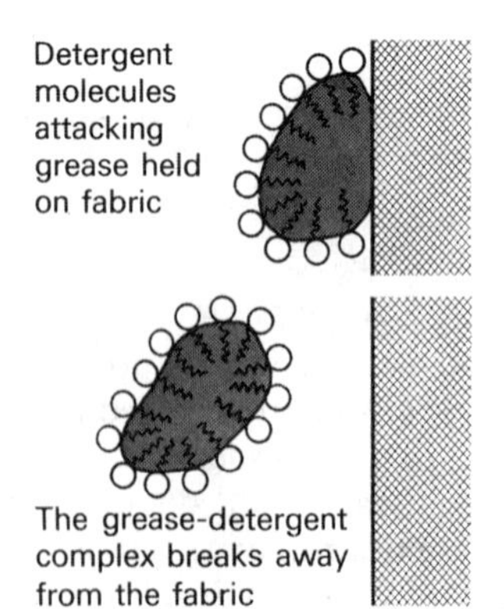

Figure 9.6 The action of detergents on grease (simplified). The ionic heads are carried away with the water and take the grease with them

The Differences Between Soapy Detergents and Soapless Detergents

As we said earlier, soaps or soapy detergents are one subdivision of the class of materials called detergents. The other main subdivision is soapless detergents.

Soapy detergents are manufactured, as described earlier, by the hydrolysis of vegetable oils or animal fats by caustic alkali. They were the first known detergents and are divided into soft soaps (usually potassium salts), hard soaps (sodium salts), soap flakes, soap powders, and toilet soaps (soft soaps with perfume and other additives). They all consist of salts of long chain carboxylic acids, i.e. containing the carboxyl (COOH) group, attached to a long hydrocarbon chain.

In Section 4.4 it was explained how vegetable oils superseded animal fats in the manufacture of margarine when it became possible to harden the oils by addition of hydrogen. This technique was also important in the manufacture of soap, for vegetable oils could be hardened and then converted into soap by hydrolysis with caustic alkali. However, as the population of the world increased, so did the demand for margarine and other foods derived from vegetable oils, and the proportion of oil available for the manufacture of soap diminished rapidly.

Materials which possessed the properties of soap but could be made from substances other than oils and fats were urgently required, and research led to the development of what are now known as soapless detergents. The essential characteristics are the same. Soapless detergents have a long hydrophobic chain of hydrogen and carbon atoms, and an ionic hydrophilic group. The difference is that the hydrocarbon chain is usually obtained from petroleum and the ionic group has to be introduced synthetically, whereas the carboxylic group in acids derived from fats is present naturally. Hydrocarbons are, of course, very readily obtained from petroleum, and the ionic groups inserted are usually either sulphate ($-O-SO_3^-$) or sulphonate ($-SO_3^-$), concentrated sulphuric acid being required for either process.

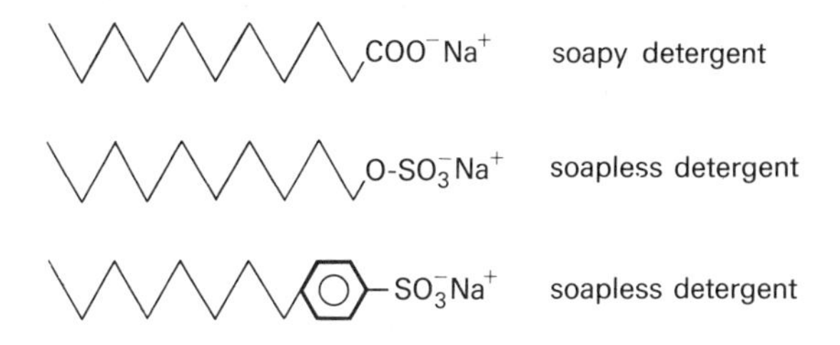

The Advantages of Soapless Detergents

Economically, the use of petroleum products as raw materials relieves the pressure on the demand for vegetable oils and animal fats needed for food substances. Practically, the variety of possible synthetic soapless detergents far exceeds the number of soaps. This means that specific soapless detergents can be manufactured for a particular purpose by using a certain kind of hydrocarbon chain or by inserting more than one ionic group. There are thousands of soapless detergents now in use; obvious examples include washing powders (often graded according to the fabrics they are best suited for) and washing up liquids. As a complete contrast, research workers are concerned with the development of detergents which disperse crude oil discharged at sea but do not endanger marine life. Other modern developments include the use of enzyme powders, which are specially formulated to attack stains containing biological protein, such as blood and sweat. This increased efficiency in cleansing properties has been offset to some extent by certain problems which have arisen due to the wide-spread use of soapless detergents. Huge foam deposits have appeared on

Specific soapless detergents can be manufactured for a particular purpose

rivers and at sewage works because the hydrocarbon chains, unlike those of soapy detergents which are derived from food substances, cannot be broken down by bacteria and dispersed. Nowadays, biodegradable soapless detergents are manufactured, the hydrocarbon chains of which are readily broken down by bacteria so that foam does not accumulate in effluent. Soapless detergents now have a 45 per cent share of the British detergent market and their use is expanding constantly.

Although soapless detergents function in the same way as soapy detergents in lowering the surface tension of water and removing dirt and grease, they have one other very important advantage apart from those outlined earlier. The calcium and magnesium salts of soapless detergents are soluble in water whereas the corresponding salts of soapy detergents are not. This means that soapless detergents can be readily used in hard water; there is no precipitate (scum), nor is any detergent wasted in removing the magnesium and calcium ions before the detergent operation can begin (Figure 9.7). This explains why the surface tension of hard water in Experiment 9.7 was lowered immediately by the application of a soapless detergent, but not by the action of soap.

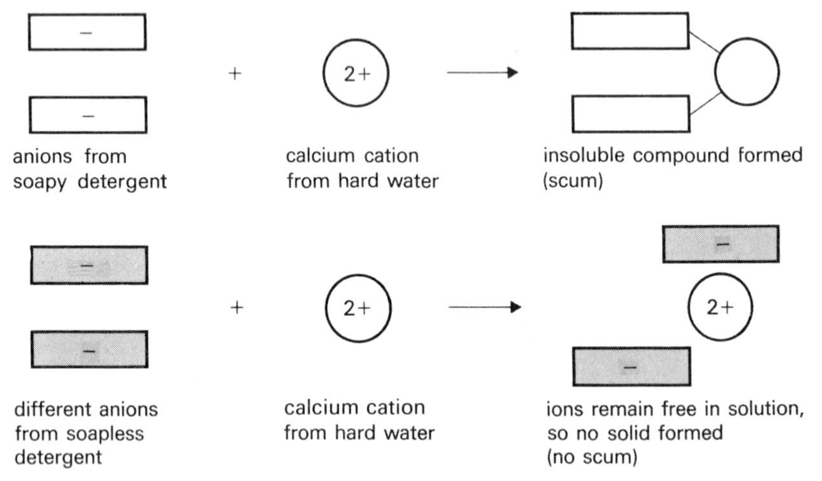

Figure 9.7 The action of soapy and soapless detergents on hard water

9.3 HYDROGEN PEROXIDE

Properties of Hydrogen Peroxide

Physical Properties

Hydrogen peroxide is always used in the form of its aqueous solution as it is extremely unstable when pure. The pure liquid is miscible with water in all proportions, and the concentration of a hydrogen peroxide solution is usually measured by its 'volume strength'. When hydrogen peroxide decomposes oxygen is formed:

$$2H_2O_2(aq) \rightarrow 2H_2O(l) + O_2(g)$$

and the volume strength corresponds to the number of volumes of oxygen potentially obtainable from one volume of solution. Thus ten volume hydrogen peroxide is of such a strength that 1 cm^3 of it can be decomposed to yield 10 cm^3 of oxygen; one litre would yield ten litres of oxygen, and so on.

Chemical Properties

Decomposition into water and oxygen Aqueous solutions of hydrogen peroxide, although they are much more stable than the pure liquid, are unstable. Bright sunlight and heat accelerate the decomposition into water and oxygen, hence the use of brown bottles for storage.

The solutions are made more stable by the addition of a negative catalyst such as sodium stannate or a dilute acid. The decomposition is accelerated at room temperatures by the addition of a positive catalyst such as manganese(IV) oxide and the reaction provides a very convenient preparation of oxygen (Section 4.3).

Hydrogen peroxide as an acid Pure hydrogen peroxide is a weak acid and has the usual effects on indicators, but dilute aqueous solutions often appear to be neutral. Salts of hydrogen peroxide are called peroxides.

Hydrogen peroxide as an oxidizing agent Hydrogen peroxide is a powerful oxidizing agent. This is explained by its readiness to decompose into oxygen (and water) and by its ability to accept electrons, particularly in acid solution. Revise the definitions of oxidation in Chapter 8.

$$H_2O_2(aq) \rightarrow H_2O(l) + [O](g)$$

or

$$\underset{}{H_2O_2(aq)} + \underset{\text{from acid}}{2H^+(aq)} + \underset{\text{from substance being oxidized}}{2e^-} \rightarrow 2H_2O(l)$$

Experiment 9.8
The Action of Hydrogen Peroxide on Lead Sulphide ***

Apparatus
Teat pipette, microscope slide or tile, tongs.
Access to a hydrogen sulphide generator (in fume cupboard), lead ethanoate (acetate) paper, ten volume hydrogen peroxide solution.

Procedure
(a) Moisten a piece of lead ethanoate paper with water, hold it in a pair of tongs, and expose it to hydrogen sulphide in the fume cupboard.

(b) Place the treated paper on a tile or microscope slide, and spot a few drops of hydrogen peroxide on to the paper from a teat pipette. Record what happens.

Points for Discussion
1. The action of hydrogen sulphide on soluble lead salts such as lead ethanoate is used as a test for the gas. A double decomposition takes place and a black precipitate of lead sulphide forms, in this case on the paper.

$$Pb^{2+}(aq) + S^{2-}(aq) \rightarrow PbS(s)$$

This is just a convenient way of preparing a little lead sulphide and is not connected with the properties of hydrogen peroxide.

2. The product of the reaction between hydrogen peroxide and lead(II) sulphide is white lead(II) sulphate, $PbSO_4$. Devise an equation for the reaction and explain why it shows hydrogen peroxide to be an oxidizing agent. Indicate clearly what has been oxidized and reduced during the reaction. You may use any definition of oxidation, including oxidation number.

3. This reaction is utilized in the restoration of old oil paintings. In view of what you have learned in this experiment, what do you think is being restored in such cases, and why does it become necessary?

4. Another example of the use of hydrogen peroxide as an oxidizing agent is the oxidation of iron(II) to iron(III), Experiment 10.16.

Hydrogen peroxide as a bleaching agent Hydrogen peroxide is a good bleaching agent. It will bleach most varieties of (grease free) human hair and both synthetic and natural fibres. These are also further examples of oxidation.

The Test for Hydrogen Peroxide If hydrogen peroxide, dilute sulphuric acid, potassium chromate(VI) solution, and pentanol (care, flammable) are mixed and shaken together, a blue colour is formed in the top layer.

Uses of Hydrogen Peroxide

100 per cent hydrogen peroxide is sometimes used to assist oxidation of rocket fuels. As previously mentioned, hydrogen peroxide is a good bleaching agent, and as such it is used in dilute solution in toothpastes, laundries, and hair bleaches. It has antiseptic properties which are utilized in cleaning cuts and in mouthwashes.

QUESTIONS CHAPTER 9

1. It is very difficult to form a lather with ordinary soap in sea-water. (a) Why is this so? (b) What would be formed instead of the lather? (c) What type of cleanser would form a lather with sea-water? Why is this type of substance effective when soap is not? (J. M. B.)

2. Differentiate between 'temporarily' and 'permanently' hard water.

3. Describe carefully what happens when carbon dioxide is passed through calcium hydroxide solution (limewater) until no further change occurs. Give equations.

4. Explain how stalactites and stalagmites are formed. Why does one usually form opposite the other?

5. Explain the terms 'efflorescent', 'deliquescent' and 'hygroscopic', and give examples to illustrate your answer.

6. Which one of the following *cannot* be used to soften permanently hard water?
A Calcium hydroxide
B Distillation
C Sodium carbonate
D An ion exchange resin
E Calgon

7. A hygroscopic substance
A is too small to be seen with the naked eye
B dissolves in moisture from the air
C absorbs moisture from the air
D absorbs hydrogen very rapidly
E reacts rapidly with hydrogen

8. Which of the following is not true about the hydrated and anhydrous forms of copper(II) sulphate?
A One is blue and the other is colourless (white).
B One is dry and the other is wet.
C One is crystalline and the other is powdered.
D They have different formula masses.
E Energy changes are involved in their inter-conversions.

9. (a) Calcium carbonate, present in rocks or soil, is one of the causes of hardness of water. Explain why this is so.
(b) Explain the use of (i) calcium hydroxide, and (ii) sodium carbonate, in the softening of hard water.
(c) Why does the presence of dissolved sodium carbonate not make water hard?
(d) A copper boiler used in the preparation of distilled water is encrusted with a layer of white scale caused by the hardness in the water used. Explain how this scale was formed from the hard water.
If supplies of dilute sulphuric, hydrochloric and nitric acids were available, which of these acids would you use to remove the scale from the boiler? Give reasons for your choice. (C.)

10. Name four ions present in sea-water and describe how you would show the presence of each if provided with about 1000 cm^3 of sea-water. Washing with soap in sea-water is difficult, but, if a synthetic detergent is used, washing becomes very easy. Explain why these differences are found. (J.M.B.)

11. Discuss the importance of water in everyday life.

12. 'It is easy to forget that water reacts chemically with many substances'. Discuss this statement, and include as many examples as possible.

13. Rain water changes considerably before it eventually returns to the sea. Describe some of these changes and explain how they occur.

14. Explain how a detergent facilitates the removal of dirt and grease from fabrics.

15. Describe what happens when a soapy detergent is first added to a sample of hard water. Why does a lather eventually form? What similarities and differences are there between soapy and soapless detergents?

10 The Metals and their Compounds

Introduction

The Importance of Metals and their Compounds

Metals and metallic compounds are among the most important substances that you are likely to study, both in the chemistry laboratory and in your daily life. Metals were so vital to people of early civilizations that whole eras of history have been named after those metals which were predominant at the time, for example, the Bronze Age, the Iron Age (page 175).

Modern civilization is dependent, to a very high degree, on metals, from steel with its numerous forms and thousands of uses to the more specialized metals such as tantalum, titanium and zirconium which have been developed and used in space missiles, jet aircraft and nuclear reactors. At the present rate of consumption a number of metals will soon be exhausted and recycling will be necessary.

You do not perhaps realize how many metal compounds form part of your daily life. Such mundane things as soap and common salt are metallic compounds, while if you look around the shelves of a chemical laboratory you will find that well over half the bottles contain substances which are compounds of metals. In this chapter we shall consider the preparation and properties of some important metallic compounds. The extraction of metals from their ores will be dealt with later in Chapter 14. You have already met some metals and their compounds when the periodicity of elements and the relationship of the alkali metals was dealt with in Chapter 5. In the following sections the general chemistry of the compounds of the metals is considered first, followed by a more detailed account of the metals and their familiar compounds.

10.1 GENERAL PREPARATIONS AND PROPERTIES OF METALLIC COMPOUNDS

Oxides

An oxide may be defined as a compound of oxygen with one other element, i.e. element + oxygen → oxide.

The Preparation of Metallic Oxides

Direct combination It *may* be possible to prepare an oxide by direct combination of the metal with oxygen; oxygen is a very reactive element and when a metal burns in air or oxygen the reaction often evolves much heat. However, this method is not a good way of preparing metallic oxides because some metals rapidly become coated with a protective layer of oxide which then stops the reaction. A poor yield of oxide results unless the metal is finely divided, i.e. the surface area is increased, and even then reaction may be slow. Refer to your work on heating metals in air (Section 4.3).

Points for Discussion

1. Which metals form strongly protective films of oxide on the surface?

2. Which metals form oxides when heated in air?

3. Iron is capable of forming three oxides, namely iron(II) oxide (FeO), iron(III) oxide (Fe_2O_3) and tri-iron tetroxide (Fe_3O_4). If iron is heated in excess of oxygen, which is the least likely to be formed?

4. When a metal which forms a number of oxides is heated in oxygen, the oxide produced is that in which the metal is in a high oxidation state, e.g. Fe_3O_4. This method therefore cannot be used for the preparation of an oxide in which such a metal is in a lower oxidation state, e.g. FeO.

The action of heat on metallic compounds

Experiment 10.1

† The Action of Heat on the Carbonate, Hydroxide and Nitrate of Copper **

Apparatus

Test-tubes and rack, Bunsen burner, asbestos square, spatula, test-tube holder, delivery tube, rubber bung, access to fume cupboard, splints.
Calcium hydroxide solution, universal indicator paper, dilute sulphuric acid, copper(II) carbonate, hydroxide and nitrate, cobalt chloride paper.

Procedure

(a) Note the appearance of each chemical before, during, and after heating.

(b) Place about two spatula measures of copper(II) carbonate into a test-tube. Heat it gently at first and then more strongly.

(c) Test any gas evolved with moist indicator paper, cobalt chloride paper, a glowing splint, and calcium hydroxide solution.

(d) Allow the contents of the tube to cool, and then add 2 cm³ dilute sulphuric acid. Warm the tube.

(e) Repeat procedures (b), (c) and (d) separately with the other copper(II) compounds but omit the test with calcium hydroxide solution, and heat the nitrate in the fume cupboard. The coloured gas evolved when the nitrate is heated is nitrogen dioxide NO_2, which has an unpleasant odour and is poisonous. Another colourless gas is also evolved in this reaction although you may not be able to detect it.

(f) Copy the following table in your notebook and complete the details.

Compound	*Original colour*	*Colour of evolved gases*	*Nature of gas*	*Colour of solid product*

Points for Discussion

1. What type of substance remained in each tube after heating?

2. Which of the following do you think is the more reasonable conclusion?

A When heated, copper(II) nitrate, hydroxide and carbonate decompose liberating one or more gases, and form copper(II) oxide.

B When heated, copper(II) nitrate, hydroxide and carbonate decompose liberating one or more gases and form copper(I) oxide.

Summary You will have no doubt noticed that each of the three copper(II) compounds chosen contains oxygen in the 'anion group'. When these compounds are heated some of the bonds are broken (but not all or we should end with copper), and the end product suggests that some of the oxygen remains bonded to the copper atoms as copper(II) oxide. There are other copper compounds which contain oxygen, e.g. copper(II) sulphate, but under the conditions of the experiment copper(II) sulphate does not break down completely to the oxide.

The action of heat on sulphates may yield the oxide but a great deal of heat is required, and it is difficult to carry out such a decomposition in a school laboratory. However iron(III) oxide may be prepared by heating iron(II) sulphate at the normal temperature of the Bunsen flame. Note the iron(II) is oxidized to iron(III) in the process.

$$2FeSO_4, 7H_2O(s) \longrightarrow 2FeSO_4(s) \longrightarrow Fe_2O_3(s) + SO_3(g) + SO_2(g)$$

Green crystalline substance	White powder	Red brown powder	Colourless gases

Note that many metallic compounds contain water of crystallization so that when they are heated, water is often driven off to form the anhydrous salt. Further decomposition may then occur. You should familiarize yourself with the appropriate colour changes that take place in these cases. For example, copper(II) nitrate appears to melt when heated, but in fact the water of crystallization driven off dissolves the salt to form a dark green solution. Further decomposition occurs after the water has been completely driven off. Similarly, when copper(II) sulphate is heated, water of crystallization is first evolved.

$$CuSO_4, 5H_2O(s) \rightarrow CuSO_4(s) + 5H_2O(g)$$

Blue crystals	White powder

These experiments suggest that the action of heat on a metallic carbonate, hydroxide or nitrate is a suitable method for making an oxide.

$$CuCO_3(s) \rightarrow CuO(s) + CO_2(g)$$

$$2Cu(NO_3)_2(s) \rightarrow 2CuO(s) + 4NO_2(g) + O_2(g)$$

$$Cu(OH)_2(s) \rightarrow CuO(s) + H_2O(g)$$

This may be so for copper(II) oxide, but what of other oxides?

The lower the position of the element in the electrochemical series, the more easily do the compounds decompose. This is illustrated in the examples which follow. With respect to metallic hydroxides, copper hydroxide decomposes easily to the oxide. Sodium hydroxide is not decomposed under ordinary laboratory heating conditions; it simply melts. Calcium hydroxide forms the oxide, but more heat is required than for magnesium hydroxide etc. so the order of increasing ease of decomposition is

$$Ca(OH)_2,\ Mg(OH)_2,\ Al(OH)_3,\ Zn(OH)_2,\ Fe(OH)_3,\ Pb(OH)_2,\ Cu(OH)_2$$

Try to write equations for the action of heat on aluminium hydroxide, iron(III) hydroxide, zinc hydroxide and lead hydroxide.

Can you name another metal hydroxide which does *not* decompose when heated?

The action of heat on carbonates can be used as a general method for preparing some oxides, but it is not suitable for preparing sodium oxide. This is to be expected from their position in the electrochemical series. [Note: aluminium and iron(III) carbonates do not exist.]

$$MgCO_3(s) \longrightarrow MgO(s)+CO_2(g)$$

$$PbCO_3(s) \longrightarrow PbO(s)+CO_2(g)$$

The action of heat on metallic nitrates is suitable for the preparation of metallic oxides, provided the metal is below calcium in the electrochemical series.

$$2X(NO_3)_2(s) \rightarrow 2XO(s)+4NO_2(g)+O_2(g)$$

The nitrates of sodium and potassium do not react in this way when heated (page 248).

The oxides of weakly electropositive metals are unstable to heat even if momentarily formed by any of these reactions and so decompose to the metal, liberating oxygen. For example, the carbonates and nitrates of mercury(II) and silver decompose on heating to form the respective metals, for the oxides initially formed are unstable.

$$2AgNO_3(s) \rightarrow 2Ag(s)+2NO_2(g)+O_2(g)$$

$$Hg(NO_3)_2(s) \rightarrow Hg(l)+2NO_2(g)+O_2(g)$$

$$2Ag_2CO_3(s) \rightarrow 4Ag(s)+2CO_2(g)+O_2(g)$$

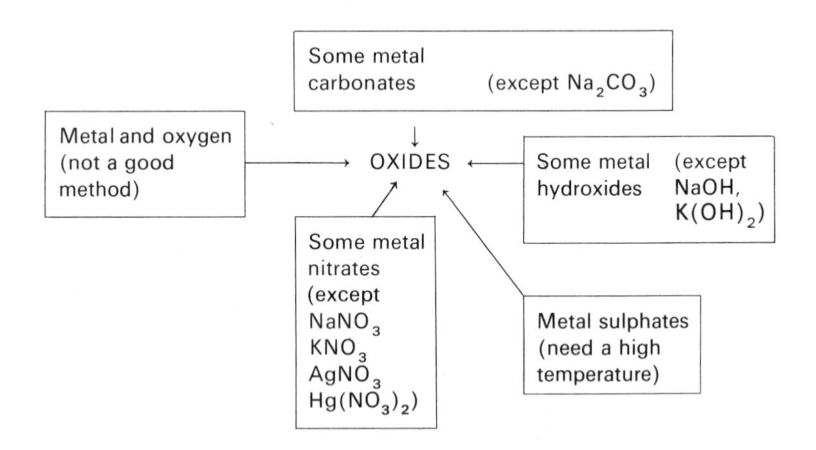

Properties of Metallic Oxides

As you already know, metals in general form oxides which are basic, but a few form amphoteric oxides. You should know at least one of these. Transition metal oxides containing a high proportion of oxygen may be acidic (e.g. CrO_3). Some metal oxides react with water to form alkaline solutions. This reaction is considered again under the next heading. Other basic properties and the appearance of specific oxides are considered in more detail with the individual metals.

Hydroxides

These are ionic compounds composed of a metal ion and one or more hydroxyl groups as the anion part. Some heavy metal hydroxides, however, are unlikely to be true hydroxides but are more likely to be hydrated oxides.

The Preparation of Hydroxides

The action of water on metallic oxides Refer back to Experiment 5.4 where metallic oxides were shaken with water and the liquid tested with universal indicator. In a few cases an alkaline solution resulted. This suggests the presence of OH^- ions. A divalent metal oxide which reacts with water forms a hydroxide:

$$MO(s)+H_2O(l) \rightarrow M(OH)_2(aq)$$

or ionically

$$O^{2-}+H_2O(l) \rightarrow 2OH^-(aq)$$

This method of making hydroxides is limited to oxides of metals near the top of the electrochemical series, such as calcium oxide.

Double decomposition If an insoluble metal hydroxide can be formed by mixing two solutions of soluble compounds, then it will be precipitated (Section 3.5). Thus to prepare an insoluble metallic hydroxide $M(OH)_x$, make a solution of a soluble salt of M, and add a solution of a soluble hydroxide. General ionic equation

$$M^{x+}(aq)+x(OH)^-(aq) \rightarrow M(OH)_x(s)$$

Which of the following is the most reasonable method for the preparation of magnesium hydroxide? (This is only slightly soluble.)

(a) Mixing solutions of magnesium chloride and sodium nitrate
(b) Dissolving magnesium oxide (sparingly soluble) in water
(c) Mixing solutions of magnesium chloride and sodium hydroxide

Experiment 10.2
†The Preparation of Insoluble Metallic Hydroxides*

Principle
Most metal hydroxides are insoluble and can be prepared by precipitation. Sodium hydroxide solution or ammonia solution are normally used as sources of hydroxide ions.

Apparatus
Rack of test-tubes, teat pipette, test-tube holder, Bunsen burner, stirring rod.
Dilute solutions of *any* soluble salts of calcium, magnesium, aluminium, iron(II), iron(III), zinc, lead, copper. Sodium hydroxide and ammonia solutions. (Remember that compounds of zinc, lead, and copper are poisonous.)

Procedure
(a) Pour about 1 cm^3 of each salt solution into a separate test-tube.

(b) Add about 1 cm^3 of sodium hydroxide solution to each tube. Note the appearance of any precipitate.

(c) Add a further 1 cm^3 of sodium hydroxide solution drop-by-drop, stirring after each addition. Warm the contents of each tube.

(d) Repeat the above procedures but use ammonia solution instead of sodium hydroxide.

(e) Record your observations in a table as follows.

Solution	*Appearance of precipitate (if any) with sodium hydroxide*	*Effect of excess sodium hydroxide*	*Appearance of precipitate (if any) with ammonia solution*	*Effect of excess ammonia solution*

Points for Discussion

1. List those hydroxides which can be prepared from sodium hydroxide or potassium hydroxide without complication.

2. List those which can be prepared from either sodium hydroxide or potassium hydroxide provided an excess of alkali is not used.

3. List those which can be made from ammonia solution without complication.

4. List those where ammonia solution may be used providing it is not in excess.

5. Using the knowledge gained from this experiment, devise tests for distinguishing between solutions of (a) sodium chloride and magnesium chloride, (b) iron(II) nitrate and iron(III) nitrate, (c) aluminium chloride and zinc chloride.

6. Write ionic equations for all the precipitates you made in the experiment.

Summary It should be obvious from your results that many metallic hydroxides are insoluble and can thus be prepared by precipitation. However, unless care is taken in the addition of sodium hydroxide or ammonia solution to solutions of aluminium, lead, zinc or copper salts, the precipitate formed will redissolve in excess of the hydroxide solution. For example, aluminium hydroxide dissolves in excess sodium hydroxide solution according to the equation.

$$Al(OH)_3(s) + NaOH(aq) \rightarrow \underset{\text{Sodium aluminate}}{NaAl(OH)_4(aq)}$$

In the case of zinc hydroxide, sodium zincate is formed:

$$Zn(OH)_2(s) + 2NaOH(aq) \rightarrow Na_2Zn(OH)_4(aq)$$

The formulae for sodium aluminate and sodium zincate are only approximations. The use of potassium hydroxide solutions would yield the corresponding potassium compounds.

Aluminium hydroxide and zinc hydroxide react in this way because they are amphoteric.

Ammonia solution precipitates metallic hydroxides in the same way but some redissolve when excess is added, because they form soluble complex salts with ammonia, *not* because they are amphoteric. For example zinc hydroxide dissolves in excess ammonia solution to form a complex tetrammine zinc ion. Copper hydroxide also dissolves in excess ammonia solution to form a complex ion, which gives a deep blue solution.

When sodium hydroxide or ammonia solution is added to a solution of an iron(II) salt, the dark green precipitate of iron(II) hydroxide very quickly turns brown at the surface where it is in contact with air, because it has been oxidized to an iron(III) compound.

Properties of Hydroxides

Metallic hydroxides which dissolve in water form alkaline solutions because they furnish hydroxide ions to the solution. Both soluble and insoluble metallic hydroxides react as bases and form salts with acids. They are also often decomposed by heat to form the oxide as indicated in the previous section. Write equations to show the effect of heat and acids on the common metallic hydroxides.

Carbonates

The Preparation of Carbonates

From carbonic acid (really carbon dioxide and water) You may suggest that carbonates can be made by using carbonic acid but you should recall (Section 3.1) that carbonic acid is too unstable to keep. It may be prepared by passing carbon dioxide gas into water, but it is a very weak acid and does not readily form carbonates with solutions of metallic salts. However, it will react with solutions of alkalis and with suspensions of insoluble bases in water, but the end product depends upon the conditions of the experiment. Carbonic acid is a dibasic acid and thus forms two series of salts, the carbonates and hydrogen carbonates. Two products are therefore possible in such reactions; for example, if a suspension of magnesium oxide reacts with carbon dioxide a clear solution of magnesium hydrogen carbonate is eventually formed. The sequence of reactions is:

$$MgO(s) + CO_2(g) \rightarrow MgCO_3(s)$$

$$MgCO_3(s) + CO_2(aq) + H_2O(l) \rightarrow Mg(HCO_3)_2(aq)$$

$$Mg(HCO_3)_2(aq) \rightarrow MgCO_3(s) + H_2O(l) + CO_2(g)$$

The carbonate is first formed, but because there is an excess of carbon dioxide, the reaction goes further to produce the acid salt, which contains a greater proportion of carbon dioxide. This last stage can be reversed if the carbon dioxide is driven off by heating so that the carbonate is reformed.

To make magnesium carbonate by this method, the hydrogen carbonate is first prepared and its solution then heated to dryness. If the white solid formed (magnesium carbonate) is heated further, it decomposes to form the oxide.

A similar reaction occurs in the case of calcium hydroxide. When carbon dioxide is passed into a solution of calcium hydroxide (limewater), insoluble calcium carbonate is first formed which gives a milky appearance to the solution. This is the test for carbon dioxide.

$$Ca(OH)_2(aq) + CO_2(g) \rightarrow CaCO_3(s) + H_2O(l)$$

When excess carbon dioxide is passed through the suspension, the milkiness disappears and the soluble hydrogen carbonate is formed. This reaction is dealt with in more detail in Section 9.2.

Experiment 10.3
Double Decomposition as a Method for Preparing Metallic Carbonates *

Apparatus
Rack of test-tubes.
Solutions of soluble salts of sodium, calcium, magnesium, aluminium, zinc, iron(II), iron(III), lead, copper(II) and silver. Sodium carbonate solution. (Compounds of zinc, lead, copper, and silver are poisonous.)

Procedure
(a) Pour about 1 cm^3 depth of each solution into separate test-tubes.

(b) Add about 2 cm^3 of sodium carbonate solution to each tube.

(c) Note the appearance of each solution before and after the addition of sodium carbonate solution.

(d) Tabulate your results as follows.

Solution	*Appearance*	*Appearance after addition of sodium carbonate*

Points for Discussion
1. Most metallic carbonates are insoluble in water (two notable exceptions being those of sodium and potassium) and may be prepared by double decomposition as in Experiment 10.3. However, some of the precipitates you have prepared are not pure carbonates, because solutions of sodium carbonate always contain some hydroxide ions as well as carbonate ions and if both the metallic carbonate and hydroxide are insoluble, the precipitate will contain both, i.e. it is a *basic carbonate,* e.g. $CuCO_3$, $Cu(OH)_2$.

2. Write equations for the reactions, assuming for convenience that 'pure' carbonates are formed.

3. You should have noticed that precipitates were formed with both aluminium and iron(III) ions, and yet their carbonates do not exist. What do you think the precipitates were in these cases? The information in point 1 should help.

4. In the instances where basic carbonates are formed the 'pure' carbonate may be obtained by using a solution of sodium hydrogen carbonate in place of sodium carbonate.

$$Mg^{2+}(aq) + 2HCO_3^-(aq) \rightarrow MgCO_3(s) + CO_2(g) + H_2O(l)$$

Properties of Carbonates

Experiment 10.4
† The Distinction between Sodium Hydrogen Carbonate and Sodium Carbonate *

Apparatus
Rack of test-tubes, Bunsen burner, asbestos square, delivery tubes, rubber bungs.
Solutions of sodium carbonate, sodium hydrogen carbonate, magnesium sulphate and calcium hydroxide.

Procedure
(a) Pour about 1 cm depth of sodium carbonate solution, and sodium hydrogen carbonate solution, into separate test-tubes. Fit each tube with a delivery tube.

(b) Heat each solution until it boils and continue heating for a few minutes. Test for carbon dioxide by passing any evolved gas through calcium hydroxide solution.

(c) To separate 1 cm^3 portions of sodium carbonate and sodium hydrogen carbonate add 1 cm^3 of magnesium sulphate solution.

Points for Discussion

1. Was carbon dioxide evolved by boiling either the carbonate or hydrogen carbonate?

2. Did both solutions behave in a similar way with magnesium sulphate solution?

3. As a result of these two tests you should be able to describe *two* ways in which sodium hydrogen carbonate can be distinguished from sodium carbonate. State the two methods.

4. A third more general difference is that most carbonates are insoluble in water, whereas all hydrogen carbonates are soluble, and in fact, with the exception of those of potassium and sodium, hydrogen carbonates cannot be obtained in the solid state. They are only stable in solution.

5. The equations for the reactions discussed are:

$$2NaHCO_3(aq) \rightarrow Na_2CO_3(aq) + CO_2(g) + H_2O(l)$$

or, ionically

$$2HCO_3^-(aq) \rightarrow CO_3^{2-}(aq) + CO_2(g) + H_2O(l)$$

$$Na_2CO_3(aq) + MgSO_4(aq) \rightarrow Na_2SO_4(aq) + MgCO_3(s)$$

or, ionically

$$CO_3^{2-}(aq) + Mg^{2+}(aq) \rightarrow MgCO_3(s)$$

6. What reaction is common to both hydrogen carbonates and carbonates?

Other reactions of carbonates When treated with dilute hydrochloric acid, they yield carbon dioxide gas, e.g. in the laboratory preparation of carbon dioxide:

$$CaCO_3(s) + 2H_3O^+(aq) \rightarrow Ca^{2+}(aq) + CO_2(g) + 3H_2O(l)$$

As seen earlier, carbonates decompose when heated to yield the oxide and carbon dioxide gas. Exceptions are potassium carbonate and sodium carbonate.

The Test for a Carbonate

To the suspected carbonate add a little dilute hydrochloric acid. Vigorous effervescence and a milky appearance when the evolved gas is passed into calcium hydroxide solution indicates the presence of a carbonate of hydrogen carbonate. The two can be distinguished as described earlier.

Nitrates

As nitrates are salts it should be possible to prepare them by the methods you used in Section 3.5. However you should realize that it is important to know whether a substance is soluble in water before you decide upon a specific method of preparation. It so happens that *all* nitrates are soluble.

The Preparation of Nitrates

Revise the principles of the preparation of salts in Section 3.5. In general, nitric acid reacts with metals, insoluble bases, alkalis and carbonates to produce nitrate salts (but see Experiment 10.5).

There are problems caused by the fact that nitric acid is also a powerful oxidizing agent. For example, if a metal exhibits variable valency, the nitrate produced by the action of nitric acid will be one in which the metal is in a high oxidation state. Thus iron(II) nitrate must be produced by, for example, interaction of barium nitrate solution and iron(II) sulphate solution.

$$FeSO_4(aq) + Ba(NO_3)_2(aq) \rightarrow Fe(NO_3)_2(aq) + BaSO_4(s)$$

$$Ba^{2+}(aq) + SO_4^{2-}(aq) \rightarrow BaSO_4(s)$$

Cautious evaporation of the filtrate from this reaction yields green crystals of the hexahydrate. These however readily form a basic iron(III) nitrate.

Experiment 10.5
The Action of Dilute Nitric Acid on Aluminium *

Apparatus
Test-tube, Bunsen burner, asbestos square.
Aluminium foil, dilute nitric acid.

Procedure
(a) Place a piece of aluminium foil about 1 cm^2 into the test-tube. Add about 1 cm^3 depth of dilute nitric acid.

(b) If there is no reaction, warm the test-tube and its contents.

Points for Discussion
1. When the dilute nitric acid was added to the aluminium foil was there any immediate reaction? If not, was there any reaction after a minute or so?

2. If you warmed the test-tube and its contents was there any reaction?

3. You know that aluminium is protected by a layer of oxide, and oxidizing acids such as nitric acid make the oxide layer even more effective so that the metal is not attacked. Although aluminium has little action with other dilute acids, it is attacked by concentrated hydrochloric acid and concentrated sulphuric acid, but not by concentrated nitric acid. Thus aluminium nitrate cannot be made by this method. The metal is said to be made *passive* by the acid, i.e. a chemically resistant oxide layer is formed. Iron is also passive to concentrated nitric acid.

Properties of Nitrates

Experiment 10.6
†Tests for Nitrates ***

Apparatus
Rack of test-tubes, test-tube holder, spatula, Bunsen burner, asbestos square. Sodium nitrate and copper nitrate solids, Devarda's alloy, concentrated sulphuric acid, dilute hydrochloric acid, dilute sodium hydroxide solution, copper turnings, deionized water, red and blue litmus papers.

Procedure
(a) Place 1 spatula measure of each salt into separate test-tubes. Add two or three copper turnings, and then cautiously add two or three drops of concentrated sulphuric acid. Heat the tube gently. Record your observations.

(b) Repeat procedure (a) but use dilute hydrochloric acid in place of concentrated sulphuric acid.

(c) Dissolve about half a spatula measure of each solid in 1 cm^3 deionized water in two separate test-tubes. To each add 1 cm^3 of dilute sodium hydroxide and 0.5 g Devarda's alloy. Warm the test-tube gently, and cautiously smell any evolved gas. Test any evolved gas with damp red and blue litmus papers.

Points for Discussion
1. Which of the following statements is the most reasonable conclusion?

A Sodium nitrate with copper turnings and concentrated sulphuric acid evolves brown fumes, but copper nitrate does not.

B Both solids evolve brown fumes when heated with copper turnings and concentrated sulphuric acid.

C Copper nitrate treated in this way evolves brown fumes but sodium nitrate does not.

2. Referring to procedure (c), which statement is best?

A Vigorous effervescence occurred, and an acid gas was evolved.

B No reaction occurred until the mixture was heated, and then vigorous effervescence occurred, an alkaline gas being evolved.

C Moderate reaction occurred, which became vigorous on heating the tube. Ammonia gas was evolved.

3. The significance of these tests is explained in the summary.

The action of heat on nitrates You have learned that when some metallic nitrates are heated they decompose to give brown fumes of nitrogen dioxide, oxygen and the oxide of the metal, and this is a useful method for preparing metallic oxides; there are exceptions however, and some of these have been referred to already. For example, the nitrates of sodium and potassium yield the nitrites and oxygen:

$$2NaNO_3(s) \rightarrow 2NaNO_2(s) + O_2(g)$$

while ammonium nitrate yields dinitrogen monoxide:

$$NH_4NO_3(s) \rightarrow N_2O(g) + 2H_2O(g)$$

Summary

All nitrates are soluble in water, and with the exception of potassium, sodium and ammonium nitrates, yield the oxide, nitrogen dioxide and oxygen on heating. Their presence in a solid may be indicated by the production of brown fumes when the solid is heated with copper turnings and concentrated sulphuric acid. A better test is the reduction of nitrates in alkaline solution by Devarda's alloy to yield ammonia gas.

Sulphates and Hydrogen Sulphates

These are salts of the dibasic acid, sulphuric acid H_2SO_4. The hydrogen sulphates contain the hydrogen sulphate ion HSO_4^-, and as this is capable of forming H_3O^+ ions in aqueous solution, they tend to be acids. However, we are more concerned here with the normal salts, i.e. the sulphates which contain the sulphate ion, SO_4^{2-}. Later reference to sodium hydrogen sulphate will serve to emphasize the distinction further.

The Preparation of Sulphates

(a) Most common sulphates are soluble in water, with the exception of barium and lead sulphates and the slightly soluble calcium sulphate. You should now be able to suggest how you could prepare samples of barium, lead, and calcium sulphates. Suggest two suitable reagents for *each* sulphate, the procedure you would adopt, and write ionic equations for the reactions which occur.

(b) Soluble sulphates are more conveniently prepared by methods such as the action of dilute sulphuric acid on insoluble bases, alkalis, carbonates and fairly reactive metals (Section 3.5). Can you name some metals which do *not* react with dilute sulphuric acid? Why are the salts formed in different oxidation states when dilute nitric and sulphuric acids are added to separate samples of iron metal? Why is calcium sulphate not prepared in quantity by the action of dilute sulphuric acid on calcium carbonate? Note that aluminium oxide must be freshly prepared before it will react with dilute sulphuric acid.

(c) In Section 3.4 you learned how it was possible to prepare a salt by titrating an alkali with an acid. Sulphuric acid can be used in this way, but it is a dibasic acid, so that *two* salts can be made according to the proportions of acid and alkali used. Refer to your work on the preparation of carbonates and hydrogen carbonates before beginning the next experiment.

Experiment 10.7
†Preparation of Salts by Titrating a Dibasic Acid with an Alkali *

Apparatus
Burette and stand, burette funnel, 25 cm^3 or 10 cm^3 pipette, pipette filler, white tile, three conical flasks, evaporating basin, tripod, gauze, Bunsen burner, asbestos square, filter funnel and stand, filter paper. Approximately 0.1 M solutions of sodium hydroxide and sulphuric acid, phenolphthalein indicator.

Procedure
Determine the respective volumes of acid and alkali required to give a neutral solution as described in Section 3.4, using phenolphthalein as indicator. Add the acid to a known volume of alkali.

Points for Discussion
1. As the indicator is purple in alkaline solution the alkali is completely neutralized when the colour of the indicator *just* disappears. The product could be either or both of the salts $NaHSO_4$ or Na_2SO_4. Write down the equations for the preparation of both salts. Study them and decide which of them could be made whilst there is still some alkali left in the solution, and which is made when all of the alkali is used up. Which salt has been made by complete neutralization?

2. How can we obtain a sample of the other salt, sodium hydrogen sulphate? Look again at the two equations. Suppose you found that 25.0 cm^3 of dilute sodium hydroxide were completely neutralized by 20.0 cm^3 of sulphuric acid. How much alkali would you need to mix with 20.0 cm^3 of sulphuric acid in order to make a sample of sodium hydrogen sulphate? Alternatively you could predict how much sulphuric acid you would need to mix with 25.0 cm^3 of alkali in order to make a sample of sodium hydrogen sulphate. The ratio of the two reagents in both your answers should be the same.

3. Devise a procedure whereby you could prepare a sample of sodium hydrogen sulphate from sodium hydroxide solution and sulphuric acid. Carry out the experiment, transfer the sample to the evaporating basin and crystallize the sodium hydrogen sulphate as described in Section 3.4.

4. It should not have been necessary to use the indicator in 3. Why?

5. You should revise the steps you have taken in order to prepare samples of the two (sodium) salts of a dibasic acid. This procedure is adopted whenever such samples are required, and the starting materials, alkali and acid, are in aqueous solution.

The Test for Soluble Sulphates and Hydrogen Sulphates

Revise the test for soluble sulphates on page 89. Hydrogen sulphates also give this test but may be distinguished from the sulphates as their aqueous solutions are acidic, and by their different behaviour when heated.

Properties of Sulphates

There are no very important properties of sulphates but the action of heat on iron(II) sulphate has already been described (page 234), and calcium sulphate is used in the manufacture of sulphuric acid.

Chlorides

These are salts of the strong monobasic acid, hydrochloric acid.

Experiment 10.8
† To Examine the Solubility of Some Metallic Chlorides *

Apparatus
Rack of test-tubes, test-tube holder, spatula, rubber bungs, labels.

Powdered samples of the chlorides of sodium, calcium, magnesium, iron(III), zinc, lead(II), copper(II), mercury(I), silver.

Procedure
(a) Put a spatula measure of each sample into a separate labelled dry test-tube, and add about 10 cm^3 deionized water.

(b) Insert the bung and shake the tube and its contents for about a minute. Place the tube in the rack and record the appearance of the contents of each tube.

(c) Remove the bung from each tube and add a further two spatula measures of the salt named on the labels. Insert the bung and repeat procedure (b).

(d) If any appear to be insoluble in water, boil the suspension for about a minute. Allow the tube to cool and examine the contents again.

Points for Discussion
1. What is your general conclusion about the solubility of most metallic chlorides in water?

2. Which chlorides are exceptions to the general rule and do not dissolve in cold water? You should already know that silver chloride (formed in the test for soluble chlorides) is insoluble; so is copper(I) chloride.

3. Which chloride is a notable exception in that it is insoluble in cold water but dissolves in hot water?

4. Which chlorides can you make by double decomposition? State the reagents you would use and write ionic equations.

5. Soluble chlorides can be made by the usual methods for preparing salts, using hydrochloric acid.

6. Some chlorides, particularly anhydrous aluminium chloride and anhydrous iron(III) chloride and many non-metallic chlorides, react with water, and so they must be prepared by *dry* methods. The usual salt preparations involve aqueous solutions and so cannot be used for such anhydrous salts which react with water, although the hydrated salts are readily obtained. Dry methods for the preparation of chlorides involve the action of either chlorine or hydrogen chloride on the heated element (page 257).

The Properties of Metallic Chlorides

Experiment 10.9
† Some Reactions of Chlorides ***

Apparatus
Rack of test-tubes, test-tube holder, Bunsen burner, asbestos square, spatula. Sodium chloride, manganese(IV) oxide, concentrated sulphuric acid, litmus papers, ammonia solution.

Procedure
(a) Put a spatula measure of sodium chloride into a test-tube. Carefully add about 0.5 cm^3 concentrated sulphuric acid. Blow across the top of the tube and test the gas with moist blue litmus paper. Hold the stopper of the ammonia solution bottle near the top of the tube.

(b) Put half a spatula measure of sodium chloride in a test-tube and add half a spatula measure of manganese(IV) oxide and 0.5 cm^3 concentrated sulphuric acid. Warm the tube and its contents (in a fume cupboard) and test any evolved gas with moist blue litmus, and moist red litmus papers.

Points for Discussion

1. What do you think the gas produced in (a) is? This reaction is used in the preparation of the gas (page 301). The equation is

$$NaCl(s) + H_2SO_4(aq) \rightarrow NaHSO_4(aq) + HCl(g)$$

2. The reactants in (b) are similar to those in (a) plus the addition of manganese(IV) oxide. What difference does the manganese(IV) oxide make to the reaction? This reaction may be used to prepare chlorine.

3. Revise the tests for soluble chlorides on page 90. Note that (a) and (b) are tests which can be used whether a chloride is soluble or insoluble.

4. There are no important general reactions of chlorides apart from these.

5. Some hydrated metallic chlorides form basic chlorides when heated, and this is why we cannot prepare these anhydrous forms by heating the hydrated salt, e.g. $AlCl_3, 6H_2O$.

Sulphides

These are the salts of the weak dibasic acid, hydrogen sulphide. The pure substance is a gas, but in aqueous solution it acts as an acid. Again, two series of salts, the normal sulphides X_2S and the hydrogen sulphides XHS are possible.

The Preparation of Sulphides

Some metallic sulphides may be prepared by direct combination. You have already prepared iron(II) sulphide in this way (Experiment 2.24). Many sulphides are insoluble in water and these may be prepared by double decomposition, but in some cases the conditions are important as shown by the next experiment.

Experiment 10.10

†To Prepare Zinc Sulphide and Copper(II) Sulphide by Precipitation (Double Decomposition)***

Apparatus

Rack of test-tubes, hydrogen sulphide generator, access to fume cupboard. Solutions of zinc sulphate and copper(II) sulphate, dilute ammonia solution, dilute hydrochloric acid. Hydrogen sulphide gas is very poisonous. It must only be used in a fume cupboard, or in a well-ventilated room.

Procedure

(a) Pour 1 cm^3 of zinc sulphate solution into each of three test-tubes.

(b) To one test-tube add 0.5 cm^3 ammonia solution, and add 0.5 cm^3 dilute hydrochloric acid to a third tube.

(c) Place the three test-tubes in the rack and pass hydrogen sulphide from the generator through the solutions. Record your observations.

(d) Repeat the above procedures with the copper(II) sulphate solution, but use sufficient ammonia solution to dissolve the initial precipitate.

(e) Repeat using a solution containing both zinc sulphate and copper(II) sulphate.

Points for Discussion

1. When hydrogen sulphide dissolves in water some sulphide ions are produced. These can react with metal ions and precipitate insoluble sulphides.

$$H_2S(g) + 2H_2O(l) \rightarrow 2H_3O^+(aq) + S^{2-}(aq)$$

$$Zn^{2+}(aq) + S^{2-}(aq) \rightarrow ZnS(s)$$

$$Cu^{2+}(aq) + S^{2-}(aq) \rightarrow CuS(s)$$

However, the pH of the solution is important. Some sulphides do not form in acid solution, and are precipitated only in alkaline solution.

2. Which alternative is correct?

A Hydrogen sulphide precipitates zinc sulphide and copper(II) sulphide from solutions of Zn^{2+} and Cu^{2+}.

B Hydrogen sulphide precipitates copper(II) sulphide from acidic and alkaline solutions of copper(II) sulphate. It does not precipitate zinc sulphide from acid solution.

C Hydrogen sulphide precipitates copper(II) sulphide from solutions of copper(II) salts but does not precipitate zinc sulphide from solutions of zinc salts.

3. Another sulphide which is precipitated in acid or alkaline solution is lead(II) sulphide. Iron(II) sulphide precipitates only in alkaline solution.

4. The explanation for 2 and 3 is as follows. When hydrogen sulphide dissolves in water an equilibrium is set up.

$$H_2S(aq) \rightleftharpoons 2H^+(aq) + S^{2-}(aq)$$

The equilibrium may be changed by addition of acid or alkali. How will it change if acid is added? How will it change if alkali is added? How does either addition affect the concentration of sulphide ion (S^{2-})? Some sulphides are very insoluble and precipitate if only a very small concentration of S^{2-} is present. Some metallic sulphides precipitate only if higher concentrations of S^{2-} are present. To which of these categories does zinc sulphide belong?

Properties of Sulphides

The sulphides of the more reactive metals are hydrolysed by water, and solutions of these sulphides are unstable. When a metallic sulphide is treated with dilute hydrochloric acid and the mixture warmed, hydrogen sulphide gas is evolved. The gas can be detected (a) by its smell (like rotten eggs), but remember it is very poisonous, and (b) by its reaction with moist lead ethanoate paper. A dark brown or black coloration is produced. (What is it?) Most lead salts are insoluble in water, so that only lead(II) nitrate and lead(II) ethanoate (acetate) can be used for this test.

10.2 SOME REACTIVE METALS. SODIUM, CALCIUM, MAGNESIUM AND ALUMINIUM

Sodium and its Compounds

Occurrence and Properties of the Metal

The metal is very reactive and does not occur in the free state. It is found as sodium ions in compounds such as sodium chloride, sodium aluminosilicates in clay, and as sodium nitrate in the rock 'caliche'. The metal was prepared first by Sir Humphrey Davy in 1807, by the electrolysis of molten sodium hydroxide. By the end of the nineteenth century the Castner cell using the electrolysis of molten sodium hydroxide was the main route to sodium. The modern method of manufacture is by means of a modified version of the Down's cell (page 386).

Sodium is a soft silvery metal, m.p. 371 K (98 °C) and b.p. 1156 K (883 °C). It tarnishes rapidly in moist air forming first a film of the monoxide which deliquesces to form sodium hydroxide solution. This in turn reacts with the carbon dioxide in the atmosphere, ultimately forming a solid on the surface (sodium carbonate). This effloresces to form the monodrate. Because of these reactions the metal is stored under a liquid hydrocarbon. The sequence of reactions may be represented:

$$4Na(s)+O_2(g) \rightarrow 2Na_2O(s)$$

$$Na_2O(s)+H_2O(l) \rightarrow 2NaOH(aq)$$

$$2NaOH(aq)+CO_2(g) \rightarrow Na_2CO_3(aq) \rightarrow Na_2CO_3, 10H_2O(s) \rightarrow Na_2CO_3, H_2O(s)$$

Sodium reacts with many elements on heating to form compounds, e.g. sodium chloride, sodium hydride, sodium phosphide and sodium oxide. It reacts vigorously with water (page 127) and with ammonia gas at 573–673 K (300–400 °C) to form sodamide:

$$2Na(s)+2NH_3(g) \rightarrow 2NaNH_2(s)+H_2(g)$$

Most compounds of sodium are colourless because the sodium ion is colourless. Can you suggest why sodium chromate, Na_2CrO_4, is coloured? Sodium compounds are ionic because the ionization potential for sodium is low, and a sodium atom easily transfers its outer valency electron to form an ionic bond. Refer back to its electron structure (page 19) and its position in the electrochemical series (Section 6.2).

Sodium Monoxide (Na_2O)

This is prepared by burning sodium in a limited supply of air. It is a typical basic oxide, and reacts violently with water to form sodium hydroxide.

Sodium Peroxide (Na_2O_2)

This oxide is formed as a yellow solid when sodium is burnt in excess air (Experiment 4.9). It is classed as a peroxide because it contains the peroxide link –O–O– and reacts with cold water to form hydrogen peroxide. The reaction with water is exothermic and oxygen is evolved. Can you suggest why sodium peroxide is stored in air-tight tins? It reacts with cold dilute hydrochloric acid to form hydrogen peroxide, but if the acid is warm, oxygen is then evolved. Write equations for these reactions. It reacts with carbon dioxide as follows:

$$2Na_2O_2(s)+2CO_2(g) \rightarrow 2Na_2CO_3(s)+O_2(g)$$

Can you suggest why it was formerly used in submarines to prolong the period for which the submarine could be submerged?

Sodium Hydroxide (NaOH)

Experiment 10.11
A Laboratory Preparation of Sodium Hydroxide **

Apparatus
Two 100 cm^3 beakers, tripod, gauze, Bunsen burner, asbestos square, rack of test-tubes, filter funnel and stand, filter paper, teat pipette, stirring rod, 50 cm^3 measuring cylinder.
Hydrated sodium carbonate, calcium hydroxide, dilute hydrochloric acid, carbon dioxide generator.

Procedure
(a) Dissolve about 5 g hydrated sodium carbonate in about 50 cm^3 deionized water in a 100 cm^3 beaker.

(b) Add about 3 g of calcium hydroxide, and heat the solution until it boils. Continue heating for one to two minutes with continuous stirring.

(c) Filter off about 1 cm^3 of the solution into a test-tube and add two or three drops of dilute hydrochloric acid. If effervescence occurs, add a further 0.5 g calcium hydroxide to the contents of the beaker, boil the solution and repeat the procedure until there is no further effervescence with hydrochloric acid. Filter off the solution into the other 100 cm^3 beaker and keep the filtrate for the next experiment.

Points for Discussion

1. You have made a solution of sodium hydroxide by a double decomposition reaction, for the other product, calcium carbonate, is insoluble in water. Write the equation for the reaction in ionic terms.

2. Why was the filtrate tested for effervescence by the addition of dilute hydrochloric acid?

3. The filtrate is not pure sodium hydroxide solution. What will it still contain? This process is used in an industrial manufacture of sodium hydroxide called the Gossage (Lime–Soda) process. Most sodium hydroxide is now made by other methods (Section 14.3).

Properties of sodium hydroxide As sodium hydroxide is a typical laboratory alkali it gives all the reactions of bases and alkalis discussed in Chapter 3.

Experiment 10.12

† The Reaction of Carbon Dioxide with Sodium Hydroxide **

Apparatus

Rack of test-tubes, test-tube holder, carbon dioxide generator.
Filtrate from Experiment 10.11.

Procedure

Pour about 5 cm^3 of the filtrate from Experiment 10.11 into a test-tube and pass carbon dioxide from the generator through the solution until no further reaction is observed.

Points for Discussion

1. Which of the following statements best describes your observations?

A No visible reaction occurs.

B A white precipitate is formed at first which dissolves when excess carbon dioxide is passed through the solution.

C No visible reaction occurs at first, but gradually a white precipitate forms which does not dissolve when excess carbon dioxide is passed through the solution.

2. Carbon dioxide dissolves in water to produce a solution which behaves as the weak acid, carbonic acid (H_2CO_3). This can form two salts with sodium hydroxide, namely sodium hydrogen carbonate ($NaHCO_3$) and sodium carbonate (Na_2CO_3). Revise the preparation of salts from sulphuric acid (Experiment 10.7) and the action of carbon dioxide on calcium hydroxide solution (page 75) before answering the next question. Which of the two sodium salts will be formed by an excess of the acid (i.e. CO_2+H_2O)? Could this be the white precipitate? Write an equation for the reaction.

3. From consideration of the following data and your answer to 2, suggest reasons for the fact that calcium hydroxide is used for *testing* for carbon dioxide but not for *absorbing* it, whereas potassium hydroxide is preferred to both sodium and calcium hydroxides for absorbing large quantities of carbon dioxide, but not for detecting it.

Calcium hydroxide is only slightly soluble. Calcium carbonate is insoluble. Potassium and sodium carbonates are both soluble. Sodium hydrogen carbonate is fairly soluble Potassium hydrogen carbonate is soluble.

Sodium Chloride ($NaCl$)

Sodium chloride is a typical chloride which occurs abundantly in nature as rock salt and in sea-water. It contains a small percentage of hygroscopic magnesium chloride, so magnesium carbonate is added to table salt to prevent caking.

Sodium Sulphate (Na_2SO_4, $10H_2O$)

A white efflorescent crystalline solid, which is a typical sulphate in its reactions (page 241), it crystallizes from water as the decahydrate. Methods of distinguishing between sulphates and hydrogen sulphates are described on page 242.

Sodium Nitrate ($NaNO_3$)

This is a typical nitrate in most of its reactions, but the action of heat on the salt is atypical. Many nitrates form the oxide (page 234) but this salt decomposes only as far as the nitrite:

$$2NaNO_3(s) \rightarrow 2NaNO_2(s) + O_2(g)$$

Sodium Carbonate (Na_2CO_3) and Sodium Hydrogen Carbonate ($NaHCO_3$)

Both these compounds are prepared on a large scale by the ammonia–soda (Solvay) process (Section 14.3). The carbonate crystallizes from water as colourless crystals of the decahydrate Na_2CO_3, $10H_2O$. These crystals effloresce on exposure to air to form a white powder consisting of the monohydrate Na_2CO_3, H_2O. Sodium hydrogen carbonate is a fine white powder which decomposes on heating to form sodium carbonate:

$$2NaHCO_3(s) \rightarrow Na_2CO_3(s) + CO_2(g) + H_2O(s)$$

Both compounds are important because of the variety of uses made of them. They may be distinguished as shown in Experiment 10.4.

The Sodium Ion (Na^+)

A non-luminous Bunsen flame is coloured an intense yellow by the vapour of a sodium salt.

Uses of Sodium and its Compounds

Metal: Preparing lead tetraethyl; manufacture of sodium peroxide and of sodium cyanide; the reduction of titanium chloride to titanium; a coolant in nuclear reactors.
Peroxide: Oxidizing agent.
Hydroxide: In the laboratory as a strong base; industrially in the manufacture of paper, artificial silk and of soap.
Carbonate: Glass making, paper making, detergents, and manufacture of sodium hydroxide.
Hydrogen carbonate: Baking soda and baking powder.
Chloride: Food industry, for glazing pottery, for the manufacture of sodium, sodium hydroxide, sodium hypochlorite and sodium chlorate.
Sodium nitrate: Manufacture of sodium nitrite which is used in dyestuffs.
Sulphide: Dyestuffs.
Sulphate: Glass making and paper making.

Calcium and its Compounds

Occurrence and Properties of the Metal

The metal does not occur in the free state as it is reactive and easily forms the calcium ion Ca^{2+}. This is explained by its electronic structure (2, 8, 8, 2.) and its position in the electrochemical series. Its compounds are ionic because of this ability to lose electrons easily, and most of them are colourless. The most common compounds are calcium carbonate ($CaCO_3$) which is found as marble, chalk, and limestone, and calcium sulphate which is found as anhydrite ($CaSO_4$) and gypsum ($CaSO_4$, $2H_2O$). Other naturally occurring compounds are calcium fluoride (fluorspar CaF_2) and calcium phosphate ($Ca_3(PO_4)_2$). It is extracted by electrolysis of the molten chloride.

As you will have observed earlier it is a light silvery metal which reacts vigorously with water, oxygen and chlorine.

$$Ca(s) + 2H_2O(l) \rightarrow Ca(OH)_2(aq) + H_2(g)$$

$$2Ca(s) + O_2(g) \rightarrow 2CaO(s)$$

It tarnishes in air forming a film of the oxide on the surface which eventually forms the carbonate as illustrated by the sequence:

$$Ca \xrightarrow{O_2} CaO \xrightarrow{H_2O} Ca(OH)_2 \xrightarrow{CO_2} CaCO_3$$

Calcium Oxide (CaO)

This is a white amorphous powder which has the reactions of a typical metallic oxide. It may be prepared by the usual methods for preparing oxides, but it is not easy to decompose the carbonate or hydroxide by heating under laboratory conditions.

Experiment 10.13

† To Prepare a Sample of Calcium Oxide by Heating Calcium Carbonate *

Apparatus

Tripod, gauze, Bunsen burner, asbestos square, mouth blowpipe, balance.
Marble chips.

Procedure

(a) Weigh a piece of marble chip.

(b) Place it on a clean gauze supported on the tripod, and heat it strongly for about five minutes. Allow the chip to cool and reweigh it. Reheat for a further five minutes. Cool and reweigh. Record your results.

(c) Repeat procedures (a) and (b) with another piece of marble chip of approximately the same mass as the first. This time heat it by means of the Bunsen burner and mouth blowpipe. Allow it to cool on an asbestos mat. Weigh it and use it immediately for Experiment 10.14.

Points for Discussion

1. Was there any significant difference in the mass of the chip before and after heating without the blowpipe?

2. Did the fiercer heating with the blowpipe produce any significant difference?

3. Why does the carbonate lose mass when it is heated strongly?

4. Which of the following statements better describes your observations?

A On heating, marble chips easily decompose liberating carbon dioxide gas.

B Marble decomposes with difficulty when heated, and liberates carbon dioxide gas.

5. Write an equation for this reaction.

6. Calcium oxide (quicklime) is made on a large scale by heating limestone in a limekiln at about 1273K (1000 °C).

Properties of calcium oxide Although it is a typical basic oxide there is one rather unusual reaction of importance, as the next experiment will show.

Experiment 10.14
†To Investigate the Effect of Adding Water to Calcium Oxide *

Apparatus
Teat pipette.
Piece of aluminium foil 6 cm × 6 cm, piece of calcium oxide from Experiment 10.13 (it will still be mainly calcium carbonate, but will be covered with a film of the oxide), deionized water, universal indicator paper.

Procedure
(a) Fold up the edges of the aluminium foil to form a shallow tray. Place it on the palm of your hand and then place the lump of 'calcium oxide' in the centre of the foil.

(b) Very carefully and slowly add deionized water, drop-by-drop to the calcium oxide by means of the teat pipette. Hold the foil in your hand over the sink, and if it becomes necessary, drop the foil into the sink. *Do not allow the calcium oxide to come into direct contact with the skin.*

(c) Watch and listen carefully for any signs of reaction as the water meets the solid.

Points for Discussion
1. Which observation is correct?

A The calcium oxide does not react with water.
B The calcium oxide reacts, crumbles to a voluminous powder, and becomes very hot, liberating steam.
C The calcium oxide crumbles to a voluminous powder and becomes very cold.

2. What do you think is formed? Write the equation for the reaction. Test the product with universal indicator paper.

3. When you heated the marble chip with the blowpipe, you may have noticed that the portion of chip being heated gave off a bright white light (incandescence). This was the basis of an old form of stage lighting in which a cylinder of calcium oxide (quicklime) was heated by means of an oxyhydrogen blowpipe. The intensely white light produced was called 'lime-light', hence the derivation of the term 'being in the limelight'.

Calcium Hydroxide ($Ca(OH)_2$) This is a fine white powder, sparingly soluble in water. The suspension in water is called 'milk of lime' and the solution is 'limewater'. It is a typical hydroxide and a useful base.

Calcium Carbonate ($CaCO_3$) This has the reactions of a typical carbonate, but see Experiment 10.13. It reacts with dilute sulphuric acid, but the reaction soon slows down because of the formation of a film of insoluble calcium sulphate on the surface of the calcium carbonate. Suggest two pairs of solutions you could mix together in order to produce a precipitate of calcium carbonate and write ionic equations for the separate reactions.

Calcium Chloride ($CaCl_2$) This has the reactions of a typical chloride. When crystallized from aqueous solution the solid obtained is the hexahydrate $CaCl_2, 6H_2O$. When heated it forms the anhydrous chloride, which has a strong affinity for water and is a good drying agent, e.g. in desiccators, guard tubes for the preparation of anhydrous materials such as aluminium chloride and iron(III) chloride. However it is not used for drying ammonia gas because it absorbs the gas to form the compound $CaCl_2, 8NH_3$.

Calcium Sulphate ($CaSO_4$) In its anhydrous form it occurs as anhydrite $CaSO_4$. The dihydrate $CaSO_4, 2H_2O$ occurs as gypsum. The salt is sparingly soluble in water and is one of the main causes of permanently hard water (Section 9.2). When gypsum is heated to 393 K (120 °C) it loses some of the water of crystallization and forms Plaster of Paris [$(CaSO_4)_2, H_2O$].

$$2CaSO_4, 2H_2O(s) \rightarrow (CaSO_4)_2, H_2O(s) + 3H_2O(g)$$

When mixed with water, it reforms the dihydrate. As this change occurs with slight expansion the material is ideal for making plaster casts. As its name suggests it is also widely used for plastering walls and making plasterboard.

Test for the Calcium Ion (Ca^{2+})

Volatile calcium compounds which contain the calcium ion impart a brick red coloration to the Bunsen flame.

Uses of Calcium and its Compounds

Metal: Deoxidizer to remove oxygen from metal castings; to reduce uranium fluoride to the metal; alloyed with lead as a cable covering.
Carbonate: Manufacture of calcium oxide; as marble and limestone in building; blast furnace and ammonia–soda process.
Chloride: Anhydrous salt as a drying agent.
Hydroxide: In the Solvay process; softening of temporarily hard water; manufacture of mortar and cement; cheap industrial base.
Sulphate: Gypsum, $CaSO_4, 2H_2O$ in the manufacture of Plaster of Paris; anhydrite, $CaSO_4$ in the manufacture of cement, sulphuric acid, and ammonium sulphate fertilizer.

Magnesium and its Compounds

Occurrence and Properties of the Metal

The metal is very reactive and is not found free in nature. It is found in the mixed carbonate, dolomite ($CaCO_3$, $MgCO_3$) and in carnallite, KCl, $MgCl_2, 6H_2O$. Magnesium halides are found in sea-water, and the metal is an essential constituent of the chlorophyll in green plants. Its electronic structure is 2, 8, 2. and it loses two electrons fairly readily to form the magnesium ion, Mg^{2+}. This ion is colourless and most magnesium compounds are colourless. Can you think of any which are coloured? The metal dissolves readily in dilute hydrochloric and sulphuric acids, liberating hydrogen gas and forming the appropriate magnesium salts. As usual no hydrogen is evolved with dilute nitric acid, but it may be evolved if very dilute nitric acid is used.

Magnesium Oxide (MgO)

It is a white refractory solid, i.e. it has a very high melting point, and it has the properties of a typical basic oxide.

Magnesium Hydroxide ($Mg(OH)_2$)

It may be prepared by double decomposition because it is only sparingly soluble in water. It is a typical base and a weak alkali.

Magnesium Carbonate ($MgCO_3$)

As explained on page 238 a white precipitate is obtained when solutions containing the carbonate ion, CO_3^{2-}, and the magnesium ion, Mg^{2+}, are mixed, but the precipitate is a basic magnesium carbonate. The carbonate is obtained if a solution of a soluble hydrogen carbonate is used in place of a soluble carbonate. Water containing dissolved carbon dioxide which passes over ground containing magnesium carbonate dissolves it to form soluble magnesium hydrogen carbonate. The resulting solution is a temporarily hard water.

Magnesium Chloride ($MgCl_2$)

This crystallizes from aqueous solution as the hexahydrate, $MgCl_2, 6H_2O$. This salt is deliquescent.

Magnesium Sulphate ($MgSO_4$, $7H_2O$)

This salt occurs naturally as the heptahydrate, Epsom salts. On heating it first loses water of crystallization and then sulphur trioxide to form the oxide. This is in contrast to the sulphates of sodium and calcium.

Tests for the Magnesium Ion (Mg^{2+})

If a magnesium solution is slightly acidified with dilute hydrochloric acid and then treated with a few drops of magneson reagent followed by excess sodium hydroxide solution, a *blue* precipitate is produced.

Uses of Magnesium and its Compounds

1. Magnesium is alloyed with other metals. Can you name two alloys of magnesium? Which property of magnesium makes them useful?
2. Why do you think magnesium is used in flashlight powders, fireworks and flares?
3. The metal is also used to reduce the fluorides of uranium and plutonium to the metal.
4. What two properties of magnesium oxide are important in its use as a furnace lining?
5. A suspension of magnesium hydroxide in water is called Milk of Magnesia. Why is it used medicinally?
6. Magnesium sulphate is commonly called 'Epsom Salts'. Can you suggest how it got its name? What is it used for?

Aluminium

Occurrence and Properties of the Metal

The metal occurs naturally in combination with other elements. It is the most abundant metal in the Earth's crust, and is found mainly in the form of alumino-silicates. The oxide is found as corundum and emery, and in the hydrated form as bauxite (Al_2O_3, $2H_2O$). The extraction of the metal is dealt with later (Section 14.3).

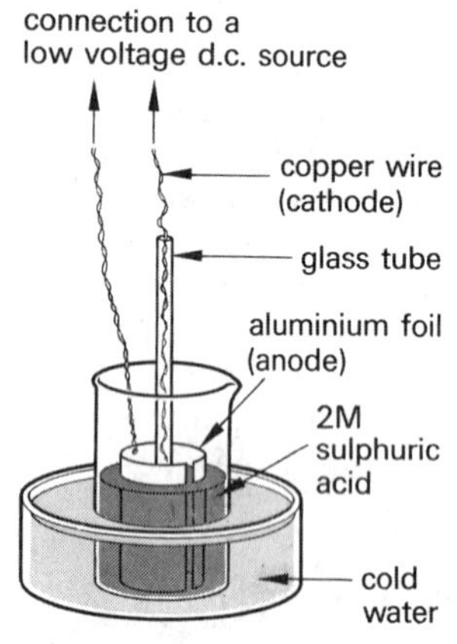

Figure 10.1 Anodizing aluminium foil

The electronic structure of aluminium is 2, 8, 3. and this indicates that the valency of aluminium is three. It is difficult to remove all three outer electrons to give the simple Al^{3+} ion, and the formation of this simple ion is rare. However, the hydrated ion $[Al(H_2O)_6]^{3+}$ occurs both in aluminium salt hydrates and in aqueous solutions of aluminium salts. Generally in the anhydrous compounds bonding tends to covalency. The metal itself is silver in colour, malleable, ductile, and of low density. It appears not to be very reactive although its position in the electrochemical series would suggest otherwise. This is because a very thin tenacious film of oxide is formed on the surface of the metal which protects it from further attack.

Aluminium readily dissolves in concentrated or dilute hydrochloric acid liberating hydrogen gas. It does not react readily with dilute sulphuric acid but will dissolve in hot concentrated sulphuric acid. Nitric acid does not dissolve aluminium. Note that the metal also dissolves in sodium hydroxide solution, and hydrogen gas is liberated,

$$2NaOH(aq) + 2Al(s) + 6H_2O(l) \rightarrow 2NaAl(OH)_4(aq) + 3H_2(g)$$

$$2OH^-(aq) + 2Al(s) + 6H_2O(l) \rightarrow 2Al(OH)_4^-(aq) + 3H_2(g)$$

Anodizing

If a piece of aluminium is made the anode of a cell at which oxygen is liberated during electrolysis, the oxygen reacts with the aluminium to thicken the thin film of oxide already present (Figure 10.1). This layer of oxide readily adsorbs a dye. Such a process is used to colour the surface, rendering the metal more resistant to corrosion by the thicker oxide layer.

Aluminium Oxide (Al_2O_3)

This is not easily prepared by burning aluminium in air. The carbonate does not exist nor is the nitrate obtained easily, so the oxide is best prepared by precipitating the hydroxide and heating the precipitate strongly. This method is used to make the aluminium oxide required for the industrial extraction of the metal. The oxide is amphoteric (Section 4.3).

Aluminium Hydroxide ($Al(OH)_3$)

This may be prepared by double decomposition but the use of excess sodium or potassium hydroxide solutions should be avoided, as the precipitated hydroxide is amphoteric and will redissolve.

Aluminium Chloride ($AlCl_3$)

The anhydrous chloride may be prepared by passing dry chlorine or dry hydrogen chloride over heated aluminium using the apparatus as in Figure 10.3 (page 257). When the solid is heated it sublimes and the vapour contains Al_2Cl_6 molecules which are covalent. In this respect it is an atypical metallic chloride.

Anhydrous aluminium chloride is hydrolysed by water. The terms hydrolysis and hydration are often confused. Hydration involves the combination of a substance and water to produce only *one* product, e.g.

$$CuSO_4(s) + 5H_2O(l) \rightarrow CuSO_4, 5H_2O(s)$$

hydrated copper(II) sulphate. Hydrolysis on the other hand is a process involving water whereby two or more products are formed, e.g. in the case of anhydrous aluminium chloride the reaction may be represented as:

$$AlCl_3(s) + 3H_2O(l) \rightarrow Al(OH)_3(s) + 3HCl(g)$$

Aluminium will dissolve in hydrochloric acid to form a solution of aluminium chloride, but the aluminium ion in this solution is the hydrated aluminium ion $Al(H_2O)_6{}^{3+}$. The anhydrous aluminium chloride cannot be obtained by crystallizing out the hydrated salt and then heating it to drive off water, because on heating the aluminium chloride reacts with the water of crystallization, i.e. it is hydrolysed.

Aluminium Sulphate ($Al_2(SO_4)_3$)

This may be made by dissolving freshly made aluminium hydroxide in dilute sulphuric acid and crystallizing the salt from the solution. The aqueous solution is acidic to litmus as a result of partial hydrolysis.

$$[Al(H_2O)_6]^{3+}(aq) \rightarrow [Al(H_2O)_5\,OH]^{2+}(aq) + H^+(aq)$$

Alums, Double and Complex salts

Substances which crystallize in similar forms are said to be isomorphous. Aluminium sulphate is isomorphous with a number of salts, and a solution of aluminium sulphate and one of these salts when crystallized yields a double salt. Such double salts are called *alums*. They have a general formula which may be represented as follows: X_2SO_4, $Y_2(SO_4)_3$, $24H_2O$ where X and Y are cations with charges of 1^+ and 3^+ respectively. A better representation is $X^+Y^{3+}(SO_4^{2-})_2$, $12H_2O$. The better known alums are potash alum $K^+Al^{3+}(SO_4^{2-})_2$, $12\,H_2O$ and chrome alum

$$K^+Cr^{3+}(SO_4^{2-})_2, 12H_2O.$$

A *double salt* is composed of two simple salts, and in aqueous solution all the ions are 'free' and retain their characteristic properties. Thus in solution potash alum 'ionizes' as follows,

$$K^+Al^{3+}(SO_4^{2-})_2, 12H_2O(s) \rightarrow K^+(aq) + 2SO_4^{2-}(aq) + (Al(H_2O)_6)^{3+}(aq)$$

The solution gives the appropriate tests for each of these ions. A *complex salt* contains a complex ion, i.e. two atoms, ions or molecules joined together so that the whole unit behaves as one ion and has a charge. An example of such a salt is tetrammine copper(II) sulphate which contains the complex tetrammine copper(II) ion, $Cu(NH_3)_4^{2+}$. If this solution is electrolysed the complex ion as a whole moves to the cathode. In aqueous solution the salt does not give the tests for the simple copper(II) ion, nor those for the ammonia molecule.

Uses of Aluminium and its Compounds

1. Why is aluminium used in aircraft bodies and the moving parts of machinery?
2. Storage tanks, milk churns, telescope mirrors are made of aluminium. Suggest why.
3. You are probably familiar with the brightly coloured aluminium kettles and saucepans used in the kitchen. What properties of the metal are made use of here?
4. The metal is a very good conductor of electricity. As such, what use is made of it?
5. Aluminium is widely used for packaging and as a covering material. Can you think of three examples? Name any properties of the metal which determine how it can be used in this way.

10.3 SOME LESS REACTIVE METALS. ZINC, IRON, LEAD, AND COPPER

Zinc

Occurrence and Properties of the Metal

The most common form in which zinc is found is as zinc sulphide (sphalerite) associated with other compounds such as lead sulphide (galena). Zinc carbonate (calamine) is also fairly widespread. The ore is first processed to give a zinc concentrate of some 60–70 per cent zinc, and in the case of sulphide deposits, with up to 30 per cent of sulphur. In the Imperial Smelting Process a charge of hard sinter, containing zinc, lead, and coke at 1073 K (800 °C) is fed into the top of a blast furnace, and air at 1223 K (950 °C) at the bottom. Reduction occurs and zinc vapour, carbon monoxide and carbon dioxide leave the furnace. The temperature is maintained in excess of 1273 K (1000 °C) to prevent reoxidation of zinc. The zinc vapour is condensed in a lead splash condenser where a solution of zinc in lead is formed, the zinc being regained by cooling the lead, outside the condenser, to a temperature below that at which zinc metal separates out, i.e. 713 K (440 °C). The purity of the metal is between 98.5–99 per cent.

Much zinc is also produced by electrolysis. The ore is heated and dissolved in dilute sulphuric acid. After purification the solution is electrolysed between lead–silver anodes and aluminium cathodes. Temperature control is essential to prevent hydrogen evolution at the cathode. The zinc produced is 99.5–99.995 per cent pure.

The atomic structure of zinc is 2, 8, 18, 2. and this suggests that in reaction it could lose two electrons to form a divalent ion, Zn^{2+}, and this is in fact the case. Compounds of zinc are mainly ionic but in the anhydrous salts the bonding does have some covalent character. The compounds are colourless in most cases unless the other ion present, e.g. the permanganate ion, confers a colour. As you may remember from your work on the electrochemical series, zinc comes about midway in the series but this only gives a qualitative idea of its reactivity. The metal does not readily dissolve in dilute acids, but the addition of a little copper(II) sulphate solution helps to speed up the reaction because a large number of minute copper–zinc cells are set up. The metal is also attacked by hot alkaline solution to form a zincate, and hydrogen gas is liberated.

$$Zn(s) + 2OH^-(aq) + 2H_2O(l) \rightarrow Zn(OH)_4^{2-}(aq) + H_2(g)$$

Zinc compounds are poisonous when taken internally.

Zinc Oxide (ZnO)

The oxide is an amphoteric oxide (Section 4.3). It is made by burning zinc in air. Remember that the oxide is white when cold, but yellow when hot.

Zinc Hydroxide ($Zn(OH)_2$)

This is an amphoteric hydroxide, and is sparingly soluble in water. It may be made by double decomposition, but in such cases it is important that excess alkali is not used, otherwise the precipitated hydroxide will redissolve to form a zincate.

$$Zn(OH)_2(s) + 2OH^-(aq) \rightarrow Zn(OH)_4^{2-}(aq)$$

Zinc hydroxide will dissolve in ammonia solution but instead of forming a zincate, a soluble complex tetrammine zinc(II) ion is formed,

$$[Zn(NH_3)_4]^{2+}$$

Zinc Carbonate ($ZnCO_3$)

When sodium carbonate solution and a solution of a soluble zinc salt are mixed, the resulting precipitate is basic zinc carbonate. The use of a solution of sodium hydrogen carbonate in place of sodium carbonate yields the normal carbonate.

The basic carbonate:

$$2Zn^{2+}(aq) + 2OH^-(aq) + CO_3^{2-}(aq) \rightarrow ZnCO_3, Zn(OH)_2(s)$$

The normal carbonate:

$$Zn^{2+}(aq) + 2HCO_3^-(aq) \rightarrow ZnCO_3(s) + CO_2(g) + H_2O(l)$$

The carbonate has the reactions of a typical carbonate, and is insoluble in water.

Zinc Chloride ($ZnCl_2$)

The anhydrous salt is prepared by passing dry chlorine or dry hydrogen chloride over heated zinc. It is very deliquescent. The anhydrous salt is soluble in ethanol and ether, and this solubility in organic solvents indicates a certain degree of covalent character. It also has a low melting point which further suggests covalent character. Crystallization of the chloride from aqueous solution yields the hydrated zinc chloride, $ZnCl_2, 6H_2O$. In this salt ions do exist $[Zn(H_2O)_6]^{2+}, 2Cl^-$.

Zinc Sulphate ($ZnSO_4$)

This is the commonest of all the zinc salts, and it is isomorphous with the corresponding sulphates of iron(II), magnesium and nickel. The hydrated salt has the formula $ZnSO_4, 7H_2O$.

Zinc Nitrate ($Zn(NO_3)_2$)

This salt may be prepared by the usual methods. It is a white deliquescent solid and has the typical reactions of a metallic nitrate.

Zinc Sulphide (ZnS)

Unlike the majority of metallic sulphides, zinc sulphide is white in colour. The use of zinc salts in paints makes use of this fact, because if hydrogen sulphide is present in the atmosphere or environment, the formation of zinc sulphide does not discolour the paint, whereas the presence of lead and other salts may cause dark coloured sulphides to form.

Uses of Zinc and its Compounds

1. Why is zinc used as a coating for iron? Galvanizing and sherardizing are two methods for producing this coat. What is the difference between them?
2. What is the most important alloy of zinc?
3. What use is made of zinc in electric batteries?
4. What property of zinc makes it useful for die-casting?
5. The oxide is used in paints and ointments. The carbonate is used in calamine lotion.
6. Mixed with barium sulphide, zinc sulphate forms 'Lithopone' which is used in paints.

Iron

Occurrence and Properties of the Metal

This is the second most abundant metal in the Earth's crust, and it occurs as various oxides (haematite, Fe_2O_3 and magnetite, Fe_3O_4) and as the disulphide FeS_2. The pure element is difficult to obtain. The large-scale extraction of the metal is dealt with in Section 14.3. This process produces cast iron which can be converted into wrought iron and steel. It may be appropriate at this point to make a brief reference to alloys and their properties before discussing the various alloys of iron.

An alloy is usually composed of two or more metals, although it may be made up of a metallic element, and elements such as carbon, silicon, nitrogen and phosphorus. Alloys are usually prepared by melting metals together and allowing the melt to cool. The atoms in a metal are close packed (Section 7.1) and the metal can be deformed by causing one plane of atoms to slide over another. However, the introduction of some larger metal atoms into the structure makes it more difficult to dislocate the layers by such deformation. On the other hand the presence of very small atoms, such as carbon or nitrogen which can fit into the small spaces between the metal atoms, will also affect the physical properties of the metal (Figure 10.2). Alloys are usually less malleable and ductile with lower melting points and electrical conductivities than the pure metal.

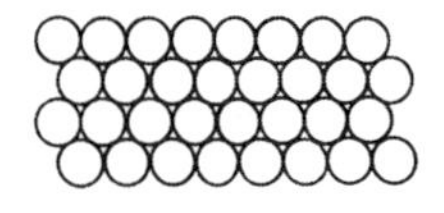
(a) pure metal

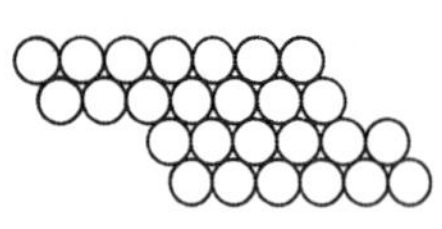
(b) deformation

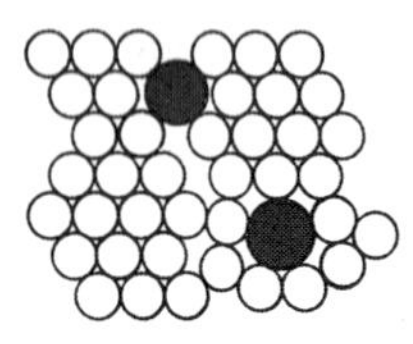
(c) large impurity atoms

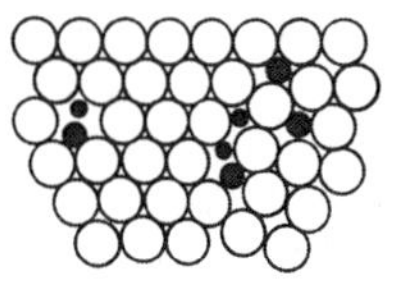
(d) small impurity atoms

Figure 10.2 Illustration of the effect of the presence of foreign atoms on the properties of a pure metal. (a) Uniform arrangement of atoms in the pure metal, (b) deformation of the metal by dislocation, (c) the effect on the atomic arrangement of large impurity atoms, and (d) the effect of small impurity atoms.

Cast iron	*Wrought iron*	*Steel*
Contains about 4 per cent carbon with a small percentage of silicon, phosphorus and sulphur. This alloy of iron expands on solidification. It is brittle and is used where strength is not important.	Much purer form obtained, for example, by heating cast iron with iron(III) oxide. This is a softer form and can be worked, e.g. made into chains etc. It is purer because it contains less sulphur, phosphorus and silicon but it still contains some carbon.	This is an alloy of iron containing between 0.15 and 1.5 per cent carbon. Its properties can be altered in several ways: (a) by altering the percentage of carbon, (b) by heat treatment, (c) by adding other metals, such as chromium, manganese, etc. The use of steel is limited only by its high density and tendency to corrosion.

On exposure to moist air, the metal rusts, and this electrochemical phenomenon is dealt with in Section 6.3. The metal reacts with steam to form tri-iron tetroxide and hydrogen gas is liberated:

$$3Fe(s) + 4H_2O(g) \rightleftharpoons Fe_3O_4(s) + 4H_2(g)$$

Iron dissolves in dilute sulphuric acid and dilute hydrochloric acid to form iron(II) sulphate and iron(II) chloride respectively. Dilute nitric acid yields iron(III) nitrate. Concentrated nitric acid does not react with iron. Can you suggest why?

The electronic structure for iron is 2, 8, 14, 2. which suggests that it is divalent; in fact it has two oxidation states, +2 and +3, and forms two series of compounds.

The compounds are ionic, although anhydrous iron(III) chloride, like aluminium chloride, possesses covalent character. The compounds are coloured, and hydrated iron(II) compounds are pale green in colour, whereas iron(III) compounds are brown, reddish-brown or yellow.

Iron Oxides (FeO; Fe_2O_3; Fe_3O_4)

Iron(II) oxide, FeO, is spontaneously oxidized in air to iron(III) oxide. It is a basic oxide and black in colour. Iron(III) oxide is a reddish-brown basic oxide. It may be prepared by heating iron(III) hydroxide, or, rather unusually, by heating iron(II) sulphate (page 234). Note the change in the oxidation number of iron. Tri-iron tetroxide, Fe_3O_4, behaves as though it was a compound oxide, and is produced when red hot iron is quenched in cold water, or when steam is passed over heated iron.

Iron Hydroxides ($Fe(OH)_2$; $Fe(OH)_3$)

Iron(II) hydroxide can be precipitated as a white solid, if sodium hydroxide dissolved in freshly boiled water is added to pure iron(II) sulphate (also dissolved in freshly boiled water) into which is passed a stream of hydrogen gas. In the absence of air (e.g. in freshly boiled water) there is less chance of oxidation of Fe^{2+} ion to Fe^{3+} ion. If air is present this oxidation proceeds and the precipitate gradually changes from green to a brownish colour. Iron(III) hydroxide may be prepared by double decomposition. Try to think of some of the materials you could use. Write an ionic equation for the reaction.

Iron Chlorides ($FeCl_2$; $FeCl_3$)

Anhydrous iron(II) chloride is obtained if *dry* hydrogen chloride is passed over heated iron filings. It is a white crystalline solid. If in a similar manner dry chlorine is passed over heated iron filings the product is anhydrous iron(III) chloride (Figure 10.3). The iron(III) chloride has similar

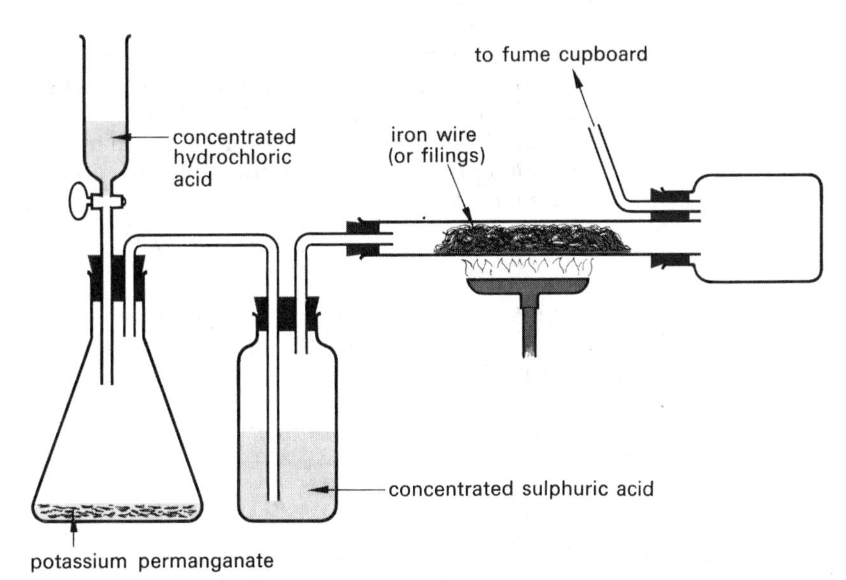

Figure 10.3 The preparation of iron(III) chloride by passing chlorine gas over heated iron filings

characteristics to aluminium chloride as both are metallic chlorides and readily sublime on heating. Its aqueous solution is acidic because of hydrolysis. These two anhydrous salts cannot be made by heating the hydrated salts; some oxidation of the iron(II) chloride occurs on heating, and the iron(III) chloride is hydrolysed by the water of crystallization. When a metal has two or more oxidation states the conditions necessary for production of a particular compound must be carefully considered. The use of hydrogen chloride produces a chloride in which the metal is in a lower oxidation state.

$$Fe(s) + 2HCl(g) \rightarrow FeCl_2(s) + H_2(g)$$

The use of chlorine in place of hydrogen chloride produces a compound in which the metal is in the highest oxidation state, in this case iron(III) chloride. Chlorine, which is very electronegative (Section 5.5), is a good 'grabber' of electrons and takes as many electrons as the metal is prepared to give, i.e. it oxidizes the iron to iron(III):

$$2Fe(s) + 3Cl_2(g) \rightarrow 2FeCl_3(g)$$

Sublimes

or ionically

$$2Fe(s) + 3Cl_2(g) \rightarrow 2(Fe^{3+}3Cl^-)(s)$$

Point for Discussion Does it matter whether hydrogen chloride or chlorine is used in the preparation of anhydrous aluminium chloride? Explain your answer.

Iron Sulphates ($FeSO_4$, $Fe_2(SO_4)_3$)

Iron(II) sulphate is the best known iron(II) salt. You have already prepared this compound (Experiment 3.14). It decomposes on heating to form iron(III) oxide, and liberates the two oxides of sulphur.

$$2FeSO_4, 7H_2O(s) \rightarrow 2FeSO_4(s) \rightarrow Fe_2O_3(s) + SO_2(g) + SO_3(g)$$

This provided the basis of one of the earliest known methods for the preparation of sulphuric acid. How do you think the acid was made? The salt is a pale green crystalline solid which contains seven molecules of water of crystallization. What colour change takes place when the salt is gently heated to about 373 K (100 °C)?

Iron(III) sulphate can be prepared by oxidation of a solution of iron(II) sulphate in the presence of nitric acid and sulphuric acid. It decomposes to iron(III) oxide on heating,

$$Fe_2(SO_4)_3(s) \rightarrow Fe_2O_3(s) + 3SO_3(g)$$

Iron(II) Sulphide(FeS)

This may be prepared by direct combination of iron and sulphur (Experiment 2.24). It is used in the laboratory for preparing hydrogen sulphide gas (page 290).

The Distinction Between Iron(II) and Iron(III) Salts

This distinction may be observed by the following series of reactions.

Experiment 10.15
To Distinguish between Iron(II) and Iron(III) Salts *

Apparatus
Rack of test-tubes, teat pipette.
0.01M solutions of iron(II) sulphate, iron(III) sulphate or chloride, potassium hexacyanoferrate(II) (ferrocyanide), potassium hexacyanoferrate(III) (ferricyanide), dilute ammonia solution. (The iron(II) sulphate solution should be made in freshly boiled water.)

Procedure
(a) Pour about 1 cm^3 of iron(II) sulphate solution into each of three test-tubes. To one of these tubes add two or three drops of potassium hexacyanoferrate(II) solution. To another tube add two or three drops of potassium hexacyanoferrate(III) solution. To the third tube add two or three drops of ammonia solution.
(b) Repeat the procedure in (a) with the iron(III) sulphate solution. Record your observations in a table.

Points for Discussion

1. From your observations would it be possible to distinguish between iron(II) and iron(III) salts by the use of potassium hexacyanoferrate(II) solution only? Explain your answer.

2. Could this be done with potassium hexacyanoferrate(III) solution? Explain your answer.

3. Is ammonia solution effective in distinguishing between iron(II) and iron(III) salts?

The interconversion of iron(II) and iron (III) ions. The relationship between the iron(II) and the iron(III) ion may be illustrated as shown:

$$Fe^{2+} \rightleftharpoons Fe^{3+} + e^-$$

From your work on oxidation and reduction (Chapter 8) you will see that the conversion of iron(II) ions to iron(III) ions involves oxidation, i.e. loss of electrons. The converse change is reduction, i.e. gain of electrons. How may these interconversions be carried out?

Experiment 10.16
The Oxidation of Iron(II) Sulphate to Iron(III) Sulphate *

Apparatus

Rack of test-tubes, Bunsen burner, asbestos square.

Freshly prepared solution of iron(II) sulphate in boiled water, dilute sulphuric acid, hydrogen peroxide solution (twenty volume), potassium hexacyanoferrate(II) solution, potassium hexacyanoferrate(III) solution.

Procedure

(a) Pour about 1 cm^3 of iron(II) sulphate solution into a test-tube and add two or three drops of potassium hexacyanoferrate(II) solution. Your observations in Experiment 10.15 should tell you if there is any iron(III) ion present. This blank test is essential, otherwise it would be difficult to prove that any iron(III) ion present after reaction was not already present before the reaction.

(b) Pour about 1 cm^3 of iron(II) sulphate solution into a test-tube. Add about 0.5 cm^3 hydrogen peroxide and 1 cm^3 dilute sulphuric acid. Boil the contents of the tube. Cool and add two or three drops of potassium hexacyanoferrate(II) solution.

Points for Discussion

1. Has the acidified hydrogen peroxide oxidized the iron(II) ion? Explain your answer.

2. In this reaction the iron(II) sulphate was converted into iron(III) sulphate.

$$2FeSO_4 \rightarrow Fe_2(SO_4)_3$$

Is the 'equation' balanced? Can you suggest why the hydrogen peroxide was acidified with *sulphuric* acid? The reaction $Fe^{2+} \rightarrow Fe^{3+} + e^-$ requires the presence of an oxidizing agent, hence the addition of hydrogen peroxide. Ionically the reaction may be represented as:

$$2Fe^{3+}(aq) + H_2O_2(aq) + 2H_3O^+(aq) \rightarrow 2Fe^{2+}(aq) + 4H_2O(l)$$

3. Other oxidizing agents such as concentrated nitric acid, potassium permanganate, and potassium dichromate may be used. Why would the use of hydrogen peroxide or concentrated nitric acid be preferable to the use of potassium manganate(VII) in the preparation of iron(III) sulphate?

4. It is easy to convert iron(II) chloride to iron(III) chloride by passing chlorine gas through an aqueous solution of iron(II) chloride. The chlorine oxidizes the iron(II)

to iron(III) by accepting electrons and also provides the 'extra' chlorine to 'balance' the reaction,

$$2FeCl_2(aq) + Cl_2(g) \rightarrow 2FeCl_3(aq)$$

or ionically

$$2Fe^{2+}(aq) + Cl_2(g) \rightarrow 2Fe^{3+}(aq) + 2Cl^-(aq)$$

5. The conversion of iron(II) oxide or hydroxide to the iron(III) compound is spontaneous, and iron(II) compounds when left exposed to the air tend to form iron(III) compounds. When iron(II) sulphate crystals are left in air they turn brown but there are insufficient sulphate ions present to completely form iron(III) sulphate, and some of the iron(II) ions are converted to the hydrated iron(III) oxide which is brown in colour. You may have observed that iron(II) sulphate solution if left in the laboratory for any length of time forms a brown deposit, and the solution itself may appear brownish. In this case it is likely that a basic iron(III) sulphate may be formed.

6. The reverse change, i.e. conversion of iron(III) ions into iron(II) ions may be accomplished as follows.

Experiment 10.17
The Reduction of Iron(III) Sulphate to Iron(II) Sulphate **

Apparatus

Rack of test-tubes, hydrogen generator. Iron filings, dilute sulphuric acid, iron(III) sulphate solution, potassium hexacyanoferrate(III) solution, potassium hexacyanoferrate(II) solution.

Procedure

(a) Pour about 1 cm^3 of the iron(III) sulphate solution into a test-tube and add two or three drops of potassium hexacyanoferrate(III) solution. This will show if there are any iron(II) ions present before reaction.

(b) Pour about 1 cm^3 of iron(III) sulphate solution into a test-tube. Add about 1 cm^3 dilute sulphuric acid, and bubble hydrogen gas from the generator through the solution for about five minutes. Divide the solution into two portions. To one portion add potassium hexacyanoferrate(III) solution. To the other portion add potassium hexacyanoferrate(II) solution.

(c) Pour about 1 cm^3 iron(III) sulphate solution into a test-tube and add 1 cm^3 dilute sulphuric acid and some iron filings. Allow the evolved hydrogen to bubble through the solution until there is no further change of colour. Add two or three drops of potassium hexacyanoferrate(III) solution.

Points for Discussion

1. Does ordinary gaseous hydrogen reduce iron(III) ions to iron(II) ions?

2. Your observations should suggest that the hydrogen generated *in situ* by procedure (c) reduces iron(III) to iron(II), whereas ordinary hydrogen (procedure (b)) does not. Actually the hydrogen formed by the second reaction is rather a red herring and obscures the more fundamental electron transfer. As the metal reacts with the acid, electrons are liberated:

$$Zn(s) \rightarrow Zn^{2+}(aq) + 2e^-$$

Some of these react with iron(III) ions and reduce them to iron(II):

$$Fe^{3+}(aq) + e^- \rightarrow Fe^{2+}(aq)$$

Other electrons react with the hydronium ions from the acid to produce hydrogen gas.

$$2H_3O^+(aq) + 2e^- \rightarrow 2H_2O(l) + H_2(g)$$

3. Other reducing agents which will bring about this reduction are magnesium and sulphuric acid, zinc and sulphuric acid and hydrogen sulphide.

4. To obtain a pure sample of iron(II) sulphate from iron(III) sulphate why is it necessary to use (i) sulphuric acid rather than hydrochloric acid and, (ii) iron instead of zinc?

5. What reagents would you use to reduce iron(III) chloride to iron(II) chloride? Test your prediction experimentally, and write equations for the reactions occurring.

Uses of Iron and its Compounds

1. Cast iron was traditionally used for a variety of purposes, many of which have been superseded by modern plastics. Can you name some of these uses? Why was cast iron preferred to other forms of iron or steel in these cases?
2. Wrought iron can be worked (wrought). What use is made of it?
3. Steel as a constructional material is very versatile. Can you suggest some properties which produce this versatility? What disadvantage has steel as such a material?

Lead

Occurrence and Properties of the Metal

The principal ore of lead is galena (lead sulphide, PbS). The metal itself has been known from ancient times. It is easily prepared in the laboratory by heating lead(II) oxide with carbon, e.g. on a charcoal block. The freshly prepared metal is bright and silvery but soon tarnishes in air due to the formation of a coating of the basic carbonate. A possible sequence of reaction is:

$$2Pb(s)+2H_2O(l) \rightarrow 2Pb(OH)_2(s)$$

$$Pb(OH)_2(s)+CO_2(g) \rightarrow PbCO_3(s)+H_2O(l)$$

and

$$Pb(OH)_2(s)+PbCO_3(s) \rightarrow PbCO_3, Pb(OH)_2(s)$$

It is a soft dense metal. Lead is placed just above hydrogen in the electrochemical series and is not attacked to any extent by acids (except nitric acid) because of the formation of a protective coating of the insoluble lead salt.

Lead Oxides

Lead(II) oxide (PbO) is a yellow or orange powder prepared by the usual methods for making an oxide. It dissolves in both acids and alkalis, i.e. it is amphoteric,

$$PbO(s)+2H_3O^+(aq) \rightarrow Pb^{2+}(aq)+3H_2O(l)$$

$$PbO(s)+2OH^-(aq)+H_2O(l) \rightarrow Pb(OH)_4^{2-}(aq) \text{ (a hydroxyplumbite).}$$

Lead(IV) oxide (PbO_2) is a dark brown powder and is prepared by heating tri-lead tetroxide with dilute nitric acid and filtering off the undissolved lead(IV) oxide which is left as a residue. In this reaction tri-lead tetroxide *behaves* as if it were a mixture of lead(II) oxide and lead(IV) oxide (cf. Fe_3O_4). The lead(II) oxide which is a typical base reacts with the nitric acid to form soluble lead nitrate in solution, leaving the insoluble lead(IV) oxide.

Tri-lead tetroxide (Pb_3O_4) is prepared by heating lead(II) oxide in air to a temperature of about 743 K (470 °C). If the temperature is raised above 773 K (500 °C) the tri-lead tetroxide decomposes, reforming lead(II) oxide and oxygen is evolved.

$$6PbO(s)+O_2(g) \rightleftharpoons 2Pb_3O_4(s)$$

The tri-lead tetroxide is a red powder hence its old name, red lead. Its reaction with nitric acid is used in the preparation of lead(IV) oxide. Tri-lead tetroxide behaves as though it was a mixed oxide.

Experiment 10.18
† To Investigate the Reactions of the Oxides of Lead **

Apparatus
Rack of test-tubes, test-tube holder, Bunsen burner, asbestos square, spatula, splints.
Lead(II) oxide, lead(IV) oxide, tri-lead tetroxide, dilute hydrochloric acid, litmus papers, concentrated hydrochloric acid.

Procedure
(a) Put one spatula measure of each oxide into separate test-tubes, and heat each tube strongly. Test for the presence of oxygen by means of a glowing splint.

(b) Put half a spatula measure of each oxide into separate test-tubes. Pour into each test-tube approximately 1 cm^3 of dilute hydrochloric acid. Warm each tube gently and test any evolved gas with moist litmus papers.

(c) Repeat procedure (b) but use concentrated hydrochloric acid (care!) instead of the dilute acid.

Points for Discussion

1. Which of the following statements best describes your observations for (a)?

A The three oxides of lead decompose on heating, oxygen gas being liberated.
B Only tri-lead tetroxide decomposes on heating to yield oxygen.
C Lead(II) oxide and tri-lead tetroxide decompose on heating to yield oxygen.

2. From your answers to point 1 and the colour changes involved, write equations for these reactions.

3. The lead(II) oxide, a typical basic oxide, reacts with dilute hydrochloric acid to form the salt, lead(II) chloride.

$$PbO(s) + 2H_3O^+(aq) + 2Cl^-(aq) \rightarrow PbCl_2(s) + 3H_2O(l)$$

As the salt is insoluble in cold water the reaction soon stops.

4. Lead(IV) oxide reacts with concentrated hydrochloric acid in a similar way to manganese(IV) oxide. What is this resemblance? The equation for the reaction is:

$$PbO_2(s) + 4H_3O^+(aq) + 4Cl^-(aq) \rightarrow PbCl_2(s) + 6H_2O(l) + Cl_2(g)$$

5. You should have found that tri-lead tetroxide resembles lead(IV) oxide in this reaction but this is not surprising for it behaves as though it is a mixed oxide, and thus as though it contains lead(IV) oxide.

Lead Carbonate ($PbCO_3$)

This insoluble, white compound is prepared by double decomposition, but the use of sodium carbonate is not suitable as it precipitates a basic carbonate. Before you suggest which soluble lead compounds may be used for this double decomposition perform the next experiment.

Experiment 10.19
† To Investigate the Reactions of Lead(II) Compounds

Apparatus
Rack of test-tubes, Bunsen burner, asbestos square, test-tube holder, teat pipettes.
Lead(II) nitrate solution, dilute hydrochloric acid, dilute sulphuric acid, sodium carbonate solution, sodium hydroxide solution, potassium iodide solution.

Procedure
(a) Pour about 1 cm^3 of lead(II) nitrate solution into each of five test-tubes.

(b) Add two or three drops of dilute sulphuric acid to one of the tubes.

(c) To another tube add two or three drops of dilute hydrochloric acid.

(d) Repeat this procedure by adding to the third tube sodium carbonate solution; to the fourth tube add sodium hydroxide solution and to the fifth tube add potassium iodide solution. Boil each solution and then allow it to cool.

(e) Record your observations in a table as follows.

	HCl (aq)	*H_2SO_4 (aq)*	*Na_2CO_3 (aq)*	*NaOH (aq)*	*KI (aq)*
First addition					
Boiling					
Cooling					

Points for Discussion

1. What conclusions can you draw as to the respective solubilities of lead(II) chloride, lead(II) sulphate, lead(II) carbonate, lead(II) hydroxide and lead(II) iodide in water?

2. Name a lead salt which is an exception to this generalization.

3. What is characteristic about the solubility of lead(II) chloride and lead(II) iodide in hot and cold water?

4. Write ionic equations for each of the reactions shown in the table.

Lead(II) Nitrate ($Pb(NO_3)_2$)

This is the most important of the soluble lead(II) salts and may be prepared by the usual methods. Lead(II) nitrate crystallizes without water of crystallization and it is a convenient source for the preparation of nitrogen dioxide. Other metallic nitrates also yield nitrogen dioxide when heated, but water of crystallization is evolved which dissolves some of the gas. The presence of water is undesirable if the gas is to be liquefied.

Uses of Lead and its Compounds

1. Lead is a very dense, malleable metal which is resistant to corrosion. What uses of the metal make use of these properties?
2. Why is lead of particular value in work involving radioactive material?
3. How are the metal and lead(IV) oxide used to produce electricity?
4. Tri-lead tetroxide, lead(II) carbonate and lead(II) chromate are used in paints. What type of paints? Why are these lead compounds used in this way? Why is the use of lead based paints for children's toys prohibited in this country?

Copper

Occurrence and Properties of the Metal

This was one of the earliest metals known to man, and its alloys, brass and bronze, are familiar materials. The metal occasionally occurs free in nature, and is found combined with other elements in compounds such as copper pyrites ($CuFeS_2$), malachite ($CuCO_3, Cu(OH)_2$) and cuprite (Cu_2O). The metal is comparatively soft, malleable, ductile and a good conductor of electricity. The pure metal is reddish-brown in colour but it tarnishes in air, becoming covered with a greenish patina of the basic carbonate. Have you noticed this characteristic colour on weather vanes, lightning conductors, etc. which have become 'weathered'? The electronic structure of copper is 2,8,18,1., but this does not indicate the fact that copper can exist in two oxidation states, +1 and +2. The latter state is found in copper(II) compounds which are the most familiar compounds of copper. Copper is placed below hydrogen in the electrochemical series, and it does not react with dilute acids unless they are also oxidizing agents. Thus it dissolves in dilute nitric acid, liberating nitrogen monoxide gas:

$$3Cu(s) + 8HNO_3(aq) \rightarrow 3Cu(NO_3)_2(aq) + 2NO(g) + 4H_2O(l)$$

If the acid is concentrated, the gas evolved is nitrogen dioxide.

$$Cu(s) + 4HNO_3(aq) \rightarrow Cu(NO_3)_2(aq) + 2NO_2(g) + 2H_2O(l)$$

These reactions are important because they afford a method of obtaining a solution containing copper(II) ions from which other copper(II) compounds can be prepared.

Copper forms two series of compounds, copper(II) compounds and copper(I) compounds. As copper(II) compounds are much more important we deal with these first.

Copper(II) Oxide (CuO)

This is a black powder which has the properties of a typical basic oxide. It may be prepared by the usual methods although direct combination is not very suitable. It is a useful starting point for the preparation of many copper salts by reaction with acids.

Copper(II) Hydroxide ($Cu(OH)_2$)

This bluish-green compound may be prepared by double decomposition using a solution of a soluble copper(II) salt and sodium hydroxide solution. Note that the addition of ammonia solution will initially precipitate copper(II) hydroxide, but this dissolves in excess of the ammonia solution to form a deep blue solution containing the tetrammine copper(II) ion (page 283). Copper(II) hydroxide is basic in character.

Copper(II) Carbonate ($CuCO_3$)

This is prepared as a basic carbonate when sodium carbonate solution is added to a solution of a copper(II) salt. The normal carbonate is obtained if sodium hydrogen carbonate is used in place of sodium carbonate. Copper(II) carbonate is green and is a typical carbonate.

Copper(II) Chloride ($CuCl_2$)

The hot metal combines with dry chlorine to form a dark brown anhydrous copper(II) chloride. The hydrated salt is prepared by the action of dilute hydrochloric acid on copper(II) oxide, copper(II) hydroxide or copper(II) carbonate. The hydrated salt is green in colour and has the formula $CuCl_2, 2H_2O$. The anhydrous salt cannot be obtained by heating the hydrated salt.

Copper(II) Nitrate ($Cu(NO_3)_2$)

This blue salt may be prepared by dissolving the metal, copper(II) oxide, hydroxide or carbonate in dilute nitric acid, and crystallizing the solution. When the hydrated salt, $Cu(NO_3)_2, 3H_2O$, is heated, the water of crystallization is released and the salt dissolves in it, giving a green solution which then loses water and the solid nitrate formed decomposes to form copper(II) oxide.

$$2Cu(NO_3)_2, 3H_2O(s) \rightarrow 2Cu(NO_3)_2(s) \rightarrow 2CuO(s) + 4NO_2(g) + O_2(g)$$

Copper(II) Sulphate ($CuSO_4, 5H_2O$)

This is probably the most familiar of all the copper(II) compounds. It is prepared by dissolving copper(II) oxide, hydroxide or carbonate in dilute sulphuric acid and crystallizing the solution. It is a blue crystalline solid. On heating to about 393 K (120 °C) (Section 10.1) it loses four-fifths of the water of crystallization, and the remainder at 523 K (250 °C). This difference in behaviour is because the first four molecules of water lost are those co-ordinated on to the copper(II) ion; the other molecule of water of crystallization is held by hydrogen bonding to the sulphate ion.

Copper(II) Sulphide (CuS)

This is precipitated as a black solid when hydrogen sulphide gas is passed into a solution of a copper(II) salt.

$$Cu^{2+}(aq) + S^{2-}(aq) \rightarrow CuS(s)$$

Copper(I) Compounds

Copper does form copper(I) compounds but many are unstable in the presence of dilute acids and water. Copper(I) oxide and copper(I) chloride can be prepared by reduction of the corresponding copper(II) compounds under suitable conditions, as shown in the next experiment.

Experiment 10.20
The Preparation of Copper(I) Oxide *

Apparatus

Test-tubes, Bunsen burner, asbestos square, filter funnel and stand, filter paper, spatula, test-tube holder.

Copper(II) sulphate solution, glucose, sodium hydroxide solution (approximately M), Fehling's solutions A and B.

Procedure

(a) Pour about 1 cm^3 of copper(II) sulphate solution into a test-tube.

(b) Add 1 spatula measure of glucose, and approximately 1 cm^3 of sodium hydroxide solution.

(c) Heat the test-tube until the liquid boils, and continue heating for a minute or so.

(d) Filter off the precipitate of copper(I) oxide.

(e) Mix 1 cm^3 of Fehling's solution A with 1 cm^3 of Fehling's solution B in a test-tube. Add 1 spatula measure of glucose. Heat the tube until the liquid boils.

(f) Record your observations.

Points for Discussion

1. What colour was the precipitate when sodium hydroxide solution was added? Explain this change and write an ionic equation for it.

2. What would you expect to be formed if this precipitate was heated? Give an equation.

3. What colour would you expect this precipitate to be? Was this the final product? What could the glucose have done?

4. Glucose is a 'reducing' sugar and the reaction discussed in 3 can be used to test for sugars of this kind. A mixture of Fehling's solutions A and B contains copper(II) ions and these are reduced by certain sugars to copper(I) ions. The final product is copper(I) oxide. Copper(I) oxide is an orange-yellow solid but in the above reaction the colour may quickly change to a reddish-brown as the particle size increases.

Test for the Copper(II) Ion

This should be carried out on an aqueous solution. The addition of ammonia solution first precipitates copper(II) hydroxide, but this re-dissolves in excess of ammonia solution to give a deep blue solution.

Uses of Copper and its Compounds

1. Can you suggest what properties of copper account for its wide use in (a) the electrical industry, (b) boilers and central heating pipes? Why is it that the much cheaper iron is not used for this purpose?
2. Can you name some of the alloys of the metal? What uses are made of them?
3. Copper(I) oxide is used in the manufacture of red glass.
4. Copper(II) sulphate is used in electroplating. It is also used in 'Bordeaux Mixture'. What is this, and what is it used for?
5. Why is copper used to such a large extent in decorative and ornamental work?

QUESTIONS CHAPTER 10

1. Which of the following metals is the most electropositive?

A Magnesium
B Lead
C Copper
D Sodium
E Aluminium

2. Which of the following gives a white precipitate with barium chloride solution, and a brick-red flame test?

A Sodium sulphate
B Copper sulphate
C Calcium sulphate
D Ammonium sulphate
E Calcium chloride

3. Sodium and potassium are members of a family of elements called

A the alkaline earth metals
B the transition metals
C metalloids
D alkali metals
E the unreactive metals

4. Sodium metal is

A hard
B denser than water
C white in colour
D a metal melting below 373 K (100 °C)
E a metal melting above 373 K (100 °C)

5. Sodium metal is kept under

A water
B ethanol
C nitric acid
D paraffin oil
E mercury

6. The object of a diaphragm in the diaphragm cell for the electrolysis of sodium chloride solution is

A to prevent production of hypochlorite
B to assist production of hypochlorite
C to keep up the resistance of the cell
D to prevent depletion of the solution around the anode
E to prevent a rise in temperature

7. Iron is galvanized by coating it with

A copper
B tin
C zinc
D aluminium
E lead

8. A white crystalline compound exposed to air changed into a white powder. The crystalline compound may have been

A potassium hydroxide
B calcium oxide
C sodium nitrate
D sodium carbonate
E potassium sulphate

9. The raw materials fed into the blast furnace for making iron are

A iron(II) oxide, calçium carbonate and coke
B iron(III) oxide, calcium oxide and coke
C iron(III) oxide, calcium carbonate and coke
D tri-iron tetroxide, calcium hydroxide and coke

10. In the Solvay or ammonia–soda process, the raw materials are

A sodium hydroxide, calcium oxide and ammonia
B sodium carbonate, calcium hydroxide and ammonia
C sodium carbonate, calcium carbonate and ammonia
D sodium sulphate, calcium carbonate and ammonia
E sodium chloride, calcium carbonate and ammonia

11. Zinc oxide is

A unaffected by heat
B decomposed to the metal by heat
C turned permanently yellow by heat
D turned temporarily yellow by heat
E blackened by heat

12. When sodium hydroxide is added to a solution of zinc sulphate it produces

A no visible change
B a white precipitate insoluble in excess alkali solution
C a white precipitate soluble in excess alkali solution
D a greenish precipitate insoluble in excess alkali solution
E a green precipitate soluble in excess alkali solution

13. The protective film of oxide on the surface of aluminium metal may be strengthened by

A cathodizing
B anodizing
C galvanizing
D hydrolysing
E sherardizing

14. Anhydrous aluminium chloride may be prepared from

A aluminium and hydrochloric acid
B aluminium and hydrogen chloride
C aluminium oxide and hydrochloric acid
D aluminium and chlorine water
E aluminium hydroxide and hydrochloric acid

15. Tri-lead tetraoxide is

A a basic oxide
B an acidic oxide
C an amphoteric oxide
D a peroxide
E a mixed or compound oxide

16. A mixture consists of a soluble salt A and an insoluble salt B. When the mixture is heated it turns black and a colourless gas which turns calcium hydroxide solution milky is evolved.

If the mixture is shaken with water and filtered, it gives a colourless solution and leaves a green residue.

The solution gives the following results on testing: (a) with sodium carbonate solution there is a white precipitate; (b) a flame test gives a brick-red coloured flame, (c) when dilute nitric acid is added followed by silver nitrate solution, there is a curdy white precipitate.

When the residue on the filter paper is dissolved in the minimum of dilute nitric acid, there is effervescence and a green solution is formed which gives: (i) a brownish-black precipitate with hydrogen sulphide, (ii) a light blue precipitate with sodium hydroxide, (iii) a light blue precipitate with ammonium hydroxide, soluble in excess of the latter to give a deep blue solution.

Identify the salts A and B and explain all the reactions described above, giving equations. (A. E. B.)

17. Describe what you would observe when the following compounds are heated, naming the product and giving the equation:

(a) potassium nitrate, (b) lead(IV) oxide, (c) sodium hydrogen carbonate (sodium bicarbonate), (d) ammonium chloride, (e) mercury(II) oxide (mercuric oxide). (A. E. B.)

18. Name a common ore of zinc and describe how the metal is obtained from it on the industrial scale.

Quote two chemical properties of zinc which show that it is a metallic element. State and explain all that would be observed in the following experiments: (a) sodium hydroxide solution is added, in turn, to separate solutions of zinc sulphate and copper(II) sulphate until an excess of alkali is present, (b) zinc foil is left for some time in a solution of lead(II) nitrate, (c) granulated zinc is added to a solution of iron(III) chloride which has been mixed with an equal volume of dilute hydrochloric acid. (W. J. E. C.)

19. From the alternatives given, name with reasons the substances you consider most suitable for the preparation of a solid sample of each of the following:

(a) Copper(II) sulphate (cupric sulphate) using copper foil, or solid copper(II) nitrate (cupric nitrate), or copper(II) oxide (cupric oxide
(b) Sodium sulphate using sodium, or molar sodium hydroxide solution, or molar sodium chloride solution
(c) Lead(II) sulphate using lead foil, or lead(II) oxide (lead monoxide), or solid lead(II) nitrate

Describe, with all essential experimental details, the preparation of any two of these sulphates. (W. J. E. C.)

20. A metal M forms two oxides in which it has valencies of 5 and 3 respectively. Write equations to represent

(a) the formation of the pentoxide by heating the metal in oxygen,
(b) the formation of the trioxide by heating the pentoxide in hydrogen. (J. M. B.)

21. Describe the effect of (a) strongly heating a piece of marble in a Bunsen flame, (b) moistening the product from (a) with water.

Give the common name of the product in each case.

Starting with a piece of marble, how would you prepare a pure dry sample of calcium sulphate? For what reason is slaked lime added to the soil in the garden? Why is it inadvisable to lime the soil shortly after applying ammonium sulphate as a fertilizer? (J. M. B.)

22. Samples are taken from bottles labelled washing soda, calcium chloride, potassium nitrate, iron(II) sulphate (ferrous sulphate), calcium sulphate, manganese(IV) oxide (manganese dioxide), lead(II) oxide (lead monoxide), iron(III) oxide (ferric oxide).

Identify five of these samples from the following descriptions: (a) green crystals, (b) damp white lumps, (c) a black powder, (d) large colourless crystals covered in white powder, (e) a red-brown powder. (J. M. B.)

11 The Chemistry of some Important and Characteristic Non-metals

Introduction. General Chemistry of Non-metallic Elements

In the previous chapter a variety of metals and their compounds were studied; in this chapter we shall study a number of non-metals and their compounds.

Out of the hundred or so elements, about seventy-five are metals, and only about sixteen non-metals, the remaining ones being *metalloids* which have properties intermediate between metals and non-metals (e.g. germanium, arsenic). Two non-metals, oxygen (46.4 per cent) and silicon (27.8 per cent) make up about three-quarters by mass of the earth's crust.

We made a simple study of the characteristic properties of non-metals in Section 2.3, and have also made reference to some of their other properties in various sections of the book. Do non-metals have high or low electronegativities? Where in the Periodic Table do they occur? With the exception of the noble gases, atoms of non-metallic elements are covalently bonded together to form molecular structures, because they need to gain electrons in order to achieve noble gas configurations. Consequently non-metals (except graphite) are poor conductors of heat and electricity, have low melting points and boiling points (except carbon and silicon which exist as giant atomic structures, page 192), and are often gases at room temperature. You should also remember that oxides of non-metals, if 'soluble' in water, give acidic solutions (page 118). Non-metallic elements never liberate hydrogen from acids, but often combine with hydrogen to form stable hydrides (e.g. hydrogen sulphide, H_2S, and ammonia, NH_3). The ions formed by non-metals (except hydrogen) are always negatively charged and are liberated at the anode during electrolysis.

11.1 CARBON AND SILICON

Introduction

The atoms of carbon and silicon both have four electrons in their outer energy levels and so these elements are in Group 4 of the Periodic Table. As it is almost impossible for an atom to gain or lose *four* electrons these elements do not form ions but are linked with other elements always by means of covalent bonding. The fact that in the elements themselves, the atoms are covalently joined into giant atomic structures, means that their physical properties are not always those of typical non-metals. Their chemical behaviour is characteristically non-metallic, e.g. they form oxides which are acidic and the elements do not liberate hydrogen from acids. Some physical constants of carbon and silicon are listed in Table 11.1.

Table 11.1 Some physical constants of carbon and silicon

Property	*Carbon*	*Silicon*
Melting point, K (°C)	3773 (3500)	1686 (1413)
Boiling point, K (°C)	4473 (4200)	2628 (2355)
Density (g cm^{-3})	3.52 (diamond) 2.52 (graphite)	2.49
Atomic radius (nm)	0.077	0.117
Electronegativity	2.5	1.8

Carbon — The Element

Forms of the element

Free carbon in the form of graphite and diamond occurs only rarely in nature, and the element is found mainly in an impure form as coal and in a combined state in petroleum and various carbonate minerals such as calcium carbonate.

Over 90 per cent of the world's *diamonds* come from South Africa where they are found in the craters of extinct volcanoes, having been formed when carbon was subjected to enormous pressures and high temperatures. Attempts to synthesize diamonds have proved only partially successful despite the use of pressures in excess of 100 000 atmospheres and temperatures in the region of 2273 K (2000 °C). Very small synthetic diamonds have been prepared by this method and these are used as cutting tools in industry, but diamonds large enough to be cut into gemstones have not yet been synthesized. Diamond is the hardest known natural substance (although it is not as hard as the synthetically produced boron nitride) and is chemically unreactive. In addition to its use as a gemstone and in cutting tools, it is also used in rock drilling equipment.

s to synthesize dia- have proved only successful

The other allotrope of carbon, *graphite,* occurs to a small extent in many countries, particularly Ceylon, and it was mined in Borrowdale in Cumberland. Nowadays, graphite is usually prepared synthetically by the Acheson process, which involves heating impure carbon with sand in an electric furnace. In contrast to diamond, graphite is one of the softest solids known and is used as a lubricant, in electrodes, as a 'moderator' in atomic reactors and, when mixed with clay, as 'lead' in pencils.

The very different properties of diamond and graphite can be explained by the structures of these compounds (page 193). Another form of carbon, in which the atoms did not appear to be arranged in any regular manner, was called 'amorphous' (without shape) carbon and was used to describe forms of the element such as coke, charcoal, soot, lampblack and coal. These forms have many uses, e.g. activated charcoal has good adsorptive power and is used to purify materials (Experiment 2.4). It is now believed that 'amorphous' carbon consists of very small fragments of graphite crystals.

Another interesting form of carbon is *carbon fibre,* which is made by the controlled thermal degradation of a textile fibre such as viscose rayon or polyacrylonitrile. The orientation of the carbon atoms in the original fibre is retained in the product. Carbon fibre is particularly useful where a material of high durability, strength, and lightness is required. Its commercial potential has not yet been exploited but it is beginning to find applications in reinforcing plastics used in turbo-engines, racing car bodies, ships' hulls, etc.

The Carbon Cycle In addition to being present in carbon dioxide in the atmosphere, carbon is an essential constitutent of all living organisms. There is a balance between the carbon dioxide liberated into the atmosphere and that used up from the atmosphere. This is conveniently summarized in the *carbon cycle* (Figure 11.1). You may like to find out for yourself how the amount of

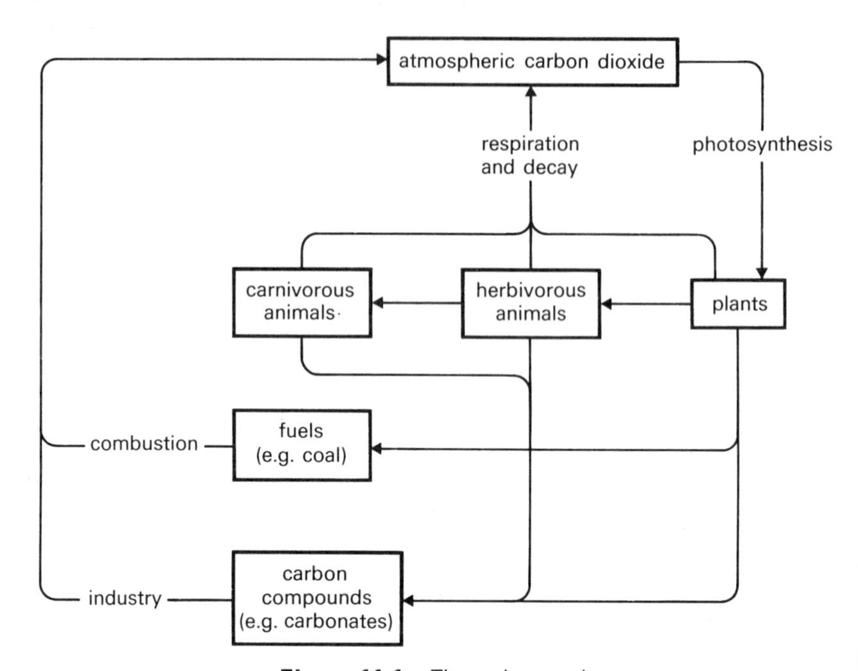

Figure 11.1 The carbon cycle

carbon dioxide then in the atmosphere may have contributed to the climate of the Ice Ages. It is important to realize that virtually all the carbon atoms on earth have been present since the earth was formed, and are constantly 'circulating' in nature. For example, it is possible that some of the carbon atoms in your body once formed part of a tree in a primeval forest, then part of a dinosaur, then part of a Viking ship, part of Shakespeare's body, etc.

Chemical Reactions of Carbon All forms of carbon, if heated to a sufficiently high temperature in a plentiful supply of air, give carbon dioxide, but carbon monoxide is formed if the supply of air is limited.

Carbon reduces many metallic oxides (those of the less vigorous metals) to the metal when the two are heated together.

Silicon—The Element

Silicon is the second most abundant element in the earth's crust (about 28 per cent by mass). Can you remember which is the most abundant? Silicon is not found in an uncombined state but occurs widely in silicates (e.g. calcium silicate, $CaSiO_3$) or silicon dioxide (silica, SiO_2) which can exist in various forms, e.g. quartz, sand, flints, etc. Most rocks contain silicates which on weathering produce particles of silicon dioxide and these eventually accumulate as sand. Chemical weathering of silicates leads to the formation of clay which contains aluminium silicate (Al_2O_3, $2SiO_2$, $2H_2O$).

Silicon can be prepared in the laboratory by heating *dry* sand with magnesium powder,

$$SiO_2(s) + 2Mg(s) \rightarrow 2MgO(s) + Si(s).$$

The industrial method of preparation is similar but uses an electric furnace, and carbon instead of magnesium as the reducing agent. You may be shown the laboratory preparation, in which a little magnesium silicide (Mg_2Si) is also formed. When hydrochloric acid is added to this compound a mixture of volatile silicon hydrides (of general formula $Si_nH_{(2n+2)}$, e.g. SiH_4, monosilane) is evolved. These hydrides, unlike their corresponding carbon (alkane) analogues, are very reactive and spontaneously flammable.

Properties and Uses of Silicon

Silicon is a hard and brittle solid, and all its forms have a diamond-like crystal structure, but many outwardly different forms exist which are built up from this same basic 'unit'. Large crystals are chemically unreactive, but powdered silicon when heated is attacked by halogens and by oxygen. Silicon is used in the making of silicon steel (a chemically resistant steel) and in the preparation of silicones (*Further Topics*).

Some Compounds of Carbon

Carbon forms an enormous number of different compounds, the vast majority of which are organic (Chapter 12). We will now consider some of its important inorganic compounds.

Carbon Dioxide (CO_2)

Preparation This gas is produced whenever carbon or any of its compounds are completely burnt in an *excess* of air or oxygen. The usual method of preparing carbon dioxide in the laboratory is to add dilute hydrochloric acid to marble chips (a pure form of calcium carbonate).

Points for Discussion

1. Write an equation for this reaction and devise a suitable apparatus using the information given in Section 4.1 and Table 4.1.

2. Some acid spray will be carried over with the gas. How would you remove this if necessary? (Remember that carbon dioxide is itself an acid gas.)

3. Carbon dioxide is also liberated by the action of heat on carbonates and hydrogen carbonates (page 234), and from fermentation (Experiment 12.7). It is produced industrially by heating calcium carbonate in a limekiln (Figure 11.2).

Figure 11.2 A vertical limekiln

Physical properties Carbon dioxide is a colourless, virtually odourless gas. It is denser than air as can be shown by the porous pot diffusion experiment (Experiment 4.11). How must the apparatus be modified in view of the facts that carbon dioxide is more dense than air and hydrogen less dense? Carbon dioxide is slightly soluble in water, but under pressure a large volume can be made to dissolve and produce a solution called 'soda water'. This principle is used in making 'fizzy' drinks (e.g. lemonade). Carbon dioxide is easily liquefied and solidified under pressure. The solid sublimes at 195 K (−78 °C) and is used as a refrigerant; it is called 'Drikold' or 'Dri-ice' as it leaves no residue after sublimation.

Chemical properties

(a) *Action with metals*
Carbon dioxide is not very chemically reactive, but metals high in the electrochemical series burn in it if heated sufficiently, as is shown by the next experiment.

Experiment 11.1
†Reaction of Magnesium with Carbon Dioxide *

Apparatus
Carbon dioxide generator, gas jar and slide, tongs, Bunsen burner, asbestos square.
Magnesium ribbon.

Procedure
(a) Collect a gas jar of carbon dioxide by downward delivery and cover the open end of the jar with the slide.

(b) Wrap a piece of magnesium ribbon around the tongs so that a piece about 12 cm long hangs from the tongs.

(c) Ignite the magnesium ribbon, remove the slide from the gas jar, and quickly plunge the burning magnesium into the gas jar. (Do not look directly at the burning magnesium.)

(d) Record all your observations.

Points for Discussion
1. What evidence have you that a reaction took place? From your observations you should be able to name one of the products formed.

2. It is probable that the initial exothermic burning of magnesium in the carbon dioxide gas decomposes the gas. What other gas will be formed by this decomposition? Does magnesium burn in this other gas? If so what compound is formed? Does one of the reaction products correspond to this compound?

3. Write an equation for the reaction.

(b) *Carbonic acid*
When carbon dioxide dissolves in water about 1 per cent of the dissolved gas also *reacts* with the water to form the weak acid, 'carbonic acid' (page 237), i.e.

$$CO_2(g)+H_2O(l) \rightleftharpoons H_2CO_3(aq)$$

or ionically

$$CO_2(g)+3H_2O(l) \rightleftharpoons 2H_3O^+(aq)+CO_3^{2-}(aq)$$

It is impossible to isolate carbonic acid but a solution of carbon dioxide in water is a weak acid as shown by its reaction with indicators. It reacts with alkalis to form either carbonates, e.g.

$$2KOH(aq) + CO_2(g) \rightarrow K_2CO_3(aq) + H_2O(l)$$

or, with excess carbon dioxide, hydrogen carbonates, e.g.

$$K_2CO_3(aq) + H_2O(l) + CO_2(g) \rightarrow 2KHCO_3(aq)$$

Point for Discussion

Can you remember why potassium hydroxide is used to absorb carbon dioxide whereas calcium hydroxide is used to detect it?

(c) *Other chemical properties*

The role of carbon dioxide in the production of hard water (page 217) and in photosynthesis, combustion, and respiration (Section 4.2) has already been discussed.

...ich baking powder ...add?

Uses of carbon dioxide Apart from its use in making 'fizzy' drinks and as a refrigerant, carbon dioxide is used in fire extinguishers. Suggest why. Industrially carbon dioxide is used in the Solvay ammonia–soda process (Section 14.3) and also to cool certain types of atomic reactors. Carbon dioxide is produced from baking powder during the baking of cakes, bread, etc. Baking powder consists of a dry mixture of sodium hydrogen carbonate ($NaHCO_3$) and an acidic substance, e.g. potassium hydrogen tartrate, $KHC_4H_4O_6$ (cream of tartar); carbon dioxide is evolved when the mixture is moistened, and the evolution becomes more rapid on heating. These bubbles of carbon dioxide cause the cake mixture to 'rise'. Can you suggest why carbon dioxide is not evolved until water is added?

Carbon Monoxide (CO)

Preparation Carbon monoxide is formed when carbon and various fuels are burnt in a *limited* supply of air or oxygen. It is usually prepared in the laboratory as a demonstration only, because of its poisonous nature, the preparation being carried out by adding a solution of methanoic acid (formic acid), dropwise, to warm concentrated sulphuric acid. The latter dehydrates the methanoic acid according to the equation:

$$H.COOH(aq) \rightarrow CO(g) + H_2O(l)$$

You may like to devise an apparatus for this preparation in your notebook (Section 4.1), bearing in mind the properties of carbon monoxide and that the two main impurities are likely to be carbon dioxide and sulphur dioxide. Commercially carbon monoxide is produced in the form of water gas and producer gas.

Physical properties Carbon monoxide is a colourless, odourless gas, insoluble in water and slightly less dense than air. It is extremely poisonous at a concentration of less than 1 per cent, and is particularly dangerous as it has no odour. It combines with the haemoglobin in the red blood corpuscles in preference to oxygen and so prevents the supply of this essential element to the body cells.

Chemical properties

(a) *Combustion*
Carbon monoxide burns very easily in air with a characteristic blue flame, forming carbon dioxide. Write an equation for this reaction. Often the supply of air to a normal household coal fire is limited and carbon monoxide is produced. Why do you think there are not a large number of carbon monoxide poisonings in the home?

(b) *As a reducing agent*
It is a good reducing agent and is sometimes used to reduce oxides of metals (except these near the top of the electrochemical series) to the metal. This is the principle in the extraction of iron from iron ore in the blast furnace (page 388).

(c) *Other properties*
It is a neutral oxide and so has no effect on acids and alkalis under normal conditions, although it does show very slight acidic tendencies by reacting with concentrated sodium hydroxide solution at 433 K (160 °C) and under pressure to form the salt, sodium methanoate:

$$CO(g) + NaOH(aq) \rightarrow H.COONa(aq)$$

Uses of carbon monoxide Carbon monoxide is used mainly as a fuel and as a reducing agent.

Carbonates and Hydrogen Carbonates

The chemistry of carbonates and hydrogen carbonates has been fully considered in Section 10.1.

Compounds of Silicon

Some properties of silicon dioxide (SiO_2), silicon tetrachloride ($SiCl_4$), and monosilane (SiH_4) have previously been considered (Section 5.3). We shall now study silicon dioxide in more detail, and also consider the formation of silicates, including glass.

Silicon Dioxide (SiO_2)

Silicon dioxide, or silica as it is more usually called, occurs naturally in the several forms previously mentioned. In its crystalline forms, it has a giant atomic structure (page 192) in which each silicon atom is bonded to four oxygen atoms and each oxygen atom to two silicon atoms in such a way that each silicon atom is at the centre of a regular tetrahedron of oxygen atoms, i.e. represented two-dimensionally,

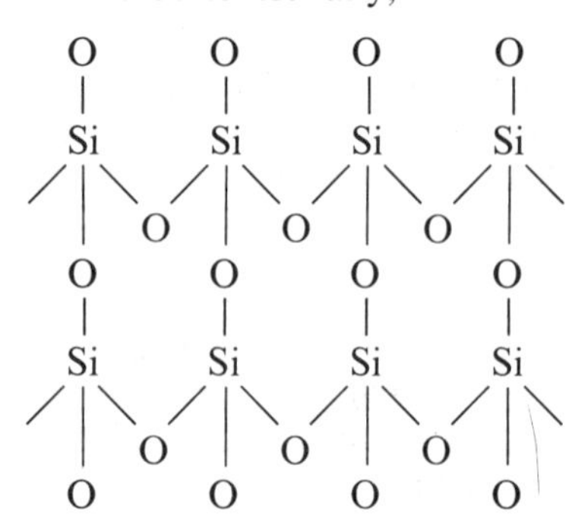

This structure is in complete contrast to the structure of the previous Group 4 dioxide we considered, i.e. carbon dioxide, which exists as individual molecules at room temperature. This results in the two oxides having completely different physical properties. Silicon dioxide is a solid of high melting point and high boiling point, whereas carbon dioxide is a gas with a low melting point and boiling point.

Properties Silica is a fairly unreactive acidic oxide with a high melting point [about 1873 K (1600 °C)]. When silica is fused, *silica glass* is formed. This has advantages over ordinary glass in that (a) it can be heated to about 1773 K (1500 °C) before it softens, (b) it has a very low coefficient of expansion and can be heated to white heat and then plunged into cold water without cracking, (c) it is more transparent than ordinary glass to visible, infra-red, and ultra-violet light. Consequently such glass is used in making glassware required to withstand attack by acids or extremes of temperature, and also for the construction of lenses and prisms.

Glass

Ordinary glass (sometimes called soda glass) is a mixture of sodium silicate and calcium silicate and is made by heating a mixture of sand, sodium carbonate, and calcium carbonate in a furnace at 1673 K (1400 °C).

$$Na_2CO_3(s) + SiO_2(s) \rightarrow Na_2SiO_3(s) + CO_2(g)$$

$$CaCO_3(s) + SiO_2(s) \rightarrow CaSiO_3(s) + CO_2(g)$$

The molten glass is then machined to specifications. For example, sheet glass can be made by pressing a horizontal rod into molten glass and lifting the rod vertically. The glass clings to the rod and a sheet is 'pulled' out of the molten mass. Another, more recent, process involves floating the molten glass on liquefied tin (Figure 11.3). The sheet is then drawn along by rollers (Figure 11.4). Shaped pieces of glassware (e.g. bottles)

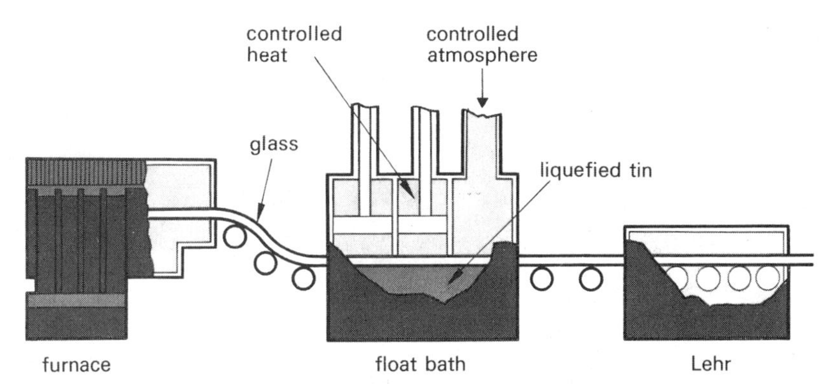

Figure 11.3 Diagrammatic representation of the Float process

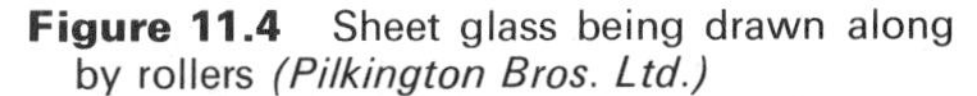

Figure 11.4 Sheet glass being drawn along by rollers *(Pilkington Bros. Ltd.)*

are usually moulded or blown into shape from small quantities of molten glass. Glass products are always cooled slowly, because if molten glass is cooled too rapidly, internal strains are set up which make the glass liable to crack easily. This slow cooling process is called *annealing,* and when the large glass mirror for the 500 cm telescope at Mount Palomar was cast, it took six months to anneal.

Cheap glass is often tinted green or yellow due to iron compounds which are present as impurities in the sand, and *coloured glass* is made by deliberately adding traces of transition metal oxides which form coloured silicates, e.g. iron(II) (green), copper(I) (red), cobalt (blue), etc. Addition of aluminium or boron oxides gives the glass a low coefficient of expansion, suitable for vessels which are to be heated, e.g. 'Pyrex'.

Glass does not melt at a definite temperature but gradually becomes more and more plastic. This is because glass is not a true solid, but is a supercooled liquid. The molecules are not arranged in any definite patterns as they would be in a true solid, but retain the random positions they had in the liquid state. An old window will have thicker glass at the bottom than at the top because the glass slowly 'flows' down.

Glass is attacked by hydrofluoric acid, but wax is not. How could you use these substances to make a design on a piece of glass? This process is called *etching*.

Figure 11.5 The Micro-plus 500 cabin cruiser, made from Cellobond ester resins and glass fibre *(B.P.)*

Fibre-glass is made by forcing molten glass through very small holes and allowing the resulting fine threads to set. It can then be woven into fabric, and used to make fireproof curtains, and other heat resistant products. An ever increasing amount of the fibre is used to reinforce plastics and so produce a strong, lightweight material which can act as a substitute for wood and metals, e.g. in making small boat hulls (Figure 11.5), car frames, etc.

11.2 NITROGEN AND PHOSPHORUS

Introduction

Nitrogen and phosphorus are in Group 5 of the Periodic Table as their atoms have five electrons in their outer shells. For reasons mentioned on page 36 it is impossible for either of these elements to lose five electrons and thus form an ion of the type X^{5+}, although they can gain three electrons and ions of the type X^{3-} are occasionally present in compounds (e.g. magnesium nitride (Mg_3N_2) and calcium phosphide (Ca_3P_2)). The majority of compounds formed by nitrogen and phosphorus contain covalent bonds. Also phosphorus, but not nitrogen, can increase its valency to five in such compounds as phosphorus pentachloride (PCl_5) and phosphorus(V) oxide (P_4O_{10}).

Nitrogen and phosphorus are typical non-metals and thus form acidic oxides etc., but they differ in physical appearance. Can you account for the fact, in terms of structure, that nitrogen is a gas at room temperature whereas phosphorus is a solid? Some physical constants of these elements are given in Table 11.2.

Table 11.2 Some physical constants of nitrogen and phosphorus

Property	*Nitrogen*	*Phosphorus*
Melting point, K (°C)	63 (−210)	317 (44) (white)
Boiling point, K (°C)	77 (−196)	553 (280) (white)
Density (g cm^{-3})	—	1.82 (white) 2.20 (red)
Atomic radius (nm)	0.07	0.11
Ionic radius (nm) (of X^{3-} ion)	0.17	0.21
Electronegativity	3.0	2.1

Nitrogen—The Element

Nitrogen was first isolated (from air) in 1772 by the Swedish chemist Carl Wilhelm Scheele. As you learnt in Section 4.2 it forms about 78 per cent by volume of air, and it is also present in a number of compounds, particularly nitrates, which occur in regions having dry climates, e.g. Chile. Why do you think nitrate deposits are only found in dry areas?

The Nitrogen Cycle

Nitrogen is an essential constituent of all living organisms, both plant and animal, and is present in their tissues in compounds called proteins. Plants acquire 'nitrogen' by absorption of nitrates and ammonium salts in solution through their root systems and use them to form proteins; herbivorous animals then consume these plants and so obtain their necessary nitrogen, and carnivorous animals consume the proteins made by these animals. These processes are part of the nitrogen cycle

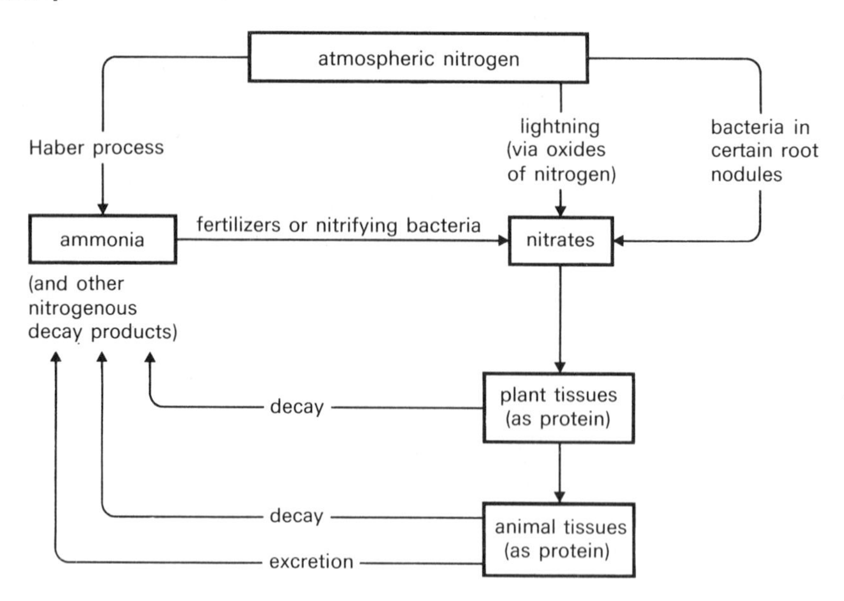

Figure 11.6 The nitrogen cycle

(Figure 11.6). The vast majority of plants can only absorb nitrogen in the form of nitrates or ammonium salts, and nitrogen which is present in a nitrate group, or in some form which is readily convertible to a nitrate, is known as *fixed nitrogen.* The supply of fixed nitrogen, essential for the life of plants and ultimately of animals, is replenished naturally by (a) the nitric acid formed in the atmosphere by electrical discharges during thunderstorms and washed down by rain, (b) the oxidation of ammonia produced during the breakdown of animal and plant tissue by bacteria in the soil, (c) the leguminous plants (e.g. peas, beans, clover) which have nodules on their roots containing colonies of bacteria capable of fixing atmospheric nitrogen directly. Plant growth can also be stimulated by increasing the natural nitrate content of the soil either by addition of manure (containing nitrogenous animal products) or of fertilizers such as ammonium nitrate. Fertilizers are usually ammonium compounds manufactured from ammonia made by the Haber process (page 381). As the nitrogen in this process is obtained from the air, man is fixing the nitrogen. It is now essential that nitrogen is fixed in this synthetic way as modern civilization, cropping, output of sewage into the sea, etc. have unbalanced the natural cycle which is no longer self-sustaining. The utilization of nitrogen by living things is a cyclic process (Figure 11.6), comparable to the carbon cycle, and in a similar way nitrogen atoms present in your body tissue will also have had an interesting history.

Point for Discussion

Why is it preferable to use a ton of ammonium nitrate, NH_4NO_3 (molecular mass 80), rather than a ton of ammonium sulphate, $(NH_4)_2SO_4$ (molecular mass 132), as a nitrogenous fertilizer?

Preparation of Nitrogen

Nitrogen can be prepared in the laboratory from air (Experiment 4.5) and is prepared industrially by the fractional distillation of liquid air (page 394). Pure nitrogen is obtained in the laboratory by gently warming a mixture of concentrated solutions of ammonium chloride and sodium nitrite. The sequence of reactions can be represented by:

$$NH_4Cl(aq) + NaNO_2(aq) \rightarrow NH_4NO_2(aq) + NaCl(aq)$$

$$NH_4NO_2(aq) \rightarrow N_2(g) + 2H_2O(l)$$

(Ammonium nitrite)

[*Note:* In this experiment ammonium nitrite is constantly being formed and decomposed; the action of heat on ammonium nitrite alone is often dangerously explosive.]

Point for Discussion

From a knowledge of the physicial properties of nitrogen and reference to Section 4.1, devise an apparatus for the preparation and collection of a pure dry sample of nitrogen.

Properties of Nitrogen

Nitrogen is a colourless, odourless, non-poisonous gas which is slightly soluble in water (Table 4.1) forming a neutral solution.

The chief chemical characteristic of nitrogen is its inertness. Nitrogen neither burns nor supports combustion and does not react unaided with any substance at room temperature. It will combine directly with oxygen (e.g. when heated in the presence of a suitable catalyst), forming nitrogen monoxide and then nitrogen dioxide

$$N_2(g) + O_2(g) \rightarrow 2NO(g) \qquad 2NO(g) + O_2(g) \rightarrow 2NO_2(g),$$

with hydrogen under specific conditions forming ammonia (page 381), and with some metals (e.g. calcium, magnesium, and aluminium) forming nitrides ($3Mg(s) + N_2(g) \rightarrow Mg_3N_2(s)$).

Uses of Nitrogen

The Haber process converts large quantities of atmospheric nitrogen into ammonia which is used to produce fertilizers and nitric acid. Nitrogen provides the necessary inert atmosphere for certain chemical reactions, and liquid nitrogen is used as a refrigerant.

Phosphorus—The Element

Manufacture

In nature, phosphorus is mainly found combined in the form of phosphate salts (e.g. calcium phosphate, $Ca_3(PO_4)_2$). It is prepared on a commercial scale by heating a mixture of phosphate rock, sand, and coke in an electric furnace at 1673 K (1400 °C). The reactions which take place using calcium phosphate for example, are:

$$2Ca_3(PO_4)_2(s) + 6SiO_2(s) \rightarrow 6CaSiO_3(s) + P_4O_{10}(g)$$

$$P_4O_{10}(g) + 10C(s) \rightarrow 4P(g) + 10CO(g)$$

The phosphorus obtained by cooling this vapour is called *white* phosphorus.

Physical Properties

Phosphorus is a solid element existing in two main allotropic forms at room temperature (i.e. 'white' phosphorus, m.p. 317 K (44 °C), b.p. 554 K (281 °C) and 'red' phosphorus, sublimes at 689 K (416 °C)). When phosphorus vapour is cooled the *metastable* white phosphorus is produced, which reverts, extremely slowly at room temperature but rapidly above 523 K (250 °C), to the stable allotrope, red phosphorus. (A metastable allotrope is one which shows little tendency to change under conditions in which it is theoretically unstable.) White phosphorus is composed of tetrahedral P_4 molecules and it is believed that these units link together in a haphazard way in red phosphorus, forming a giant atomic structure.

Chemical Properties

White phosphorus is chemically very reactive and is also poisonous, and it is not usual to study this allotrope experimentally at an introductory level. It is spontaneously flammable in air and has to be stored under water, with which it does not react. Even the heat of the hand is often sufficient to ignite it, thus it must never be allowed to come into contact with the skin. White and red phosphorus burn in an abundant supply of air to form phosphorus(V) oxide (Experiment 4.10). Write an equation for this reaction. White phosphorus also ignites spontaneously in chlorine and reacts violently with other halogens. Strong oxidizing agents oxidize it to phosphoric acid (H_3PO_4) and with strong alkali, phosphine (PH_3) is formed. If white phosphorus is boiled in water, the steam evolved is luminous in the dark. This phenomenon is called *phosphorescence* and is a simple if rather dangerous test for white phosphorus.

Uses of Phosphorus

Phosphorus is mainly used to make fertilizers (e.g. superphosphate of lime, $Ca(H_2PO_4)_2 + CaSO_4$) to replace phosphorus compounds in the soil, as it is an essential constituent of living matter. Ammonium phosphate $(NH_4)_3PO_4$, is a particularly useful fertilizer as it supplies both nitrogen and phosphorus to the soil. Small amounts of phosphorus are used as rat poisons and in the preparation of special bronze alloys. Match heads were once made from white phosphorus but this is now prohibited (why?). A sulphide of phosphorus (P_4S_3) is used in making the heads of 'strike anywhere' matches.

Compounds of Nitrogen

Nitrogen is present in a variety of compounds, notably ammonia and its related salts, nitric and nitrous acids and their salts, and a number of compounds with oxygen and the halogens. The industrial production of ammonia and nitric acid are considered in Section 14.3, and we discussed the properties of dilute nitric acid and metal nitrates in Sections 3.1 and 10.1 respectively. We shall now deal with ammonia and its related compounds in more detail, discuss nitric acid, and briefly mention some of the oxides of nitrogen.

Ammonia (NH_3)

Preparation Ammonia is formed when any ammonium salt is heated with an alkali, either solid or in solution (page 79). This can be represented ionically as

$$NH_4^+(aq) + OH^-(aq) \rightarrow NH_3(g) + H_2O(l)$$

The usual laboratory preparation is to heat an intimate mixture of solid ammonium chloride (NH_4Cl) and dry calcium hydroxide ($Ca(OH)_2$).

Experiment 11.2
† Laboratory Preparation of Ammonia **

Apparatus and Procedure

Using the information given in Section 4.1 and Table 4.1 devise an apparatus and procedure for the production of a dry sample of ammonia gas. Remember ammonia gas must be dried by passing it through calcium oxide.

Points for Discussion

1. Why do you think *dry* calcium hydroxide was used in this experiment? Would solid sodium hydroxide be as suitable?

2. Why cannot ammonia be dried using concentrated sulphuric acid or calcium chloride?

Physical properties Ammonia is a colourless gas, less dense than air, with a characteristic pungent odour and is *very* soluble in water (Table 4.1 and Experiment 11.3) forming ammonia solution. Ammonia is easily liquefied under pressure and is usually transported in this form. Ammonia molecules are pyramidal in shape (page 200).

Experiment 11.3

† The Fountain Experiment ***

Apparatus

Ammonia generator, *dry* round bottom flask and other apparatus indicated in Figure 11.7.

Universal indicator solution.

Procedure

(a) Fill the round bottom flask with dry ammonia gas. By passing the gas through the flask for some time. Stop the flow of gas and tighten both screw clips tightly to prevent the gas escaping.

(b) Set up the apparatus as shown in Figure 11.7 colouring the water green with a few drops of universal indicator.

(c) With the open end of the rubber tubing well below the surface of the water, unscrew the clip and observe what happens.

Points for Discussion

1. Why is this experiment called the fountain experiment? Can you explain what causes this effect?

2. What colour change does the universal indicator undergo? Why does this change take place?

3. Ammonia is very soluble in water and when the screw clip is opened some of the gas immediately dissolves. This reduces the pressure in the flask and the larger (atmospheric) pressure on the surface of the water in the beaker forces it up the tube, where it emerges from the fine jet in the form of a fountain. Ammonia continues to dissolve rapidly, and so the pressure in the flask is being continually reduced and the fountain is maintained.

4. Did all the ammonia dissolve? Explain your answer.

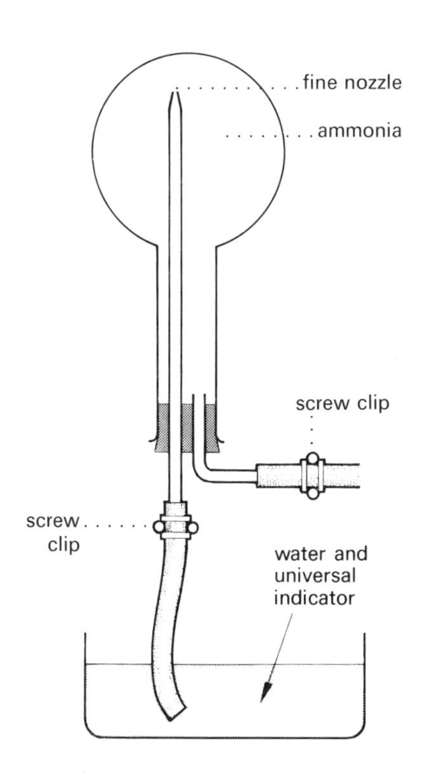

Figure 11.7 The fountain experiment

Chemical properties Ammonia is the only common alkaline gas and neutralizes acid solutions and acid gases (Experiment 1.13) to form ammonium salts.

Ammonia reduces the heated oxides of metals below iron in the electrochemical series, e.g.

$$2NH_3(g) + 3CuO(s) \rightarrow 3Cu(s) + 3H_2O(g) + N_2(g)$$

Ammonia also reacts with the halogens, the products depending on whether (a) ammonia or (b) the halogen is in excess, i.e.

(a) $$8NH_3(g) + 3Cl_2(g) \rightarrow N_2(g) + 6NH_4Cl(s)$$

(b) $$NH_3(g) + 3Cl_2(g) \rightarrow NCl_3(l) + 3HCl(g)$$

The tests for ammonia have previously been described (Experiment 3.7).

Ammonia will not burn in air, but burns in oxygen to produce nitrogen and steam. This can be demonstrated by the following experiment.

Experiment 11.4
† Burning Ammonia in Oxygen ***

Apparatus

As illustrated in Figure 11.8, access to fume cupboard.

Cylinders, or alternative supplies, of dry ammonia and oxygen.

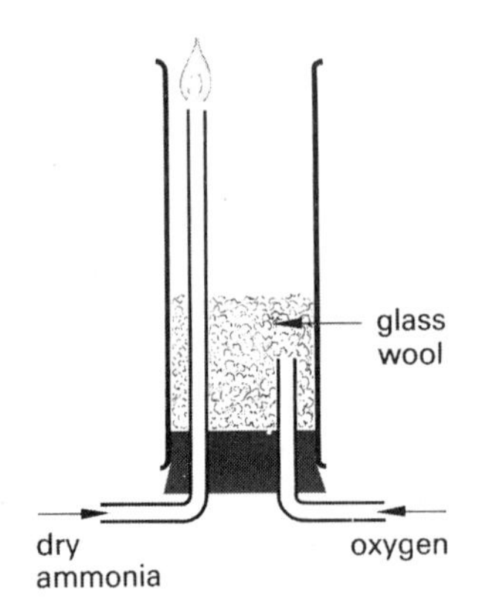

Figure 11.8 Combustion of ammonia in oxygen

Procedure

(a) Assemble the apparatus shown (Figure 11.8) in a fume cupboard.

(b) Adjust the rate of flow of the two gases until a flame can be lit at the outlet of the ammonia tube. Record any observations.

Points for Discussion

1. What colour was the flame?

2. What is the purpose of the glass wool?

3. Did you notice any evidence for the formation of water vapour? Write an equation for the combustion.

4. In the presence of a catalyst (e.g. platinum), ammonia is oxidized by air to give nitrogen monoxide:

$$4NH_3(g) + 5O_2(g) \rightarrow 4NO(g) + 6H_2O(g)$$

This reaction is the first step in the industrial production of nitric acid from ammonia (Section 14.3).

Uses of ammonia Ammonia is chiefly produced industrially (Haber process) as an intermediate in the preparation of fertilizers or for oxidation into nitric acid. It is also used to prepare urea (used in the production of plastics, in addition to being a fertilizer), and as a refrigerant.

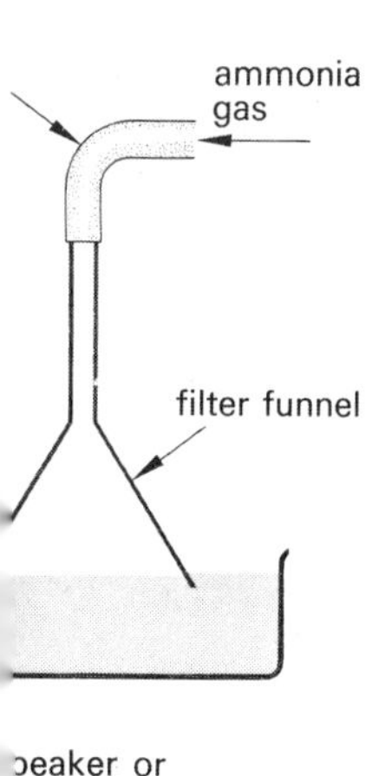

11.9 Preparation of
ation of ammonia

Ammonia solution When ammonia dissolves in water the ammonia solution formed contains the ions $NH_4^+(aq)$ and $OH^-(aq)$. Because of this the formula NH_4OH is sometimes wrongly ascribed to this solution.

Will the solution finally formed be alkaline or acidic? Give a reason for your answer. The most concentrated solution of ammonia has a density of 0.880 g cm^{-3} and is commonly called concentrated ammonia or '880' ammonia.

When ammonia is passed into water, special precautions must be taken to prevent the water 'sucking back' into the ammonia generator, as ammonia is extremely soluble (Experiment 11.3). Sucking back can usually be prevented by attaching an inverted filter funnel to the end of the delivery tube and placing the rim of the funnel just below the surface of the water (Figure 11.9). Why would sucking back occur if the delivery tube dipped directly into the water? How does the use of a filter funnel prevent this? Can you think of another advantage of using this method to dissolve a gas in water?

Properties of ammonia solution When ammonia solution is warmed, ammonia gas is liberated and this provides a useful way of 'preparing' the gas for laboratory purposes.

Ammonia solution is an alkaline solution (Section 3.2) and although not as strong an alkali as sodium hydroxide solution, it has similar reactions. For example, it neutralizes acids to give ammonium salts, (e.g. $NH_3(aq) + HCl(aq) \rightarrow NH_4Cl(aq)$) and precipitates insoluble hydroxides by double decomposition,

$$Fe^{2+}(aq) + 2OH^-(aq) \rightarrow Fe(OH)_2(s)$$

Some hydroxides redissolve after precipitation if excess of ammonia solution is added (even though the hydroxides are not amphoteric) due to the formation of soluble complexes (Experiment 10.2) e.g. initially

$$Cu^{2+}(aq) + 2OH^-(aq) \rightarrow \underset{\text{Light blue precipitate}}{Cu(OH)_2(s)}$$

followed by

$$Cu(OH)_2(s) + 4NH_3(aq) \rightarrow \underset{\text{Deep blue solution}}{[Cu(NH_3)_4]^{2+}(aq) + 2OH^-(aq)}$$

Ammonium Salts

Preparation Ammonium salts are usually prepared in the laboratory by adding ammonia solution to the appropriate acid, and industrially by passing ammonia gas into the acid.

Points for Discussion

1. Why do you think the laboratory preparation uses ammonia solution and the industrial preparation ammonia gas?

2. Write equations for the preparation of ammonium sulphate and ammonium nitrate by both of these methods.

3. Why is ammonium carbonate not prepared by the previous methods? It is usually prepared by heating a dry mixture of ammonium sulphate and calcium carbonate; the product (very impure) is separated by sublimation and purified by recrystallization from water.

Properties All ammonium salts are soluble in water and produce ammonia gas when heated with an alkali (page 79). Ammonium salts always decompose when heated, producing different nitrogenous products, and some also sublime (ammonium chloride, page 56, and ammonium carbonate). The salts which sublime also decompose as can be shown by the next experiment.

Experiment 11.5
Effect of Heat on Ammonium Chloride ***

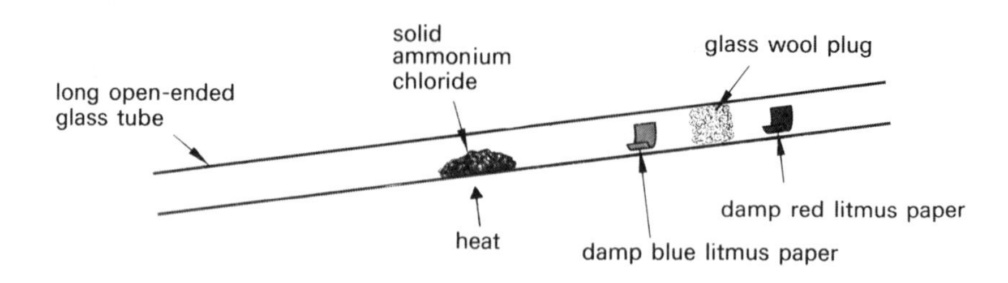

Figure 11.10 The action of heat on ammonium chloride

Apparatus
As in Figure 11.10.

Procedure
(a) Assemble the apparatus (Figure 11.10), taking care to place the damp blue and red litmus papers as shown.

(b) Gently heat the ammonium chloride and note any changes taking place.

Points for Discussion
1. What changes did you see taking place in the apparatus? What must have been produced to account for these changes? Where must these 'products' have come from?

2. Name the 'products' and write an equation to show how they are formed.

3. How do you account for the fact that one piece of litmus turns red, while the other turns blue?

4. In this experiment ammonium chloride is decomposed by heat, i.e.

$$NH_4Cl\ (s) \rightleftharpoons NH_3\ (g) + HCl\ (g)$$

The products, when cool, reform the original compound, and thus the reaction is reversible. Such a reaction is called a *thermal dissociation*.

5. Other ammonium compounds decompose on heating, but do not reform on cooling (i.e. the reaction is not reversible). Reactions of this nature are called *thermal decompositions,* e.g.

$$NH_4Cl(s) \rightleftharpoons NH_3(g) + HCl(g)$$
$$NH_4NO_2(s) \rightarrow N_2(g) + 2H_2O(g)$$
$$NH_4NO_3(s) \rightarrow N_2O(g) + 2H_2O(g)$$
$$(NH_4)_2SO_4(s) \rightarrow NH_3(g) + NH_4HSO_4(s)$$

Uses The uses of some typical ammonium compounds are listed in Table 11.3.

Nitric Acid (HNO_3)

Preparation Nitric acid is prepared by heating approximately equal masses of potassium nitrate and concentrated sulphuric acid at a temperature of about 403 K (130 °C). The temperature is kept as low as possible to avoid thermal decomposition of the acid. The equation for the reaction is:

$$H_2SO_4(l) + KNO_3(s) \rightarrow HNO_3(l) + KHSO_4(s)$$

Table 11.3 Uses of some common ammonium compounds

Chemical name	*Common name*	*Formula*	*Use*
Ammonia solution	Ammonium hydroxide	NH_3 (aq)	In the household for washing and cleaning
Ammonium carbonate	—	$(NH_4)_2CO_3$	In smelling salts
Ammonium chloride	Sal-ammoniac	NH_4Cl	In dry batteries, Leclanché cells and in soldering
Ammonium nitrate	—	NH_4NO_3	As a fertilizer (Nitram) and in explosives
Ammonium sulphate	Sulphate of ammonia	$(NH_4)_2SO_4$	Widely as a fertilizer

An all glass apparatus is used as cork and rubber are rapidly attacked by the acid vapour. A suitable arrangement, using a retort, is shown in Figure 11.11. The nitric acid produced by this method condenses as a liquid containing more than 90 per cent nitric acid and is called *fuming nitric acid* as it fumes in contact with atmospheric moisture. Concentrated nitric acid contains about 68 per cent nitric acid.

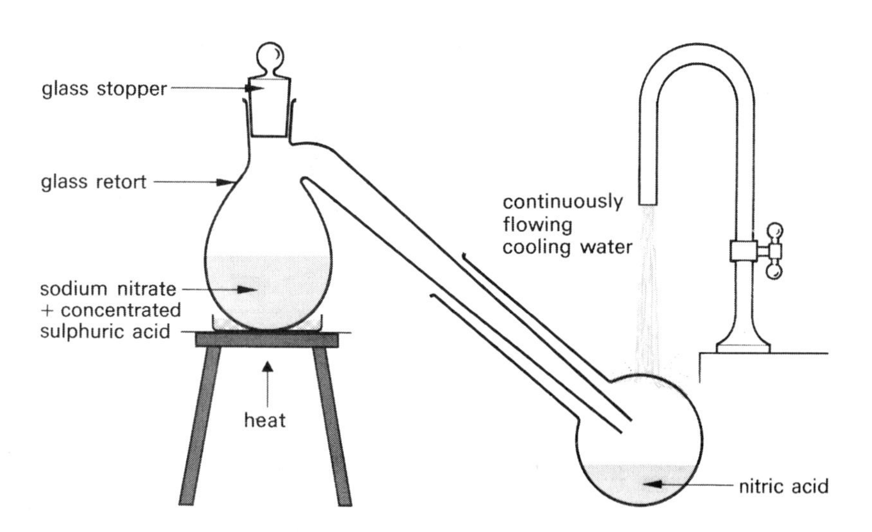

Figure 11.11 Laboratory preparation of nitric acid

Commercially, large quantities of nitric acid are made by the catalytic oxidation of ammonia (Section 14.3).

Physical properties Pure nitric acid is a colourless liquid, m.p. 231 K (−42 °C), b.p. 359 K (86 °C) and density 1.52 g cm^{-3}. The concentrated acid is often yellowish in colour due to the presence of oxides of nitrogen, particularly nitrogen dioxide. The colour deepens on exposure to daylight because it causes slow decomposition of the acid:

$$4HNO_3(aq) \rightarrow 4NO_2(aq) + 2H_2O(l) + O_2(g)$$

This is why concentrated nitric acid is stored in dark bottles.

Chemical properties The chemical properties of nitric acid can be conveniently divided into four groups.

(a) Acidic properties
Dilute nitric acid is a typical strong acid (except in its reaction with metals) whose reactions have been studied in Section 3.1.

(b) Thermal decomposition
Concentrated nitric acid when heated undergoes thermal decomposition, liberating oxygen:

$$4HNO_3(aq) \rightarrow 4NO_2(g) + 2H_2O(l) + O_2(g)$$

This can be shown experimentally by using the apparatus illustrated in Figure 11.12. Fairly strong heating is required and rubber and cork cannot be used as they are attacked by nitric acid. The nitrogen dioxide produced dissolves in the water and the presence of oxygen in the collection tube can be confirmed in the usual way.

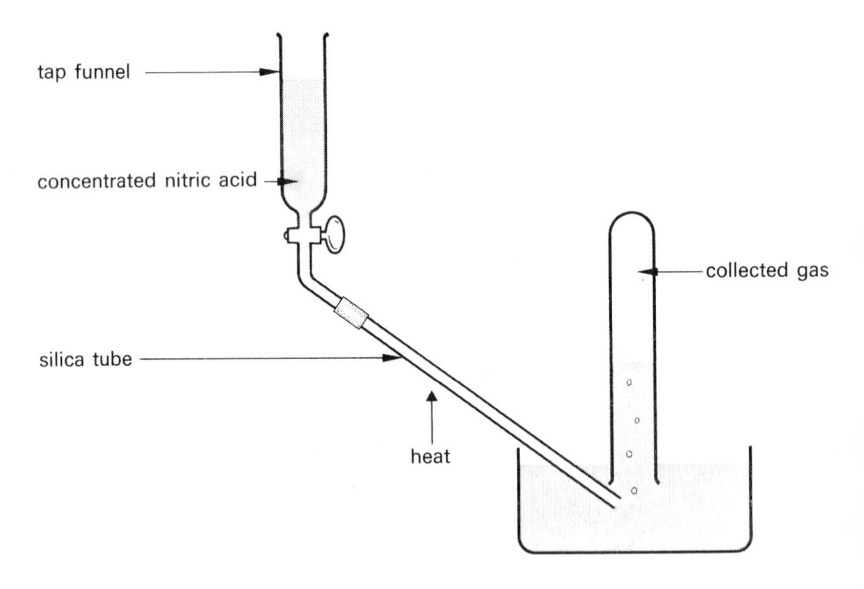

Figure 11.12 Thermal decomposition of nitric acid

(c) As a nitrating agent in organic chemistry
Nitration is a reaction in which a nitro (NO_2) group is introduced into a molecule. Nitric acid in water ionizes to give H_3O^+ and NO_3^- ions, i.e.

$$HNO_3 + H_2O \rightleftharpoons H_3O^+ + NO_3^-$$

However, if *concentrated* nitric acid is mixed with concentrated sulphuric acid the mixture ionizes as follows:

$$HNO_3 + 2H_2SO_4 \rightleftharpoons NO_2^+ + H_3O^+ + 2HSO_4^-$$

and the NO_2^+ ion (the nitronium ion) produced brings about the nitration reactions in organic chemistry. The compounds so produced are called nitro-compounds. For example, benzene (C_6H_6) can be nitrated to nitrobenzene ($C_6H_5NO_2$). Nitro-compounds are used as intermediates in the production of certain dyes and others are used in explosives, e.g. T.N.T. (trinitrotoluene).

(d) Oxidizing properties
Concentrated nitric acid is a powerful oxidizing agent, and we have already considered many reactions illustrating this, for example the reaction of metals with the acid. Most metals, including those below hydrogen in the electrochemical series, are oxidized by concentrated nitric acid to form a metal nitrate. In cases where the metal would normally liberate hydrogen from the acid, this hydrogen is also oxidized, to water. Some metals are rendered passive with concentrated nitric acid (Section 10.1). Some non-metals (e.g. carbon) are also oxidized by the concentrated acid.

The nature of the products formed in nitric acid oxidations depends on the substance being oxidized. Thus it is possible to produce a range of decomposition products depending on the strength of the reducing agent, e.g.

compound	nitric acid	nitrogen dioxide	nitrogen monoxide	dinitrogen monoxide	nitrogen	ammonia
formula	HNO_3	NO_2	NO	N_2O	N_2	NH_3
oxidation number of the nitrogen	+5	+4	+2	+1	0	−3

increasing strength of reducer →

However in most of the oxidizing reactions, the nitric acid behaves in the following way:

$$2HNO_3 \rightarrow 2NO_2 + H_2O + [O]$$

Experiment 11.6
†Oxidizing Reactions of Concentrated Nitric Acid ***

Apparatus
Hydrogen sulphide generator, gas jar and slide, teat pipette, test-tube, access to fume cupboard.
Concentrated nitric acid, dilute potassium iodide solution.

Procedure
(a) Working in a fume cupboard, collect a gas jar of hydrogen sulphide by downward delivery. Add a few drops of concentrated nitric acid to the jar and observe any changes.

(b) Pour a little potassium iodide solution into a test-tube and add a few drops of concentrated nitric acid. Note any changes taking place.

Points for Discussion

1. What did you observe in procedure (a)? What is this product? Write a simple equation for the reaction representing the nitric acid as [O].

2. What change took place in procedure (b)? What caused this change?

3. By reference to your work in Chapter 8 state, in each of the above cases, which substance has been oxidized and which substance has been reduced.

Uses Nitric acid is used for the manufacture of fertilizers and explosives and as an intermediate in the formation of dyes.

Oxides of Nitrogen Nitrogen forms six oxides, N_2O, NO, NO_2, N_2O_3, N_2O_5, and NO_3. They are usually encountered as reaction products of nitric acid on metals (Section 3.1) and when nitrates are heated (Section 10.1).

Point for Discussion

Using the information in Section 4.1 and Table 4.1, devise a suitable apparatus for the preparation and collection of nitrogen dioxide, made by heating lead nitrate crystals (pages 234 and 263).

11.3 OXYGEN AND SULPHUR

Introduction

Oxygen and sulphur are in Group 6 of the Periodic Table as both have six electrons in their outer shells. The elements are electronegative and thus readily form divalent ions (X^{2-}) in addition to forming divalent covalent compounds. Sulphur, like phosphorus, can increase its valency beyond that expected from its electronic configuration, and forms compounds in which the covalency may increase to six, e.g. sulphur hexafluoride, SF_6. Both oxygen and sulphur exhibit allotropy, oxygen forming diatomic (O_2) and triatomic (O_3, ozone) molecules, while sulphur forms a variety of chain and ring structures, both in the element itself and in sulphur compounds (Section 7.3). Both sulphur and oxygen form linkages of the type —O—O— (in peroxides, page 119) and —S—S—.

Compounds containing oxygen and hydrogen give rise to many properties which do not occur in similar compounds containing sulphur and hydrogen. This is because in the former case hydrogen bonding can take place (see page 213).

Some physical constants of sulphur and oxygen are given in Table 11.4.

Oxygen and its Compounds

The chemistry of oxygen has been fully considered in Section 4.3 and related compounds such as water (Section 9.1), hydrogen peroxide (Section 9.3) and metallic oxides (Section 10.1) have also been studied. The use of oxygen as an oxidizing agent is also considered in Chapter 8.

Sulphur—The Element

Extraction Sulphur occurs free in nature and also combined in sulphide ores (page 261) and sulphate ores (page 248). Over 80 per cent of the world's supply of sulphur is extracted from the deposits of naturally occurring sulphur in Texas and Louisiana. This free sulphur is thought to have been formed

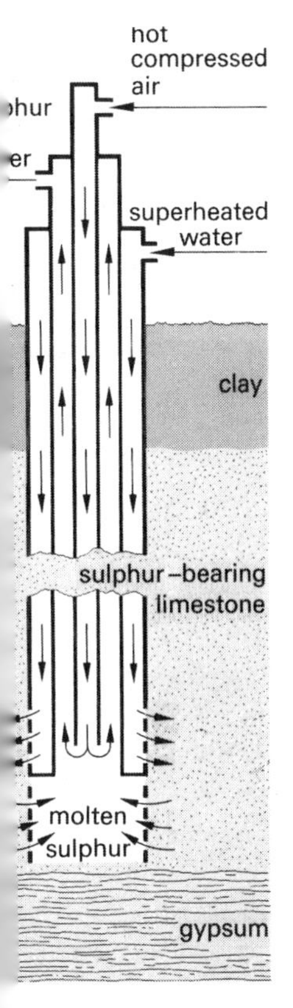

re 11.13 Sulphur ex-
ction by the Frasch
ocess

Table 11.4 Some physical constants of oxygen and sulphur

Property	*Oxygen*	*Sulphur*
Melting point, K (°C)	54 (−219)	392 (119)
Boiling point, K (°C)	90 (−183)	717 (444)
Density ($g\ cm^{-3}$)	—	2.06 (rhombic)
Atomic radius (nm)	0.066	0.104
Ionic radius (nm) (of X^{2-} ion)	0.132	0.174
Electronegativity	3.5	2.5

by the bacterial reduction of gypsum ($CaSO_4$, $2H_2O$) which is often found with it. The American deposits are about 30 metres thick; each consists of a dome-shaped layer of sulphur and limestone on top of gypsum. The sulphur-bearing strata lie under quicksands at depths greater than 160 metres and are extracted by an ingenious method known as the *Frasch process.* A metal tube about 15 cm in diameter and containing two concentric inner tubes is sunk into the top of the sulphur deposit (Figure 11.13). Water, superheated to 443 K (170 °C) under pressure, is forced, still under pressure, down the outer tube, melting the sulphur, which collects in a pool. Hot compressed air is forced down the centre tube, producing a light froth of molten sulphur which rises up the middle tube and is collected in large containers. After the sulphur has solidified, the sides of the containers are dismantled and the blocks of sulphur are broken up. The sulphur obtained by this method is usually over 99.5 per cent pure.

Physical Properties

Sulphur is a yellowish solid and is used in the laboratory either as roll sulphur (made by casting molten sulphur into 'sticks') or as flowers of sulphur (a yellow powder made by condensing sulphur vapour). It is insoluble in water but soluble in some solvents (e.g. carbon disulphide, Experiment 2.23). It shows the typical physical properties of a non-metal, i.e. it has a low melting point and boiling point and is a poor conductor of heat and electricity. The changes which take place when sulphur is heated have already been discussed, as have the various allotropes of the element (Section 7.3).

Chemical Properties

Chemically sulphur is quite reactive. It burns in air or oxygen to form sulphur(IV) oxide (SO_2) plus a small amount of sulphur(VI) oxide (SO_3) (Experiment 4.10). It combines directly on heating, with hydrogen, the halogens (except iodine), carbon, phosphorus and with all the common metals (e.g. Experiment 2.24), to form sulphides. It is oxidized to sulphur(IV) oxide by concentrated nitric acid, and by hot concentrated sulphuric acid.

Uses of Sulphur

Most of the world's supply of sulphur is used in the manufacture of sulphuric acid (Section 14.3). Small amounts are used in the manufacture of matches, gunpowder, sulphur-based insecticides, certain drugs, paper and in the vulcanization of rubber.

Compounds of Sulphur

Sulphur forms a variety of compounds, but we shall restrict our attention to a consideration of the hydride (H_2S), the oxides (SO_2 and SO_3), the acids (H_2SO_4 and H_2SO_3) and their corresponding salts.

Hydrogen Sulphide (H_2S)

Preparation Hydrogen sulphide gas is evolved when an acid reacts with a metallic sulphide. The usual laboratory preparation is to add dilute hydrochloric acid to iron(II) sulphide:

$$FeS(s) + 2HCl(aq) \rightarrow FeCl_2(aq) + H_2S(g)$$

You have already carried out this reaction as part of Experiment 2.24. The gas prepared by this method is not pure even if the hydrogen chloride vapour is removed by passing through water. This is because iron sulphide always contains some free iron which reacts with the acid to produce hydrogen. It is often convenient to have supplies of hydrogen sulphide readily available in a laboratory and a Kipp's apparatus (Figure 4.1) can be used for this purpose.

Physical properties Hydrogen sulphide is a colourless, extremely poisonous gas with an odour like rotten eggs. It is denser than air and is slightly soluble in water.

Chemical properties It burns with a blue flame, forming sulphur(IV) oxide and steam in a plentiful supply of air, and sulphur and steam in a limited supply of air. Write equations for these combustions.

An aqueous solution of hydrogen sulphide is weakly acidic (page 245) and passage of the gas through solutions of metallic salts often precipitates the sulphides of these metals (Section 10.1). The ability of hydrogen sulphide to act as a powerful reducing agent has been considered in Chapter 8, and the test for hydrogen sulphide using lead ethanoate paper was given on page 229.

Sulphur(IV) Oxide (Sulphur Dioxide) (SO_2)

Preparation Sulphur(IV) oxide is usually prepared in the laboratory by heating copper turnings with concentrated sulphuric acid:

$$Cu(s) + 2H_2SO_4(aq) \rightarrow CuSO_4(aq) + SO_2(g) + 2H_2O(l)$$

Using the apparatus given in Section 4.1 and the information in Table 4.1 devise an apparatus for the preparation and collection of a dry sample of sulphur(IV) oxide. Industrially the gas is usually prepared either by burning sulphur in air or by roasting sulphide ores in air. Laboratory supplies are often conveniently obtained from a syphon of liquefied sulphur(IV) oxide.

Physical properties Sulphur(IV) oxide is a colourless, dense, poisonous gas with a suffocating odour. It is easily liquefied under a small pressure (three atmospheres) and is usually stored and transported in this form. It is quite soluble in water (Table 4.1) and the formation of an acidic solution indicates that it is an acidic oxide.

Chemical properties Sulphur(IV) oxide is a reducing agent, e.g. with warm lead dioxide:

$$SO_2(g) + PbO_2(s) \rightarrow PbSO_4(s)$$

However, most of its reducing reactions take place in aqueous solution or at least require the presence of moisture. These are better regarded as reactions of *sulphurous acid* which is formed when some of the dissolved sulphur(IV) oxide reacts with the water:

$$SO_2(g) + 3H_2O(l) \rightleftharpoons 2H_3O^+(aq) + SO_3^{2-}(aq)$$

The following experiment illustrates some redox reactions of sulphur(IV) oxide in aqueous solution.

Experiment 11.7
†Some Redox Reactions of Sulphur(IV) Oxide in Aqueous Solution***

Apparatus
Cylinder of sulphur(IV) oxide or sulphur(IV) oxide generator, two 50 cm^3 beakers, teat pipettes, two gas jars and slides.
Dilute hydrochloric acid, barium chloride solution, ten volume hydrogen peroxide solution, small beakers containing solutions of bromine water, acidified potassium permanganate, acidified potassium dichromate and hydrogen sulphide generator.

Procedure
All these experiments should be performed in a fume cupboard.
(a) Bubble sulphur(IV) oxide into a beaker half full of water in order to form an aqueous solution of the gas. Divide the solution into two portions and to one portion add a few drops of dilute hydrochloric acid followed by a few drops of barium chloride solution. Note any change. Add a little more dilute hydrochloric acid and again note any change taking place. To the second portion add a few cm^3 of ten volume hydrogen peroxide solution, followed by a few drops of dilute hydrochloric acid and a few drops of barium chloride solution. Note any change and then add a little more dilute hydrochloric acid until no further change occurs.
(b) Bubble sulphur(IV) oxide into the beaker containing bromine water. Note any change taking place.
(c) Repeat (b) using (i) the solution of potassium permanganate, and (ii) the solution of potassium dichromate.
(d) Collect a gas jar of sulphur(IV) oxide and one of hydrogen sulphide. Place the gas jar of sulphur(IV) oxide on top of the gas jar of hydrogen sulphide so that their covered ends are in contact. Remove the slides and record any observations.

Points for Discussion
1. Using the information on page 89 explain why a white precipitate formed in procedure (a) when hydrochloric acid and barium chloride solution were added to the aqueous solution of sulphur(IV) oxide. This precipitate dissolved in excess hydrochloric acid, but when hydrogen peroxide solution was added to the second portion of solution and the test repeated, the precipitate was insoluble in excess hydrochloric acid. Explain these observations.

2. In aqueous solution, sulphur(IV) oxide reacts with water to produce SO_3^{2-} (sulphite) ions (page 290). These ions are responsible for the *reducing* properties of the solution, and in bringing about reductions, they are themselves oxidized to sulphate ions, i.e.

$$SO_3^{2-}(aq) + [O] \rightarrow SO_4^{2-}(aq)$$

(in terms of addition of oxygen)

or

$$SO_3^{2-} + H_2O - 2e^- \rightarrow SO_4^{2-} + 2H^+$$

(in terms of electron transfer)

or

$$S^{IV}O_3^{2-} \rightarrow S^{VI}O_4^{2-}$$

(in terms of oxidation number)

These principles were fully discussed in Chapter 8.

3. In the first experiment the sulphite ions were oxidized by hydrogen peroxide to sulphate ions:

$$H_2O_2(aq) + SO_3^{2-}(aq) \rightarrow SO_4^{2-}(aq) + H_2O(l)$$

and although both ions form a precipitate with barium ions, barium sulphite is soluble in excess hydrochloric acid but barium sulphate is not.

4. In procedure (c) the two oxidizing agents used provide the oxygen to oxidize the sulphite ion and in the process become reduced. The reduction can be simply expressed as

$$SO_3^{2-}(aq) + \underset{\text{from oxidizing agent}}{[O]} \rightarrow SO_4^{2-}(aq)$$

The colour changes you observed in each case were due to the following changes:

(i) $MnO_4^-(aq)$ manganate(VII) ion (purple) → $Mn^{2+}(aq)$ Manganese(II) ion (colourless) and

(ii) $Cr_2O_7^{2-}(aq)$ → $Cr^{3+}(aq)$
Dichromate(VI) ion (orange) → Chromium(III) ion (green)

[Note: these reactions must be carried out in acidic solution.]

5. In procedure (b) bromine molecules (brown) are converted to bromide ions (colourless) according to the following equation:

$$Br_2(aq) + SO_3^{2-}(aq) + H_2O(l) \rightarrow 2Br^-(aq) + SO_4^{2-}(aq) + 2H^+(aq)$$

Which substance is being reduced and which one is being oxidized?

6. In the examples so far considered sulphur(IV) oxide has been acting as a reducing agent, and although its reaction with potassium dichromate(VI) solution is used as the *test* for the gas it is important to remember that this and the other reactions considered are given by many reducing agents, including alkenes (page 314).

7. In procedure (d) two reducing agents, sulphur(IV) oxide and hydrogen sulphide, were allowed to mix. Did any reaction occur? Give a reason for your answer. Name one of the products formed, and write an equation for the reaction. Although you may have concluded that the two gases readily react together, this is not strictly the case as at least a trace of moisture must be present for the reaction to occur. A little of the sulphur(IV) oxide reacts with this moisture to produce SO_3^{2-} ions which then react as follows:

$$2H_2S(g) + SO_3^{2-}(aq) + 2H^+(aq) \rightarrow 3H_2O(g \text{ or } l) + 3S(s)$$

(The reaction continues as the water vapour formed reacts with sulphur(IV) oxide to produce more SO_3^{2-} ions.) In this reaction, which substance is being oxidized and which is being reduced? You should have concluded that the sulphur(IV) oxide (really the SO_3^{2-} ions) has been reduced and that it must therefore act as an *oxidizing* agent in the presence of hydrogen sulphide. This is because hydrogen sulphide is a more powerful reducing agent than sulphur(IV) oxide. Both the hydrogen sulphide and sulphur(IV) oxide react to form sulphur, and this is illustrated in the following scheme:

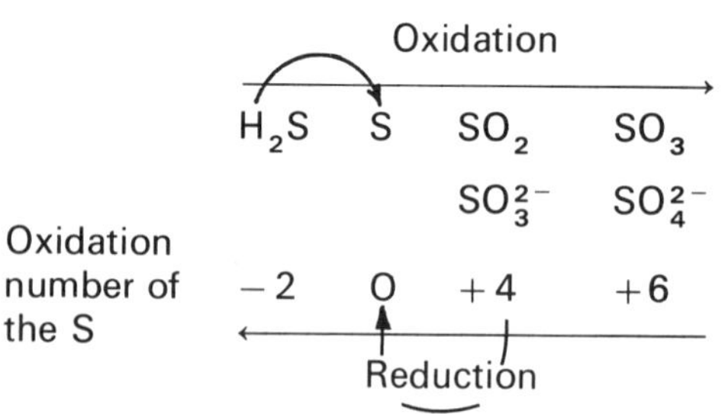

Uses Sulphur(IV) oxide is an intermediate in the preparation of sulphuric acid (Section 14.3). It is also used as a preservative for canned fruit and fruit juices.

Liquid sulphur(IV) oxide is used to purify the light oil fraction of petroleum and as a non-aqueous solvent. The following experiment illustrates a further use of sulphur(IV) oxide.

Experiment 11.8
†Action of Sulphur(IV) Oxide on Various Materials**

Apparatus
Cylinder of sulphur(IV) oxide or alternative supply, and as required by procedure (a). Coloured articles of wool, silk, and straw. Flowers.

Procedure
(a) Devise an apparatus or system whereby the articles can be left in contact with moist sulphur dioxide for a period of time.

(b) After treatment with sulphur(IV) oxide leave the articles in air and preferably also in sunlight for a few days. Note any changes which take place throughout the experiment.

Points for Discussion

1. What happened to the coloured articles after exposure to moist sulphur(IV) oxide? What kind of compound does this suggest sulphur(IV) oxide is?

2. From your observations in (b), what conclusions can you draw as to the permanency of the change in (a)?

3. The bleaching action of moist sulphur(IV) oxide is thought to be due to its ability, as a reducer, to remove oxygen from the colouring compound and make it colourless. The wood pulp used for making newspapers is bleached in this way. The bleaching, however, is not permanent and on exposure to air and light the original colour slowly returns due to aerial oxidation. In addition to your experiments, you will perhaps also have noticed this with old newspapers.

Sulphur(VI) Oxide (SO_3)

Preparation Sulphur(VI) oxide was first prepared by heating iron(III) sulphate:

$$Fe_2(SO_4)_3(s) \rightarrow Fe_2O_3(s) + 3SO_3(g)$$

Industrially it is produced by the direct combination of sulphur(IV) oxide and oxygen in the presence of a heated catalyst (platinum or vanadium pentoxide, page 384). The following experiment illustrates the industrial principle on a laboratory scale.

Experiment 11.9

†Preparation of Sulphur(VI) Oxide***

Apparatus

Cylinders of sulphur(IV) oxide and oxygen, or alternative supplies, and the apparatus and reagents indicated in Figure 11.14, small paint brush.

Fairly concentrated sulphuric acid, soap solution.

Procedure

(a) Assemble the apparatus illustrated in Figure 11.14 but do not yet place the ice bath in position.

(b) Check if the glass–rubber joints are air-tight by applying a little soap solution to each joint using a small paint brush.

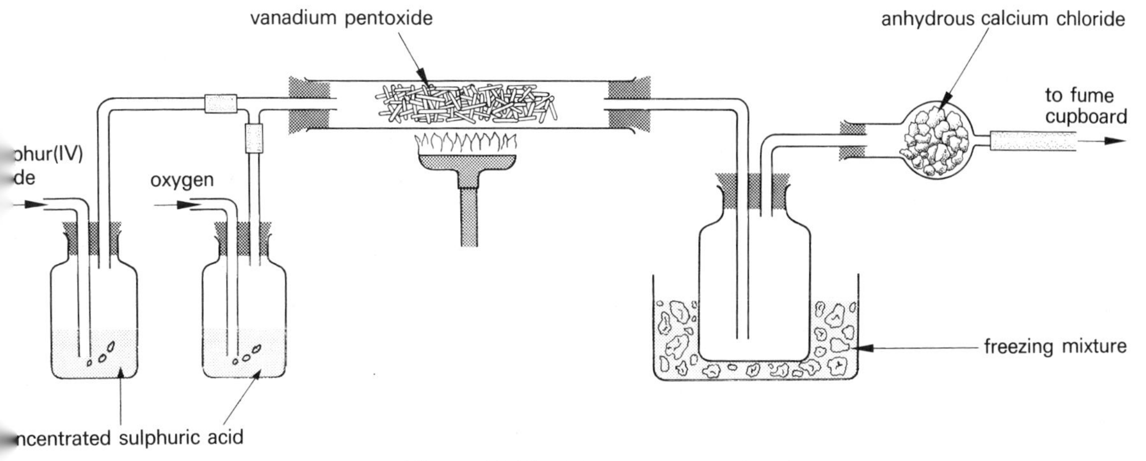

Figure 11.14 Laboratory preparation of sulphur(VI) oxide

(c) Purge the apparatus free of moist air by passing dry oxygen through it for half a minute.

(d) Place the ice bath in position and heat the tube containing the catalyst, using a 'spreader Bunsen' if available.

(e) Pass sulphur(IV) oxide and oxygen over the heated catalyst and adjust their rates of flow so that they are approximately equal.

(f) Continue to pass the gases over the heated catalyst until a reasonable amount of the product has formed in the cooled receiver.

(g) Observe this product. Carefully add a few drops of water to the sulphur(VI) oxide and record any observations. Dissolve the remainder of the sulphur(IV) oxide in fairly concentrated sulphuric acid.

Points for Discussion

1. How did the test you performed using soap solution help you to decide if a 'joint' was leaking?

2. Why were the gases dried before reaction? Why was any moist air removed from the apparatus before reaction? What was the purpose of the calcium chloride tube?

3. Why was an ice bath placed around the receiving vessel? (See Table 4.1.)

4. How could you tell when the flow rates of sulphur(IV) oxide and oxygen were approximately equal?

5. In what state was the product collected? What did it look like?

6. Why was the sulphur(VI) oxide 'removed' by dissolving in fairly concentrated sulphuric acid and not water?

Properties and uses of sulphur(VI) oxide Sulphur(VI) oxide consists of white, silky needle-like crystals which melt at 290 K (17 °C) and boil at 317 K (44 °C). It is an acidic oxide and it fumes strongly in moist air to form sulphuric acid:

$$SO_3(s) + H_2O(l) \rightarrow H_2SO_4(aq)$$

The reaction with water is violent and a mist of acid droplets forms rather than a solution. This mist remains suspended in the gas and it is then virtually impossible to absorb the sulphur(VI) oxide completely. This difficulty can be overcome by dissolving the sulphur(VI) oxide in fairly concentrated sulphuric acid.

Virtually all the sulphur(VI) oxide produced is used immediately, in the vapour state, in the preparation of sulphuric acid.

Sulphuric Acid (H_2SO_4)

The industrial preparation of sulphuric acid is considered in detail in Section 14.3. The laboratory preparation is as in procedure (g) of Experiment 11.9.

Physical properties Sulphuric acid is probably the most important chemical that is manufactured. About 30 million tons are produced annually throughout the world. Pure sulphuric acid is a colourless, viscous liquid, which on heating loses sulphur(VI) oxide to form the concentrated acid which contains 98 per cent of the acid. The concentrated acid has a strong affinity for water and evolves a great amount of heat when added to it. Because of this, the acid is always added to water and *not* the water to the acid because in the latter case the first portion of water is quickly vaporized to steam and 'spits out' corrosive droplets of the acid.

Chemical properties The chemical properties of the acid can be conveniently divided into four groups.

(a) Acidic properties
Dilute sulphuric acid is a typical strong acid, giving the characteristic colour changes with indicators, liberating carbon dioxide from carbonates and hydrogen carbonates and neutralizing alkalis. It reacts with metals above hydrogen in the electrochemical series, liberating hydrogen and forming the appropriate salt. (The reaction with sodium or potassium is explosive.) It is a dibasic acid and gives rise to two series of salts; sulphates and hydrogen sulphates (page 88). Concentrated sulphuric acid displaces acids of lower boiling points from their salts when the two are heated together, e.g.

$$Cu(NO_3)_2(aq) + H_2SO_4(aq) \rightarrow 2HNO_3(g) + CuSO_4(aq)$$

This in fact is used in one of the tests for the nitrate grouping (page 240), and in the laboratory preparations of nitric acid (page 284) and hydrochloric acid (page 301).

(b) As a drying agent
Because of its strong affinity for water, *concentrated* sulphuric acid is used for drying many gases, e.g. sulphur dioxide, oxygen, chlorine, hydrogen chloride, etc. It cannot be used to dry ammonia or hydrogen bromide—why not?

(c) As a dehydrating agent

Experiment 11.10
† Reaction of Concentrated Sulphuric Acid and Sugar ***

Apparatus
Evaporating basins, asbestos square, spatula.
White sugar, concentrated sulphuric acid, hydrated copper(II) sulphate.

Procedure
(a) Place a few spatula measures of sugar into an evaporating basin and stand the basin on an asbestos square.

(b) *Carefully* pour a small amount of concentrated sulphuric acid on to the sugar and record your observations.

(c) Repeat (a) and (b) but use hydrated copper(II) sulphate instead of the sugar.

Points for Discussion
1. What happened when the concentrated acid was poured on to the sugar? What element did the resulting product remind you of?

2. The formula of cane sugar is $C_{12}H_{22}O_{11}$. What must have been removed from the compound in order to leave only this element?

3. The equation for the reaction is

$$C_{12}H_{22}O_{11}(s) \xrightarrow[H_2SO_4]{\text{concentrated}} 12C(s) + 11H_2O(g)$$

4. Can you explain what happened to the hydrated copper(II) sulphate?

5. In these experiments *concentrated* sulphuric acid has *extracted the elements of water* from the sugar and chemically combined water from the hydrated salt. In such cases it is known as a *dehydrating agent*. You must make sure you understand the difference between a drying agent and a dehydrating agent.

6. The preparation of carbon monoxide from methanoic acid (page 273) is another example of concentrated sulphuric acid acting as a dehydrating agent.

7. Concentrated sulphuric acid gives severe skin burns because it removes water and the elements of water from the skin tissue which is consequently destroyed. Such 'burns' should be dealt with *immediately* by first washing with water and then treating with a moist paste of sodium hydrogen carbonate.

(d) As an oxidizing agent
The use of *concentrated* sulphuric acid as an oxidizing agent has previously been considered (Chapter 8). You will remember that in such reactions the sulphuric acid is reduced to sulphur dioxide, i.e.

$$H_2SO_4 \rightarrow SO_2 + [O] + H_2O$$

The oxidation of copper by concentrated sulphuric acid is used as the laboratory preparation of the gas.

Uses Sulphuric acid has a wide variety of uses, the main ones being the manufacture of fertilizers, rayon, dyes, plastics, drugs and explosives. It is also used in the refining of petroleum products and cleaning metal surfaces before electroplating (a process known as 'pickling'). When sulphur(VI) oxide is added to concentrated sulphuric acid, a form of the acid known as *oleum* is produced. It is extremely corrosive and is used to sulphonate aromatic compounds (i.e. substitute a hydrogen atom by a sulphonic ($-SO_3H$) group). Sulphonated compounds are used in the manufacture of drugs, detergents and dyes.

Sulphurous Acid (H_2SO_3)

This acid is formed when sulphur(IV) oxide dissolves in water (page 290). As with carbonic acid, only a small percentage of the dissolved gas reacts and the pure acid cannot be isolated. The use of sulphurous acid as a reducing agent was considered in Experiment 11.7. The use of aqueous sulphur(IV) oxide as a bleaching agent makes use of the fact that it reduces the dye by extracting oxygen from it, leaving a colourless compound (i.e. dye $-$ oxygen $\rightarrow$ colourless (page 293)). Contrast this with the bleaching action of chlorine by oxidation.

Sulphates and Hydrogen Sulphates

These salts of sulphuric acid have been fully considered in Section 10.1.

Sulphites and Hydrogen Sulphites

These are the two salts of the dibasic acid, sulphurous acid, and contain the ions SO_3^{2-} and HSO_3^- respectively. These salts can be prepared by passing sulphur(IV) oxide through aqueous solutions containing the cation of the salt to be prepared, e.g. (for sodium salts),

$$2Na^+(aq) + SO_2(g) + 3H_2O(l) \rightarrow Na_2SO_3(aq) + 2H_3O^+(aq)$$

followed by

$$2Na^+(aq) + SO_2(aq) + SO_3^{2-}(aq) + H_2O(l) \rightarrow 2NaHSO_3(aq)$$

A solution of the normal salt (sodium sulphite) is first formed and this is then converted to a solution of the acid salt (sodium hydrogen sulphite) by excess sulphur(IV) oxide. If the normal salt is required, the acid salt is first prepared and an equimolecular amount of sodium hydroxide is added to the solution:

$$NaHSO_3(aq) + NaOH(aq) \rightarrow Na_2SO_3(aq) + H_2O(l)$$

Sulphites give the same reduction and oxidation reactions as sulphurous acid because both contain the sulphite ion (SO_3^{2-}). In addition, on boiling a solution of a sulphite with sulphur a thiosulphate is produced,

$$SO_3^{2-}(aq) + S(s) \rightarrow \underset{\text{(Thiosulphate ion)}}{S_2O_3^{2-}(aq)}$$

(e.g. sodium thiosulphate, $Na_2S_2O_3$).

Sulphites are used for bleaching, disinfecting, and preserving, and for removing traces of chlorine from bleached cotton and linen.

Soluble sulphites form a white precipitate with barium chloride solution:

$$Ba^{2+}(aq) + SO_3^{2-}(aq) \rightarrow BaSO_3(s)$$

Sulphates also give this test, but the sulphite can be distinguished by the fact that the precipitate is soluble in dilute hydrochloric acid. Barium sulphate remains insoluble under such conditions. Why do you think dilute hydrochloric acid is added to the barium chloride solution when testing for a sulphate?

11.4 THE HALOGENS, WITH PARTICULAR REFERENCE TO CHLORINE

Introduction

The family of elements in Group 7 of the Periodic Table is known as the Halogens. The name halogen means 'salt producer' and is derived from the fact that all halogens form ionic salts of the type M^+X^- (e.g. Na^+Cl^-).

As each halogen atom has seven electrons in its outer shell, they either form ions of the type X^-, or show a covalency of one. In addition, after fluorine, the elements can also have an apparent valency greater than one. The elements themselves exist as simple covalent molecules (e.g. Cl_2, Br_2) and also form diatomic interhalogen compounds (e.g. ClBr). IF does not exist.

Some physical constants of the halogens are given in Table 5.8. Many of their chemical properties have already been discussed in Section 5.5. The next experiment illustrates another important property of the halogens.

Experiment 11.11

† Halogen Displacement Reactions ***

Apparatus
Chlorine generator, rack of test-tubes, spatula, access to fume cupboard when using chlorine.
Bromine water, solutions of potassium chloride, potassium bromide, and potassium iodide, iodine.

Procedure
Note any colour changes taking place in each of the procedures.

(a) Bubble chlorine into potassium bromide solution.

(b) Bubble chlorine into potassium iodide solution.

(c) Add a little bromine water to potassium iodide solution, and allow the two solutions to mix.

(d) Add a small crystal of iodine to a solution of (i) potassium chloride, (ii) potassium bromide.

Points for Discussion

1. In (a) and (b) the reactants were both virtually colourless. What did you conclude from your observations when they were mixed? Does the colour produced suggest what one of the products might be, in each case?

2. Is a similar reaction occurring in (c)? If so, what do you think has happened?

3. Do similar reactions also take place in (d)? Explain your answer.

4. Considering only the three halogens used in this experiment, which halide ions can be displaced from an aqueous solution by (i) chlorine, (ii) bromine, and (iii) iodine? Which halide ions would you expect fluorine to displace from aqueous solution? Unfortunately, although your prediction may be theoretically correct, the reaction of fluorine with aqueous solution is complicated because fluorine reacts with water.

5. You should have concluded that chlorine will displace bromide ions and iodide ions from aqueous solution, that bromine will displace iodide ions from aqueous solution but not chloride ions, and that iodine will displace neither chloride nor bromide ions from aqueous solution.

This order of halogen displacement is the same as that of the reactivity of the halogens (page 145). The reactivity of a halogen atom depends on its ability to gain an electron, and this can be predicted from its electronegativity value (Section 5.5). Thus the more electronegative chlorine is a better 'grabber' of electrons than bromine, which in turn is better than iodine. Therefore, when chlorine is added to a solution containing bromide ions, the chlorine 'grabs' electrons from the bromide ions which are converted into bromine molecules:

$$Cl_2(g) + 2Br^-(aq) \rightarrow 2Cl^-(aq) + Br_2(l)$$

However, when iodine is added to a solution containing bromide ions, there is no reaction because the bromine, which is a better 'grabber' of electrons, already has possession of the electrons and will not lose them to the iodine, i.e. $I_2 + 2Br^- \rightarrow$ no reaction. Write ionic equations, where appropriate, for the other reactions you performed.

In the remainder of this section we shall consider other properties of the halogens by studying a typical halogen (chlorine) in more detail.

Chlorine—The Element

Davy first identified the gas as an element in 1810 and gave it the name of chlorine, from the Greek, chloros, greenish-yellow.

Preparation

In the laboratory chlorine is prepared by the oxidation of chloride ions in acidic solution. The usual method is the oxidation of concentrated hydrochloric acid, using manganese(IV) oxide or potassium manganate(VII) as the oxidizing agent. The latter reagent is the most convenient as no heating is required. The simplified equation for the reaction is

$$2HCl(aq) + \underset{\left[\begin{array}{c}\text{From}\\\text{oxidizing}\\\text{agent}\end{array}\right]}{\text{'O'}} \rightarrow H_2O(l) + Cl_2(g)$$

Using the information given in Section 4.1 and in Table 4.1, devise a suitable apparatus for the preparation and collection of a pure dry sample of chlorine gas. Remember that hydrogen chloride vapour will be present in the chlorine gas evolved and should be removed, and that chlorine reacts with alkaline solutions (page 301).

Industrially chlorine is obtained as a by-product in many electrolytic reactions, e.g. from the Kellner–Solvay cell (page 392), and the Down's cell (page 386).

Physical Properties

Chlorine is a dense, greenish-yellow gas, with a choking smell. It is a poisonous gas, and was used as such in World War I. In addition to being slightly soluble in cold water (Table 4.1), it is easily liquefied under pressure (seven atmospheres), and is thus stored and transported in the liquid state in steel cylinders.

Chemical Properties

Direct combination Chlorine is a very reactive gas and most elements, when heated, react directly with it to form chlorides. Notable exceptions are carbon, nitrogen, and oxygen. The reactions of metals with chlorine were mentioned in Section 10.1.

Experiment 11.12
† The Reactions of Metals with Chlorine ***

Apparatus
Chlorine generator, two gas jars and slides, sharp knife, filter paper, combustion spoon, asbestos paper, Bunsen burner, access to fume cupboard. Dutch metal, sodium.

Procedure
(a) Working in a fume cupboard, collect two gas jars of chlorine by downward delivery and cover them with the slides.

(b) Place a small piece of asbestos paper into the 'cup' of a combustion spoon and on to this paper place a small piece of freshly cut sodium. Heat the sodium in the spoon and then plunge it into a gas jar of chlorine. Record any observations you make.

(c) Carefully add some Dutch metal to a gas of chlorine and record your observations.

Points for Discussion
1. From the evidence of the two metals used, would you say that chlorine combines readily with metals?

2. What compounds will be formed in each case? Write the equation for the reaction with sodium.

3. You will remember that the direct combination of a metal with chlorine is used (i) to prepare anhydrous chlorides e.g. aluminium chloride (page 253), and (ii) to prepare the higher chlorides of metals having more than one valency state e.g. iron(III) chloride (page 257).

Affinity for hydrogen Chlorine has a particular affinity for hydrogen and, in addition to reacting explosively with the element in sunlight (page 145), it reacts with many compounds containing hydrogen as illustrated by the next experiment.

Experiment 11.13
† Affinity of Chlorine for Hydrogen ***

Apparatus
Chlorine generator, combustion spoon, two gas jars and slides, Bunsen burner, filter paper, test-tube and holder, access to fume cupboard.
Wax taper (or small candle), turpentine.

Procedure
(a) Working in a fume cupboard, collect two gas jars of chlorine.

(b) Light a wax taper. or place a small length of candle on to a combustion spoon and light it. Plunge the taper (or spoon) into a gas jar of chlorine and record your observations.

(c) Pour a little turpentine into a test-tube and warm it gently (care, turpentine is flammable). Fold a piece of filter paper so that it will fit into the test-tube, and dip it into the tube so that it becomes moistened with warm turpentine. Then drop the filter paper into a gas jar of chlorine. Again record your observations.

Points for Discussion
1. Did chlorine react vigorously with the taper, and turpentine? Name one of the products formed in each case. Write equations for these reactions. (The formula of turpentine can be taken as $C_{10}H_{16}$, and that of the taper (or candle) can be considered to be C_xH_y.)

2. You will note that in both cases the chlorine has removed the hydrogen, to form hydrogen chloride, and left carbon (in the form of soot). It is thus acting as an oxidizing agent. Predict how chlorine will react with hydrogen sulphide (H_2S), and water (H_2O). Check your predictions with the information given on pages 204 and 300.

As an oxidizing agent The reactions of chlorine as an oxidizing agent were considered in Chapter 8 and the previous experiment illustrates further examples.

Chlorine water When chlorine is bubbled into water, part of the chlorine which dissolves in water also reacts with the water to give a mixture of hydrochloric and hypochlorous acids:

$$Cl_2(g) + H_2O(l) \rightarrow HCl(aq) + HClO(aq)$$

This solution is known as *chlorine water*, and the presence of the hypochlorous acid makes its properties different from those of dry gaseous chlorine. Hypochlorous acid is a strong oxidizing agent and has another characteristic property as shown by the next experiment.

Experiment 11.14

† A Characteristic Property of Chlorine Water ***

Apparatus

Chlorine generator, 5×250 cm^3 beakers, burette, squat beaker, spills.

Selection of cloth dyed with vegetable dyes, coloured flowers, grass, etc., blue litmus paper, two pieces of paper (one printed and one freshly written on with ordinary ink).

Procedure

(a) Working in a fume cupboard, bubble chlorine into five beakers, each about half full of water.

(b) Into the first beaker place coloured flowers and/or grass, etc. Into the second beaker place the pieces of dyed cloth. Into the third beaker place the two pieces of paper. Into the fourth beaker place a piece of blue litmus paper. Record any changes taking place in each beaker.

(c) Using the solution from the remaining beaker, *completely* fill a burette and pour the remaining solution into a squat beaker. Place your finger over the open end of the burette, invert it, and stand it in the solution in the squat beaker (Figure 11.15). Leave the burette exposed to sunlight for a few days. Test any gas collected with a glowing splint.

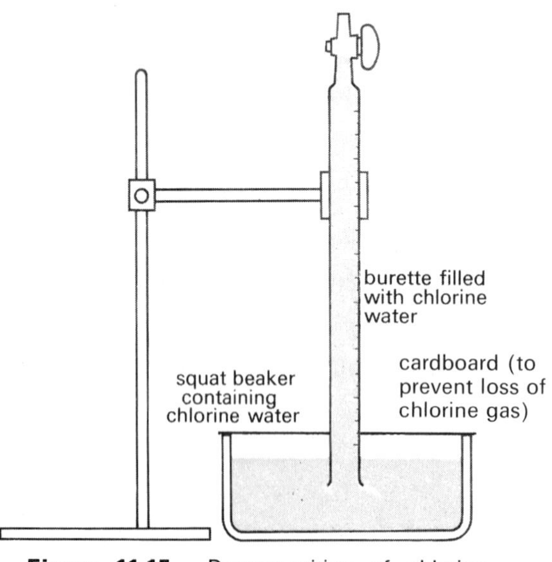

Figure 11.15 Decomposition of chlorine water by sunlight

Points for Discussion

1. What effect did the chlorine water have on the various articles and materials placed in it?

2. What gas did you collect in procedure (c)?

3. When chlorine dissolves in water the hypochlorous acid (HClO) formed slowly decomposes in the presence of light to liberate oxygen:

$$2HClO(aq) \rightarrow 2HCl(aq) + O_2(g)$$

Thus chlorine water is an oxidizing solution, and it bleaches materials by oxidizing the colour matter in such substances to colourless compounds, e.g.

$$\underset{\text{Coloured}}{HClO(aq) + Dye(s)} \rightarrow \underset{\text{Colourless}}{HCl(aq) + DyeO(s)}$$

or ionically,

$$OCl^-(aq) + Dye \rightarrow Cl^-(aq) + \text{oxidized Dye.}$$

Compare this reaction with the bleaching action of aqueous sulphur dioxide (page 293) which is a reducing reaction. The ease with which it loses oxygen makes hypochlorous acid a powerful antiseptic and disinfectant, as it kills bacteria by oxidation.

4. Dry chlorine will not affect blue litmus paper, but moist chlorine or chlorine water first turns it red due to the presence of hydrochloric acid. The slower bleaching action then removes this colour and leaves the paper colourless. This reaction is used as a *test for chlorine*.

Action with alkalis As chlorine in aqueous solution is acidic, it is not surprising that chlorine reacts readily with alkalis. The reaction depends upon the conditions. With *cold dilute* alkali, a chloride and hypochlorite are formed, whereas with *hot concentrated* alkali a chloride and chlorate are formed, e.g. (with sodium hydroxide):

$$Cl_2(g) + 2NaOH(aq) \rightarrow NaCl(aq) + \underset{\text{Sodium hypochlorite}}{NaClO(aq)} + H_2O(l)$$

and

$$3Cl_2(g) + 6NaOH(aq) \rightarrow 5NaCl(aq) + \underset{\text{Sodium chlorate(V)}}{NaClO_3(aq)} + 3H_2O(l)$$

or ionically

$$Cl_2(g) + 2OH^-(aq) \rightarrow Cl^-(aq) + ClO^-(aq) + H_2O(l)$$

and

$$3Cl_2(g) + 6OH^-(aq) \rightarrow 5Cl^-(aq) + ClO_3^-(aq) + 3H_2O(l)$$

Hypochlorites are salts of hypochlorous acid and as, like the acid, they produce OCl^- ions in solution, they are also good oxidizing and bleaching agents. As hypochlorites are more stable than the free acid they are used extensively as bleaches and antiseptics. Common products such as 'Domestos', 'Parazone', 'Milton', etc. are essentially solutions of sodium hypochlorite.

Uses

Modern industry is making an ever increasing demand for chlorine and about three million tons are produced annually. A large proportion of this is used in chlorinating organic compounds to make rubbers, plastics (e.g. P.V.C.), and solvents (e.g. tetrachloromethane, and chloroethenes). Chlorine is also used for sterilizing water and sewage, as a bleaching agent, in the preparation of drugs, insecticides, and dry-cleaning fluids, and in the production of hydrochloric acid.

Compounds of Chlorine

Chlorine forms chlorides (Section 10.1) and also forms a series of oxides and oxyacids which are comparatively unimportant. By far the most important compound of chlorine is hydrogen chloride.

Hydrogen Chloride (HCl)

Preparation Hydrogen chloride is prepared in the laboratory by the action of concentrated sulphuric acid on sodium chloride. As you will remember from page 295, sulphuric acid can be used to displace a more volatile acid from one of its salts and use was made of this fact in the preparation of nitric acid from sodium nitrate (page 284). The laboratory preparation of hydrochloric acid also illustrates this principle. It is usual to warm the reaction mixture after the initial vigorous reaction between the two reactants eases. The equation for the reaction is

$$NaCl(s) + H_2SO_4(aq) \rightarrow NaHSO_4(s) + HCl(g)$$

Using the information given in Table 4.1 and Section 4.1, devise a method of preparing and collecting a dry sample of hydrogen chloride gas. Why do you think *concentrated* sulphuric acid is used in this reaction?

On a large scale hydrogen chloride is produced by the same reaction, but the heating is continued to red heat when more hydrogen chloride is produced due to the sodium hydrogen sulphate initially formed reacting with excess sodium chloride:

$$NaCl(s) + NaHSO_4(s) \rightarrow Na_2SO_4(s) + HCl(g)$$

Where chlorine and hydrogen are both readily available (e.g. as by-products from the Kellner–Solvay cell), hydrogen chloride is made by direct union of the gases in the presence of activated charcoal (and the absence of direct sunlight, which would produce an explosive combination of the gases).

Physical properties Hydrogen chloride is a colourless gas, slightly denser than air. It has a strong, irritating odour and fumes strongly in moist air forming hydrochloric acid droplets. It is *very* soluble in water (Table 4.1), as can be shown by the 'fountain experiment' (page 281).

Chemical properties Dry hydrogen chloride is not particularly reactive at ordinary temperatures, although very reactive metals such as sodium 'burn' in it to form the chloride and hydrogen, e.g.

$$2Na(s) + 2HCl(g) \rightarrow 2NaCl(s) + H_2(g)$$

Other metals (above hydrogen in the electrochemical series) when heated will react with the gas. If a metal can form two chlorides (e.g. iron) then it will react with hydrogen chloride to form the chloride in which the metal is in its lower valency state (e.g. $FeCl_2$).

Hydrogen chloride forms 'steamy' fumes with ammonia gas (page 12) and this reaction together with the usual test for a chloride (page 90) is used to identify the gas.

Hydrochloric Acid ($HCl(aq)$)

As hydrogen chloride is extremely soluble in water, the solution must be prepared using a funnel to prevent 'sucking back' (page 283). The solution produced is acidic due to the formation of hydrochloric acid:

$$HCl(g) + H_2O(l) \rightarrow H_3O^+(aq) + Cl^-(aq)$$

Hydrochloric acid is a typical strong acid whose properties have already been discussed in Section 3.1. The acid gives rise to chloride salts, which have also been considered earlier (Section 10.1).

QUESTIONS CHAPTER 11

1. A solid element on burning in oxygen forms a gas at room temperature. The element could be:

A sodium
B sulphur
C phosphorus
D hydrogen
E silicon

2. Which of the following compounds would make the best nitrogenous fertilizer? (H = 1, N = 14, O = 16, Na = 23, Mg = 24, S = 32, Cl = 35.5)

A NH_4Cl
B $(NH_4)_2SO_4$
C $NaNO_3$
D NH_4NO_3
E $Mg(NO_3)_2$

3. Br_2 can be produced by the reaction of Br^- with

A I_2, B Cl_2, C I^-, D Cl^-, E HCl

4. Dehydration of a substance is best described as

A the removal of moisture
B the addition of water
C the removal of water
D the addition of moisture
E the removal of the elements of water

5. Sulphur

A forms two alkaline oxides
B is spontaneously flammable
C burns with a blue flame
D conducts electricity in the molten state
E is usually stored in the form of sticks in water

6. Hydrochloric acid

A is present in the stomach
B can be reduced to chlorine
C is a covalent compound
D bleaches litmus paper
E forms both normal and acid salts

7. Ammonium sulphate is widely used as a fertilizer because

A it provides nitrogen for the plants
B it provides sulphur for the plants
C it provides oxygen for the plants
D it poisons harmful bacteria
E it helps to break up the soil

8. Liquid ammonia, b.p. 239.6 K (−33.4 °C), is used as a cheap nitrogenous fertilizer, and is injected about 12 cm below the soil surface when conditions are suitable.

(a) Why is the liquid ammonia not sprayed directly on to the soil?
(b) What are the suitable conditions of the soil needed before injecting the liquid?
(c) Why has this 'fertilizer' not found widespread application?

9. Name a suitable drying agent for each of the following gases.

(a) Chlorine
(b) Hydrogen sulphide
(c) Ammonia (J. M. B.)

10. Name *two* crystalline forms of carbon and give *two* uses of each form. (J. M. B.)

11. Name *one* acidic and *one* alkaline gas which, in each case, is very soluble in water. Draw a diagram to show how a very soluble gas can be safely dissolved in water contained in a beaker and explain briefly how the apparatus avoids the danger of 'sucking back'. (J. M. B.)

12. When dilute hydrochloric acid is warmed with each of the following compounds, a gas is evolved. Name this gas and give a chemical test to establish its identity.

(a) Sodium carbonate (c) Sodium sulphite
(b) Sodium sulphide. (J. M. B.)

13. (a) Neither water nor dry sulphur trioxide crystals alone turn dry blue litmus paper red. However, the litmus paper does turn red when it is added to a solution of sulphur trioxide crystals in water. Explain this observation.

Write an ionic equation to show what happens when sulphur trioxide dissolves in water.

(b) Describe what is seen when blue copper(II) sulphate crystals are added to (i) dilute sulphuric acid, (ii) concentrated sulphuric acid. (J. M. B.)

14. Bottles containing three white powders, sodium carbonate, calcium hydroxide and ammonium chloride respectively, have lost their labels. Using only water and dilute hydrochloric acid as additional reagents, what experiments would you perform in order to re-label the bottles correctly? (J. M. B.)

15. Ammonia can be prepared by heating an ammonium salt with an alkali.

(a) Name a pair of reagents suitable for this reaction.
(b) Give the simplest ionic equation for the reaction.

Ammonia is (i) very soluble in water and (ii) less dense than air. How does each of these properties determine the way in which ammonia is collected in a gas jar? (J. M. B.)

16. What are the components of (a) water gas, (b) producer gas? How is each of these gases made? Which is the better fuel? Why is the less effective fuel gas used at all?

17. Name *two* common drying agents that are *not* suitable for drying ammonia gas, and give the reason in each case. Give the name, and formula, of the chemical generally used to dry ammonia. Name and give the formula of *two* ammonium salts and give *one* different use for *each*. (C.)

18. Give reasons for the following.

(a) In the preparation of nitric acid from solid sodium nitrate, sulphuric acid is used rather than hydrochloric acid.
(b) Chlorine water is not stored in sunlight.
(c) Yellow phosphorus is stored under water.
(d) Concentrated ammonia solution is kept in a tightly stoppered bottle.

19. Describe briefly the manufacture of nitric acid from ammonia, giving the essential chemistry but *no* technical details. Give *two* uses of nitric acid and *one* physical property. Describe *two* experiments that illustrate the oxidizing power of concentrated nitric acid. Include in your accounts evidence that oxidation has occurred. (C.)

20. (a) Draw a labelled diagram of the apparatus you would use to prepare and collect a sample of hydrogen chloride, starting from sodium chloride. Write an equation for the reaction and state the reaction conditions.
(b) Explain why solutions of ammonia and hydrogen chloride in water are alkaline and acidic respectively whereas dry ammonia and dry hydrogen chloride have no effect on dry red or blue litmus paper.
(c) Describe and explain the changes that take place when a little ammonium chloride is heated in a test-tube. (C.)

21. Describe and illustrate the laboratory preparation and collection of sulphur dioxide from sulphuric acid. Explain how you would convert sulphur dioxide into sulphur trioxide. How do the two oxides differ physically?

Give a chemical test to distinguish between the solutions obtained when each of these oxides is dissolved in water.

The result of the test should be given for each solution. (J. M. B.)

22. How may pure dry chlorine be prepared and collected in the laboratory? How is the gas usually identified? When red-hot iron wire is plunged into a gas jar of chlorine, dense brown fumes are formed and the iron continues to glow. Write:

(a) the name of the product formed
(b) the equation of the reaction which takes place
(c) the name of an element which will ignite in chlorine without being previously heated (J. M. B.)

23. How would you prepare and collect hydrogen sulphide in the laboratory? What is the usual chemical test for this gas? Give the equation for the reaction, and state what is observed in each case, when hydrogen sulphide is passed through an aqueous solution of (a) copper(II) sulphate, (b) sulphur dioxide, (c) chlorine. (J. M. B.)

24. How would you prepare and collect a sample of dry ammonia? State *two* uses of the gas. How does it react with hydrogen chloride? Describe an experiment which shows that ammonia is a reducing agent. (J. M. B.)

25. Draw a fully labelled diagram to illustrate a laboratory method for the preparation and collection of chlorine, including a method of drying the gas.

Describe experiments, *one* in each case, which show that chlorine can oxidize (a) by removing hydrogen from a compound, (b) in the presence of water by adding oxygen to a compound. What is the effect of passing chlorine into hot sodium hydroxide solution until it is saturated? (J. M. B.)

26. By means of a labelled diagram and an equation, show how a sample of carbon dioxide can be made and collected in the laboratory. Without giving any details of the apparatus used, describe briefly how you would convert carbon dioxide into pure carbon monoxide. A journalist in a motoring magazine wrote:

'On a busy roadway, the proportion of carbon monoxide has varied from 6 parts per million to 180 parts per million.'

(a) At what time of day would you expect the concentration of carbon monoxide to be high?
(b) By what reaction is carbon monoxide formed?
(c) What is the effect of carbon monoxide on blood, and why does this make the gas so poisonous? (J. M. B.)

27. A grey-black compound, A, reacted with dilute hydrochloric acid to liberate a gas, B, which had an unpleasant smell and which blackened a lead ethanoate paper.

A white crystalline sodium compound, C, reacted with warm dilute hydrochloric acid to liberate a gas D, which had a sharp pungent smell and which turned an orange potassium dichromate paper green.

What do you think the compounds A, B, C, and D, might be? Explain the reactions and observations. Describe the effect of the two gases B and D on iron(III) chloride solution. What would be the effect of mixing the two gases B and D together? (J. M. B.)

12 An Introduction to Organic Chemistry

Organic chemistry has been defined as the chemistry of the compounds of carbon. Carbon is a unique element in that its atoms possess the remarkable ability to join up with one another, apparently indefinitely, to form chains (which may be straight or branched), or ring structures or mixtures of the two. This results in the known existence of nearly 3 million organic compounds compared with about 60 000 of the inorganic type.

The classifying of this vast number of organic compounds is simpler than you might think because they tend to fall into ordered sequences known as *homologous series*. All members of such a series have similar properties as they all possess the same *functional group*.

An organic molecule usually consists of a fairly unreactive hydrocarbon portion (i.e. hydrogen and carbon atoms) joined to another atom or group of atoms (e.g. a chlorine atom, a hydroxyl group) known as the functional group, which is largely responsible for the reactions of the compound, i.e.

Compounds which form a sequence such that the hydrocarbon parts of successive members differ by the factor —CH_2— and which all contain the same functional group, are said to be members of a homologous series, e.g. CH_3—OH, CH_3CH_2—OH, $CH_3CH_2CH_2$—OH etc. Members of such a series can be represented by a general formula, e.g. $C_nH_{(2n+1)}OH$. In the next two sections we shall be studying a few typical homologous series. Nomenclature, homologous series, and functional groups are considered in more detail in the *Supplementary Text*.

12.1 HYDROCARBONS. ALKANES, ALKENES AND ALKYNES

Introduction

Hydrocarbons, as the name implies, are compounds composed of hydrogen and carbon only. There are four important series of hydrocarbons, three of which are composed of aliphatic compounds and exist as 'straight' or branched chains, and one aromatic series based on benzene.

1. *Alkanes:* each carbon atom is joined to an adjacent carbon atom by means of a *single bond*, the remaining carbon atom valencies being satisfied by hydrogen atoms (e.g. CH_3—CH_2—CH_2—CH_3).
2. *Alkenes:* two adjacent carbon atoms at some point in the chain are linked together by a *double bond* (e.g. CH_3—CH_2—CH=CH_2).
3. *Alkynes:* two adjacent carbon atoms at some point in the chain are linked together by a *triple bond* (e.g. CH_3—CH_2—C≡CH).
4. *Aromatics:* most contain a benzene ring (i.e. ⌬).

Other series of hydrocarbons exist, but these are unimportant, except to specialists in hydrocarbon chemistry.

Thus alkanes are characterized by having single bonds only, whereas alkenes and alkynes contain *one* carbon–carbon double bond and one carbon–carbon triple bond respectively.

Alkanes (Paraffins)

General Characteristics

The alkanes form an homologous series of general formula $C_nH_{(2n+2)}$. Thus when n = 1, the alkane is CH_4, methane, and has the graphical formula

```
     H
     |
  H—C—H
     |
     H
```

When n = 2 the alkane is C_2H_6, ethane, of graphical formula

```
     H  H
     |  |
  H—C—C—H
     |  |
     H  H
```

The names and some physical constants of ten members of the series are given in Table 12.1.

Table 12.1 Physical properties of some alkanes

Name of alkane	*Formula*	*Melting point K (°C)*	*Boiling point K (°C)*	*Density (g cm⁻³)*	*Physical state at room temperature*
Methane	CH_4	91 (−182)	109 (−164)	—	Gas
Ethane	C_2H_6	90 (−183)	184 (−89)	—	Gas
Propane	C_3H_8	85 (−188)	231 (−42)	—	Gas
Butane	C_4H_{10}	135 (−138)	272 (−1)	—	Gas
Pentane	C_5H_{12}	144 (−129)	309 (36)	0.63	Liquid
Hexane	C_6H_{14}	178 (−95)	342 (69)	0.66	Liquid
Heptane	C_7H_{16}	182 (−91)	371 (98)	0.68	Liquid
Pentadecane	$C_{15}H_{32}$	283 (10)	543 (270)	0.77	'Liquid'
Hexadecane	$C_{16}H_{34}$	291 (18)	560 (287)	0.77	Solid
Heptadecane	$C_{17}H_{36}$	295 (22)	575 (302)	0.78	Solid

As is general in any homologous series the melting points and boiling points rise as the value of 'n' in the general formula increases. This is because the intermolecular attraction between molecules increases as the molecules become larger. The liquid alkanes have low specific gravities and, being immiscible with water, float on its surface.

Occurrence

Alkanes are formed in nature by the decomposition of vegetable and animal matter in the absence of air. Over millions of years nature has built up vast stocks of coal and oil from such decompositions and large quantities of alkanes are among the products obtained when coal and oil are distilled in the absence of air. Methane is always found associated with

coal and oil deposits, and it is indeed the main constituent of natural gas (>78 per cent) which is used widely as a fuel in Britain, the U.S.A., Canada, Germany, and Italy. The recent discoveries of accumulations of this gas under the North Sea has revolutionized Britain's gas industry. Methane is also obtained as a by-product of the distillation of coal and oil and at sewage works. It can also be observed bubbling in ponds and marshy regions where vegetable matter is decomposing and for this reason the gas was originally named 'marsh gas'. Animals also break down vegetable matter during digestion and so produce methane, and it has been estimated that a cow produces over 500 000 cm^3 per day. Methane is also the main constituent of the atmospheres of the planets Jupiter and Saturn.

w produces half a mil- cm³ of methane per day!

Experiment 12.1

† Laboratory Preparation of Methane ***

Apparatus
As in Figure 12.1, mortar and pestle, spatula, rack of test-tubes fitted with rubber bungs.
Sodium ethanoate, soda lime.

Procedure

(a) Grind a small amount of fused sodium ethanoate (sodium acetate) to a powder in a mortar, and then mix it thoroughly with about twice its bulk of soda lime (a non-deliquescent form of sodium hydroxide).

(b) One-third fill a hard glass test-tube (150×25 mm) with the mixture and set up the apparatus as shown in Figure 12.1.

(c) Heat the test-tube in the middle at first and then gradually move the flame towards the closed end of the tube.

(d) Discard the first two test-tubes of gas collected and then collect four more test-tubes of gas. Stopper these samples and place them in the test-tube rack for the next experiment.

Points for Discussion

1. Why was it necessary to grind up the solid sodium ethanoate before mixing it with the soda lime?

2. Why do you think it is necessary to start heating the reacting mixture as shown and then gradually move the flame as indicated?

3. Why is it necessary to discard the first two test-tubes of gas collected?

4. The equation for the preparation of methane is:

$$\underset{\text{Sodium ethanoate}}{CH_3COONa(s)} + \underset{\text{Soda lime}}{NaOH(s)} \rightarrow \underset{\text{Methane}}{CH_4(g)} + \underset{\text{Sodium carbonate}}{Na_2CO_3(s)}$$

This method of preparation (decarboxylation) which involves heating together soda lime (a non-deliquescent form of

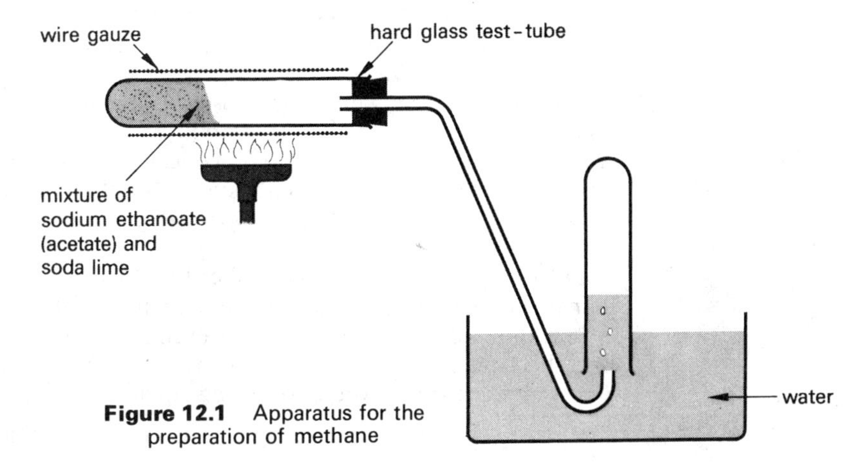

Figure 12.1 Apparatus for the preparation of methane

sodium hydroxide) and the anhydrous sodium salt of a carboxylic acid, is a general method of preparing alkanes, (i.e. as alkanes are members of an homologous series, they have common methods of preparation). Which compound would you have to heat with soda lime in order to produce (i) ethane, (ii) butane? If these compounds are available it may be possible for you to prepare samples of ethane and butane.

Properties of Alkanes

Experiment 12.2
† Reactions of Methane **

Apparatus
Bunsen burner, asbestos square, spills.
Rack of stoppered test-tubes of methane, red and blue litmus papers, solution of bromine water.

Procedure
Carry out the following tests on samples of methane and record your observations.

(a) Cautiously sniff a small amount of the gas.

(b) Test the gas with (i) damp red litmus paper and (ii) damp blue litmus paper.

(c) Ignite a test-tube of the gas using a lighted splint (care!).

(d) Add a small amount of bromine water to a test-tube of methane, quickly replace the stopper, and shake the tube so that the contents are well mixed.

Points for Discussion

1. Some of the *physical properties* of a gas can be remembered by answering the following questions: colour? odour? density? solubility (in water)?

2. Is methane an acidic gas, an alkaline gas, or a neutral gas?

3. What did you observe when a lighted splint was applied to methane?

4. Was there any apparent reaction between bromine water and methane? What criterion did you use to decide this?

5. Would you expect the results of procedures (b), (c), and (d) to be similar or different for other members of the homologous series? If you prepared (or have available) samples of ethane and butane you will be able to find out if your answer is correct, by performing the three tests using these homologues.

Chemical Properties of Alkanes

You should have noticed that shaking methane with a solution of bromine water did not remove the orange colour of the solution, or at most only removed it slowly, i.e. the bromine reacted very slowly with the methane.

[Note: bromine is often used in organic chemistry to establish ideas about different kinds of compounds, and it is easy to see if it has reacted as the orange colour will disappear. Bromine water is more convenient to use than bromine vapour.]

The slow reaction with bromine is typical of methane as it is chemically unreactive, and this chemical stability is characteristic of the whole alkane series. In fact the only chemical reactions methane and other alkanes undergo are (a) combustion, (b) thermal decomposition, (c) slow substitution reactions (e.g. with halogens).

Combustion You probably noticed that when you ignited methane, it burnt with a slightly luminous flame, or if you had allowed some air to mix with it before ignition you may even have noticed a slight explosion. In either case the methane burns and is oxidized by the air to carbon dioxide and water according to the following equation:

$$CH_4(g) + 2O_2(g) \rightarrow CO_2(g) + 2H_2O(g)$$

If insufficient oxygen is present for complete combustion (as is the case in underground mine workings) then poisonous carbon monoxide is produced instead of carbon dioxide and this is the deadly 'after damp' that remains in coal mines after explosions. A similar situation also exists in the cylinders of petrol engines (petrol is a mixture containing some alkanes), and when incomplete combustion occurs the exhaust fumes contain carbon monoxide and occasionally even carbon. This is why it is dangerous to run a car engine inside a garage and why petrol engines occasionally need a 'de-coke'.

Decomposition by heat Methane is more stable than most organic compounds due to the strength of the C—H bonds, and a large amount of heat energy has to be used to break these bonds. At temperatures approaching 1273 K (1000 °C) the gas breaks down into carbon and hydrogen, i.e.

$$CH_4(g) \xrightarrow[(1000\,^\circ C)]{1273\ K} C(s) + 2H_2(g)$$

Higher alkanes (i.e. those containing a larger number of carbon atoms) tend to break up more readily into simpler hydrocarbons when heated, especially in the presence of a catalyst. This process is known as *'cracking'*.

Reaction with halogens Methane reacts slowly with bromine in diffused daylight, and there is no reaction if the two are mixed in the dark.

The reason for this situation is that for a bromine molecule to react with a molecule of methane, it is necessary to break bonds. The Br—Br bond is weaker than a C—H bond and so will break first, but only when sufficient energy is available in the system. The thermal energy of the molecules at room temperature is insufficient to bring about reaction and it is necessary to supply additional energy either by heating or by using the energy of light rays. This is why no reaction takes place in the dark, but a slow reaction takes place in diffused daylight. In direct sunlight where more energy is available an explosive reaction occurs. The reactions of a halogen with an alkane can be summarized by the following equations,

$$CH_4 + Br_2 \rightarrow CH_3Br + HBr$$
Monobromomethane

$$CH_3Br + Br_2 \rightarrow CH_2Br_2 + HBr$$
Dibromomethane

$$CH_2Br_2 + Br_2 \rightarrow CHBr_3 + HBr$$
Tribromomethane

$$CHBr_3 + Br_2 \rightarrow CBr_4 + HBr$$
Tetrabromomethane

These reactions occur simultaneously and a mixture of products always results. A similar and faster reaction takes place when chlorine is used instead of bromine. Iodine has virtually no reaction with methane.

You will notice that in each step of the reaction in diffused daylight a hydrogen atom of the original methane is substituted by a bromine atom. Such a reaction in which an atom of one compound is *substituted* by another atom is known as a *substitution reaction*. *Substitution* can be defined as a reaction between an organic compound and another element or compound of the molecular type X—Y, during which X enters the organic molecule, displacing an atom which then combines with Y.

In alkanes the four valencies of each carbon atom are used to the greatest combining power and the compounds are said to be *saturated.* Compounds which contain a double or triple bond between two adjacent carbon atoms are said to be *unsaturated.* Saturated compounds can only react either by substitution reactions or by disruption of the molecules (e.g. cracking, combustion), whereas unsaturated ones usually react by addition reactions (page 313). Because substitution reactions first involve breaking covalent bonds, they tend to take place much more slowly than the rapid reactions which take place when unsaturated compounds undergo addition reactions. This difference in the rates of the reactions between bromine and the two classes of compounds is used as a distinguishing test.

Predicting the Reactions of Other Alkanes

As all the alkanes belong to an homologous series they will all react in a similar manner. Write the equations for (a) the combustion of propane in air, (b) the thermal decomposition of butane, and (c) the substitution reactions of ethane with chlorine in the presence of diffused daylight. This latter reaction again occurs in steps; write an equation for each step. You will find that after one chlorine atom has been substituted the second one may substitute either on the same carbon atom as the first or on the other carbon atom. Both compounds exist and are isomeric (page 200).

Summary of the General Properties of Alkanes

Alkanes are saturated hydrocarbons belonging to an homologous series of general formula $C_nH_{(2n+2)}$. They exist as gases, liquids, or solids depending upon the length of the carbon atom chain. From butane onwards they show an increasing degree of isomerism. All members dissolve in organic solvents, but are insoluble in water, on which the liquid members float as they have low densities. Chemically they are unreactive, and used to be called paraffins (from the Latin, parum, 'little' and affinis, 'activity'), the solid members being known as paraffin waxes. Alkanes burn readily in air and the gaseous and liquid members are used extensively as fuels, for example, natural gas (mainly methane), 'calor' gas (mainly butane) and engine fuels (mainly liquid alkanes). The higher alkanes are also used as solvents and as raw materials in the plastics industry and in the manufacture of a number of chemicals. Solid alkanes (paraffin waxes) have a wide variety of uses, e.g. vaseline is paraffin wax and oil.

Test for an Alkane See page 321.

Alkenes (Olefines)

General Characteristics

Alkenes are unsaturated hydrocarbons characterized by having a carbon–carbon double bond at some point in the molecule. They form an homologous series of general formula C_nH_{2n} and are named by replacing the ending *-ane* of the corresponding alkane by *-ene,* which signifies the presence of the double bond in the compound. Thus when n = 2 the alkene is C_2H_4, ethene (ethylene) and has the graphical formula

```
H       H
 \     /
  C=C
 /     \
H       H
```

Similarly when n = 3, C_3H_6, propene,

```
H     H    H
 \    |   /
H—C—C=C
 /        \
H          H
```

is formed. When n = 4 the alkene is called butene. This alkene has three isomeric forms. See if you can work out what they are.

You will notice that there is no alkene corresponding to n = 1; this is because all alkenes contain a $>C=C<$ functional group, i.e. at least two carbon atoms are needed.

Some physical constants for the first six alkenes are given in Table 12.2. Again observe the gradual increase in boiling point with increasing number of carbon atoms. Alkenes are insoluble in water, but dissolve in organic solvents. They are not found to any great extent in nature; coal gas and natural gas contain small amounts of ethene but the bulk of that used industrially is produced by the 'cracking' of petroleum.

Table 12.2 Physical constants of some alkenes

Name and formula of alkene	*Melting point K (°C)*	*Boiling point K (°C)*	*Density (g cm⁻³)*	*Normal physical state*
Ethene $CH_2{=}CH_2$	104 (−169)	169 (−104)	—	Gas
Propene $CH_3{-}CH{=}CH_2$	88 (−185)	226 (−47)	—	Gas
But-1-ene $CH_3{-}CH_2{-}CH{=}CH_2$	88 (−185)	267 (−6)	—	Gas
Pent-1-ene $CH_3{-}CH_2{-}CH_2{-}CH{=}CH_2$	135 (−138)	303 (30)	0.64	Liquid
Hex-1-ene $CH_3{-}CH_2{-}CH_2{-}CH_2{-}CH{=}CH_2$	133 (−140)	336 (63)	0.67	Liquid
Hept-1-ene $CH_3{-}CH_2{-}CH_2{-}CH_2{-}CH_2{-}CH{=}CH_2$	154 (−119)	367 (94)	0.70	Liquid

Preparation of Alkenes

Experiment 12.3
†Laboratory Preparation of Ethene ***

Apparatus

As in Figure 12.2, rack of test-tubes and corks to fit, small gas jar and cover, spatula.

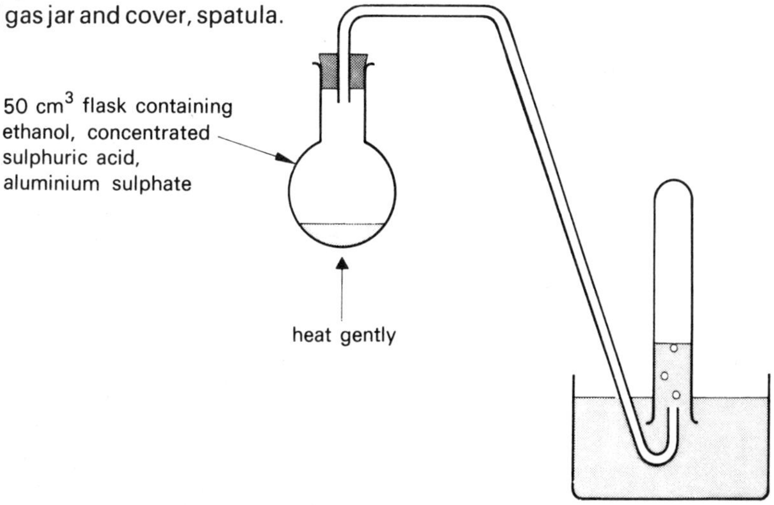

Figure 12.2 Apparatus for the preparation of ethene

Procedure

(a) Pour about 10 cm^3 of ethanol into the round bottom flask and add approximately 4 cm^3 of concentrated sulphuric acid, followed by a spatula measure of aluminium sulphate.

(b) Arrange the apparatus as shown in Figure 12.2 so that the gas evolved is collected over water.

(c) Heat the flask, controlling the heat so that a steady evolution of gas occurs. Collect six test tubes and one small gas jar of gas but discard the first two tubes collected. Stopper these samples and keep them for the next experiment.

Points for Discussion

The equation for the preparation of ethane is

$$C_2H_5OH(g) \rightarrow C_2H_4(g) + H_2O(g)$$

Ethanol vapour — Ethene — Steam

This method, i.e. dehydration of the appropriate alcohol, is a general method of preparing alkenes. The dehydration can be carried out using a variety of dehydrating agents. The usual ones are hot, unglazed porcelain (using ethanol vapour), concentrated sulphuric acid, aluminium oxide, or 'syrupy' phosphoric(V) acid. Which alcohol would you dehydrate if you wished to prepare propene? Write an equation for this reaction.

Properties of Alkenes

Experiment 12.4
† Reactions of Ethene **

Apparatus

Bunsen burner, asbestos square, spills, teat pipette, small gas jar, access to fume cupboard.

Rack of stoppered test-tubes of ethene and small gas jar of ethene, red and blue litmus papers, bromine, solution of bromine water, a dilute solution of potassium manganate(VII) acidified with dilute sulphuric acid.

Procedure

(a) Cautiously sniff a small amount of the gas.

(b) Test the gas with (i) damp red litmus paper, (ii) damp blue litmus paper.

(c) Ignite a test-tube of the gas using a lighted spill. (Care; ethene/air mixtures are explosive.)

(d) Add a small amount of bromine water to a test-tube of ethene, quickly replace the stopper, and shake the tube so that the contents are well mixed.

(e) Repeat procedure (d) using acidified potassium manganate(VII) instead of bromine water.

(f) Working in a fume cupboard, place two drops of bromine into a small gas jar by means of a teat pipette. Allow *all* the bromine to vaporize and then invert this gas jar over the gas jar of ethene. Record your observations.

Points for Discussion

1. You should now be able to write down at least four *physical properties* of ethene.

2. Is ethene an acidic gas, an alkaline gas, or a neutral gas?

3. What do you observe when a lighted spill is applied to ethene?

4. From your observations in procedure (f) can you suggest what could have happened when bromine gas and ethene are mixed? Could a similar reaction have taken place in procedure (d)?

5. Would you expect the results of procedures (b), (c), (d), (e) and (f) to be similar or different for other members of the alkene homologous series? If you have samples of other alkenes you will be able to test your predictions.

Chemical Properties of Alkenes

Combustion You would notice that when ethene was ignited it either gave a minor explosion or burnt with a very luminous flame which may have been smoky. Can you suggest any reason why the flame may appear smoky? Write an equation for the combustion of ethene in a plentiful supply of air.

Reactions with halogens When ethene was mixed with bromine vapour or shaken with bromine water, the orange coloration *rapidly* disappeared, suggesting that the bromine had reacted with the ethene. Although you could not see the ethene 'disappear' you should have noticed the appearance of some oily streaks on the sides of the gas jars. These streaks indicate the formation of a new compound which could only have arisen from the reaction of bromine and ethene. This new compound is formed according to the following equation:

$$C_2H_4(g) + Br_2(g) \rightarrow C_2H_4Br_2(l)$$

or graphically,

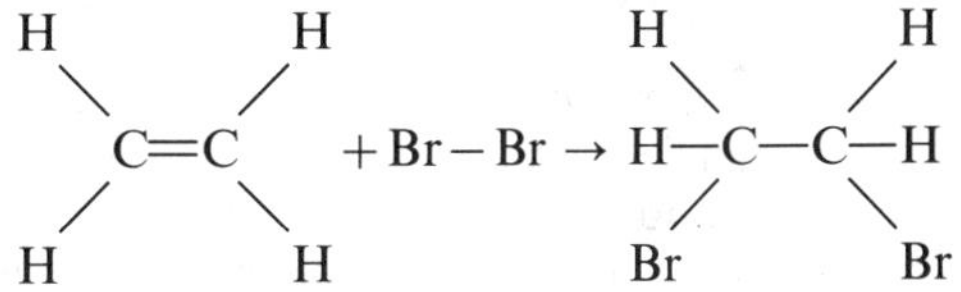

$$\begin{array}{ccc} H & & H \\ & C{=}C & \\ H & & H \end{array} + Br{-}Br \rightarrow \begin{array}{ccc} H & & H \\ H{-} & C{-}C & {-}H \\ Br & & Br \end{array}$$

(1,2-dibromo-ethane)

As you will notice a bromine molecule has *added* on to an ethene molecule to form a single new molecule. Such a reaction is known as an *addition reaction*. An addition reaction only occurs when a compound is *unsaturated* (e.g. it contains carbon–carbon double or triple bonds), and this enables the compound to combine directly with other substances to form a single new compound. This can be thought of simply as the compound having 'room' to accommodate more atoms, i.e.

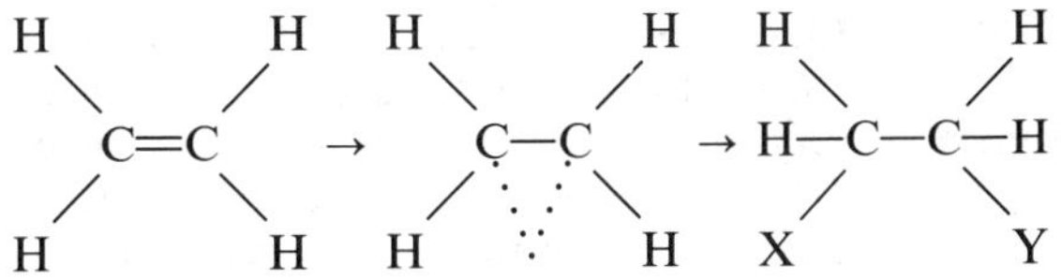

$$\begin{array}{ccc} H & & H \\ & C{=}C & \\ H & & H \end{array} \rightarrow \begin{array}{ccc} H & & H \\ & C{-}C & \\ H & \because & H \end{array} \rightarrow \begin{array}{ccc} H & & H \\ H{-} & C{-}C & {-}H \\ X & & Y \end{array}$$

[Note: the C=C bond is more reactive than the C—H bond and thus unsaturated compounds always undergo addition reactions in preference to substitution reactions. Compare alkanes which can only undergo substitution reactions.]

A similar reaction to that between bromine and ethene occurs between chlorine and ethene. Why do you think this is so? Would you expect the reaction with chlorine to take place more or less readily than with bromine? Explain your answer.

Reaction with acidified potassium manganate(VII) By reference to your work on oxidation in Chapter 8 you should be able to say why the colour changes observed with acidified potassium manganate(VII) took place. In the reaction the ethene was oxidized and the reaction can be represented by the following equation:

$$\underset{\text{(H}_2\text{C=CH}_2\text{)}}{\text{H}_2\text{C=CH}_2} + H_2O + \underset{\text{From potassium manganate(VII)}}{[O]} \longrightarrow \underset{\text{Ethan-1,2-diol (ethylene glycol)}}{\text{H–C(H)(HO)–C(H)(OH)–H}}$$

```
H       H
 \     /
  C=C      + H2O + [O]
 /     \             From
H       H            potassium
                     manganate(VII)

             H        H
              \      /
  ----------> H—C—C—H
              /      \
            HO        OH
            Ethan-1,2-diol
            (ethylene glycol)
```

Again note that the reaction is an addition reaction. Can you understand why the compound formed is named ethan-1,2-diol? This compound is used as an anti-freeze and in the preparation of Terylene.

Other addition reactions Ethene also reacts additively with a number of other compounds, for example, hydrogen (in the presence of finely divided nickel at 413 K (140 °C)) and hydrogen bromide (a concentrated aqueous solution at 373 K (100 °C)). Write equations for these two reactions and name the products formed in each case.

Polymerization Ethene can be liquefied by applying pressure, and a specialized form of addition reaction takes place when the liquid ethene, still under pressure, is heated in the presence of a trace of oxygen as catalyst. The ethene molecules add on to each other and form a long chain which can vary in length from about 100 to 1000 carbon atoms depending on the temperature and pressure used, i.e.

```
                                 CH2      CH2
                             \  /  \    /  \   /
'n' (CH2=CH2)        →        CH2    CH2     CH2
'n' ethene molecules         Part of one polyethene molecule
```

A reaction of this kind is called a *polymerization* and the white solid formed in this case is Polythene (from poly, 'many', ethene) which is an important plastic. See the *Supplementary Text* for further details.

Predicting the Reactions of Other Alkenes

Other alkenes react in a similar manner to ethene. Why? Write equations for (a) the combustion of hexene in air, (b) the reaction of butene with chlorine (c) the reaction of propene with (i) hydrogen chloride, (ii) Baeyer's reagent, (iii) hydrogen, (d) the polymerization of propene. In each case state the type of reaction taking place, the conditions needed, and the name of the products formed.

Summary of the General Properties of Alkenes

Alkenes are unsaturated hydrocarbons belonging to an homologous series of general formula C_nH_{2n}. They exist as gases, liquids or solids depending upon the length of the carbon atom chain. Members of the series from butene upwards occur in isomeric forms, the three isomers of butene being,

$$CH_3{-}CH_2{-}CH{=}CH_2 \text{ (but-l-ene)}$$

$$CH_3{-}CH{=}CH{-}CH_3 \text{ (but-2-ene)}$$

and

$$(CH_3)_2C{=}CH_2 \text{ (2-methyl propene)}$$

All alkenes are soluble in organic solvents, but are insoluble in water.

As all alkenes contain the same functional group, i.e. a carbon–carbon double bond, they all have similar chemical properties. The presence of this double bond results in the compounds being very reactive by comparison with the alkanes, and in their undergoing a large number of addition reactions.

Uses of Alkenes

Alkenes can be obtained in plentiful supply and comparatively cheaply by the cracking of petroleum. Their ability to react with a wide variety of compounds makes them very important starting materials in the industrial production of a large range of organic chemicals (Figure 12.3), including ethanol, 'glycol' (used as a coolant in aircraft engines, and in 'anti-freeze'), detergents, and in margarine production. Due to their ability to undergo addition polymerization, their most important use is in the manufacture of plastic materials. Such plastics as polythene, polystyrene, polyvinyl chloride (P.V.C.), polypropylene, synthetic rubbers etc. are becoming increasingly important in the modern world.

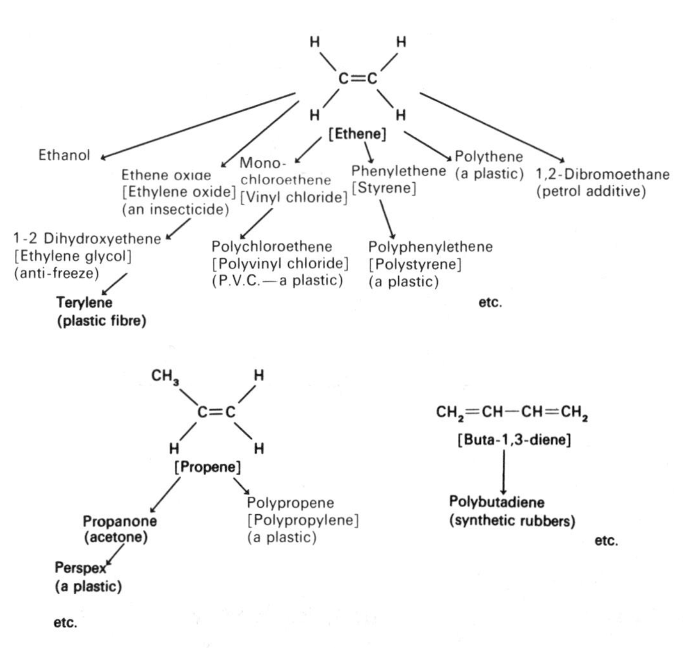

Figure 12.3 Some of the organic compounds produced from alkenes

Tests for Alkenes

(a) The rapid decolorization of a solution of bromine water

(b) The rapid decolorization of acidified potassium manganate(VII)

Together these two tests show that a compound is unsaturated, but not necessarily that it is an alkene as the tests are also given by alkynes. The method of distinguishing between alkenes and alkynes is given on page 321.

Alkynes

General Characteristics

Alkynes are unsaturated hydrocarbons characterized by having a carbon–carbon triple bond at some point in the molecule. They form an homologous series of general formula $C_nH_{(2n-2)}$ and are named by replacing the ending *-ane* of the appropriate alkane by *-yne* which signifies the presence of the triple bond in the compound. Thus when n = 2 the alkyne is C_2H_2, ethyne (acetylene), and has the graphical formula H—C≡C—H. Similarly when n = 3, C_3H_4, propyne,

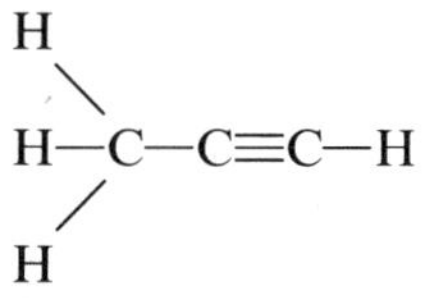

```
     H
      \
   H—C—C≡C—H
      /
     H
```

is formed. What are the names and formulae of the next two alkynes? These alkynes have two and three isomers respectively; see if you can work out what they are.

Some physical constants for the first six alkynes are given in Table 12.3. Again note the gradual increase in boiling point with increasing number of carbon atoms. Alkynes are insoluble in water but dissolve in organic solvents. They are usually too reactive to be found free in nature. The industrial methods of preparing ethyne are discussed after its laboratory preparation.

Table 12.3 Physical constants of some alkynes

Name and formula of alkyne	*Melting point K (°C)*	*Boiling point K (°C)*	*Density (g cm^{-3})*	*Normal physical state*
Ethyne CH≡CH	Sublimes	189 (−84)	—	Gas
Propyne CH_3—C≡CH	172 (−101)	250 (−23)	—	Gas
But-1-yne CH_3—CH_2—C≡CH	147 (−126)	281 (8)	—	Gas
Pent-1-yne CH_3—CH_2—CH_2—C≡CH	183 (−90)	313 (40)	0.69	Liquid
Hex-1-yne CH_3—CH_2—CH_2—CH_2—C≡CH	141 (−132)	344 (71)	0.72	Liquid
Hept-1-yne CH_3—CH_2—CH_2—CH_2—CH_2—C≡CH	192 (−81)	373 (100)	0.73	Liquid

Preparation of Alkynes

Experiment 12.5
† Laboratory Preparation of Ethyne ***

Apparatus
As in Figure 12.4, rack of test-tubes and bungs to fit, spatula, spills.

Procedure
(a) Place some calcium carbide into the conical flask and assemble the apparatus as shown (Figure 12.4).

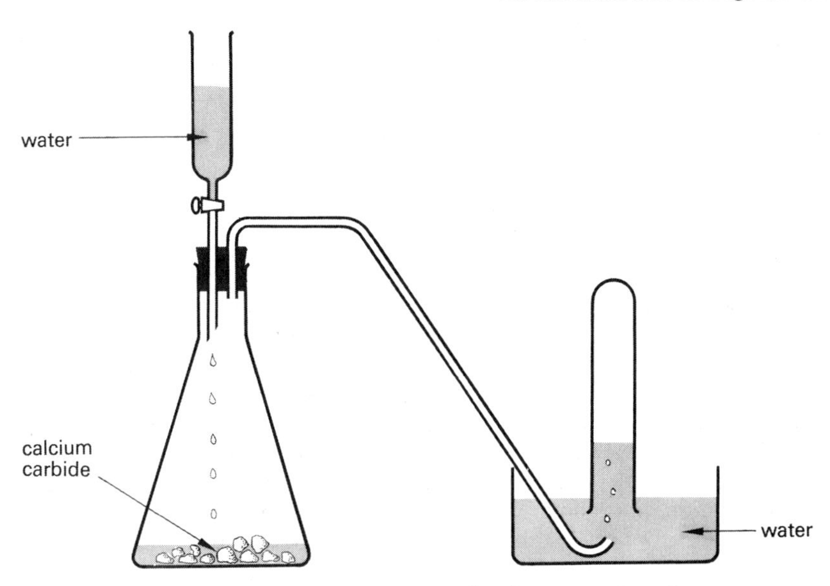

figure 12.4 Apparatus for the preparation of ethyne

(b) Slowly drop water on to the calcium carbide and collect small samples of the gas evolved in test-tubes. Test these samples with a lighted splint (take care; hydrogen/air mixtures are explosive), and when the gas burns quietly, without exploding, collect about six test-tubes of ethyne. Stopper these samples and keep them for the next experiment.

Points for Discussion

1. Why was it necessary to test small samples of ethyne with a lighted splint before a final collection of the gas was made?

2. The reaction between water and calcium carbide is normally used to prepare ethyne. The equation for the reaction is:

$$CaC_2(s) + 2H_2O(l) \rightarrow Ca(OH)_2(aq) + C_2H_2(g)$$

Do you think this method could be adopted as a general method of preparing alkynes or can it only be used for ethyne? A general method of preparing alkynes is to boil a 1,2-dibromoalkane with alcoholic potassium hydroxide solution, e.g.

$$\underset{\text{(1,2-Dibromobutane)}}{CH_3.CH_2.CHBr.CH_2Br(l)} + 2KOH \rightarrow \underset{\text{(But-1-yne)}}{CH_3.CH_2.C{\equiv}CH(g)} + 2KBr(aq) + 2H_2O(l)$$

3. Ethyne is obtained industrially (i) as a by-product in the cracking of various petroleum fractions and (ii) by the partial oxidation of natural gas:

$$4CH_4(g) + 3O_2(g) \rightarrow 2C_2H_2(g) + 6H_2O(g)$$

In countries where electrical power is cheap, calcium carbide is manufactured in an electrical furnace from coke and quicklime (calcium oxide):

$$3C(s) + CaO(s) \xrightarrow{(2273\ K)}_{(2000\ °C)} CaC_2(s) + CO(g)$$

and water is added to the carbide formed, to generate ethyne.

Properties of Alkynes

Experiment 12.6

†Reactions of Ethyne ***

Apparatus

Bunsen burner, asbestos square, tripod, gauze, filter funnel and stand, small gas jar, spatula, teat pipette, spills, filter paper, access to fume cupboard.

Rack of stoppered test-tubes of ethyne, red and blue litmus papers, solution of bromine water, Baeyer's reagent, bleaching powder, dilute hydrochloric acid, calcium carbide fragments, ammoniacal solutions of copper(I) chloride and silver nitrate.

Procedure

(a) Repeat procedures (a) to (e) in Experiment 12.4, using your samples of ethyne instead of ethene.

(b) Working in a fume cupboard, generate a little chlorine by placing a spatula measure of bleaching powder into a small gas jar and adding a little dilute hydrochloric acid. Test for chlorine gas (care) by its bleaching action on moist litmus paper, and when it has filled the jar drop a few small pieces of calcium carbide into the jar and record your observations.

(c) Using two further test-tubes of ethyne, add to one tube about 2 cm^3 of ammoniacal copper(I) chloride solution and to the other about 2 cm^3 of ammoniacal silver nitrate solution. Shake both of the tubes so that the gas and solution mix. Filter the resulting precipitates and place each filter paper on a gauze supported by a tripod. *Very carefully* warm the filter papers by gently heating the gauze with a small Bunsen flame. Record all your observations.

Points for Discussion

1. Write down at least four *physical properties* of ethyne.

2. Is ethyne an acidic gas, an alkaline gas, or a neutral gas?

3. What did you observe when ethyne was tested with a lighted splint?

4. Did both bromine and chlorine react with ethyne? Was the order of their activities what you would have predicted? Explain your answer.

5. What evidence have you that ethyne is an unsaturated compound?

Chemical Properties of Alkynes

It was probably apparent to you that ethyne was more reactive than ethene, and is in fact very often explosively so. This fact is partially due to ethyne having a large positive heat of formation (Section 13.2), i.e.

$$2C+H_2 \rightarrow C_2H_2 \quad (\Delta H = +201 \text{ kJ mole}^{-1})$$

and partially due to the presence of the triple bond in the molecule. Because of its high degree of unsaturation ethyne forms addition compounds very readily. The conversion usually occurs in two stages, the triple bond changing first to a double bond, and then to a single bond, i.e.

$$-C\equiv C- \dashrightarrow \;>C=C<\; \dashrightarrow \;-\overset{|}{\underset{|}{C}}-\overset{|}{\underset{|}{C}}-$$

Combustion You would notice that a test-tube of ethyne either burnt in air with a luminous smoky flame, or ignited explosively, because certain ratios of ethyne to oxygen form explosive mixtures. Write an equation for the combustion of ethyne in a plentiful supply of oxygen. The products of this equation are both colourless gases; why then do you think that the flame in your experiment was smoky? It may help you to know that ethyne contains 92.3 per cent by weight of carbon.

Reaction with halogens Ethyne reacts explosively with chlorine. You should have noticed one of the products which formed in the gas jar. What was this product? Using this knowledge write an equation for the reaction. If the reaction is performed in the presence of a negative catalyst (Section 14.2), e.g. antimony pentachloride $SbCl_5$, then, by contrast, a smooth *addition* reaction occurs in *two stages* according to the following equations:

$$\mathrm{H{-}C{\equiv}C{-}H(g) + Cl{-}Cl(g) \rightarrow} \begin{array}{ccc} \mathrm{H} & & \mathrm{H} \\ & \mathrm{C{=}C} & \\ \mathrm{Cl} & & \mathrm{Cl} \end{array} \quad (\mathrm{l})$$

1,2-Dichloroethene

$$\begin{array}{ccc} \mathrm{H} & & \mathrm{H} \\ & \mathrm{C{=}C} & \\ \mathrm{Cl} & & \mathrm{Cl} \end{array} (\mathrm{l}) + \mathrm{Cl{-}Cl(g)} \rightarrow \begin{array}{ccc} \mathrm{H} & & \mathrm{H} \\ & \mathrm{Cl{-}C{-}C{-}Cl} & \\ \mathrm{Cl} & & \mathrm{Cl} \end{array} (\mathrm{l})$$

1,1,2,2-Tetrachloroethane

The reaction between bromine and ethyne, as expected, is less vigorous and stepwise additions take place under normal conditions. Write equations for these addition reactions and name the compounds formed.

Other addition reactions Ethyne also reacts stepwise and additively with a number of other compounds, for example hydrogen (in the presence of a finely divided nickel catalyst at 413 K (140 °C) and the hydrogen halides (e.g. hydrogen chloride) at room temperature. Write equations for these reactions. You will notice that after the first addition of hydrogen chloride to ethyne, the compound monochloroethene (vinyl chloride),

$$\begin{array}{ccc} \mathrm{H} & & \mathrm{H} \\ & \mathrm{C{=}C} & \\ \mathrm{H} & & \mathrm{Cl} \end{array}$$

is formed. Subsequent addition of another hydrogen chloride molecule could theoretically result in either of the following products:

$$\begin{array}{ccc} \mathrm{H} & & \mathrm{H} \\ & \mathrm{H{-}C{-}C{-}Cl} & \\ \mathrm{H} & & \mathrm{Cl} \end{array} \quad \text{or} \quad \begin{array}{ccc} \mathrm{H} & & \mathrm{H} \\ & \mathrm{H{-}C{-}C{-}H} & \\ \mathrm{Cl} & & \mathrm{Cl} \end{array}$$

1,1-Dichloroethane 1,2-Dichloroethane

V. V. Markownikoff found that in such situations the hydrogen atom becomes attached to the carbon atom already bonded to the largest number of hydrogen atoms. Which of the two compounds would be formed in this case?

Reactions of other alkynes The only important member of the alkyne series is ethyne. Other alkynes react in a similar manner to ethyne.

Summary of the General Properties of Alkynes

Alkynes are very reactive compounds due to their large positive heats of formation and the presence of triple bonds. In cases where the reactions are not explosive, addition reactions usually occur and take place in two stages. Ethyne is so unstable that it cannot be stored safely under compression alone; it is dissolved in propanone under pressure and stored in steel cylinders. To further reduce the risk of explosion the cylinders are usually packed with a porous material (e.g. asbestos) to absorb the solution.

Uses of Ethyne

When ethyne undergoes stepwise addition reactions the compound produced by the first step is always an alkene derivative. If the amount of reactant used is limited, then the reaction will not go to completion and the alkene derivative can be separated and polymerized. For example, when a limited amount of hydrogen chloride reacts with ethyne the intermediate product is monochloroethene (vinyl chloride), $CH_2{=}CHCl$, and on polymerization this compound forms the well known plastic polyvinyl chloride (P.V.C.). Thus alkynes can also be used as starting materials in the manufacture of a variety of plastics, by varying the compound added to the alkyne.

Ethyne (acetylene) has been used for many years in oxy-acetylene burners to generate the very high temperatures (well over 2000 K) which are used in welding and metal cutting (Figure 12.5). The increasing import-

Figure 12.5 A welder using an oxy-acetylene torch for cutting steel

ance of the compound is, however, due to the large range of organic compounds for which it acts as a starting material—such compounds as ethanoic (acetic) acid (used in preparing aspirin, cellophane and 'celanese' fibre), ethanoic (acetic) anhydride (used in the preparation of dyestuffs and Rayon), polyvinyl chloride (the plastic, P.V.C.), polyvinyl acetate (a plastic used in Triplex safety glass), 1,1,2,2-tetrachloroethane ('Westron', a solvent for waxes, grease, rubber, oil etc.), and synthetic rubbers.

Test for Ethyne and the Distinction Between Alkanes, Alkenes and Alkynes

Due to its unsaturated nature ethyne reacts in a characteristic way with acidified potassium manganate(VII) and bromine water, although the latter reaction sometimes takes place more slowly than expected. However, as similar results are also obtained with alkenes, a further test must be applied to distinguish between these two types of compound. This test relies on the ability of ethyne to precipitate acetylides when reacted with ammoniacal solutions of certain metal salts, e.g. copper(I) chloride, silver nitrate.

$$C_2H_2(g) + Cu_2Cl_2(aq) \rightarrow \underset{\text{Reddish-brown precipitate of copper(I) acetylide}}{Cu_2C_2(s)} + 2HCl(aq)$$

$$C_2H_2(g) + 2AgNO_3(aq) \rightarrow \underset{\text{White precipitate of silver acetylide}}{Ag_2C_2(s)} + 2HNO_3(aq)$$

In addition to the formation of a characteristic precipitate, a series of short, sharp explosions is heard when these precipitates are dried by warming over a Bunsen flame. Alkenes do not give this reaction. As alkanes are saturated they do not react with Baeyer's reagent and only react slowly with bromine water.

12.2 ALCOHOLS

General Characteristics

Alcohols are among the earliest compounds to have been prepared by man. Since ancient times they have been made by the fermentation of sugar solutions, and even now this method is still widely employed, despite the discovery of alternative methods of preparation. You are no doubt aware that beer (about 4 per cent), wines (about 10–20 per cent) and spirits (up to 40 per cent) contain 'alcohol'. The alcohol referred to is mainly ethanol (ethyl alcohol), C_2H_5OH. This is an important member of the group of aliphatic alcohols which, as mentioned in the introductory section to this chapter, form an homologous series of general formula $C_nH_{(2n+1)}OH$ and are named by substituting the ending *-ol* for the final *-e* in the name of the corresponding alkane. Thus when n = 1, the alcohol is CH_3OH, methan*ol;* when n = 2, C_2H_5OH, ethan*ol* etc.

Alcohols containing one —OH group attached to an alkyl radical are said to be *monohydric* alcohols. Other alcohols exist which contain more than one hydroxyl group, e.g. 1,2-dihydroxyethane (ethylene glycol), $CH_2OH.CH_2OH$, contains two hydroxyl groups and is said to be a *dihydric* alcohol. Similarly 1,2,3-trihydroxypropane (glycerol), $CH_2OH.CHOH.CH_2OH$, is a *trihydric* alcohol. Many other *polyhydric* alcohols are known.

Table 12.4 Some physical constants of various alcohols

Name of alcohol	*Melting point K (°C)*	*Boiling point K (°C)*	*Density (g cm^{-3})*
Methanol	179 (−94)	338 (65)	0.79
Ethanol	156 (−117)	351 (78)	0.79
Propanol	147 (−126)	370 (97)	0.80
Butanol	183 (−90)	390 (117)	0.81
Pentanol	194 (−79)	410 (137)	0.81
Hexanol	226 (−47)	431 (158)	0.81
Heptanol	239 (−34)	449 (176)	0.82

Some physical constants for a number of monohydric alcohols are listed in Table 12.4. All members of the series having less than twelve carbon atoms are liquids at room temperature and show the usual trend of increasing boiling points with increasing numbers of carbon atoms.

An alcohol molecule is made up of two parts, the hydrocarbon portion which is hydrophobic (water hating) and the hydroxyl group which is strongly hydrophilic (water loving). The first three alcohols are very soluble in water (miscible in all proportions) due to the greater influence of the hydrophilic group. However, as the length of the hydrocarbon chain increases, the solubility of the alcohol decreases, and alcohols higher in the series than hexanol are practically insoluble. As the chemical reactions of all members of the series are similar we shall confine our attention to one particular member, namely ethanol.

Ethanol

Preparation of Ethanol

Ethanol is prepared industrially either by fermentation or from ethene. The fermentation method can readily be demonstrated in the laboratory as follows:

Experiment 12.7

† Laboratory Preparation of Ethanol by Fermentation *

Apparatus

250 cm^3 conical flask fitted with bung and delivery tube as in Figure 12.6, two 100 cm^3 beakers, simple distillation apparatus (Experiment 2.7), evaporating basin, Bunsen burner, asbestos square, spatula, splints.

Cane sugar or glucose, yeast, ammonium phosphate, calcium hydroxide solution.

Procedure

(a) Dissolve approximately 15 g of cane sugar or glucose in about 50 cm^3 of water in a conical flask. Add to this solution two spatula measures of yeast and a little ammonium phosphate (a yeast nutrient).

(b) Stopper the flask and leave it in a warm room, preferably near a warm radiator.

(c) Observe the flask occasionally and when the solution begins to ferment arrange for the delivery tube to dip into a beaker of calcium hydroxide solution. Record any observations you make. Leave the solution fermenting for several days.

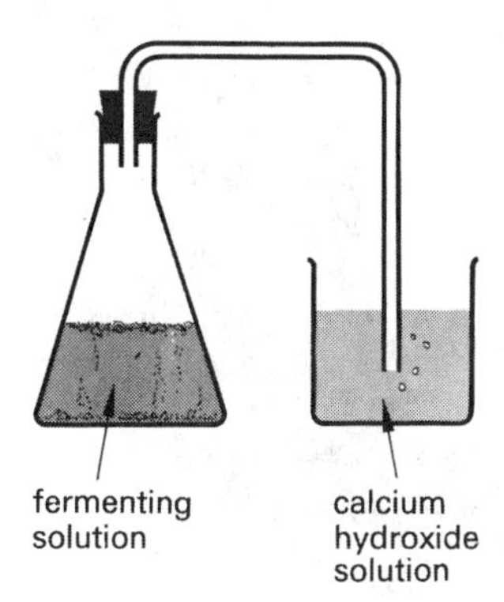

Figure 12.6 Apparatus used in the fermentation of sugar by yeast

(d) After several days decant most of the fermented solution into a round bottom flask and assemble the apparatus (as in Experiment 2.7) for the fractional distillation of this solution.

(e) Distil the solution and record the temperature at which liquid first begins to distil. Collect the first 1 or 2 cm³ of distillate in an evaporating basin and attempt to ignite this liquid using a lighted splint. Also attempt to ignite a sample of the undistilled liquid from the round bottom flask by transferring a few drops into another evaporating basin and applying a lighted splint.

Points for Discussion

1. The word fermentation is derived from the Latin word *fevere,* which means 'to boil up'. (a) From the observations you made while Experiment 12.7 was taking place, do you think this process is well named? Give a reason for your answer. (b) What evidence have you that fermentation is a chemical reaction? (c) Name *two* of the reaction products.

2. Fermentation is the term applied to chemical reactions brought about by yeast, bacteria, etc. These substances produce complex molecules called *enzymes* which can catalyse various chemical changes. In your preparation, cane sugar (or glucose) was being converted to ethanol according to the following equations,

$$\underset{\text{Cane Sugar}}{C_{12}H_{22}O_{11}(aq)} + H_2O(l) \rightarrow \underset{\text{Glucose}}{C_6H_{12}O_6(aq)} + \underset{\text{Fructose}}{C_6H_{12}O_6(aq)}$$

and/or

$$\underset{\text{Glucose or fructose}}{C_6H_{12}O_6(aq)} \rightarrow \underset{\text{Ethanol}}{2C_2H_5OH(aq)} + 2CO_2(g)$$

You will notice that the molecular formulae of glucose and fructose are the same; what name is used to describe such compounds?

3. Which two liquids are being separated during the distillation process? Which liquid distilled off first? Explain your answer using the results you obtained in procedure (e).

4. Most liquids can be separated completely from water either by distillation or by addition of a suitable drying agent (e.g. anhydrous calcium chloride). However, if an ethanol–water mixture is fractionally distilled it is impossible to obtain a percentage of alcohol greater than about 95 per cent. This mixture is commercially known as *rectified spirit* and is normally suitable for most laboratory work where ethanol is to be used.

Alcoholic Beverages

As you will discover in the next section, starch is a natural polymer which can be broken down into sugar molecules and thence to ethanol. Various compounds containing forms of starch can thus be used to make alcoholic

Figure 12.7 A malt whisky distillery at Muir-of-Ord, Perthshire

beverages (Figure 12.7). For example, beer is made from the starch in barley, and the resulting alcoholic solution is then boiled with hops to give it a bitter taste. Similarly, a wide variety of wines may be made from substances containing either starches or sugars. In addition to grapes, such substances as elderberries, beetroot, potatoes, rhubarb etc. can be used. Can you think of some others? It is estimated that over one million people in Britain make wine in their homes in this way.

When making wine you must be careful not to overexpose the wine to air in the early stages of fermentation, because oxidizing bacteria, whose spores are always present in air, may oxidize the wine to vinegar, i.e. the ethanol to ethanoic acid. This is how vinegar is made. Fortunately, when the alcoholic content of the wine is greater than about 12 per cent, the bacteria become inactive and so wines and spirits do not turn sour on exposure to air. Beer, which has a lower alcoholic content, would turn sour in air fairly rapidly and so 'stabilizers' are usually added. In contrast, when the alcoholic content of wine reaches about 17 per cent, the yeast enzymes then cease to function and further increase in the alcoholic content of the solution must be brought about by distillation, or even direct addition of alcohol, as in port.

You have probably noticed that some spirits are said to be 40° proof or 70° proof, etc. This does *not* mean that such solutions contain 40 per cent or 70 per cent alcohol. The term is derived from an old method of determining the alcohol content of a solution, as such solutions were taxed according to the amount of alcohol they contained. The method consisted of pouring the alcoholic liquor over gunpowder and then applying a flame. If the gunpowder was left dry enough to ignite, it was 'proof' that the liquor under test did not contain too much water. The liquor was then said to be *proof*. If the gunpowder was left too damp to ignite, the liquor was 'underproof'. Nowadays the Customs and Excise officer determines the amount of alcohol by the much less exciting method of measuring the density of the solution with a hydrometer.

Manufacture of Ethanol

Much of the world's supply of ethanol is still manufactured by fermenting natural starches or sugars, e.g. from starch in potatoes (Europe), from starch in rice (Asia) and from sugar in sugar cane (America). The carbon dioxide produced is usually collected and used on site in the making of lemonade and other 'fizzy' drinks. Industrially speaking, what do you think are the advantages and disadvantages of the fermentation process?

The advent of the petrochemical industry has provided a reasonably cheap alternative means of preparation, which can be carried out much more speedily. This involves using ethene produced during the 'cracking' of petroleum oils. The ethene is made to react with water according to the following equation:

$$C_2H_4(g) + H_2O(l) \rightarrow C_2H_5OH(l)$$

This reaction can be brought about in two ways: (1) by absorbing the gas in concentrated sulphuric acid at 353 K (80 °C) and 25 atmospheres pressure, and treating the resultant solution with steam; (2) by passing ethene and steam over a catalyst of phosphoric acid on kieselguhr, at a temperature of 573 K (300 °C) and a pressure of 60 atmospheres.

Properties of Ethanol

You have already observed the effect of adding ethanol to water (Experiment 1.4), and know how it reacts with dehydrating agents (page 312). The following simple experiments can be performed using 'rectified' spirit, and absolute alcohol. Other reactions of ethanol will be mentioned in the ensuing discussion.

Experiment 12.8

† Reactions of Ethanol ***

Apparatus

Teat pipettes, watch-glass or evaporating basin, Bunsen burner, asbestos square, splints, rack of test-tubes, sharp knife, filter paper, 50 cm^3 beaker, boiling tube.

Supply of ethanol (rectified spirit, and absolute alcohol), supply of sodium, fairly concentrated solution of potassium dichromate(VI), concentrated sulphuric acid, glacial ethanoic (acetic) acid, 20 per cent solution of potassium iodide, concentrated sodium hypochlorite solution.

Procedure

(a) Place a few drops of ethanol on to a watch-glass or evaporating basin and ignite the liquid using a lighted splint.

(b) Pour about 2 cm^3 of the ethanol (absolute alcohol) into a test-tube and add one small piece of freshly cut sodium about the size of a rice grain (care). Test any gas evolved with a lighted splint. When all the sodium has 'dissolved', carefully warm the solution, and then place a few drops on a microscope slide using a teat pipette. Allow the ethanol to evaporate. Record all your observations.

(c) Pour about 2 cm^3 of the potassium dichromate(VI) solution into a test-tube and *carefully* add about 1 cm^3 of concentrated sulphuric acid. To this solution add a few drops of ethanol and shake the tube well, but carefully. *Gently* warm for a few moments in a small Bunsen flame, and note any changes which occur. Also from time to time, gently waft some of the evolved vapour so that you are able to detect its odour.

(d) Pour about 1 cm^3 of ethanol into a test-tube and add about 5 cm^3 of 20 per cent potassium iodide solution, followed by a few drops of concentrated sodium hypochlorite solution. Gently warm the tube for a few minutes and record your observations.

(e) Place about 1 cm^3 of ethanol and 1 cm^3 of glacial ethanoic acid into a test-tube and carefully add about 0.5 cm^3 of concentrated sulphuric acid. Warm the tube over a small flame for a few minutes, gently shaking the contents during warming. Then pour the contents of the tube into a small beaker half full of water. Record your observations and note any detectable odour.

Points for Discussion

1. Write down as many *physical properties* of ethanol as you can.

2. What was the gas evolved when sodium was added to ethanol? What information have you about the other compound formed during this reaction?

3. Does ethanol react with (a) acidified potassium dichromate(VI), (b) a solution of potassium iodide and sodium hypochlorite? Explain your answers.

4. Did you notice anything characteristic about the compound produced in (e)? What do you think these types of compounds could be used for?

Chemical Properties of Ethanol

Ethanol undergoes a variety of reactions, mainly due to the presence of the —OH functional group in the compound. However in some cases, e.g. with chlorine, the hydrocarbon radical (C_2H_5—) also takes part in the reaction.

Combustion What colour was the flame when ethanol burnt in air? Did a similar flame result when you ignited the first 1 or 2 cm^3 of distillate in Experiment 12.7? Write an equation for the combustion of ethanol in a plentiful supply of air.

Reaction with sodium When sodium reacts with ethanol one of the products is hydrogen, the other product is sodium ethoxide, C_2H_5ONa. Write an equation for the reaction of sodium with ethanol. The evolution of hydrogen from an organic compound on addition of sodium indicates the presence of a hydroxyl group. The organic compound must be thoroughly dried before the reaction. Why is this necessary?

Reaction with acidified potassium dichromate(VI) What type of reagent is acidified potassium dichromate(VI)? Does this help you to explain why its colour changes from orange to green when warmed with ethanol (see Chapter 8)? In this reaction ethanol is oxidized to ethanal (acetaldehyde), according to the following equation:

$$\underset{\text{Ethanol}}{CH_3.CH(OH)H} + [O] \rightarrow \underset{\text{Ethanal}}{CH_3.C(=O)H} + H_2O$$

Ethanal has an unpleasant odour; did you notice it during your experiment? Further oxidation of the ethanal will result in the formation of ethanoic acid, according to the following equation:

$$\underset{\text{Ethanal}}{CH_3CHO} + [O] \rightarrow \underset{\text{Ethanoic acid}}{CH_3COOH}$$

Did you smell this acid?

The 'iodoform' reaction When ethanol is warmed with a solution of potassium iodide and sodium hypochlorite, a yellow precipitate of triiodomethane (iodoform), CHI_3, is produced by a complicated series of reactions. This reaction, commonly called the *iodoform reaction,* is given by compounds containing the structure

$$CH_3-C(=O)-X$$

(where X = hydrogen, or an alkyl or phenyl group, or a carboxylic acid group). Compounds which on oxidation produce this structure (e.g. ethanol) also give the iodoform reaction. This is a useful method of distinguishing between ethanol and methanol. Why will methanol not give this test?

Dehydration of ethanol Ethanol can be dehydrated to produce ethene either by using aluminium oxide or concentrated sulphuric acid (page 312). The temperatures at which these reactions take place are approximately 633 K (360 °C) and 453 K (180 °C) respectively. If however, the reactions are carried out at 533 K (260 °C) and 413 K (140 °C) respectively then the dehydration takes place differently and in both cases an ether, ethoxyethane (diethyl ether), is produced instead of ethene, according to the following equation:

$$\underset{\text{Ethanol}}{2C_2H_5OH(l)} \rightarrow \underset{\text{Ethoxyethane}}{C_2H_5-O-C_2H_5(l)} + H_2O(l)$$

We have referred to a similar compound before, CH_3-O-CH_3, which is isomeric with ethanol itself (page 200). This example illustrates a general principle of organic chemistry: that even a minor alteration of reaction conditions can often alter the type or percentage of products formed.

Reaction with organic acids Ethanol reacts with organic acids to form compounds called esters, and these compounds are considered under a separate heading.

Organic Acids and Esters

We have seen that oxidation of ethanol leads to the eventual production of ethanoic acid. This oxidation process also occurs naturally, by bacterial oxidation, and some wines often have a sour taste due to the formation of this acid. This acid, commonly known as acetic acid (from the Latin, acetum, 'vinegar'), is a typical organic acid. The characteristic group of an organic acid is the *carboxyl* group (—COOH) which is usually attached to an alkyl radical (R—), i.e. RCOOH. The radical in the case of ethanoic acid is the methyl radical, i.e. $CH_3.COOH$. *Carboxylic acids* show typical acidic properties such as turning blue litmus red, evolving hydrogen when treated with magnesium, zinc or iron and liberating carbon dioxide from carbonates. The structural formula of the carboxyl group is

$$-\mathrm{C}\begin{matrix}\nearrow\!\!\!\!/\,\mathrm{O} \\ \searrow \mathrm{OH}\end{matrix}$$

This structure may lead you to think that the compound contains an alcoholic grouping due to the presence of the hydroxyl group. Indeed, typical hydroxyl group reactions take place, for example with sodium. However, other alcoholic properties are absent and the compound mainly reacts as an acid, because the hydrogen atom can easily be lost from the hydroxyl group in the form of a hydronium ion in aqueous solution, i.e.

$$-\mathrm{C}(=\mathrm{O})\mathrm{OH} \rightleftharpoons -\mathrm{C}(=\mathrm{O})\mathrm{O}^- + \mathrm{H}^+$$

$$\mathrm{H}^+ + \mathrm{H_2O} \rightleftharpoons \mathrm{H_3O}^+$$

or overall:

$$\mathrm{CH_3COOH(l) + H_2O(l) \rightleftharpoons CH_3COO^-(aq) + H_3O^+(aq)}$$

in which the equilibrium lies well to the left-hand side. Thus the proportion of hydronium ions produced is small, and consequently the acid strength is low (Section 3.3).

One of the reactions you performed with ethanol was to heat it with a carboxylic acid (ethanoic acid) in the presence of a little concentrated sulphuric acid. Can you remember what happened? The equation for the reaction is:

$$\mathrm{C_2H_5OH(l) + CH_3COOH(aq) \rightleftharpoons CH_3COOC_2H_5(l) + H_2O(l)}$$

The compound produced is called ethylethanoate (ethyl acetate) and is a member of a group of compounds called *esters* which have the general formula RCOOR′, where R and R′ are radicals. An ester is formed when an alcohol reacts with an acid (usually an organic acid, but not necessarily so), i.e.

$$\text{Alcohol} + \text{Acid} \rightleftharpoons \text{Ester} + \text{Water}$$

e.g.

$$\mathrm{R'.OH + R.COOH \rightleftharpoons R.COOR' + H_2O}$$

The forward reaction is known as *esterification* and the backward reaction as *hydrolysis*. The equilibrium can be driven to the right by removing the water and this is one function of the sulphuric acid, its other function is to catalyse the reaction.

You will have found experimentally that the ester you produced was a pleasant smelling liquid which floated on the surface of a beaker of water. Esters of organic acids are well known for their pleasant odours and are therefore used in perfumes, flavouring essences, etc. It is an easy matter for you to prepare a variety of esters and compare their odours, by using different alcohols and carboxylic acids and subjecting them to the same reaction conditions as you used to prepare ethylethanoate. The fabric 'Terylene' is a polyester, i.e. formed of many ester molecules polymerized together.

This new ester gives your soap fabulous fragrance

We mentioned earlier that esters can be hydrolysed with water. If an alkali (e.g. sodium hydroxide solution) is used instead of water, the reaction takes place more quickly and the sodium salt of the carboxylic acid is formed and not the acid itself, i.e.

$$R.COOR' + NaOH \rightarrow R.COONa + R'OH$$

This reaction is known as *saponification,* and is used in the manufacture of soap.

Uses of Ethanol

A vast amount of ethanol is produced industrially because, in addition to its use in beverages, it has widespread applications as a solvent, a fuel, and in the manufacture of synthetic rubber, 'chloroform' etc. The ethanol usually encountered in school laboratories has been 'denatured' (i.e. rendered unfit for drinking!) and is called methylated spirit. It is produced either as (i) '*industrial methylated spirit'*, by adding methanol (5 per cent) to rectified spirit, or (ii) '*mineralized methylated spirit'* by adding methanol (9.5 per cent) to rectified spirit together with small amounts of unpalatable substances such as paraffin oil. The liquid is also dyed a purple colour.

Tests for Alcohols

It is usual to characterize an alcohol by the ester it forms, as these compounds have pleasant, distinctive odours. However, the esters formed by methanol and ethanol have very similar odours, and another means of distinguishing between them must be used. This is the iodoform reaction, which is given by ethanol but not by methanol (page 327). The latter by itself is not sufficient to characterize ethanol as positive results are also given by other compounds. Thus both of the tests mentioned must be used to characterize ethanol.

12.3 LARGE MOLECULES

To an inorganic chemist a 'molecule' of sulphuric acid, H_2SO_4, containing seven atoms is a large molecule. However, in organic chemistry even a molecule of glucose, $C_6H_{12}O_6$, containing twenty-four atoms is considered to be comparatively small, as certain molecules are known which contain tens of thousands of atoms. Such molecules, with a vastly different scale of molecular size, are called macromolecules and they are usually *polymers*.

Some polymers occur naturally (e.g. starch and cellulose), while others are man-made (e.g. Polythene and nylon). Most man-made polymers are called plastics and have given rise to a large rapidly expanding industry, which is discussed in more detail in the *Supplementary Text*.

How Large Molecules are Formed

The derivation of the word polymer (from the Greek, polys, 'many'; meros, 'part') gives a clue as to how such large molecules are formed. They are in fact built up by the linking together of many smaller units, called *monomers,* to form much larger units which may consist of long chains, sheets, or three-dimensional networks. This process is called *polymerization* and has previously been defined on page 314. Consider a particular monomer M; if two monomers combine together then a dimer M—M (M_2) is formed, if three monomers combine then a trimer M—M—M (M_3) is formed etc. When 'n' monomers (where n usually varies between fifty and 50 000) combine then a polymer (M_n) will be formed. Polymerization can be brought about in two ways:

(a) by the successive linking together of similar molecules or of pairs of dissimilar molecules, each of which contain one or more unsaturated bonds. This is called *addition polymerization*;

(b) by the successive linking together of two or more molecules which do not contain unsaturated bonds. In such cases small molecules (often water molecules) are eliminated and the process is called *condensation polymerization.*

Addition Polymerization

We have previously mentioned an example of this method on page 314, namely the polymerization of ethene to polythene. The reaction can be summarized as follows,

$$\text{'n' } CH_2{=}CH_2 \xrightarrow[\text{1000 atmospheres}]{\text{473 K (200 °C)}} (-CH_2-CH_2-)_n$$

Gas — + A trace of oxygen as a catalyst — White waxy solid

n can vary between 1000 and 10 000. By applying great pressure, the molecules of ethene are 'squeezed' together and undergo *addition* reactions with each other to form polyethene ('Polythene'),

$$H_2C{=}CH_2 + H_2C{=}CH_2 \text{ etc.} \rightarrow -CH_2-CH_2-CH_2-CH_2-CH_2-$$

Monomers — Polymer

Table 12.5 Some common addition polymers

Common name of polymer	*Name and formula of monomer*	*Manufacture of polymer*	*Some uses of polymer*
Polyvinyl chloride	Monochloroethene (vinyl chloride) $H_2C{=}CHCl$	Vinyl chloride is polymerized under pressure at 333 K (60 °C) using a catalyst (e.g. hydrogen peroxide)	Insulating covering for electrical cables, plastic mackintoshes, motor car upholstery, suitcase coverings, gramophone records
Polystyrene	Phenylethene (styrene) $C_6H_5(H)C{=}CH_2$	Phenylethene is stirred with a catalyst (e.g. benzoyl peroxide) until a viscous syrup is formed. The poly-merization is completed by heating this syrup	Heat insulator in buildings, as packaging material, spheres for model making. 'Expanded' form made by generating gas in syrup during polymerization
Perspex	Methylmethacrylate $H_2C{=}C(CH_3)COOCH_3$	Methylmethacrylate is stirred with a catalyst (e.g. benzoyl peroxide) until a syrupy liquid is formed. The syrup is then heated to complete the poly-merization	Glass-like transparency and resistance to weathering make it useful as a glass substitute, e.g. in aircraft windows, street light fittings, reflectors on motor vehicles, T.V. guard screens, etc.

This reaction is now usually carried out at atmospheric pressure using a special organo-metallic catalyst discovered by the German chemist, K. Ziegler, in 1953. The polythene formed by this method is sometimes known as 'low pressure' polythene and has the advantage over the polythene formed at high pressures of not softening in boiling water.

Other common polymers formed by addition polymerization are *polyvinyl chloride (P.V.C.), polystyrene,* and *Perspex* (Table 12.5). Draw a few units of each of these polymers in your notebooks.

Condensation Polymerization

Whereas the molecules linked together in the previous method were unsaturated molecules (i.e. molecules with 'spare' valencies) the molecules which take part in this type of reaction are usually saturated. They join up with one another because they possess groups which react together. For example, when a molecule containing a carboxylic acid grouping

$$-C(=O)OH$$

which is acidic, is brought into contact with a molecule containing an amine grouping

$$-NH_2$$

which is basic, a neutralization reaction occurs in which a water molecule is eliminated and a new compound formed, i.e.

$$\square-C(=O)OH + H_2N-\blacksquare \longrightarrow \square-C(=O)-N(H)-\blacksquare + H_2O$$

Such a reaction in which two or more molecules combine to form a new compound with the elimination of water or some other simple substance, is called a *condensation* reaction. As a result of the reaction a dimer has been formed. Why do you think this particular reaction cannot proceed past the dimer stage, i.e. why cannot the dimer formed react with other monomer molecules? How do you think the original monomer molecules could be modified so as to produce a large polymer molecule? The American chemist, W. H. Carothers, solved this problem in 1934 when he suggested that a large polymer would be formed if *two* of the same reactive groupings were present on each of the original monomers, i.e. each molecule should have *two* functional groups:

HOOC—□—COOH and H_2N—▨—NH_2

The first reaction would then produce

HOOC—□—CO—NH—▨—NH_2 + H_2O

which could undergo further reaction with other molecules of monomer because there are still unreacted functional groups of opposite types at the ends of the 'molecule'. Successive reactions would thus lead to the formation of a long polymeric chain compound, i.e.

HOOC—□—CO—NH—▨—NH—CO—□—~~ ~~—□—CO—NH—▨—NH_2

This is the principle used in the formation of the polymer *nylon*. As the polymer has been formed by a series of condensation reactions, it is known as a *condensation polymer*. Other common examples of condensation polymers are *Terylene, bakelite,* and *urea-formaldehyde* (Table 12.6). In the formation of Terylene an alcoholic group reacts with a carboxylic acid group and so a *polyester* is formed (Section 12.2).

Some Other Terms used in Polymer Chemistry

The following terms are among some of those commonly used when considering polymers.

Co-polymer A co-polymer is a polymer formed by the polymerization of a mixture of two or more different monomers.

Thermoplastic A thermoplastic (sometimes also called *thermosoftening plastic*) is a material which softens on heating and hardens on cooling, the process being capable of repetition. Such plastics are easily moulded into shape but are usually not very heat resistant.

Thermosetting material Thermosetting materials are those which can acquire plasticity on being heated but undergo a chemical change on continued heating, due to the formation of a molecular network structure between the long chain molecules. This produces a rigid three-dimensional network which prevents the plastic softening on reheating.

Table 12.6 Some common condensation polymers

	Common name and formula of monomers	*Common name and formula of polymer*	*Some uses of polymer*
(a)	$HOOC-C_6H_4-COOH + HO-CH_2-CH_2-OH$ 1,4-Dicarboxylbenzene (terephthalic acid) + Ethane-1, 2-diol (ethylene glycol)	$[-OC-C_6H_4-COOCH_2CH_2O-]_n$ Terylene	Mainly as a fibre (e.g. in clothing, and fishing lines)
(b)	$C_6H_5OH + H_2C{=}O + C_6H_5OH$ Hydroxy-benzene (phenol), Methanal (formaldehyde)	$[-CH_2-C_6H_3(OH)-CH_2-C_6H_3(OH)-CH_2-]_n$ Bakelite	To make buttons, knife handles, switches and distributor heads for motor vehicles, cameras, radio and telephone equipment
(c)	$O{=}C(NH_2)_2$, $O{=}CH_2$, $O{=}CH_2$ Carbamide (urea), Methanal (formaldehyde)	$[-N(C{=}O)-CH_2-N(C{=}O)-CH_2-N(C{=}O)-]_n$ cross-linked by C=O to the matching chain $-N(C{=}O)-CH_2-N-CH_2-N(C{=}O)-$ Urea–formaldehyde	Domestic kitchenware, e.g. plates, saucers, cups etc., control knobs, bottle caps

Plasticizer This is a substance used to increase the plasticity (flexibility) of a polymer. Camphor is a plasticizer for polymerized nitrocellulose, the camphor molecules penetrating between nitrocellulose molecules, enabling the latter to move more easily relative to one another.

Filler A substance (e.g. wood flour) which, when mixed with plastic, increases the bulk of the plastic and also its shock and heat resistance.

Preparation of Polymers

In the following two experiments you will prepare polymers by both of the methods we have discussed, and will become familiar with some of the terms mentioned.

Experiment 12.9

†Laboratory Preparation of Polystyrene ***

Apparatus

100 cm³ round bottom flask fitted with reflux condenser, Bunsen burner, asbestos square, tripod, gauze, oil bath, 100 cm³ and 250 cm³ beakers, measuring cylinder, 0–360 °C thermometer, spatula, filter paper, filter funnel, crucible, pipe-clay triangle.

Phenylethene (styrene), methanol, lauroyl peroxide.

Procedure

(a) Mix together about 10 cm³ of phenylethene and a pinch of lauroyl peroxide in a round bottom flask.

(b) Reflux the mixture over an oil bath (care) for about forty-five minutes, as shown in Figure 12.8. The temperature of reflux should be about 423 K (150 °C).

(c) Allow the contents of the flask to cool and pour the mixture into approximately five times its own volume of methanol in a beaker. Note the appearance of the polymer. It can be made to harden to a wax by agitating it with a spatula under the surface of a fresh supply of methanol.

(d) Isolate some of the wax formed in (c) by filtration and dry off the excess methanol using filter paper. Carefully heat a little of this solid wax in a crucible until it becomes pliable and then allow it to cool.

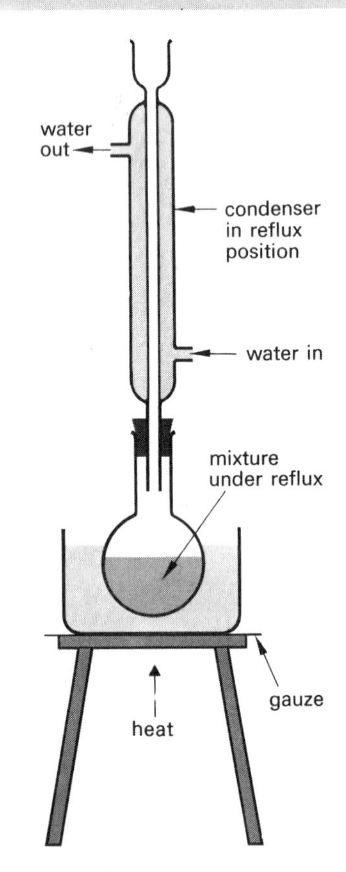

Figure 12.8 Apparatus used for refluxing mixtures

Points for Discussion

1. Are the physical states of the styrene and the final product the same? Which state contains the largest molecules? Do you think therefore that polymerization of the styrene has taken place?

2. What is the purpose of the lauroyl peroxide in this experiment?

3. From your observations in procedure (d), do you think polystyrene is a thermosoftening plastic or a thermosetting plastic?

4. Make a list of some common uses of polystyrene, by consulting books, magazines, advertisements etc. Is polystyrene used in your home? If so, for what purposes?

Experiment 12.10
†Laboratory Preparation of Nylon **

Apparatus
Two 100 cm³ beakers, 5 cm³ and 50 cm³ measuring cylinders, forceps, glass rod, crucible, pipe-clay triangle, access to fume cupboard.
Decan-1,10-dioyl chloride (sebacoyl dichloride), 1,6-diaminohexane (hexamethylene diamine), azobenzene, tetrachloromethane, approximately M sodium hydroxide solution. Gloves should be worn during this experiment, and if possible the operations should be conducted in a fume cupboard. Tetrachloromethane is toxic.

Figure 12.9 Preparation of nylon

Procedure
(a) Dissolve about 1 cm³ of decan-1,10-dioyl chloride in 10 cm³ of tetrachloromethane in a 100 cm³ beaker. Add a trace of azobenzene to colour this solution.

(b) In a separate beaker, make a solution of 1 g of 1,6-diaminohexane in 20 cm³ of M sodium hydroxide solution.

(c) Carefully pour the solution from (b) on top of the solution made in (a), taking care to avoid mixing of the two solutions.

(d) Using forceps take hold of the thin film of nylon which forms at the interface of the two solutions. Gently pull this film upwards and out of the beaker. Wrap this 'rope' of nylon around a glass rod and continue to remove nylon until the 'rope' breaks (Figure 12.9).

(e) Wash a small sample of your nylon with water, allow it to dry and then carefully heat the sample in a clean crucible to see if it melts.

Points for Discussion
1. Why do you think a trace of a colouring agent (azobenzene) was added to the solution prepared in (a)?

2. Why did the nylon only form at the interface of the two solutions? The area of this interface is comparatively small compared to the amount of nylon removed. How can this fact be explained?

3. From your observations in procedure (e) do you think nylon is a thermosoftening plastic or a thermosetting plastic?

4. Make a list of some of the common uses of nylon.

Natural Polymers

These are polymers formed in nature and include such materials as starch, cellulose, proteins, wool, silk, and rubber. When considering plastics we were concerned with the way in which these polymers were synthesized by man. In the next few pages we shall be investigating the way in which polymers are built up naturally. In the next experiment we shall attempt to break down one of nature's polymers and find out from which monomers it is made.

Breaking down Starch Molecules

As you are probably aware, starch is present in many foodstuffs such as potatoes and bread (wheat), and is an important source of bodily energy (page 343). Starchy foods are broken down in the body during digestion by enzymes such as the amylase often present in saliva. This is one way in which the starch molecules can be split up, another is to perform a similar hydrolysis, in a test-tube, using a dilute mineral acid (e.g. hydrochloric acid) instead of the enzyme. We shall use both of these methods in the next experiment.

Distinguishing between Starch and Sugars

Before we perform the experiment there are certain facts we must know. We will first attempt to find out what kind of products may be formed when starch is depolymerized. A very simple test is to chew a piece of bread for a while, so that you are actually breaking down the starch in your mouth. Is the resultant taste sweet or sour? Does this give you any clue as to the kind of substance formed? If you are studying biology you probably know already that starch molecules are broken down during digestion into various types of sugar molecules.

We can follow the rate at which the starch molecules are split up by testing small samples of the solution at regular intervals during the experiment for the presence of (a) starch and (b) sugars. These tests can be performed as follows:

Test for starch To a few drops of the solution under test add a small amount of an aqueous solution of iodine in potassium iodide. The appearance of a blue colour indicates the presence of starch.

Test for a reducing sugar To a few drops of the solution under test add a few drops of 2M sodium hydroxide solution until the solution is alkaline. To this mixture add a few drops of Fehling's solution (which is deep blue in colour), and carefully boil the mixture. The formation of a yellow precipitate which turns red on standing indicates the presence of a reducing sugar. This test is given by common sugars such as maltose ($C_{12}H_{22}O_{11}$), fructose ($C_6H_{12}O_6$), and glucose ($C_6H_{12}O_6$), but not by sucrose ($C_{12}H_{22}O_{11}$) as it is not a reducing sugar.

Carry out these tests on samples of starch and reducing sugars before you perform the main experiment. You will then know exactly what to expect when the tests are positive.

Experiment 12.11

† Breaking Down Starch Molecules *

Apparatus

Rack of test-tubes, one test-tube fitted with a bung through which passes a long (about 20 cm) piece of glass tubing, clamp and stand, Bunsen burner, asbestos square, teat pipettes.

Freshly prepared 1 per cent starch solution, iodine in potassium iodide solution, approximately 2M sodium hydroxide, approximately 2M hydrochloric acid, Fehling's solution, red litmus paper.

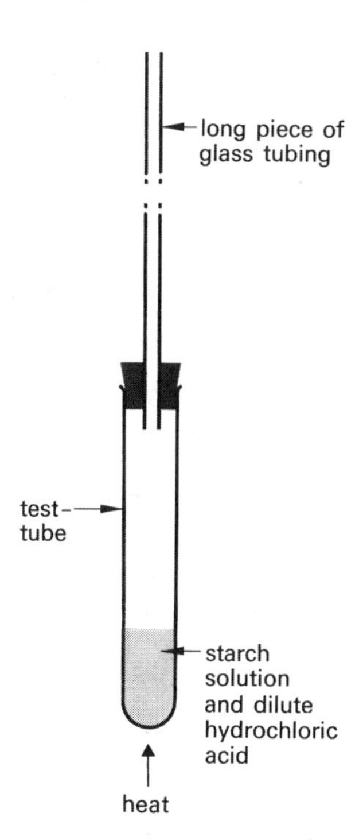

Figure 12.10 Hydrolysis of starch by dilute hydrochloric acid

Procedure

A. Hydrolysis due to enzymes present in saliva.

(a) Place about 10 cm^3 of freshly prepared starch solution into a test-tube and add to it about 1 cm^3 of saliva. Make sure the contents of the tube are well mixed.

(b) Allow the mixture to stand at room temperature and, using a teat pipette, remove two samples (five drops each) and place them in separate test-tubes. Test these two samples for the presence of (a) starch and (b) a reducing sugar using the tests previously outlined.

(c) Wash out the test-tubes and repeat procedure (b) at intervals of a few minutes.

(d) When the hydrolysis is complete, retain the remaining solution for a later experiment.

B. Hydrolysis due to the action of dilute hydrochloric acid.

(a) Pour about 10 cm^3 of freshly prepared starch solution into a test-tube and add to it about ten drops of 2M hydrochloric acid. Fit the test-tube with a bung through which is placed a long piece of glass tubing to act as a simple air condenser.

(b) Support the test-tube in a clamp (Figure 12.10), boil the solution for a few minutes and then remove two five-drop samples using a teat pipette.

(c) Test these samples for the presence of starch and a reducing sugar as in the previous hydrolysis.

(d) Boil the solution again for a few more minutes and repeat the two tests.

(e) Repeat the boiling until the hydrolysis is complete. Keep this solution for a later experiment.

Points for Discussion

1. Were starch molecules broken down in both experiments? What evidence have you to support your answer?

2. Were all the starch molecules broken down, or only some of them? Account for your answer.

3. Which hydrolysis broke down starch molecules more readily, the saliva enzymes or the dilute hydrochloric acid? Give reasons for your answer.

4. What type of compound is present in the solutions remaining at the end of each experiment?

You should have concluded that at the end of each experiment a solution containing reducing sugar was present. In the next experiment you will try to find out exactly which sugars these are.

The Identification of the Products of Hydrolysis

Experiment 12.12

† Chromatographic Identification of the Sugars left in Solution after Experiment 12.11 *

Apparatus

Bunsen burner, asbestos square, clamp and stand, teat pipette, sheet of chromatography paper, capillary tubes, paper clips, gas jar and cover, oven, spray bottle. Solutions from Experiment 12.11, approximately 2M sodium hydroxide, 1 per cent aqueous solution of glucose, 1 per cent aqueous solution of maltose, solvent, locating agent, red litmus paper.

Procedure

(a) Concentrate the solution from the first hydrolysis by evaporation.

(b) Treat the solution from the second hydrolysis in a similar manner after first neutralizing with sodium hydroxide solution.

(c) Take a rectangle of chromatography paper of such a size that when it is shaped into the form of a cylinder it stands freely in a gas jar. Test that it will do this and then open out the paper again.

(d) Rule a line about 2 cm from the bottom of the paper.

(e) By means of clean capillary tubes spot on to this line, at well spaced intervals, a drop of each of the following four solutions, (i) an aqueous 1 per cent solution of glucose, (ii) an aqueous 1 per cent solution of maltose, (iii) the enzyme hydrolysed solution (concentrated), (iv) the acid hydrolysed solution (concentrated). The spots should be about 0.5 cm in diameter, after spreading.

(f) Allow the spots to dry and then shape the chromatography paper into the form of a cylinder, making sure the edges do not overlap. Secure the cylinder by fastening the 'edges' with paper clips.

(g) Place the cylinder into a gas jar containing about 0.5 cm depth of solvent and place a cover over the jar.

(h) Allow the chromatogram to run until the solvent nearly reaches the top of the cylinder.

(i) Remove the completed chromatogram from the jar and allow it to dry. Then spray it with the locating agent, or dip it quickly into the locating agent.

(j) Allow the paper to dry in air and then heat it in an oven at 373 K (100 °C) for two or three minutes or warm cautiously above a gauze heated by a Bunsen flame. Note the colour and positions of any 'spots' produced.

Point for Discussion

Did both of the hydrolysed starch solutions contain a sugar? If so, which sugar or sugars did each solution contain? Explain your answer.

Structure of Starch You should have found that the hydrolysis of starch solution when catalysed by enzymes produced maltose, and when catalysed by an acid produced mainly glucose. These two sugars have the molecular formulae $C_{12}H_{22}O_{11}$ and $C_6H_{12}O_6$ respectively, and a molecule of maltose consists of two glucose molecules condensed together with the elimination of water:

$$\underset{\text{Glucose}}{2C_6H_{12}O_6} \rightleftharpoons \underset{\text{Maltose}}{C_{12}H_{22}O_{11}} + H_2O$$

or by simple representation,

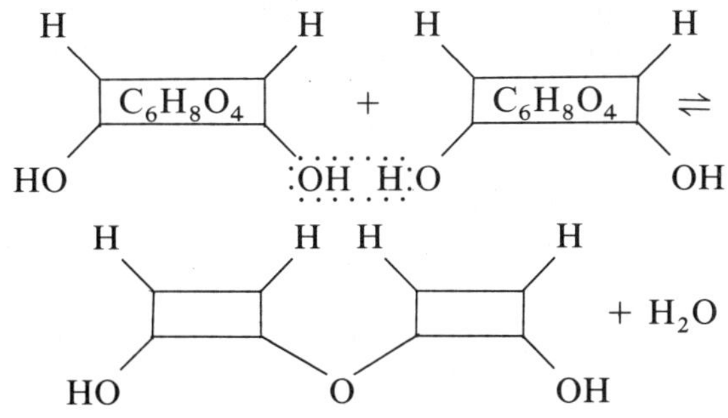

$$\mathrm{HO{-}[C_6H_8O_4]{-}OH} + \mathrm{HO{-}[C_6H_8O_4]{-}OH} \rightleftharpoons \mathrm{HO{-}[\;]{-}O{-}[\;]{-}OH} + H_2O$$

If we imagine many glucose molecules (monomers) condensing together then we would produce a very large molecule (polymer), i.e.

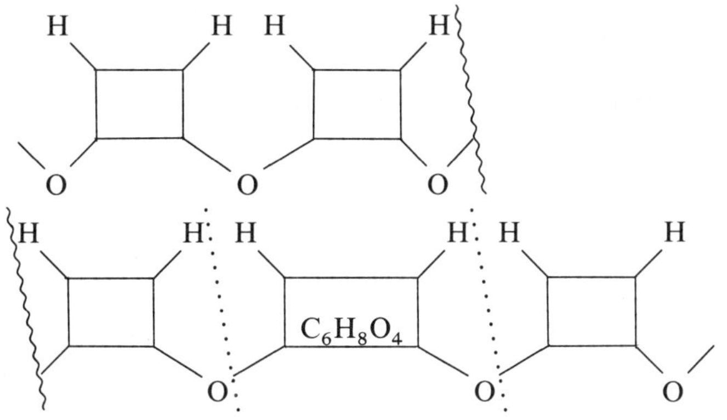

$C_6H_8O_4$

'Single unit'

A 'single unit' of this would correspond to the formula $C_6H_{10}O_5$ and the polymer would thus have the formula $(C_6H_{10}O_5)_n$. This polymer is *starch* and n has a value of about 50 000. Other polymers, (e.g. cellulose) have the same molecular formula, but the atoms are arranged in different ways.

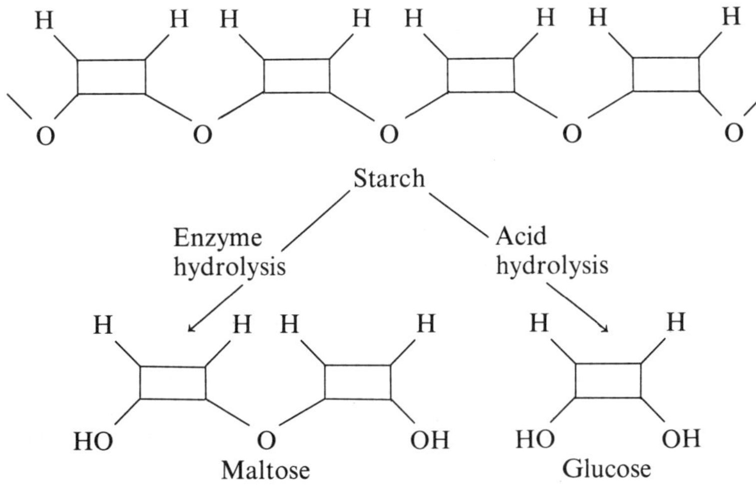

Starch

Enzyme hydrolysis

Acid hydrolysis

Maltose

Glucose

In humans, starch is broken down by various enzymes so that it can be digested. When the blood carries the sugars to the liver they are polymerized again to form glycogen (animal starch) which is stored in the liver. In between meals, or when necessary, this is broken down again and the sugars formed are used to produce energy. Why is it that humans cannot store sugars themselves, but must polymerize them to starches for storage?

Summary

In this section we have briefly considered the ways in which polymers are formed and have also referred to some of nature's polymers and man's efforts (a) to find from which monomeric units they are formed and (b) to produce identical polymers synthetically and in some cases to improve on nature's efforts.

The developments in this field of large molecules have already changed our lives and may do so even more dramatically in the future. Polymer technology could well in the foreseeable future produce inexpensive housing, inexpensive transport and even super-efficient all-plastic engines. Biochemists may not only be able to replace diseased parts of the body by synthetic ones, but may also be able to control disease, genetics, and the synthesis of life itself.

QUESTIONS CHAPTER 12

1. There are a large number of organic compounds because

A carbon atoms can join together to form long chains
B carbon reacts vigorously with many elements
C carbon has a variable valency
D carbon forms covalent bonds easily
E there are millions of living organisms

2. North Sea gas is mainly methane, which

A is heavier than air
B decolorizes Baeyer's reagent
C is used to fill meteorological balloons
D forms water vapour on burning
E forms addition compounds only

3. Alkanes
A are all gases
B have the general formula $C_nH_{2n+2}O$
C contain only carbon and hydrogen
D are usually soluble in water
E are usually reactive compounds

4. A sample of oil was vaporized and the vapour passed over a heated catalyst; a gas formed which was found to decolorize bromine water. The gas

A is ethane
B contains ethane
C contains a saturated hydrocarbon
D contains an unsaturated hydrocarbon
E contains a mixture of alkanes

5. The passage of gaseous ethene over a catalyst at high pressures to form a white flexible solid is called

A condensation
B neutralization
C esterification
D polymerization
E addition

6. Large molecules can be built up by the combination of a number of smaller molecules. These smaller molecules are called

A polymers
B allotropes
C isomers
D dimers
E monomers

7. Briefly explain what is meant by (a) a hydrophilic compound, (b) a saturated compound, (c) an unsaturated compound.

8. Give *two* uses for each of the following compounds (a) ethene, (b) ethyne, (c) ethanol.

9. A mixture of 400 cm^3 of a gas A, and 400 cm^3 of hydrogen reacted completely in the presence of a catalyst to give 400 cm^3 of a new gas, B. When 400 cm^3 of this new gas B, were mixed with 400 cm^3 of hydrogen in the presence of a different catalyst, 400 cm^3 of ethane were produced. (All volumes measured under the same conditions of temperature and pressure.)

A liquid, D, was produced when B and bromine reacted together. Addition of another catalyst to B gave a solid, E, which did not react with bromine and had a molecular mass of about 20 000. Explain these results, identify A, B, D, E and give equations for the reactions.

How might polyvinyl chloride be obtained from A? (J. M. B.)

10. Describe how you would obtain a sample of almost pure ethanol from sugar.

Outline the steps by which ethanol is manufactured from petroleum oil. Name all the chemical processes involved.

Describe *one* chemical and *one* physical test to distinguish between pure water and pure ethanol. State the result of each test on each of these substances. (J. M. B.)

11. Outline some of the main physical and chemical differences between organic and inorganic compounds.

12. (a) Which homologous series of organic compounds can be represented by the following general formulae:

Series A $C_nH_{(2n+2)}$;
Series B C_nH_{2n};
Series C $C_nH_{(2n+1)}COOH$;
Series D $C_nH_{(2n+1)}OH$?

(b) Give the name and structural formula of *one* compound in each series.

(c) Describe reactions by which

(i) a *named* compound of series B can be converted to a compound of series A;
(ii) a *named* compound of series D can be converted to a compound of series B.

(d) Name an important natural source of compounds of series A and give *two* industrial uses of such compounds. (C.)

13. Chloroethane can be formed from an alkene by an addition reaction and from an alkane by a substitution reaction. Explain the meaning of the terms alkene, alkane, addition reaction, and substitution reaction and write equations for the two reactions referred to in the first sentence.

Describe (a) *two* other addition reactions of an alkene, and (b) *one* reaction by which chloroethane can be obtained from ethanol. (C.)

14. Explain the meaning of (a) fermentation, (b) distillation. How are these processes used in obtaining ethanol from cane sugar? Explain the chemistry involved in the preparation.

Write down the structural formula of ethanol. Describe the preparation of ethyl ethanoate from ethanol, and give two other chemical reactions of ethanol, stating the conditions under which each takes place. (A. E. B.)

15. (a) Write down the empirical formula and the molecular formula of (i) ethene (ii) ethane.

(b) State two physical properties of ethene. Give equations for two reactions in which it takes part and name the products in each.

(c) Describe how ethene can be converted to polyethene.

(d) Give two chemical properties of methane which illustrate the fundamental differences between the properties of an alkane and those of an alkene, explaining what those fundamental differences are. (A. E. B.)

16. A hydrocarbon, A, and hydrogen in the presence of a catalyst gave B which, on treatment with more hydrogen and another catalyst, gave C. A and B reacted with excess of bromine to give D and E respectively. When B was dissolved in concentrated sulphuric acid and the product hydrolysed with water, it was found that ethanol was formed. Explain these results and give the formulae for A, B, C, D, and E, and names for A, B, and C.

How may vinyl chloride be prepared, and why is it such an important compound? (J. M. B.)

17. Explain with suitable examples the following terms:

(a) substitution, (b) addition, (c) polymerization, (d) esterification.

18. (a) When hydrochloric acid and silver nitrate solution are allowed to mix, reaction takes place immediately. However when ethanol and ethanoic acid are mixed an ester is formed very slowly. What explanations can you give for this difference?

(b) Write the structural formulae of the two gaseous hydrocarbons of molecular formulae C_3H_8 and C_3H_6. Given unlabelled supplies of these two gases, how would you (i) find out which was which, (ii) show experimentally that each gas contains hydrogen and carbon?

19. By drawing their structural formulae show the differences in structure between ethane and ethene. Explain how the bond between a carbon atom and a hydrogen atom in these compounds is formed.

By naming the reagents, stating the conditions, and writing an equation for each reaction, describe how ethene could be converted into (a) ethane, (b) chloroethane, (c) 1,2-dibromoethane, (d) ethanol. (J. M. B.)

13 Energy Changes in Chemistry

13.1 WHAT IS ENERGY?

Energy is easy to recognize but difficult to define. The nearest we can get to a simple definition is to say that energy is something that will move matter and so do work, and that all the many forms of energy are interconvertible.

Heat and electricity are familiar forms of energy and other examples are chemical, kinetic and potential energy. Each of these types of energy can be changed into any of the other forms. Thus the energy stored in coal (chemical energy) can be used when the coal burns (heat energy) to change water into steam and so drive the turbines in a dynamo (kinetic energy) to produce electricity (electrical energy). This in turn can be used to move a cable car up a mountain (kinetic energy) until it reaches the summit where it has energy due to its position (potential energy).

From earliest times man has used the stored chemical energy in fuels such as wood and peat to provide warmth, but until comparatively recently muscular effort was the main source of energy for any work that had to be done. (The effort expended by all the slaves during the twenty years required to build an Egyptian pyramid is approximately equal to the energy needed for two minutes of a space rocket launch.) The Industrial Revolution was due to the discovery that the chemical energy stored in coal could be eventually transformed to kinetic energy in machines, and today we use the energy stored in coal and oil to do all the heavy work in factories, to transport us in trains, ships and aeroplanes and to provide heat and light for our work and leisure.

Much of the energy from fossil fuels is converted into electrical energy which is vitally important for the modern way of life. Increasing use of electrical appliances, the needs of the developing countries, and the population explosion are producing an accelerating demand for electrical power and a consequent increase in the rate at which coal and oil are being consumed. Thus supplies of fossil fuels once thought to be limitless are in danger of being exhausted within the foreseeable future.

A great deal of research is being carried out to find alternative sources of power. Modern atomic power stations use the energy released during atomic fission to produce electricity, but ways of utilizing the tremendous stores of energy locked up in atomic nuclei have yet to be perfected. Further sources of energy being considered are fuel cells, and the sun. Although almost all our present sources of energy, from the food we eat to crude oil, are derived from the sun's energy by photosynthesis, the direct use of sunlight as a source of energy is insignificant. The problem is immediate and pressing; some new source of energy, or more efficient methods of using known sources must be discovered in the near future, otherwise civilization as we know it will come to an end.

Units

From the chemist's point of view, the majority of chemical reactions involve some observable heat change, and it is convenient to use units which are expressed in terms of heat energy. The calorie was formerly the unit of heat used in chemistry, and a calorie is defined as the quantity of heat needed to raise the temperature of one gramme of water through 1 K. The international unit of measurement of energy is the *joule*, and this is defined in heat terms as the energy required to raise the temperature of one gramme of water through 0.239 K. From this, $\frac{1}{0.239} = 4.18$ joules are equivalent to one calorie.

In this text we shall work in joules. 4.18 joules are needed to raise the temperature of one gramme of water through 1 °C.

Where Does Chemical Energy Come From?

As you know, elements tend to combine together to form stable compounds. In so doing they form bonds with each other, and when bonds are formed, there is always a release of energy. We cannot directly measure the actual energy within an atom, but only the *difference* in energy which occurs when the atom takes up a more stable state. This energy which an atom possesses but which is not measurable directly is sometimes referred to as its *intrinsic energy* (intrinsic = being within). Substances which have a high energy content are less stable than substances with a lower energy content. Chemical reactions are often processes whereby a substance forms a new material of lower intrinsic energy (i.e. becomes more stable) and gives out energy in the process. There are reactions where energy is absorbed, but these are rarely spontaneous. We can represent the combination of an atom A with an atom B by means of an energy level diagram.

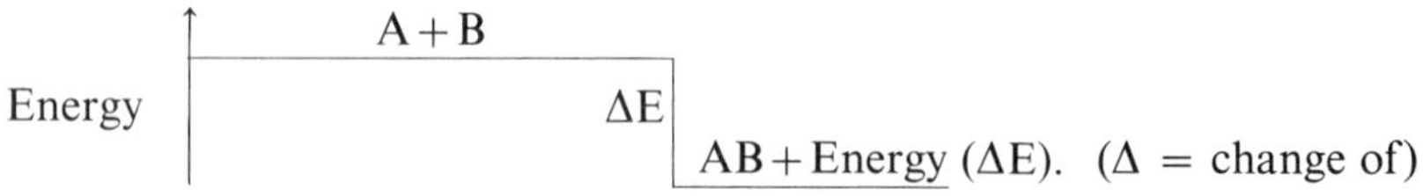

Energy ↑

A + B

ΔE

AB + Energy (ΔE). (Δ = change of)

We can see that the intrinsic energy within compound AB must be less than what was already present in A and B separately. In general, the source of any energy evolved in chemical reactions is the making of new, more stable bonds than are present in the starting material(s). It is important to realize that bonds may be broken as well as made during such processes. For example, A might have a giant structure and need to be broken into individual ions or atoms of A before they can react with B. The same may be true of B. However, if more energy is released by forming AB than is used in breaking up either or both A and B, the overall effect will be a release of energy and the reaction will be exothermic.

It does not follow however that it is easy for all exothermic reactions to occur, nor that they may be spontaneous. For example, the change diamond → graphite is exothermic, but is not spontaneous. This is because molecules will only react together when they have a certain minimum amount of energy, and this energy is called the *activation energy*. This minimum amount of energy needed by the reacting molecules differs with each reaction.

We see in Figure 13.1 for example, that the products have less energy than the reactants. Before they can be formed however, there is an initial

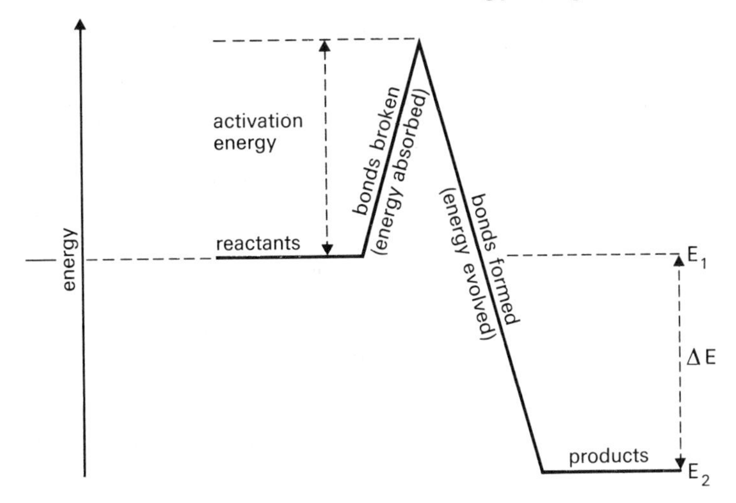

heat change = energy of products − energy of reactants

$\Delta E = E_2 - E_1$

Figure 13.1 An exothermic reaction

energy barrier or 'hill' which has to be reached and this requires expenditure of energy. During this absorption of heat (and this is why many reactions must be heated to initiate them) bonds will be broken and a greater number of molecules will attain the necessary energy for reaction to occur. Bonds are then made and energy is released, thus compensating for part or all of the energy absorbed. The height of the energy barrier indicates the activation energy. In Figure 13.1 the heat evolution exceeds the heat absorption so that overall there is an energy release and the reaction is exothermic. The energy change in the reaction (ΔE) is equal to the difference between the energy of the products (E_2) and those of the reactants (E_1).

Figure 13.2 illustrates the case where the energy of the products is higher than that of the reactants. Thus heat evolution on bond formation

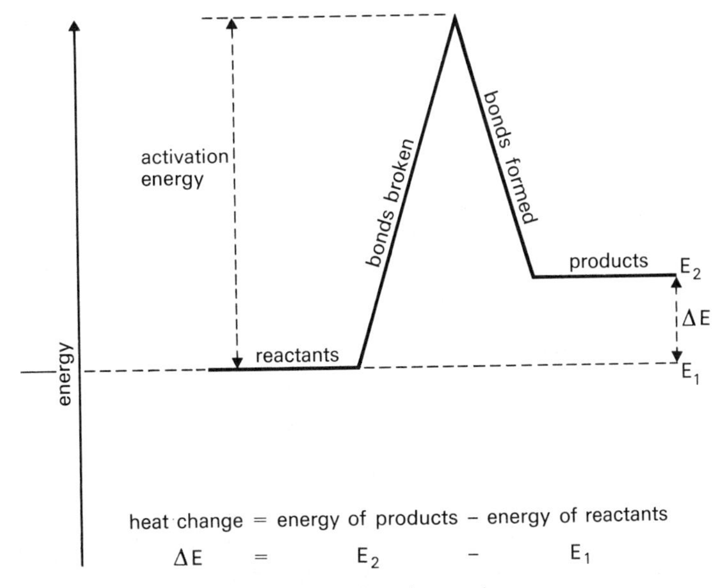

heat change = energy of products − energy of reactants

$\Delta E = E_2 - E_1$

Figure 13.2 An endothermic reaction

does not fully compensate for the activation energy, and the overall effect is heat absorption, i.e. the reaction is endothermic. Remember that bond making *releases* energy and bond breaking *absorbs* energy. The overall energy change will depend upon the number and 'strength' of the bonds concerned. In energy terms, more energy is required to break a mole of covalent bonds than a mole of weak bonds (such as intermolecular bonds). Conversely, more energy is liberated in the making of strong bonds than in the making of weak bonds.

A chemical reaction usually involves an energy change, although in some cases this energy change may not be obvious as (a) the two opposing effects (bond making and bond breaking) may be similar in magnitude, and (b) the reaction may be so slow that any heat liberated or absorbed over a short time period may be too small to be measurable.

The reaction of hydrogen and oxygen molecules to form liquid water will serve to illustrate a combination of energy changes (Figure 13.3).

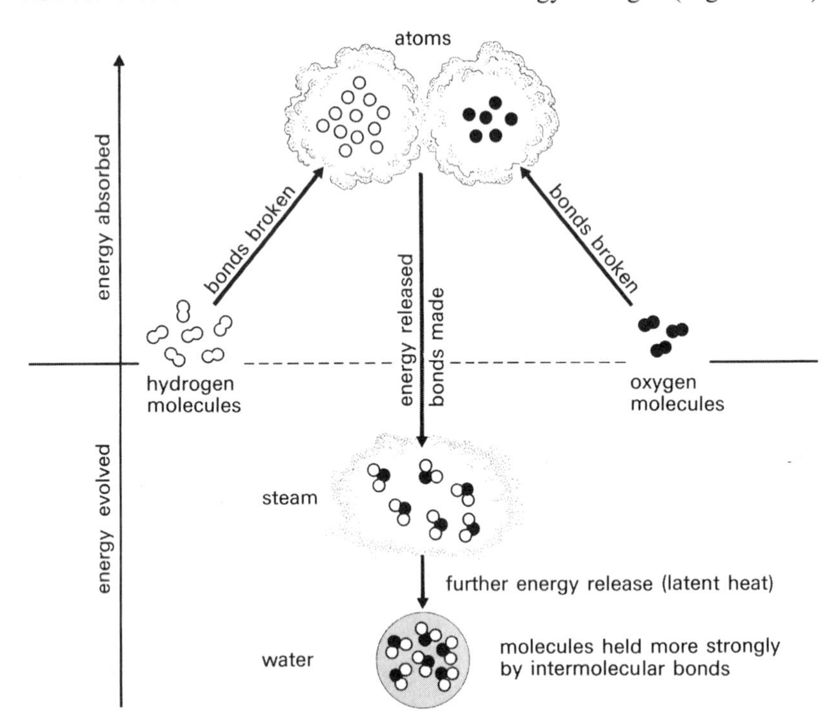

Figure 13.3 Energy changes in the formation of water from its elements

The observed effect is a release of energy, even though two of the stages involve bond breaking. This is because two single H—H and one double O=O bonds are being broken, and in this case less energy is used than is evolved when four new H—O bonds are formed.

Energy Changes Involving Melting Point, Boiling Point and Latent Heat of Vaporization

The melting point and boiling point are the temperatures at which the physical state of a substance changes. When a substance melts or boils, bonds must be broken as either change results in a more disordered state in which the particles have more freedom than before. When water is heated its temperature rises until, at standard pressure, it reaches 373 K (100 °C). At this temperature the water boils and liquid water changes to steam, but although the water may still continue to take heat energy from the heat source, there is no further rise in temperature. What then, is happening to this 'hidden' energy? What is it being used for? As you may know, it is being used to overcome the intermolecular forces between the liquid water molecules so that they separate and escape as a gas (steam) at the same temperature as the water. The energy required to bring about this change is known as the *latent heat of vaporization* and it differs for different liquids. As this energy is a measure of the degree of intermolecular attraction (i.e. the higher the intermolecular forces, the higher is the energy needed to overcome them), we can use it to compare the intermolecular forces in different liquids, but a valid comparison is only possible if we consider the *same number* of molecules of liquid in each case. Thus the molar heat of vaporization is defined as the energy required to vaporize one mole of a liquid at its boiling point. For example, the molar heat of vaporization of water is 41.2 kJ mol^{-1} and that for tetrachloromethane is 30.5 kJ mol^{-1}. As the physical change from a liquid to a vapour involves breakage of intermolecular bonds only, more energy is required to separate 6.023×10^{23} molecules of water at its boiling point than is needed to separate 6.023×10^{23} molecules of tetrachloromethane at its boiling point, because the water molecules are more strongly bound together than are the molecules of tetrachloromethane.

The ΔH Convention

Consider the reaction in which magnesium is burnt in oxygen (air). Once the magnesium is burning there is a very vigorous reaction and the breaking and making of bonds during this reaction must result in an overall release of energy. The energy is produced in the form of heat and light, and as the magnesium oxide cools to room temperature the energy is transferred to the environment. We do not know what the energy content of the magnesium or the oxygen was before the reaction, nor that of the magnesium oxide after the reaction, but we do know that as energy has been given out, the magnesium oxide must have less energy than the reactants, magnesium and oxygen (Figure 13.4). The energy or heat content of a system is called the *enthalpy* of the system and is indicated by the symbol H. We cannot measure it directly, only the change in H, and this is designated as ΔH (Δ = change of). This symbol refers to heat changes at constant pressure, i.e. under normal laboratory conditions.

Thus ΔH = energy of products (E_2)—energy of reactants (E_1).

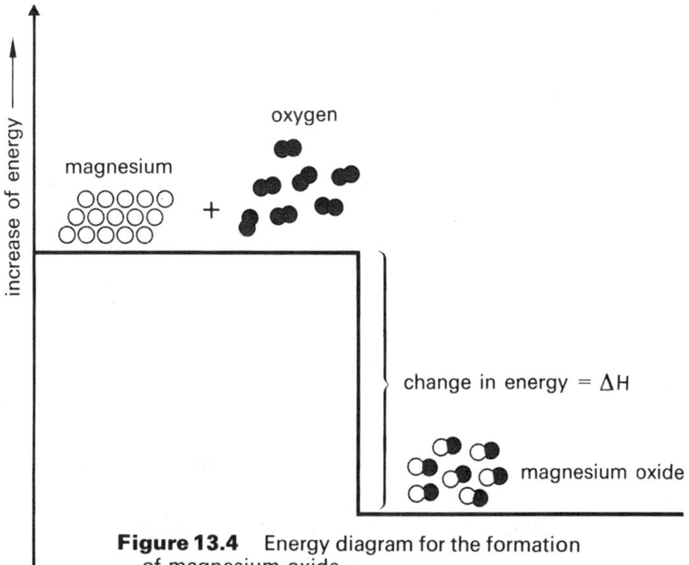

Figure 13.4 Energy diagram for the formation of magnesium oxide

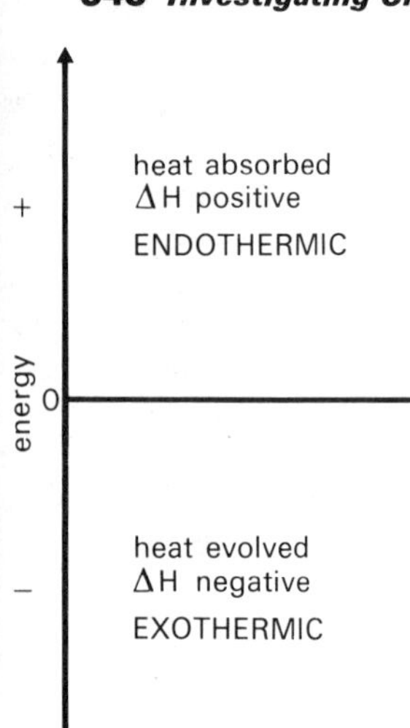

Figure 13.5 The ΔH convention

If E_1 is greater than E_2 then ΔH will be negative, and heat will be given out, i.e. the reaction is exothermic. If the energy of the reactants is less than that of the products ΔH is positive and heat will be absorbed, i.e. the reaction is endothermic. This is illustrated in an energy diagram such as Figure 13.5.

By convention ΔH is usually taken as the quantity of heat evolved or absorbed in a reaction which produces or uses one mole of a substance in the type of reaction *stated*. There are many ways of describing ΔH, depending upon the particular reaction concerned. Some of these will become more familiar to you as you work through the next section.

Note: in an exothermic reaction, the overall heat change is that needed to be taken out of the system in order to keep the temperature of the system constant, i.e. so that the products are returned to the starting temperature. The converse is true of an endothermic reaction.

Heat of Reaction This is the heat change which takes place when reaction occurs between the number of moles indicated by the equation. Note that this is one of the few cases in which ΔH is expressed in terms of the equation rather than per mole of product. The reaction between aluminium and oxygen is,

$$4Al(s) + 3O_2(g) \rightarrow 2Al_2O_3(s) \qquad \Delta H = -3360 \text{ kJ g-equation}^{-1}$$

Thus the heat of reaction in the formation of two moles of aluminium oxide is a release of 3360 kJ.

Heat of Precipitation This is the heat change which occurs when one mole of a substance is precipitated from solution. The reaction between silver nitrate and sodium chloride is,

$$AgNO_3(aq) + NaCl(aq) \rightarrow AgCl(s) + NaNO_3(aq) \qquad \Delta H = -x \text{ kJ mol}^{-1}$$

ΔH = −x kJ represents the heat liberated during the precipitation of one mole of silver chloride.

Heat of Combustion This is the heat change which occurs when one mole of substance is *completely* burnt in oxygen. The reaction between magnesium and oxygen is:

$$2Mg(s) + O_2(g) \rightarrow 2MgO(s) \qquad \Delta H = -y \text{ kJ g-equation}^{-1}$$

Therefore the heat of combustion of magnesium is *half* the overall energy change shown in the equation because two moles of magnesium are used, i.e. $\frac{1}{2}$ y kJ mol^{-1}.

Heat of Formation This is the heat change which occurs when one mole of substance is formed from its *elements:*

$$C(s) + O_2(g) \rightarrow CO_2(g) \qquad \Delta H = -394 \text{ kJ mol}^{-1}$$

The heat of formation of CO_2 is −394 kJ.

Heat of Neutralization This is the heat change which occurs when one mole of hydronium ions is neutralized by a base. The equation for the reaction between hydrochloric acid and sodium hydroxide solution may be represented as

$$HCl(aq) + NaOH(aq) \rightarrow NaCl(aq) + H_2O(l)$$

or ionically:

$$H_3O^+(aq) + OH^-(aq) \rightarrow 2H_2O(l) \qquad \Delta H = -57.5\ kJ$$

The heat of neutralization is -57.5 kJ. For sulphuric acid (a dibasic acid) however, the reaction is

$$H_2SO_4(aq) + 2NaOH(aq) \rightarrow Na_2SO_4(aq) + 2H_2O(l)$$

or

$$2H_3O^+(aq) + 2OH^-(aq) \rightarrow 4H_2O(l) \qquad \Delta H = -115\ kJ\ g\text{-equation}^{-1}$$

and the heat of neutralization ΔH is *half* the heat of reaction, because two moles of hydronium ions have been neutralized i.e. $\Delta H = -57.5$ kJ.

Heat of Solution

This is the heat change associated with the dissolving of a mole of substance (solute) in an appropriate (stated) quantity of solvent. It is important to state the quantity of solvent used, for when a solution is diluted with more of the solvent a further heat change may occur. For example, if one mole of copper(II) chloride is dissolved in 18 moles of water the heat evolved is 44 kJ. If the quantity of water is 48 moles, the heat evolved is 38 kJ. In most cases, especially elementary work, the quantity of solvent may not be stated, and this is to be taken as meaning that the dilution is such that addition of more solvent causes no further temperature change.

13.2 SOME DETERMINATIONS OF ENTHALPY CHANGES

Heat of Precipitation

Experiment 13.1

† To Determine the Heat of Precipitation of Silver Chloride ***

Apparatus

Small polythene bottle of about 60 cm³ capacity fitted with a one hole rubber bung carrying a 0–50 °C thermometer, 25 cm³ measuring cylinder.

Stoppered bottles of 0.5M solutions of silver nitrate, ammonium chloride, potassium chloride and sodium chloride. Allow these solutions to stand overnight in the laboratory to attain room temperature.

Procedure

(a) Measure out 25 cm³ of silver nitrate solution and pour it into the polythene bottle. Insert the bung fitted with the thermometer and note the temperature of the solution. It may be necessary to invert the bottle in order to cover the bulb of the thermometer with liquid. Handle the bottle so that the minimum quantity of heat is transferred from the hand to the liquid. Let the initial temperature = t_1 °C.

(b) Measure out 25.0 cm³ of 0.5M ammonium chloride solution and add it to the silver nitrate solution in the polythene bottle.

Quickly replace the bung in the bottle and shake *gently* to allow mixing of the solutions. Note the maximum temperature (t_2 °C).

(c) Repeat procedures (a) and (b) using (i) 0.5M potassium chloride and (ii) 0.5M sodium chloride in place of the ammonium chloride.

Specimen calculation

25 cm³ 0.5M silver nitrate solution react with 25 cm³ of 0.5M ammonium chloride. A molar solution contains one mole of solute dissolved in one litre of solution. Therefore 25 cm³ of 0.5M silver nitrate contain

$$\frac{25 \times 0.5}{1000} = 0.0125 \text{ mole } Ag^+$$

Similarly 25 cm^3 of 0.5M ammonium, potassium or sodium chloride solutions contain 0.0125 mole of NH_4^+ K^+ or Na^+ respectively. Now the total volume after mixing was 50 cm^3. The temperature change was $(t_2 - t_2)$ °C. Suppose in this reaction this was a rise of 3.0 °C. Assuming the specific heat of the solution to be the same as that for water, the heat change will be equal to:
specific heat of water × mass of water × change in temperature

$$= 4.18 \times 50 \times 3 \text{ joules}$$

$$= 627 \text{ joules}$$

The formation of silver chloride in this reaction may be represented

$$Ag^+(aq) + Cl^-(aq) \rightarrow AgCl(s)$$

Thus

one mole Ag^+ + 1 mole Cl^-
→ 1 mole AgCl

So

0.0125 mole + 0.0125 mole
→ 0.0125 mole

The formation of 0.0125 mole AgCl produces a heat change of 627 joules. The formation of one mole AgCl produces a heat change of

$$\frac{627 \times 1}{0.0125} \text{ joules} = 50.16 \text{ kJ mol}^{-1}$$

This means that when one mole of silver ions combine with one mole of chlorine ions to form one mole of silver chloride, 50.16 kilojoules of heat energy are liberated.

Points for Discussion

1. Calculate ΔH for each of the reactions.

2. What did you notice about your values for each reaction (to a close approximation)?

3. Can you suggest an explanation for your observations? It will help if you write an ionic equation for each reaction.

Heats of Combustion

We will now see how to determine the energy change when one mole of a substance is completely burnt in air or oxygen, i.e. its heat of combustion.

Experiment 13.2

†To Determine the Heat of Combustion of some Alcohols **

Apparatus

Small tin about 250 cm^3 capacity, small bottle of about 60 cm^3 capacity fitted with rubber bung and small glass tube, 100 cm^3 measuring cylinder, thermometer, 0–50 °C, clamp and stand, pipe cleaner, three asbestos mats.
Ethanol, methanol, propan-1-ol.

Procedure

(a) Clamp the tin in an upright position, so that it will be *just* above the flame from the wick.

(b) Measure out 100 cm^3 of water, pour it into the tin, and note the temperature of the water when it becomes steady.

(c) Half fill the small bottle with ethanol, and fit the bung and small glass tube into the bottle. Insert the pipe cleaner through the glass tube into the ethanol to act as a wick (Figure 13.6).

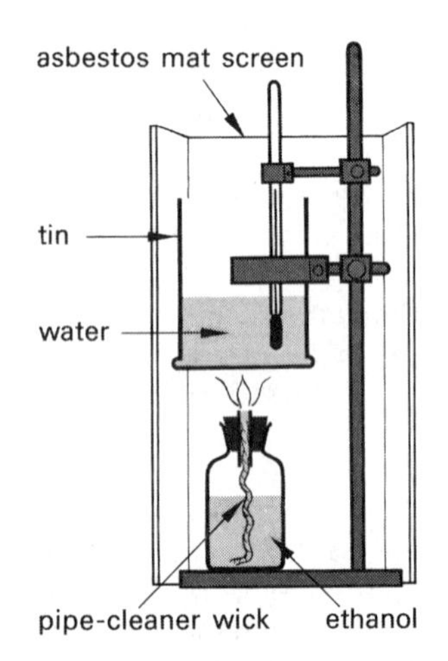

Figure 13.6 The determination of the heat of combustion of ethanol

(d) Weigh the bottle and its contents.

(e) Light the wick and immediately place the bottle beneath the tin. Arrange the asbestos mats around the tin and the lamp to shield it from draughts.

(f) Use the thermometer to stir the water and after the temperature has risen about 20 °C, extinguish the flame, note the temperature and reweigh the bottle.

(g) Repeat the procedure using (i) methanol and (ii) propan-1-ol.

Specimen calculation

Mass of bottle + ethanol (initially)	= 24.630 g
Mass of bottle + ethanol (finally)	= 24.400 g
Mass of ethanol burnt	= 0.230 g
Final temperature of water	= 33.0 °C
Initial temperature of water	= 18.0 °C
Rise of temperature	= 15.0 °C
Mass of water	= 100 g

Heat evolved
= Mass of water × specific heat of water × rise of temperature
= $100 \times 4.18 \times 15$ joules
= 6.27 kJ

Thus when 0.23 g of ethanol is completely burnt the heat evolved is 6.27 kJ. Therefore if 46 g ethanol (one g-mole) is completely burnt the heat evolved is

$$\frac{6.27 \times 46}{0.23} = 1254 \text{ kJ mol}^{-1}$$

The heat of combustion of ethanol is
$\Delta H = -1254 \text{ kJ mol}^{-1}$

Points for Discussion

1. Calculate the heats of combustion of the three alcohols. The results will be lower than the accepted values. Can you suggest any sources of error which could account for these differences?

2. The heat of combustion is the heat change when one mole of substance is completely burnt in oxygen. Why is it not necessary to burn one mole of substance in order to find this value?

3. Compare the heats of combustion of three alcohols. Their formulae are CH_3OH, CH_3CH_2OH and $CH_3CH_2CH_2OH$. Do you notice any pattern in your results? Look at the differences between the heats of combustion and compare them with the 'difference' between appropriate formulae.

4. Experiment 13.2 has dealt with the heat of combustion of liquid fuels, but it is more difficult to deal with solid materials, such as carbon, sulphur, coal etc. and the apparatus is more complicated. Industrially and in research laboratories use is made of the *bomb calorimeter*, in which a known mass of substance is burnt in excess of oxygen under pressure. The heat evolved is used to raise the temperature of a known mass of water, and hence the heat change may be calculated. In the school laboratory, use can be made of a simplified apparatus which you may see demonstrated: Figure 13.7 shows this being used for liquids.

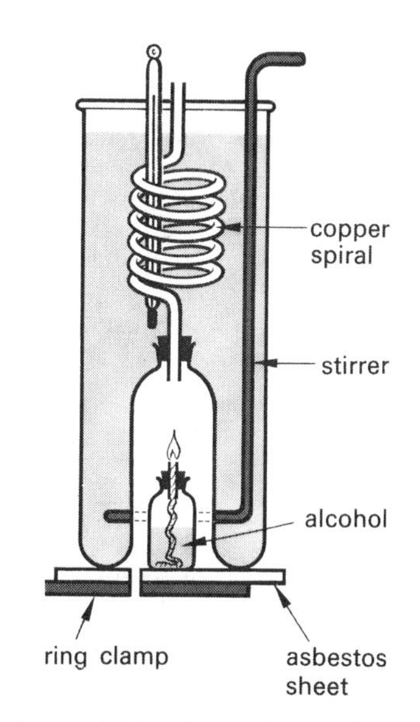

Figure 13.7 A simple calorimeter

5. When 4 g of sulphur is burnt in a bomb calorimeter, the heat evolved raises the temperature of 500 g of water by 17 °C. Calculate the heat of combustion of sulphur per mole of atoms (S = 32).

The Importance of the Determination of Heat of Combustion Values

Values of the heats of combustion of materials are important when related to the energy needs of industry or the energy needs of biological systems. Such values are referred to as *calorific values*. However, the quoted units may vary. For example, industrial values will deal with much larger quantities of fuels than we might use in a laboratory, and the units used are on a much larger scale. Two examples will serve to illustrate this point.

Gasoline	Gross calorific value	20 300 Btu lb^{-1}
Heavy fuel oil	calorific value	18 300 Btu lb^{-1}

(A British Thermal Unit (Btu) is the quantity of heat required to raise the temperature of one pound of water by 1 °F. 1 Btu = 1.05 kJ approximately.)

With respect to biological requirements such as the calorific value of food, the quantities involved are less so that values may be quoted in kilojoules $kilogramme^{-1}$, e.g.

Food	*Energy in kJ* kg^{-1}
Lettuce	672
Potatoes	3528
Bread	10 920

As you have already learnt, chemists quote the calorific value in terms of kJoules per mole.

Note that for heats of combustion, although bonds may be initially broken, more or stronger bonds are formed later, and heat is evolved, i.e. ΔH is always negative. Fuels are used to provide energy, and they normally do so by combustion processes. The greater the heat of combustion the more useful is the fuel likely to be, although other factors such as cost, toxicity, density, transport, availability, pollution etc. must be borne in mind. These factors are considered in *Further Topics*.

Heat of Neutralization

Heat changes also occur when solutions of acids and alkalis react together. In this case the heat change is referred to as the heat of neutralization. The reaction may be represented by the ionic equation:

$$H_3O^+(aq) + OH^-(aq) \rightarrow 2H_2O(l)$$

The heat of neutralization is usually expressed per mole of 'hydrogen ions' neutralized. It is important to realize that when dibasic acids such as sulphuric acid are used, one mole of acid is capable of liberating *two* moles of 'hydrogen ions', whereas one mole of hydrochloric acid is capable of liberating only one mole of 'hydrogen ions'. For the comparison to be valid we must use the same standard, and hence that chosen is one mole of 'hydrogen ions'. A simple determination of heat of neutralization may be made as follows.

Experiment 13.3

To Determine the Heat of Neutralization of Hydrochloric Acid by Sodium Hydroxide Solution *

Apparatus

Wax cup 'calorimeter' in suitably sized beaker, cotton wool, thermometer 0–100 °C, copper wire stirrer, 50 cm^3 measuring cylinder.

2M solutions of hydrochloric acid, sulphuric acid, ethanoic acid, and sodium hydroxide.

Procedure

(a) Pad the outside of the cup with cotton wool, and place it in the beaker.

(b) Measure out 50 cm^3 of 2M sodium hydroxide solution into the calorimeter. Place the thermometer in the calorimeter and note the temperature (t_1).

(c) Measure 50 cm^3 of 2M hydrochloric acid into the measuring cylinder. (If the cylinder is used to measure out the sodium hydroxide it should be washed out with water and then rinsed with 2M hydrochloric acid before measuring the 50 cm^3 of acid.)

(d) Pour the acid into the calorimeter, stir and note the highest temperature (t_2).

(e) Set out your results as follows.

Total volume of solution	100 cm^3
Initial temperature	t_1
Final temperature	t_2
Temperature difference	$t_2 - t_1 = T$
Total heat change	

$= $ Mass of water $\times$ specific heat water $\times$ temperature difference
$= 100 \times T \times 4.18$ joules
$= 418 \times T$ joules

50 cm^3 of 2M hydrochloric acid contain

$\frac{2 \times 50}{1000}$ moles $= 0.1$ mole HCl and 0.1 mole HCl is capable of liberating 0.1 mole of hydrogen ion. Therefore neutralization of 0.1 mole H_3O^+ liberates $418 \times T$ joules. Neutralization of 1 mole of H_3O^+ will liberate $\frac{418 \times T}{0.1}$ joules.

$\therefore$ Heat of neutralization of hydrochloric acid $\Delta H =$ $(-4180 \times T)$ J mol^{-1}

(f) Repeat the experiment using in turn, 2M solutions of nitric acid, sulphuric acid and ethanoic (acetic) acid.

Points for Discussion

1. Calculate the heat of neutralization for each acid used. Your results are approximations but nevertheless you should be able to draw some conclusions. Three of the results, those for hydrochloric, sulphuric and nitric acids should be very similar, but that for ethanoic acid will be different. Since the values for the first three are similar it must mean that the same reaction occurs in each case, i.e. the combination of hydronium ions from the acid with hydroxide ions from the alkali to form water.

$$H_3O^+(aq) + OH^-(aq) \rightarrow 2H_2O(l)$$

2. Hydrochloric, sulphuric and nitric acids are strong acids (Section 3.3). This means that they are, to all intents and purposes, fully dissociated into ions in dilute aqueous solution. Thus all the bond breaking required to separate the ions has been done as has all the bond making between the ions and the water. The ions are already hydrated before neutralization. This applies to the sodium hydroxide too. The only reaction which takes place on mixing is

$$H_3O^+(aq) + OH^-(aq) \rightarrow 2H_2O(l)$$

a bond making process for which the accepted value of ΔH is $= -57.5$ kJ mol $(H_3O^+)^{-1}$

3. Ethanoic acid is a weak acid, and exists in equilibrium with its ions:

$$CH_3COOH(aq) \rightleftharpoons H_3O^+(aq) + CH_3COO^-(aq)$$

During neutralization however the ethanoic acid will furnish *all* of its hydronium ions because, as each of these ions is removed, more dissociation will occur (Le Chatelier's principle, Section 14.2) until the acid is completely neutralized. In addition, the release of these 'new' free ions enables them to react with water, so that more bond making can take place. Thus ΔH for the overall reaction involving ethanoic acid is not just the energy change associated with the neutralization reaction

$$H_3O^+(aq) + OH^-(aq) \rightarrow 2H_2O(l)$$

The basic neutralization reaction is thus complicated by (a) further dissociation (bond breaking) and (b) hydration (bond

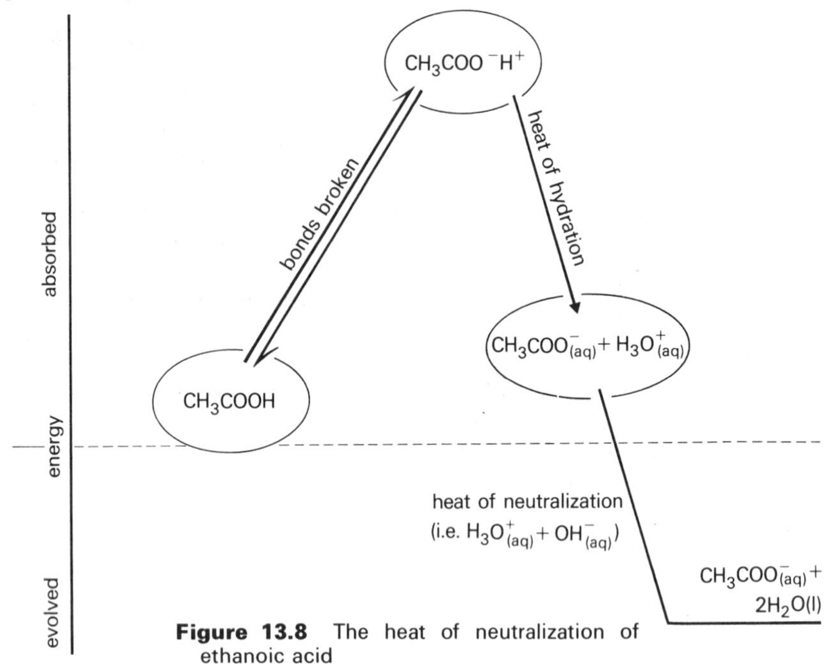

Figure 13.8 The heat of neutralization of ethanoic acid

making). The overall effect of these reactions in the case of ethanoic acid is to give an approximate value of

$$\Delta H = -55.2 \text{ kJ mol}^{-1}$$

It is just fortuitous that in this case the energy required to break the O–H bonds in ethanoic acid is almost exactly 'balanced out' by the energy given out when the ions so formed are hydrated. This results in the value for the heat of neutralization being fairly similar to that for the strong acids (–55.2 kJ mol^{-1} compared with –57.5 kJ mol^{-1}). However this is not always so. Figure 13.8 illustrates the energy changes occurring in this type of neutralization. Similar reasoning holds for other weak acids and also for weak bases.

4. Can you suggest why, if you spill a little concentrated sulphuric acid on your hand, you should immediately dilute it with water rather than neutralize it with sodium hydroxide solution?

Heat of Solution

The following experiment illustrates qualitatively the effect of dissolving various solutes in water.

Experiment 13.4
To Observe the Effect of Dissolving Various Solutes in Water *

Apparatus

Small polythene bottle of about 60 cm^3 capacity fitted with rubber bung carrying a thermometer (0–50 °C), spatula.

Anhydrous and hydrated copper(II) sulphate, ammonium chloride, deionized water.

Procedure

(a) Half fill the polythene bottle with deionized water.

(b) Replace the bung and record the temperature of the water.

(c) Remove the bung, quickly add about four spatula measures of anhydrous copper(II) sulphate. Replace the bung and after gently shaking the bottle, record the temperature.

(d) Repeat the procedure with (i) hydrated copper(II) sulphate and (ii) ammonium chloride as the solute.

Points for Discussion

1. Did you notice any difference in behaviour between the solutions? In one case an exothermic reaction occurs, and in the others an endothermic reaction.

2. Refer to lattice energies and hydration energies on page 94 before reading point 3.

3. You should have noticed in Experiment 13.4 that the dissolving of anhydrous copper(II) sulphate in water is exothermic, while the solution of hydrated copper(II) sulphate in water is an endothermic reaction. This may be explained by the fact that once the ions have become separated they become hydrated, as a result of new bonds being made with water molecules to form hydrated ions. If bond making (hydration) outweighs bond breaking (lattice energy) the overall effect is a temperature rise.

In the case of a hydrated salt these bonds with water are already formed, and the solution of the salt in water produces a temperature drop because energy is required to separate the hydrated ions, and no energy is regenerated by hydration, which has already occurred.

4. The next experiment gives a more quantitative idea of the energy changes which occur when a solute dissolves in water.

Experiment 13.5

†To Investigate the Heat of Solution of Sodium Carbonate in the Anhydrous and Hydrated Forms *

Apparatus

Waxed paper or polythene cup, beaker in which cup fits, cotton wool, thermometer 0–100 °C, copper wire stirrer, 50 cm³ measuring cylinder, watch-glass, desiccator.

Anhydrous and hydrated sodium carbonate, deionized water.

Procedure

(a) Pad the outside of the cup with cotton wool and insert in the beaker as in Experiment 13.3.

(b) Note the laboratory temperature and assume the initial temperature of the solids are the same as the room temperature.

(c) Measure 50 cm³ of deionized water into the 'calorimeter' and take the temperature of the water.

(d) Weigh out 1.06 g of anhydrous sodium carbonate on to a watch-glass.

(e) Using the thermometer as a stirrer (care) stir the water and add the anhydrous sodium carbonate. Note the maximum change of temperature.

(f) Weigh out 2.86 g of freshly powdered hydrated sodium carbonate and repeat the above procedure.

(g) Record your results in a table as follows.

Substance	Formula	Mass	Volume of water	Temperatures in °C Initial	Final	Difference ΔT
Anhydrous sodium carbonate	Na_2CO_3	1.06 g	50 cm³			
Hydrated sodium carbonate	$Na_2CO_3, 10H_2O$	2.86 g	50 cm³			

Points for Discussion

1. What is the reason for choosing these particular masses? (The formula mass of anhydrous sodium carbonate is 106, and that of the hydrated form 286.)

---

2. It is important to realize that when a solution is diluted with water a further heat effect may occur. The heat of solution then, is defined as the heat change which occurs when one mole of solute is dissolved in a *stated* quantity of water.

Specimen calculation

15.95 g of anhydrous copper(II) sulphate were dissolved in water and made up to 1 litre in a calorimeter. The temperature of the water was 17 °C and that of the solution 18.5 °C. What is the heat of solution of anhydrous copper(II) sulphate?

Both the water and the calorimeter gain heat, but we ignore the small quantity of heat taken in by the calorimeter.

Heat gained by solution

= Mass of water × specific heat of water × temperature change

= $(1000 \times 1.5 \times 4.18)$ joules

= 6.27 kJ

The molar mass of anhydrous copper(II) sulphate is 159.5 g. Now 15.95 g of anhydrous copper(II) sulphate dissolved in 1 litre of solution liberates 6.27 kJ heat energy.

Therefore 159.5 g of anhydrous copper(II) sulphate will liberate $\frac{6.27 \times 159.5}{15.95}$ kJ when dissolved in 1 litre of solution.

Heat of solution of one mole of anhydrous copper(II) sulphate is 62.7 kJ l^{-1}

$$\Delta H = -62.7 \text{ kJ mol}^{-1}$$

If we draw an energy diagram for this energy change it will look like this.

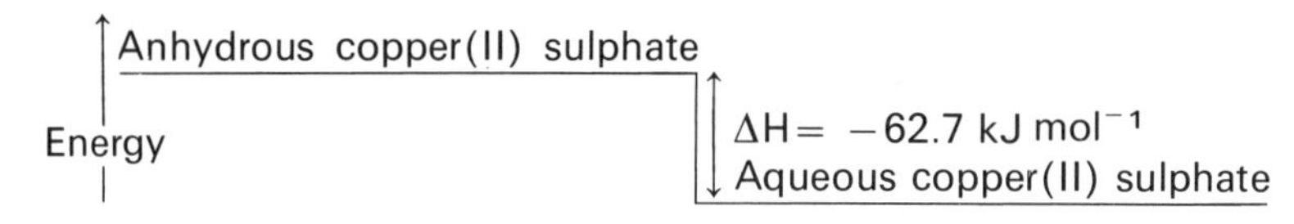

---

3. Using your results which were tabled in (f) calculate (a) the heat of solution of anhydrous sodium carbonate and (b) the heat of solution of hydrated sodium carbonate. Both these solutions are made up in 50 cm^3 of solution.

Summary

In this section we have discussed different types of energy changes which in themselves are heats of reaction, but which for convenience have been subdivided into heat of precipitation, heat of combustion, heat of formation, heat of neutralization and heat of solution. In nearly all cases the heat change is referred to one mole of product (or reactant) except in the case of heat of reaction which refers to the quantities reacting as indicated by the equation.

As all energy changes obey the Law of Conservation of Energy, then theoretically the energy of the system at the start of the reaction should equal the energy of the system at the end, plus or minus the energy lost to or gained from the environment (surroundings) in the form of heat. In practice it is not easy to measure accurately this gain or loss of heat by temperature changes, as there are bound to be some heat leakages unless very complicated apparatus is used. In the simple methods we have employed some sources of error are:

(a) inaccurate reading of the thermometer especially if graduated in intervals of 1 °C only.

(b) assuming the specific heat of the solution is the same as that for water.
(c) when solutions are mixed, there is a dilution of each solution by the other, and in some instances this dilution effect is accompanied by a temperature difference. This has been neglected here.
(d) the container itself will tend to absorb some heat and it may be necessary to allow for this. To do this, the mass and specific heat of the material of the container must be known.
(e) heat losses to the surroundings by inadequate insulation.

Can you suggest any further errors resulting from the apparatus you have used?

In this section we have considered the nature of energy changes and their relation to bond making and bond breaking. As might be expected, the energy required to break different bonds varies and depends upon the nature of the two elements making up the bond or bonds. It would be logical to expect that where multiple bonds (double or triple) are formed between atoms of the same element or atoms of two different elements, the combined strength of these bonds would be greater than for the single bond, and this is found to be so.

Table 13.1 Some average bond energies

Bond	*Energy* *(kJ mol^{-1})*
Br—Br	193
C—C	348
C=C	603
C≡C	840
C—Cl	340
C—H	414
Cl—H	432
Cl—Cl	243
Br—H	365
I—H	298
N≡N	950
O—H	466
Si—H	319
Si—O	454

Table 13.1 gives values for some average bond energies. These values must however be carefully interpreted. In Section 12.1 you found that ethene, which contains the C=C bond, very easily forms addition compounds, whilst ethane which contains the C—C bond is comparatively unreactive. It would appear that the C=C double bond is easier to break than the C—C bond, which is contrary to the values given. In fact, what happens is that only one of the C=C bonds is broken during addition and this is fairly easily accomplished with little energy change. Both bonds are *not* broken.

QUESTIONS CHAPTER 13

1. What is meant by the term combustion? Is combustion restricted to reactions involving oxygen? Why will sawdust burn more rapidly than a log of wood? Why will water (H_2O) and carbon dioxide (CO_2) not burn?

2. What is meant by the terms exothermic and endothermic reactions? Illustrate your answer by means of energy diagrams.

3. Define the term heat of neutralization. When a dilute solution of a strong acid is neutralized by a solution of a strong base, the heat of neutralization is found to be nearly the same in all cases. How can you account for this? How can you account for the cases where there is a difference?

4. Give six forms of energy. For each of the forms you give mention one example.

5. Name two examples of (a) a gaseous fuel (b) a liquid fuel and (c) a solid fuel. Can you think of any advantages or disadvantages each of these types has compared to each other?

The quantity of heat is = mass of water × specific heat of water × rise of temperature.
The specific heat of water is 4.18 joules $g^{-1}\,°C^{-1}$.

6. How many joules are required to heat

(a) 100 cm^3 of water from 10 °C to 20 °C
(b) 500 cm^3 of water from 20 °C to 50 °C
(c) 1 litre of water from 20 °C to 100 °C

7. How many joules of heat are lost when

(a) 100 cm^3 of water cool from 50 °C to 20°C
(b) 500 cm^3 of water cool from 70 °C to 30 °C

8. 40 litres of water are heated from 40 °C to 80 °C. Which of the following quantities of heat is required?

A Approximately 6.688 kJ
B Approximately 66.88 kJ
C Approximately 668.8 kJ
D Approximately 6688 kJ

9. For the reaction

$H_2(g) + Cl_2(g) \rightarrow 2HCl(g),\ \Delta H = -91\ kJ\ mol^{-1}$

This means

A the heat absorbed when 1 g-mole of hydrogen chloride is formed from its elements is 91 kJ
B the heat evolved when 1 g-mole of hydrogen chloride is formed from its elements is 91 kJ
C the heat of reaction of hydrogen and chlorine is 91 kJ
D the heat energy of hydrogen chloride is greater than for the heat energies of one molecule of hydrogen and one molecule of chlorine

10. Draw energy level diagrams for the following reactions.

(a) $P_4(s) + 5O_2(g) \rightarrow P_4O_{10}(s)$ $\quad \Delta H = -3005\ kJ$

(b) $B_2H_6(g) + 3O_2(g) \rightarrow B_2O_3(s) + 3H_2O(g)$ $\quad \Delta H = -2040\ kJ$

(c) $N_2(g) + 2O_2(g) \rightarrow 2NO_2(g)$ $\quad \Delta H = +66\ kJ$

11. State which of the following reactions are exothermic and which are endothermic.

(a) $H_2(g) + I_2(g) \rightarrow 2HI(g)$ $\quad \Delta H = +50.1\ kJ\ g\text{-equation}^{-1}$

(b) $CO(g) + H_2(g) \rightarrow H_2O(g) + C(s)$ $\quad \Delta H = -129\ kJ\ g\text{-equation}^{-1}$

(c) $Ag(s) + \frac{1}{2}Cl_2(g) + aq \rightarrow AgCl(s)$ $\quad \Delta H = -61\ kJ\ mol^{-1}$

(d) $N_2(g) \rightarrow 2N(g)$ $\quad \Delta H = +470\ kJ\ g\text{-equation}^{-1}$

If these reactions were carried out in heat insulated containers would the temperature in each container, i.e. of the contents, be higher or lower after the reaction was complete?

12. The combination of sulphur dioxide and oxygen to produce sulphur trioxide is exothermic.

$2SO_2(g) + O_2(g) \rightarrow 2SO_3(g)$ $\quad \Delta H = -187\ kJ\ g\text{-equation}^{-1}$

(S = 32 O = 16)

If 128 grammes of sulphur dioxide react completely, how many moles of sulphur trioxide would be formed?

A 0.25 mole
B 0.5 mole
C 0.75 mole
D 1 mole
E 2 moles

13. If 0.2 mole of sulphur trioxide is formed in the previous reaction what would be the heat energy evolved?

A 9.35 kJ
B 187 kJ
C 18.7 kJ
D 37.4 kJ
E 4.675 kJ

14. The formula for methanol is CH_3OH. It was found in an experiment to determine the latent heat of vaporization that to turn 8 g of methanol at its boiling point into vapour required 187 kJ of heat energy. What is the molar heat of vaporization of methanol? (Hint: calculate first the molecular mass of methanol. You are then required to determine the heat energy required to convert this mass in grammes into vapour at its boiling point.)

(C = 12, O = 16, H = 1)

15. (a) Describe and write equations for

(i) one reaction of industrial importance in which heat is absorbed
(ii) one reaction of industrial importance in which heat is produced
(iii) two reactions in which light energy is absorbed

(b) What is meant by the calorific value of a fuel? A fuel such as coal and a food fuel such as bread are used in very different ways. Explain why the calorific value is an important characteristic of the fuel and food.

(c) Give the names and approximate composition (by volume) of two gaseous fuels. (C.)

16. Explain what is meant by an exothermic reaction. Give examples of exothermic reactions occurring between (a) a gas and a solid, (b) a liquid and a solid, (c) two gases. State the conditions under which these reactions take place and the equation for the reaction.

Outline one commercial process in which the basic reaction of the process is exothermic. (A. E. B.)

17. (a) Explain how producer gas and water gas are obtained. Give equations.

(b) Indicate the approximate composition of each gas in these two mixtures.

(c) State which of the two gases has the higher calorific value and explain why this is so.

(d) Excluding those already mentioned, name two liquid fuels and one gaseous fuel that contain carbon.

(e) Outline the large-scale production of one of the fuels given in (d). (A. E. B.)

18. (a) How do the processes of respiration and the burning of fuel (i) resemble one another, (ii) differ from one another?

(b) When we burn coal, we are making use of stored energy that came originally from the sun. Explain this statement. (C.)

14 Rate of Change. The Chemical Industry

14.1 RATE OF CHANGE

What is 'Rate of Change'? Why is it Important?

By rate of change in chemistry we mean the speed of a chemical reaction, but we have to be careful how we measure it. One simple way of deciding the rate at which a reaction takes place is to find the time it takes from start to finish. Strictly speaking we are not measuring the true speed of a reaction by such a method, for you will learn later that a reaction does not proceed at a constant rate; it is quicker at certain times than others and the overall time does not indicate the rate at a certain stage, but it does give some idea of the average speed throughout the reaction.

A more scientific way of investigating reaction rates is to find the actual speed of the reaction at a given moment of time. You will be using both methods in this section.

You may be wondering how we actually measure speeds of chemical reactions. As a reaction proceeds, the starting materials are used up and so their concentration decreases. Similarly the concentration of the products increases with time. Thus by measuring the rate at which the concentration of the reactants is decreasing, or the rate at which the concentration of the products is increasing, we can find the 'speed of the reaction'.

It is important that we do not interfere with a reaction when we take the measurements which will enable us to calculate the rate at which it is proceeding, as this may alter the rate. If possible, therefore, we should use such factors as colour changes, weight loss, the volume of a liberated gas, and so on, as a measure of the reaction rate. These changes can be observed without disturbing the reaction.

It should be obvious that the rates of reactions vary enormously. You have seen reactions between solutions in which a precipitate appears apparently instantaneously, yet on the other hand the rusting of iron may take several days. It is very important that chemists should be able to understand how they can influence the rates of reactions. If it is possible to speed up a process the advantages may be considerable; in industry it is more economical to use a reaction which rapidly produces a desired chemical than one which takes several days. On the other hand there are reactions, for example certain decompositions, which occur at a faster rate than is required, and in such cases chemists seek ways of slowing down the process. The purpose of this section is to establish how we may influence the rates of chemical reactions. Factors which we will consider are concentration, pressure, temperature, light, surface area, and catalysts.

Concentration Changes

Experiment 14.1
† The 'Iodine Clock' *

Principle
When all of the specified reagents are mixed together iodine is produced. This reacts with the sodium thiosulphate until the latter is used up when the excess iodine turns the starch blue, so that the appearance of a blue colour marks the formation of a fixed mass of iodine.

Apparatus
Five 100 cm^3 beakers, teat pipette, 150 × 25 mm test-tubes in rack, white paper or tiles, two measuring cylinders. Solution X (6 g potassium iodide l^{-1} and 7.5 g sodium thiosulphate l^{-1}), twenty volume hydrogen peroxide solution, dilute sulphuric acid, and freshly prepared starch solution.

Procedure
(a) Stand the beakers side by side on white paper or tiles. Add 20 cm^3 of solution X, 10 cm^3 of dilute sulphuric acid, and about ten drops of starch solution to each beaker.

(b) Prepare five test-tubes containing 30 cm^3, 25 cm^3, 20 cm^3, 15 cm^3, and 10 cm^3 respectively of the hydrogen peroxide solution. Where necessary make the contents of each tube up to 30 cm^3 with water.

(c) Add the contents of tube 1 to beaker 1, tube 2 to beaker 2, etc., *at the same time*. (You will need help from other groups). Record your observations.

Points for Discussion
1. Which of the reagent concentrations was varied in the experiment?
2. Which of the beakers was the first to show the formation of iodine? Was the hydrogen peroxide concentration in this beaker high, low, or intermediate?
3. Which of the beakers was the last to show the formation of iodine? Was the hydrogen peroxide concentration in this high, low, or intermediate?
4. What do you conclude about the effect of increasing the concentration of one of the reactants in this example?

Experiment 14.2
† The Reaction Between Calcium Carbonate and Hydrochloric Acid **

Principle
You will remember from earlier work that one of the products of this reaction is the gas carbon dioxide. What are the other products? There is no change in the total mass of reactants and products during a chemical reaction in which no substance escapes (why not?), but if a gas is evolved there will be a loss of mass as the gas escapes. We can use this loss in mass to investigate how the rate of the reaction varies with different concentrations of, for example, the acid.

Apparatus
Direct reading balance, 100 cm^3 measuring cylinder, graph paper, 100 cm^3 conical flask, stop-clock.
Marble chips, 2M hydrochloric acid, cotton wool.

Procedure
(a) Place 20 g of marble chips in the conical flask and loosely insert a plug of cotton wool in the neck of the flask. (This is to stop acid spray emerging from the flask later in the experiment.) Weigh the flask and contents.

(b) Dilute 20 cm^3 of 2M hydrochloric acid with water to produce 40 cm^3 of M acid.

(c) Remove the flask from the balance and reset the balance to take into account the extra mass of approximately 40 g which is about to be added to the flask. Take out the cotton wool plug, quickly pour the acid into the flask, start the stop-clock, and reinsert the plug. Replace the flask on the balance. (The mass of flask and contents at time 0 is, to a very close approximation, the mass of flask, marble chips, and cotton wool +40 g, as the density of M hydrochloric acid is very nearly 1 g cm^{-3})

(d) Record the mass of flask and contents every 30 seconds for about fifteen minutes.

(e) Set out your results in a table as follows.

Time (sec)	*Initial mass of flask and contents (M_1) (g)*	*Mass at given time of flask and contents (M_2) (g)*	*Mass of carbon dioxide formed = M_1–M_2 (g)*

(f) Plot a graph of mass loss (= mass of carbon dioxide formed) on y axis against time (x axis).

(g) Repeat procedures (a) to (f) but use 40 cm^3 of 2M (instead of M) acid. If possible plot the second graph on the same axes as the first but in a different colour.

Points for Discussion

1. If the reaction proceeded at a constant speed the graph would have been a straight line (why?). Obviously this is not so. At which stage of each reaction was the rate the fastest? How did you decide?

2. Try to explain why the graph is steeper at the beginning than at the end. Why does it eventually become a horizontal straight line?

3. What can you conclude about the effect of increasing the concentration of one of the reactants (hydrochloric acid) in this reaction? Note that to make a fair comparison you will have to compare the two graphs over the same time interval.

Experiment 14.3
† The Reaction Between Sodium Thiosulphate and Dilute Hydrochloric Acid *

Principle

When sodium thiosulphate and dilute hydrochloric acid are mixed a precipitate of sulphur is formed, and this can be used to follow the speed of the reaction. This illustrates a third way of investigating reaction rates.

Apparatus

100 cm^3 conical flask, 10 cm^3 measuring cylinder, 100 cm^3 measuring cylinder, stop-clock, white paper or tile.
Approximately 2M hydrochloric acid, solution of sodium thiosulphate (approximately 40 g l^{-1}).

Procedure

(a) Mark a pencil cross on the paper or tile so that when a flask stands on it the cross can be seen through the bottom of the flask.

(b) Measure out 30 cm^3 of the sodium thiosulphate solution and mix it with 20 cm^3 of water in the flask.

(c) Pour 5 cm^3 of the acid into the flask and, at the same time, start the stop-clock. Swirl the contents of the flask and then place it over the cross.

(d) Look down at the cross through the flask and as soon as the reaction has produced enough precipitate to obscure the cross stop the clock and note the time.

(e) Predict how the procedures in (f) will affect the time taken to obscure the cross, and compare your predictions with the actual results.

(f) Repeat procedures (a) to (d) but use, in turn, 10 cm^3, 20 cm^3, 40 cm^3, and 50 cm^3 of sodium thiosulphate solution, and add water where necessary to keep the volume constant at 50 cm^3 before addition of the acid.

Points for Discussion

1. Does this experiment further confirm your conclusions to Experiments 14.1 and 14.2?

2. Before chemical reactions can take place, collisions must occur between ions or molecules of the reacting substances. (These colliding particles must also have the necessary energy, see Activation Energy page 344). Try to use this idea to explain your conclusion about the way in which concentration affects reaction rates, and also to explain why most reactions slow down as they proceed.

Surface Area

Many chemical reactions take place between a solid and a liquid, a liquid and a gas, or a solid and a gas. The reactants in such cases are in two distinct phases and the system is said to be *heterogeneous*. Systems in one phase (e.g. liquid and liquid, or gas and gas) are said to be *homogeneous*.

We cannot easily vary the area of contact between the reactants in a homogeneous system unless we increase their concentrations, but we can in a heterogeneous system. Suppose you are going to eat 2 kg of potatoes and you have the choice of eating them as boiled potatoes, chips or crisps. Obviously the surface area of each of these varieties is different, even if the mass is kept constant. Your 'rate of eating' will be affected by your choice. It would be much more difficult to eat seventy packets of crisps than the same mass of boiled potatoes, because the surface area of the crisps is so much greater.

An increase in surface area can have the opposite effect. For example, granulated sugar is used in making fruit cakes, where long, slow cooking is essential, but finely divided (caster) sugar is used in sponge cakes because it dissolves more quickly and the cake is cooked in a shorter time. Similarly, boiled potatoes are cooked more quickly if they are first cut up into small pieces so that they provide a greater area of contact with the water. How do you think that an increase in surface area of a reactant will affect the rate of a *chemical* reaction? The next experiment should provide an answer.

Experiment 14.4

†The Reaction Between Powdered Calcium Carbonate and Hydrochloric Acid **

Apparatus and Procedure

Refer to Experiment 14.2 and then repeat the operations but use 6 g of roughly crushed calcium carbonate and 40 cm³ of M hydrochloric acid. Take readings as before, and if possible plot the results on the same graph as the original one but in a different colour.

Points for Discussion

1. Compare the curve for the reaction with powdered calcium carbonate and M acid with that for the reaction between lump calcium carbonate and M acid.

2. What effect does an increase in surface area (i.e. more finely divided particles) have on the reaction rate? Was your prediction correct? Can you explain this in terms of particle collisions?

3. Why did the three graphs become horizontal at the same value on the y axis?

Temperature

Experiment 14.5

†To Investigate the Effect of Temperature Changes on Reaction Rates **

Aim

Devise a suitably modified procedure for either (or both) Experiments 14.1 and 14.3 so that you can investigate the effect varying temperatures have on reaction rates. Remember that it is possible to cool solutions as well as to heat them. Write up the experiment in the usual way and make a conclusion based upon answers to the following questions.

Points for Discussion

1. How do temperature changes affect reaction rates?

2. Can you partially account for this in terms of particle collisions?

3. A rise of 10 K approximately doubles the rate of many reactions (although this depends upon the activation energy). If a reaction takes 128 minutes at 303 K (30 °C), how long will it take at 373 K (100 °C)?

Catalysts

Catalysts have been referred to in other parts of the book (e.g. the preparation of oxygen, page 115) and you probably have some idea of what the term means. The following experiments should enable you to understand the properties of catalysts more fully.

Experiment 14.6

†The Catalytic Thermal Decomposition of Potassium Chlorate(V)***

Necessary Information

When potassium chlorate(V), $KClO_3$, is heated it eventually decomposes and produces potassium chloride and oxygen:

$$2KClO_3(s) \rightarrow 2KCl(s) + 3O_2(g)$$

This is a thermal decomposition. This reaction must be demonstrated, and a safety screen must be used.

Apparatus

Tripod, Bunsen burner, asbestos square, stop-clock, glass rod, spatula, three test-tubes (150 × 25 mm), splints, clamp stand(s), access to balance, filtration apparatus, oven.
Potassium chlorate, copper(II) oxide.

Procedure

(a) Add five spatula measures of potassium chlorate(V) to one of the tubes.

(b) Weigh accurately about one spatula measure of copper(II) oxide and record the mass. Mix the copper(II) oxide with five spatula measures of potassium chlorate(V) and place this mixture in the second tube.

(c) Place five spatula measures of copper(II) oxide in the third tube.

(d) Clamp the tube containing potassium chlorate(V) horizontally, place the Bunsen below the tube and immediately start the clock. Continuously test for the evolution of oxygen with a glowing splint. As soon as oxygen is evolved stop the clock and note the time. [Note: take care not to allow any carbon from the end of a splint to fall into the tubes.]

(e) Repeat (d) with the other two tubes in turn. Oxygen may not be evolved from both of them.

(f) After cooling add water to the tube which originally contained the mixture. Potassium chlorate(V) and potassium chloride are both soluble in water but copper(II) oxide is not. Stir and if necessary add more water until all the white solid has dissolved. Warm if necessary.

(g) Weigh a piece of filter paper and use it to filter the suspension from (f) Dry the paper and the residue of copper(II) oxide in an oven until the mass is constant. Record the result, calculate the mass of copper(II) oxide, and compare it with that of the copper(II) oxide used at the start of the experiment.

Points for Discussion

1. Which of the tubes produced oxygen?

2. Did copper(II) oxide alone liberate oxygen when heated?

3. Does copper(II) oxide help potassium chlorate(V) to decompose more easily? Explain your answer.

4. Does any copper(II) oxide appear to have been used up in the reaction? Explain your answer.

Experiment 14.7

†The Reaction Between Sodium Potassium Tartrate (Rochelle Salt) and Hydrogen Peroxide in the Presence of a Cobalt(II) Salt**

Principle

When hydrogen peroxide and sodium potassium tartrate are heated together a reaction takes place in which a gas is evolved. The rate of the reaction can be estimated by the rate of evolution of gas bubbles.

Apparatus

Spatula, test-tube (150 × 25 mm), measuring cylinder, glass rod, test-tube holder, Bunsen burner, asbestos square. Twenty volume hydrogen peroxide solution, sodium potassium tartrate, cobalt(II) chloride.

Procedure

(a) Pour 10 cm^3 of the hydrogen peroxide solution into the test-tube and add two spatula measures of sodium potassium tartrate. Stir until dissolved.

(b) Heat the contents of the tube until vigorous effervescence occurs. It may be necessary to heat the liquid almost to boiling.

(c) Cool the tube under the tap until the contents are again at room temperature. Add a few crystals of cobalt(II) chloride so that the solution is pink in colour.

(d) Warm the tube gently and slowly until vigorous effervescence occurs. Hold the tube away from the burner until no further colour change occurs. Record all your observations.

Points for Discussion

1. Does cobalt(II) chloride speed up the reaction between sodium potassium tartrate and hydrogen peroxide? It does not itself provide the gas.

2. Cobalt(II) ions are pink in colour but in other oxidation states cobalt ions have different colours. The other reactants and products are colourless. In view of the colour changes during the experiment which of the following seems to be the most reasonable conclusion?

A Cobalt(II) ions speed up the reaction between sodium potassium tartrate and hydrogen peroxide; they do not take part in the reaction and are unchanged at the end.

B Cobalt(II) ions speed up the reaction between sodium potassium tartrate and hydrogen peroxide; they take part in the reaction and are not reformed.

C Cobalt(II) ions speed up the reaction between sodium potassium tartrate and hydrogen peroxide; they take part in the reaction and are reformed.

3. Copper(II) oxide and cobalt(II) chloride are positive catalysts in these last two experiments. Consider your findings in both experiments before deciding which of the following is the most suitable definition of a positive catalyst.

A A positive catalyst changes the speed of a chemical reaction. It is unchanged chemically and in mass at the end of the reaction and so does not always appear to take part, although it actually does so and is returned to its original chemical state.

B A positive catalyst speeds up a chemical reaction. It is unchanged chemically and in mass at the end of the reaction, even though it may have taken part in the reaction.

C A positive catalyst speeds up a chemical reaction. It does not take part in the reaction and is unchanged at the end of the reaction.

4. *Negative* catalysts (inhibitors) slow down the rate of chemical reactions, but otherwise they have the same properties as positive catalysts.

5. As you have seen, manganese(IV) oxide catalyses the decomposition of hydrogen peroxide to water and oxygen. This reaction takes place, slowly, even in the absence of a catalyst and so negative catalysts are usually added to keep hydrogen peroxide solutions stable. Typical negative catalysts for this purpose are dilute acids or glycerine.

6. A situation which is probably more familiar is the use of lead tetra-ethyl ('anti-knock') in petrols. This is a negative catalyst which helps to stop pre-ignition ('knocking') which occurs when the petrol/air mixture in the cylinders ignites before it should do so. Similarly the use of rust inhibitors in anti-freeze solutions, chemicals to retard atmospheric oxidation in plastics, and antidotes in cases of poisoning are further familiar examples of negative catalysts.

7. A catalyst is a substance which *alters the rate* of a reaction and is unchanged chemically and in mass at the end. Positive and negative catalysts are specific types. (A positive catalyst offers an alternative reaction route with a lower activation energy, and conversely a negative catalyst raises the activation energy.) Sometimes in industry substances called *promoters* are used to make a catalyst more efficient.

8. If you are studying biology you will probably often refer to substances which speed up chemical processes in living organisms. Such substances are called *enzymes* and they are really a special group of catalysts.

Light

Light is a form of energy

As light is a form of energy it should come as no surprise to learn that it can speed up many chemical reactions and alter the course of others. Thus methane and chlorine react together only slowly in the dark, but in diffused daylight the reaction proceeds more rapidly (Section 12.1). In bright sunlight, however, a different reaction occurs; a mixture of the two explodes, producing carbon and hydrogen chloride instead of the substitution products obtained in diffused light (page 309). Those of you studying biology will be familiar with experiments which show that plants are unable to manufacture starch when kept in the dark.

You should be able to quote other examples of the way in which light can affect reactions. Why do you think that brown bottles are used for storing certain chemicals? Try to find examples of these.

Experiment 14.8
Colour Changes With Silver Halides *

Apparatus

Rack of test-tubes.
Dilute solutions of potassium (or sodium) chloride, bromide, and iodide, silver nitrate solution.

Procedure

(a) Pour a few cm^3 of potassium (or sodium) chloride solution into a test-tube. Add a few cm^3 of silver nitrate solution. Allow to stand.
(b) Repeat (a) using two other test-tubes and solutions of first bromide ions and then iodide ions.
(c) Record the colours of the precipitates over a period of time.
(d) Repeat (a) and (b) but immediately place the tubes in a dark cupboard. Remove and examine the tubes after ten minutes.

Points for Discussion

1. What are the precipitates formed in the tubes? Give ionic equations for the reactions.

2. Why does it not matter whether sodium or potassium salts are used?

3. What caused the colour changes? Explain your answer.

4. The colour changes are due to the formation of metallic silver. This reaction is utilized in photography. Films, papers, and plates are coated with silver halides mixed with gelatine. Areas of the material exposed to light produce silver. Development increases the formation of silver in these areas and fixing dissolves away the unused silver halides.

Rates of Reaction. A Summary

The rate of a chemical reaction can be increased in one or more of the following ways.

(a) Increase the temperature of the system. Care with enzymes; too high a temperature may destroy them.
(b) Increase the area of contact between the reactants (i.e. the surface area).
(c) Increase the concentration of one or more of the reactants. Note that in reactions between gases an increase in pressure will increase the concentrations of the gases. (Can you explain why this happens?) In such cases the rate of reaction will also be increased.
(d) Use a positive catalyst.
(e) Light also speeds up some reactions.

14.2 REVERSIBLE REACTIONS. EQUILIBRIA

Reactions Which go Both Ways

Although the principles you have learned in the previous section are generally applicable to most chemical reactions, they must be applied with care when two opposing reactions are proceeding at the same time at detectable rates.

Experiment 14.9

† The Effect of Alternate Additions of Acid and Alkali to Bromine Water *

Apparatus

Glass rod, 100 cm^3 beaker, white tile or paper, two teat pipettes, small measuring cylinder.

Bromine water, dilute sulphuric acid, dilute sodium hydroxide solution.

Procedure

(a) Pour approximately 10 cm^3 of the bromine water into the beaker and stand it on the tile or paper.

(b) Using one of the teat pipettes add dilute sodium hydroxide solution slowly, with stirring, until a colour change takes place.

(c) Using the other teat pipette, add dilute sulphuric acid, with stirring, to the solution obtained in (b) until a colour change takes place.

(d) Repeat (b) and (c) alternatively several times, using the same 'bromine' solution. Record all your observations.

Points for Discussion

1. You have seen reactions of this type before. When blue, hydrated copper(II) sulphate is heated, the water of crystallization is removed and a white powder is formed. If water is then poured on to the white powder heat is evolved and the original form of the chemical is reobtained. This can be repeated at will. Thus not only is it possible to write:

$$\underset{\text{Blue, crystalline solid}}{CuSO_4, 5H_2O(s)} \rightarrow \underset{\text{White powder}}{CuSO_4(s) + 5H_2O(g)}$$

but also:

$$\underset{\text{White powder}}{CuSO_4(s) + 5H_2O(l)} \rightarrow \underset{\text{Blue crystalline solid}}{CuSO_4, 5H_2O(s)}$$

2. Similarly, you learned in Section 9.2 that when water containing dissolved carbon dioxide reacts with calcium carbonate (e.g. limestone), soluble calcium hydrogen carbonate is formed which causes water to become temporarily hard. We can write

$$CaCO_3(s) + H_2O(l) + CO_2(aq) \rightarrow Ca(HCO_3)_2(aq)$$

But, in the same section, you learned that boiling such a solution of calcium hydrogen carbonate causes the reaction to go the other way, and that this same reverse process occurs more slowly in the formation of stalactites and stalagmites. Thus, also,

$$Ca(HCO_3)_2(aq) \rightarrow CaCO_3(s) + H_2O(l) + CO_2(g)$$

3. In Experiment 14.9 you saw another example of this kind. Reactions which can be made to go in either direction are said to be *reversible*. We do not write out the two reactions separately, as above, but instead we combine them.

$$CaCO_3(s) + CO_2(aq) + H_2O(l) \rightleftharpoons Ca(HCO_3)_2(aq)$$

$$CuSO_4, 5H_2O(s) \rightleftharpoons CuSO_4(s) + 5H_2O(g)$$

The sign $\rightleftharpoons$ means that the reaction is *reversible*. Can you think of any other reversible reactions you have encountered?

Can Both Reactions in a Reversible Reaction Take Place at the Same Time?

So far we have assumed that a reversible reaction proceeds completely either in one direction or the other. Is this always so? As both forward and backward reactions can take place it seems reasonable to suppose that they may do so *at the same time*, so that some kind of 'balance' is obtained between the two. If this is so such systems will always contain a mixture of both 'reactants' and 'products' because, although some substances are being used up by the forward reaction, they are also being continually reformed by the reverse reaction. Does this actually happen?

Experiment 14.10
† A Simple Experiment With Litmus Solution *

Apparatus
Three Petri dishes, white paper or tile, glass rod, small measuring cylinder, teat pipette.
Neutral (purple) litmus, dilute sulphuric acid, dilute sodium hydroxide solution, deionized water.

Procedure
(a) Pour 10 cm^3 of dilute sulphuric acid into Petri dish A and add approximately 1 cm^3 of neutral litmus solution. Stir.

(b) Wash the glass rod in deionized water. Pour 10 cm^3 of dilute sodium hydroxide solution into Petri dish B. Add approximately 1 cm^3 of neutral litmus solution and stir.

(c) Repeat (b) using Petri dish C but use 10 cm^3 of deionized water instead of the alkali.

(d) Stand Petri dish A on the paper or tile and place dish B on top. View the two from above. Compare the 'combined' colour of A+B with that in dish C.

Points for Discussion
1. In a neutral solution there are equal concentrations of the factors responsible for acidity (i.e. H^+ or, more accurately, H_3O^+) and for alkalinity (OH^-). The solution is then neutral because there is no excess of either of the two factors. There are only *two* forms of litmus. In the presence of acid the red variety is formed. In the presence of alkali the blue colour is formed. 'Purple' litmus is *not* a third variety.

2. The colour change of litmus is obviously reversible as the blue colour in alkaline solutions can be made to go red by addition of acid and back to blue again by adding more alkali.

$$\text{Acid form of litmus (red)} \underset{+\text{acid}}{\overset{+\text{alkali}}{\rightleftharpoons}} \text{Alkaline form of litmus (blue)}$$

Consider the information in point 1 and your observations before you answer the next question. What must be present, and in what proportions, in a neutral solution in order that the litmus appears to be purple? Explain your answer.

3. Does your answer to 2 suggest that it is possible to have a mixture of the products of both reactions in a reversible reaction? Does it seem reasonable to suppose that both reactions can proceed at the same time so that such a mixture is produced? Explain your answers.

A Theoretical Approach to Reversible Reactions

We can use the ideas learned in the last section to establish theoretically that both reactions in a reversible system can proceed at the same time, and that eventually a 'balance' state will be formed with a mixture of products and reactants.

Suppose we have a reversible reaction $A+B \rightleftharpoons C+D$. If, initially, we have only A and B then the only reaction which can proceed is the forward one. As time goes on the rate of this reaction will decrease. Why? However, the forward reaction will produce C and D as products, and as the reaction continues the concentrations of C and D will increase. It seems reasonable to suppose that the backward reaction will take place, and speed up, as more C and D are formed. Thus the reaction → is gradually slowing down and the reaction ← is gradually speeding up. Eventually the rates of the two reactions will become equal and 'balance' each other, and at this stage there will be a mixture of A, B, C, and D. We cannot alter the proportions of A, B, C, and D present in this mixture unless we alter the conditions of the experiment, e.g. by adding more substances or by changing the temperature.

Chemical Equilibrium

Equilibrium?

When both forward and backward reactions proceed at the same rate in a reversible system we say that a state of *equilibrium* has been reached. In the previous example we would say that A and B are in equilibrium with C and D; the system contains a mixture of reactants and products, and their proportions cannot be changed if the conditions of the experiment are kept constant.

It is not always possible for a reversible reaction to reach equilibrium in an open system, i.e. one in which reactants and products can be removed. If, for example, one or more of the products is constantly escaping, the reverse reaction cannot take place to any great extent and so equilibrium cannot be attained. This is why the system

$$Ca(HCO_3)_2(aq) \rightleftharpoons CaCO_3(s) + CO_2(g) + H_2O(g)$$

seems to go completely one way or the other. If a solution of calcium hydrogen carbonate is boiled in an open system the products carbon dioxide and steam can escape into the air, the backward reaction never gets established, and the forward reaction can go to completion. Why do you think the reverse process goes almost to completion in nature?

Experiment 14.11
To Establish the Idea of a Chemical Equilibrium ***

Apparatus
Two 100 cm³ separating funnels or tap funnels, two 100 cm³ beakers, 100 cm³ measuring cylinder, glass rod, white paper, access to balance.
Iodine, trichloromethane (chloroform), potassium iodide solution.

Procedure
(a) Pour 20 cm³ of trichloromethane into a beaker. Weigh out a few crystals of iodine and add them to the trichloromethane. Stir to dissolve.

(b) Weigh out a second similar portion of iodine and dissolve it in 20 cm³ of potassium iodide solution in a second beaker.

(c) Pour the solution from (a) into one of the funnels and *carefully* add 20 cm³ of *fresh* potassium iodide solution.

(d) Pour 20 cm³ of *fresh* trichloromethane solution into the second funnel and *carefully* add the solution from (b).

(e) Observe any colour changes in the two funnels over several minutes. Do not disturb the contents of the funnels.

(f) Shake each funnel vigorously until no further change appears to take place in the intensity of the colours when viewed against a white background. Compare the colours in the two funnels.

(g) Carefully drain off the lower layer from one of the funnels and replace it with a fresh 20 cm³ portion of trichloromethane. Shake vigorously until no further colour change takes place. Compare the intensity of the colours with those in the other funnel.

(h) Record all your observations.

Points for Discussion

1. What is the name given to liquids which do not mix together, such as trichloromethane and potassium iodide solution?

2. What is the colour of a solution of iodine in (a) trichloromethane, (b) potassium iodide solution?

3. How could you obtain a brown solution of iodine from a purple one, and *vice versa*?

4. Although the changes in the experiment are physical changes rather than chemical reactions, they still illustrate the principles of equilibrium. How can you account for your observations in (e) in terms of diffusion of iodine? Why does diffusion take place?

5. The 'reactions' can be expressed:

Iodine in trichloromethane $\rightleftharpoons$
(purple)
Iodine in potassium iodide solution
(brown)

How does the experiment show that the reaction is reversible?

6. Why were the 'reactions' (diffusions) rapid at first? In (f) and (g) a stage was reached in each funnel when no further colour change took place. What had then been established? Why was there no further colour change?

7. The colour intensities of each similar layer in both funnels at equilibrium (stage f) were identical because there was the same total amount of iodine in each funnel. This illustrates an important principle. The *same* equilibrium state has been reached in two *different* ways. In one funnel all the iodine was in trichloromethane to start with, but in the other funnel initially all of the iodine was in potassium iodide solution. Thus if there is a fixed number of atoms in a system, and the conditions are constant, the *same* equilibrium will be reached no matter which 'side' it is approached from.

8. The equilibrium in (f) was disturbed in (g) because some iodine was removed from the system (how?), although a *new* equilibrium was established later. The colours were then less intense because the new system contained less iodine. The *proportion* of iodine in each layer was the same, however, because the other conditions were kept constant.

When Equilibrium is Reached are There Equal Concentrations of Reactants and Products?

You may well consider that the answer to this question is 'yes' because at equilibrium the two reactions proceed at the same rate. However, it could be that the two reactions in a reversible system are only equal in rate when, say, 40 per cent of the reactants and 60 per cent of the products are present. The following simple experiment should help to provide an answer to the problem, and also to revise the ideas already established.

Experiment 14.12
† To Investigate Equilibrium Positions Using a Simple Analogy *

Apparatus

2 × 50 cm^3 measuring cylinders, two pieces of glass tubing a few centimetres longer than the height of the measuring cylinders.
(There should be several types of bore so that different groups can investigate different 'reactions'. Each group should have two dissimilar pieces of tubing.)

Procedure

(a) Pour 50 cm^3 of water into one of the measuring cylinders.

(b) Take one of the pieces of tubing (A_1) and place it in measuring cylinder A so that it touches the bottom. Place a thumb or finger tightly over the other end of the tube A_1, lift the tube out of the water and hold it over the other measuring cylinder B. Release your thumb or finger so that the water 'trapped' in the tube falls into B (Figure 14.1).

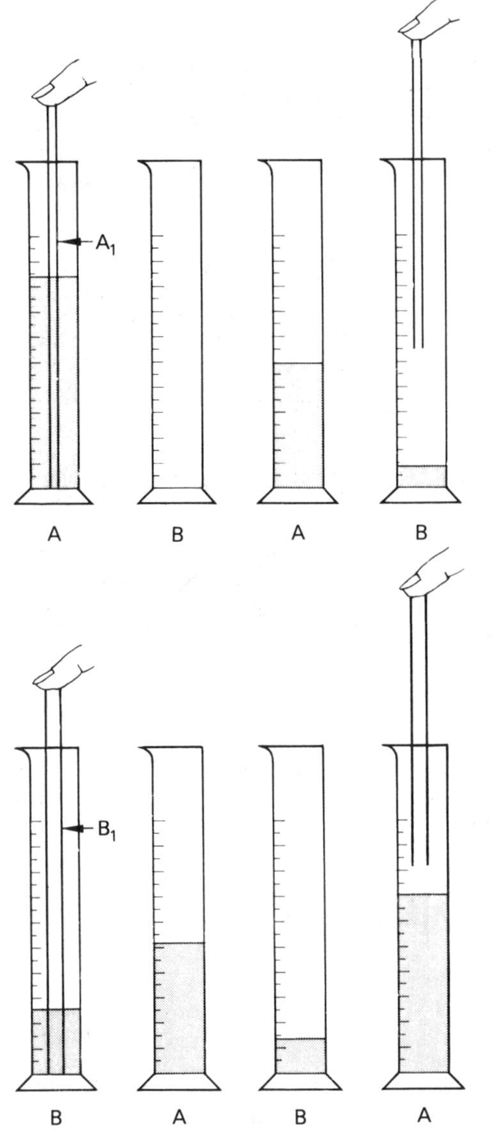

Figure 14.1 Obtaining an equilibrium

This procedure represents a chemical reaction; some of the contents of the first tube have 'reacted' (been used up) and become products in the second tube. The volume of water transferred represents the rate of the 'reaction', A → B.

(c) Using the second piece of tubing, B_1, transfer some of the water now in measuring cylinder B back into A by the same procedure as before.

This represents the 'reverse reaction', B → A. The volume of water transferred is proportional to the rate of the 'reaction' B → A. Note the volumes of water in each of the measuring cylinders.

(d) Repeat procedures (b) and (c) alternately and always in pairs (i.e. one forward reaction and one backward reaction = a pair) taking volume readings after each pair until no further change takes place in the volumes of the liquids in the tubes. This is when 'equilibrium' has been established; just as much water is being transferred from A to B as is being transferred from B to A. The rates of the two reactions are equal.

(e) Mark the level of the liquid in the two cylinders. Compare your volumes with those of other groups who have established equilibrium.

Points for Discussion

1. Plot (on the same graph) the volumes of water in cylinders A and B against the number of pairs of transfers. Comment on the shapes of your graph.
2. Why did the rate of the reaction from A to B slow down during the experiment? Why did the rate of the reaction from B to A increase?
3. Did you have equal volumes of reactants and products at equilibrium? Did other groups? Do you think that it is possible to have such a mixture at equilibrium? If so how would it be obtained using glass tubes?
4. Predict the relative volumes you would obtain at equilibrium if you were to approach it from the 'other side', i.e. starting with the same volume of water but in test-tube B (the products). Verify your prediction by doing the experiment.

Is a Chemical Equilibrium Dynamic or Static?

An equilibrium would be *static* (i.e. at rest) if the two opposing reactions stop completely at equilibrium so that no further change then takes place. In a *dynamic* ('in motion') equilibrium both reactions would reach the same rate and then continue to react at that speed, so that although there would appear to be no further reaction, both processes would in fact continue. This idea is feasible because, if you had continued both 'reactions' after establishing equilibrium in Experiment 14.12, there would have been no further change in the relative amounts of reactants and products, although both reactions were still proceeding.

It is quite difficult to prove that a certain state of equilibrium is dynamic because, as we have said, both forward and back reactions would be proceeding at the same time and at the same speed, and so there would be no outward sign that any reaction is taking place.

The situation may be compared to the movement of polystyrene spheres from one compartment of a box to another. (The barrier between the two compartments corresponds to the activation energy of the reactions.)

1.
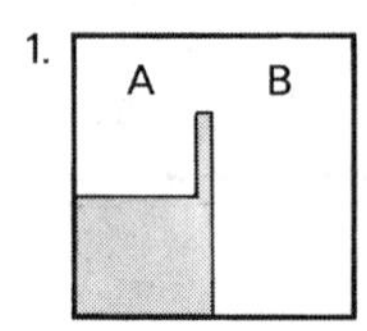

Suppose we have a glass fronted box divided into two unequal compartments A and B.

2.
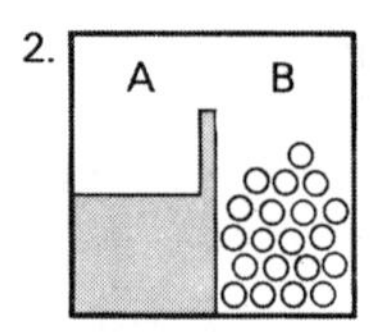

20 small polystyrene spheres are placed in compartment B.

3.
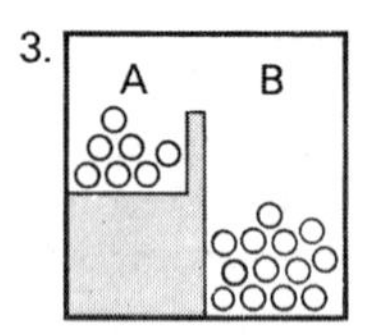

Suppose that after shaking vigorously for three minutes the spheres are arranged as shown; seven in A and thirteen in B.

4.
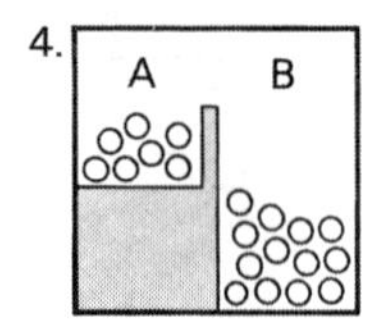

After shaking the box for a further three minutes the spheres may still be arranged 7:13, i.e. equilibrium would be established.

If the equilibrium in 4 is static the seven spheres in compartment A will be the same as the seven spheres in A after stage 3. On the other hand, if the equilibrium is dynamic, the seven spheres in A at stage 4 will be different from those in A at stage 3, because spheres will be constantly exchanged between A and B. How could you find out if the equilibrium is dynamic or static?

This problem is, of course, easy to solve in the case of the polystyrene spheres. You would just have to colour a few of them so that their progress could be followed.

1.
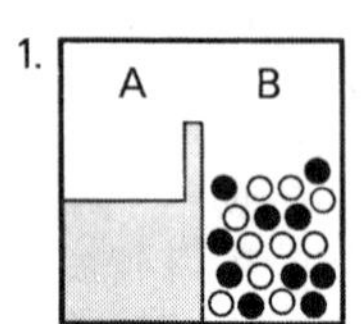

The same box as before but this time some of the spheres are coloured.

2.
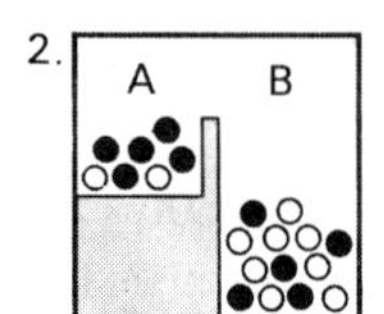

This could be the situation after shaking the box for three minutes. Five of the seven spheres in compartment A are coloured.

3.
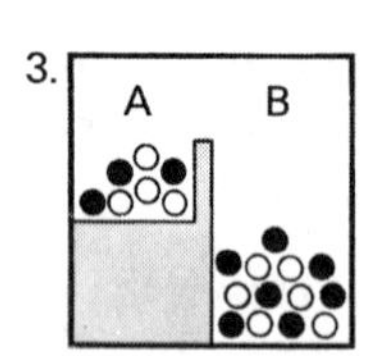

If, after shaking for a further three minutes, it is found that three of the seven spheres in A are coloured then the equilibrium is dynamic.

4.
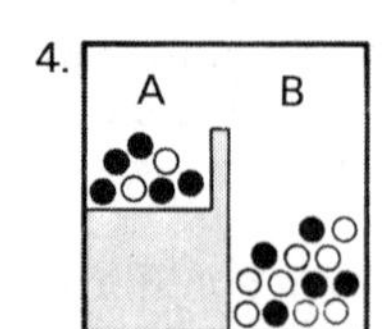

If the number of coloured spheres in A is unchanged by shaking, the equilibrium is probably static.

Unfortunately we cannot colour individual ions or molecules and so watch their progress during a reversible reaction, but there *is* a way of labelling these particles; we can use a radioactive isotope to 'label' one of the groups of particles taking part in a reversible reaction.

If we allow a system to establish equilibrium and then introduce a radioactive isotope of one of the reagents, we can follow its movements by means of a Geiger-Müller counter. If none of the radioactive material is found to be transferred from reactant to product (or *vice versa*) then the equilibrium is static. If, on the other hand, it is shown that radioactive particles have 'swopped over' after equilibrium has been established, the equilibrium must be dynamic; i.e. the reactions must be still continuing.

The equilibrium between a solid and its saturated solution is not usually regarded as chemical but this kind of system can be used to illustrate equilibrium and to show whether it is dynamic or static.

1. 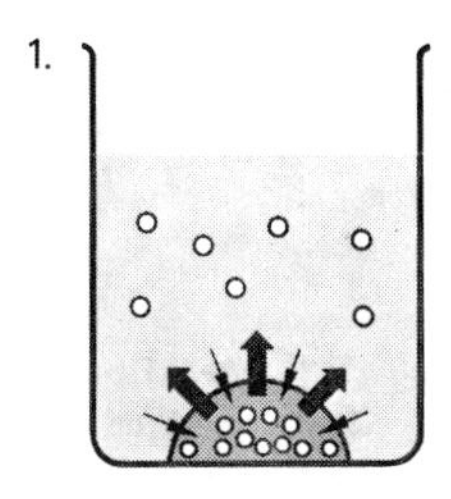

If we have a solid in contact with unsaturated solution more solid is dissolving than is being precipitated by the 'back' reaction.

2. 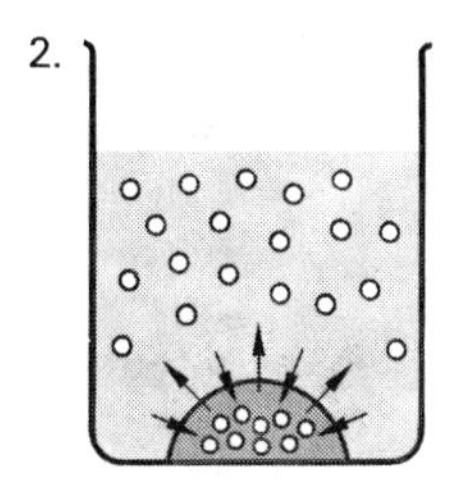

If we have a solid in contact with its saturated solution equilibrium has been established. The weights of dissolved and undissolved solid remain unchanged.

If the equilibrium is static all the particles of solute in the solution will remain there. If the equilibrium is dynamic some particles will be precipitated from the solution and an equal number of particles from the solid will dissolve.

3. 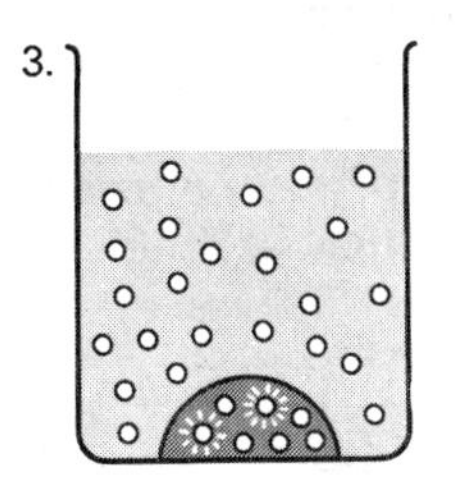

Some of the solid particles can be labelled by using a radioactive isotope, and put into contact with the saturated (non-radioactive) solution from 2.

After some time the saturated solution from 3 can be examined using a Geiger-Müller liquid counter. If the solution is radioactive then some particles must have left the solid and dissolved in the solution, even although it is already saturated. As there can be no change in the mass of the dissolved solid (why?) an equal number of solute particles must have been precipitated on to the solid. So although there is no apparent change, an exchange of particles from solid to solution, and from solution to solid, is taking place and the equilibrium is dynamic.

The equilibrium between a solution of lead(II) chloride and the solid salt is a suitable one for investigation. If a saturated solution of lead(II) chloride is formed in contact with undissolved solute the following equilibrium will be established:

$$PbCl_2(s) \rightleftharpoons Pb^{2+}(aq) + 2Cl^-(aq)$$

Natural lead is not radioactive so that the saturated solution should not be radioactive. If more solid lead(II) chloride containing a radioactive lead isotope is added to the saturated solution it will serve as our tracer or label. The already saturated solution cannot accept any *additional* dissolved lead(II) chloride (if the conditions are kept constant) but if the equilibrium is dynamic the solution will become radioactive; the total amount of dissolved salt will remain unchanged, but dissolved and undissolved lead(II) chloride will be exchanged constantly at the same rate.

Experiment 14.13

†To Examine the Nature of the Equilibrium Between a Solid and its Saturated Solution ***

Note

Make sure that you understand the principle of the experiment before it is demonstrated to you.

Apparatus

Spatula, Geiger-Müller liquid counter, scaler, mechanical stirrer, two 100 cm^3 beakers, centrifuge and appropriate test-tubes, teat pipette.

Propanone (acetone), lead(II) chloride, radioactive lead(II) chloride, deionized water.

Procedure

(a) Fill the Geiger-Müller liquid counter with deionized water and take a count of radioactivity for five minutes. This is the background count under these conditions.

(b) Add a spatula measure of 'ordinary' lead(II) chloride to about 30 cm^3 of deionized water in a beaker. Stir vigorously for several minutes so that some of the solid dissolves and a saturated solution is formed.

(c) Pour some of the *suspension* from (b) into a suitable test-tube and centrifuge. Transfer 5 cm^3 (or whatever volume the liquid counter will hold) of the clear, saturated solution to the liquid counter. Take a count for five minutes.

(d) Centrifuge the rest of the suspension from (b) and pour the clear, saturated solution into a beaker. Add a spatula measure of radioactive lead(II) chloride. Stir mechanically for at least twenty minutes.

(e) Centrifuge a sample of the suspension from (d) and pour the appropriate volume of the clear solution into the liquid counter. Count for five minutes.

Points for Discussion

1. Is 'ordinary' lead(II) chloride radioactive?

2. Is the saturated solution radioactive after being mixed with radioactive lead(II) chloride?

3. Is this typical equilibrium static or dynamic? Make sure that you fully understand the reasons for your answer.

Is it Possible to Control the Relative Proportions of Products and Reactants in a Chemical Equilibrium?

We have said that a particular equilibrium position depends upon a particular set of conditions, and in the last section you learned that changes in concentration or temperature and the addition of catalysts can alter the rate of chemical reactions. Is it possible to alter the equilibrium position in a reversible reaction by making one or more of these changes? If this is to be done the back reaction must be affected more than the forward reaction, or *vice versa*. If we were to shake the box of spheres (page 372) at the same rate as before, but with the box at an angle, we would obtain a different ratio of spheres in the two compartments (i.e. not 7:13); the equilibrium position would be changed. If it is possible to change the equilibrium position in a chemical reaction the advantages could be considerable. If we have to produce a certain substance by means of a reversible reaction it is obviously desirable to be able to modify the equilibrium position so as to obtain as large a proportion of the product as is economically feasible.

It may be that the changes we can induce in a reversible reaction will influence both forward and back reactions equally, in which case the equilibrium position will not be changed. However, even this will be useful, for if we can speed up both reactions we can at least reach the equilibrium state more quickly. We can again use our analogy of the box of spheres in order to illustrate this. Shaking the box more vigorously than before (but at the original angle) will not alter the distribution of spheres once 'equilibrium' has been attained, but the 'equilibrium' (the 7:13 arrangement) will be obtained more quickly. Similarly, equilibrium in the box will be reached more rapidly, for a given intensity of shaking (temperature), if the barrier between the compartments is lower (i.e. in chemical terms, if a positive catalyst is used).

ting for equilibrium to be blished would be very ly

In industrial processes a reversible reaction attaining an equilibrium position in which the products occupy 90 per cent of the system is of little value if it takes a fortnight to reach that position. It is better to attain equilibrium in a few hours, even if the proportion of products is small, for the process can be repeated. The ideal is to control a reaction so that it reaches equilibrium quickly and also contains a high proportion of the required product.

Concentration Changes in Reversible Reactions

Consider the reversible reaction:

$$A + B \rightleftharpoons C + D$$

If the principles we discussed in the last section can be applied to reversible reactions, an increase in the concentration of A and/or B at equilibrium should only speed up the forward reaction. Similarly, addition of C and/or D at equilibrium should only speed up the back reaction. If this is correct then it *will* be possible to change an equilibrium position, for we will be able to speed up one reaction more than the other. Thus addition of A and/or B at equilibrium should produce a new equilibrium position richer in C and D.

Experiment 14.14
†The Hydrolysis of Bismuth Trichloride *

Apparatus
Rack of test-tubes, measuring cylinder, two teat pipettes, glass rod.
Clear solution of bismuth trichloride in concentrated hydrochloric acid, concentrated hydrochloric acid, deionized water.

Necessary Information
Bismuth trichloride is hydrolysed by water in the reversible reaction:

$$BiCl_3(s) + H_2O(l) \rightleftharpoons BiOCl(s) + 2HCl(aq)$$

Bismuth oxychloride is insoluble in water and forms a white precipitate.

The solution of bismuth trichloride in concentrated hydrochloric acid contains the chemicals expressed in the equation for the hydrolysis, i.e. four different chemicals. They are in equilibrium with each other.

Procedure
(a) Add water to a sample of the equilibrium mixture in a test-tube. (Make sure that you use the same teat pipette for the bismuth trichloride solution throughout the experiment.) What is the name of the white precipitate formed? How could you make some of the white precipitate 'dissolve'? Test your theory.

(b) Try to obtain several different equilibrium positions from the same volume of original solution by using different concentrations of either one of the reactants or one of the products.

(c) Write up your procedure in the usual way.

Points for Discussion

1. Does this experiment confirm that new equilibrium positions can be produced by varying the concentration of reactants or products at equilibrium?

2. How else could you alter an equilibrium position by influencing concentrations but without *adding* anything? Give an example using the last experiment.

3. The principles you have just learned are applicable to all systems in equilibrium. They are referred to again in the summary at the end of this section.

Temperature Changes in Reversible Reactions

As an increase in temperature speeds up chemical reactions, we should be able to reach an equilibrium position more quickly by raising the temperature. It would be very useful if we could also speed up one reaction more than the other.

Any reaction involves energy changes because bonds are made and broken, and this usually results in a temperature change. One of the two reactions in a reversible system involves breaking strong bonds and making weaker ones and is therefore endothermic; the reverse reaction must be exothermic as opposite changes occur. This is true of any reversible reaction. The system thus has its own 'built in' temperature control factor, irrespective of what we supply to it. In other words, if a new equilibrium position is formed the system must either lose or gain heat energy according to whether the equilibrium position moves in the direction of the exothermic or endothermic reaction. As the two reactions are different with respect to energy (temperature) changes, perhaps a change in temperature will influence one more than the other.

The solubility of ammonium chloride Ammonium chloride dissolves in water and the process is endothermic (ΔH is $+$ve). The reverse reaction (the formation of solid ammonium chloride from ions in solution, which we shall refer to as the precipitation) is therefore exothermic, with ΔH $-$ve. If you have studied Chapter 13 you should be able to explain why these energy changes are as stated. If we make a saturated solution of ammonium chloride in contact with excess solid at, say, 273 K (0 °C) a dynamic equilibrium will exist between dissolved and undissolved solute, i.e. both dissolving and precipitating reactions will be occurring. There will be a fixed amount of solute dissolved in, say, 100 cm^3 of solution.

$$NH_4Cl(s) \underset{\text{exothermic } \Delta H - ve}{\overset{\text{endothermic } \Delta H + ve}{\rightleftharpoons}} NH_4^+(aq) + Cl^-(aq)$$

The following table shows the masses of ammonium chloride which can be dissolved in 100 cm^3 of water at various temperatures.

Temperature, K (°C)	*Solubility of ammonium chloride in water (g l^{-1})*
273 (0)	290
283 (10)	330
293 (20)	370
303 (30)	410

We can conclude that, although the rates of *both* reactions (in this case dissolving and precipitation) are increased as the temperature is raised, one reaction is affected more than the other because the amount of dissolved ammonium chloride changes with temperature. Look again at the data and make sure that you understand this conclusion.

Points for Discussion

1. Which of the two reaction rates in this example is increased most by raising the temperature?

2. Is this the reaction which is endothermic or exothermic? Does this reaction cause the system to lose heat or gain heat?

3. This rebellious behaviour is found to be true of all equilibria. It is referred to again in the summary at the end of this section.

The equilibrium between dinitrogen tetroxide and nitrogen dioxide You will meet further examples of the effect of temperature changes on equilibrium positions in the next section, but we will consider one further example here because the same reaction can also be used to study the effect of pressure changes on equilibria.

Dinitrogen tetroxide (N_2O_4) is a colourless gas which is fairly stable at temperatures just above its boiling point, 294 K (21 °C). As the temperature rises, the gas dissociates into the dark brown gas nitrogen dioxide (NO_2) and the two gases exist in equilibrium.

$$\underset{\text{Light brown}}{N_2O_4(g)} \underset{\Delta H - ve}{\overset{\Delta H + ve}{\rightleftharpoons}} \underset{\text{Dark brown}}{2NO_2(g)}$$

You may be able to explain why the energy changes are as shown. If a fixed mass of dinitrogen tetroxide is warmed gradually at constant pressure its volume will increase (why?). If the gas is 'ideal' it will obey Charles' Law (page 408), and it is thus possible to calculate the volume a fixed mass of the gas should occupy at certain temperatures. Study the following data and the equation for the reaction and try to answer the questions.

Temperature K (°C)	*Theoretical volume of a certain fixed mass of the gas if it obeys Charles' Law (cm³)*	*Actual volume of the fixed mass of gas (cm³)*
308 (35)	48	51
318 (45)	50	55
328 (55)	52	62
338 (65)	54	73

Points for Discussion

1. Why do you think that the reaction $N_2O_4 \rightarrow 2NO_2$ is endothermic and the reverse one is exothermic?

2. Does dinitrogen tetroxide obey Charles' Law?

3. Charles' Law shows how an ideal gas expands when it is heated. If the volume of a gas at a certain temperature is greater than it should be due to expansion alone, some reaction could be taking place so that a greater number of molecules is formed. These would then occupy a greater volume at the same pressure. Which of the two reactions in this reversible system could cause this to happen?

4. If both forward and back reactions are speeded up equally by an increase in temperature there would be no significant volume change (over such a small temperature range) other than that due to expansion, because the total number of molecules in the system would be unchanged. Make sure that you understand this. Are both reactions speeded up equally by an increase in temperature in this example? If not, which one is speeded up most, the endothermic one or the exothermic one? (Your answer to 3 should help you to answer this.) Does this further confirm your earlier conclusion about the effect of temperature changes on equilibria?

Pressure Changes in Reversible Reactions

The volumes of liquids and solids do not alter significantly when the pressure upon them is changed (why not?), but in a reaction between gases an increase in pressure brings the molecules closer together and thus increases their concentration. It follows that the rate of a reaction between gases (and of both reactions in a gaseous equilibrium) will be speeded up when the pressure of the system is increased, and equilibrium will thus be reached more quickly. This is really a concentration effect as mentioned earlier (page 366). Does a pressure change also affect the equilibrium position? In some reactions there is a 'built in' pressure regulator, just as there is a 'built in' temperature regulator. For example, in the reaction

$$N_2O_4(g) \rightleftharpoons 2NO_2(g)$$

an increase in the rate of the forward reaction produces more molecules. If the system is enclosed, i.e. is at constant volume, this will cause an increase in pressure.

Experiment 14.15
The Effect of Pressure on the Equilibrium Between Dinitrogen Tetroxide and Nitrogen Dioxide ***

Apparatus
100 cm³ glass syringe (lubricated with thin oil and absolutely air-tight) containing about 50 cm³ of the equilibrium mixture of dinitrogen tetroxide and nitrogen dioxide, white paper or screen.

Procedure
(a) Revise the colours of the two gases present in the equilibrium mixture. Watch carefully throughout the experiment for changes in the intensity of the brown colour. This will give some indication of the relative proportions of the two gases.

(b) Hold the syringe in front of a white screen and press the plunger in as far as possible so as to increase the pressure on the system. Hold the plunger steady for a few seconds and then release the pressure. Record your observations.

(c) Decrease the pressure by pulling out the plunger as far as it will go. Again hold the plunger steady for a few seconds before releasing it. Record your observations.

(d) The changes are summarized in Figure 14.2 so that the experiment can be discussed if you do not have a chance to do it.

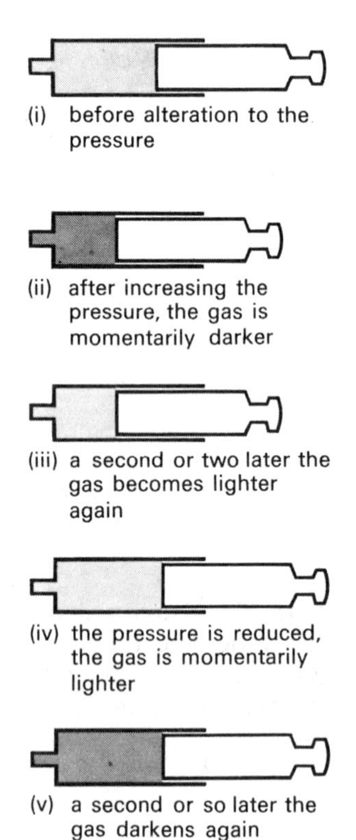

Figure 14.2 The effect of pressure changes on dinitrogen tetroxide

Points for Discussion
1. When the pressure was first increased the brown colour darkened. This is because the gas was more concentrated, and is not due to a change in the equilibrium position. What happened to the colour a few seconds later as the equilibrium adjusted to the new conditions? Which gas must have been present in a greater proportion than at the beginning?

2. Do you think that an increase in pressure has increased the rate of both reactions equally? If not, which reaction has been speeded up more than the other? Is this the reaction which produces fewer or more molecules? How will this particular reaction affect the pressure?

3. By similar reasoning, try to explain your observations when the pressure was reduced.

4. This principle is shown by all other gas reactions when there is a 'built-in' pressure regulator, i.e. where the equation shows different total numbers of molecules of reactants and products. In a reaction such as

$$I_2(g) + H_2(g) \rightleftharpoons 2HI(g)$$

where the equation shows an equal total number of reactant and product molecules (two in this case), pressure changes do not affect the equilibrium position, although they will alter the time taken to reach equilibrium provided that the substances are gases. These ideas are summarized at the end of this section.

Catalysts in Reversible Reactions

If a positive catalyst is added to a system in equilibrium it will *not* alter the equilibrium position; both reactions are speeded up to the same extent so that equilibrium is attained more quickly. In the synthesis of ammonia,

$$N_2(g) + 3H_2(g) \rightleftharpoons 2NH_3(g)$$

finely divided iron is normally used as a catalyst. This also equally speeds up the decomposition of ammonia and so does not change the proportion of ammonia at equilibrium.

A Discussion and Summary of the Factors which Affect Chemical Equilibria. The Principle of Le Chatelier

Revise your conclusions about the ways in which temperature, pressure, and concentration changes affect equilibria before you try to answer the questions.

Points for Discussion

1. If we increase the concentration of one substance, (e.g. a reactant) in a system in equilibrium, the equilibrium position changes. To bring about this change is some of the added reactant used up or is more of it produced?

2. If we increase the temperature by giving heat to a system in equilibrium, the equilibrium position changes. To bring about this change do the reactions which occur cause the system to gain or lose heat?

3. In the case of a gas reaction which involves a change in the number of molecules, if we increase the pressure the equilibrium position changes. To bring about this change do the reactions which occur produce an increase or decrease in pressure?

4. From your conclusions to 1, 2, and 3 try to formulate a general principle to describe what happens when the conditions of temperature, pressure, or concentration are altered for a system at equilibrium. This principle was first stated by H. L. Le Chatelier and it is known as Le Chatelier's principle (stated after Table 14.1).

5. The factors which affect chemical equilibria are summarized in Table 14.1.

Table 14.1 The factors which affect chemical equilibria

Factor	*Type of reversible reaction*	*Attainment of equilibrium*	*Effect on equilibrium position*
Increase in the concentration of a substance X	Any	Faster	New equilibrium position containing a lower concentration of X
Increase in temperature	Most	Both reactions speeded up so equilibrium attained more quickly	New equilibrium with a higher proportion of the substance(s) made by the endothermic reaction
Decrease in temperature	Most	Opposite to above, therefore slower	New equilibrium with a higher proportion of the substance(s) made by the exothermic reaction
Increase in pressure	Gas reactions where the equation shows an equal number of molecules of reactants and products	Faster (concentration effect)	No change
Decrease in pressure	Gas reactions as above	Slower (concentration effect)	No change
Increase in pressure	Gas reactions where the equation shows an unequal number of reactant and product molecules	Faster (concentration effect)	New equilibrium with a higher proportion of the substance(s) produced by the reaction causing a reduction in the number of molecules (and hence a reduction in pressure)
Decrease in pressure	Gas reactions as above	Slower (concentration effect)	New equilibrium with a higher proportion of the substance(s) produced by the reaction causing an increase in the number of molecules (and hence the pressure)
Positive catalyst	Most	Faster	No change
Negative catalyst	Most	Slower	No change

Le Chatelier's principle states: if one of the factors of a system at equilibrium is changed the system changes so as to *oppose* the effect, and a new equilibrium is established.

14.3 SOME IMPORTANT INDUSTRIAL PROCESSES

The Synthesis of Ammonia by the Haber Process

Most of the world's supply of ammonia is obtained by the Haber process. It proved difficult to synthesize ammonia from nitrogen and hydrogen until L. F. Haber developed a suitable catalyst. His process enabled Germany to produce her own independent route to explosives during World War I, for a large proportion of the ammonia is converted into nitric acid which in turn can be used to manufacture explosives.

The nitrogen cycle (page 278) has been disturbed by the demands of civilization; more nitrogen compounds are removed from the soil than are replaced naturally, and our way of restoring the balance is to add fertilizers to the soil. These are usually compounds of ammonia, which are soluble and contain nitrogen in a form which is readily available to plants. You should know that plants cannot use atmospheric nitrogen directly, even though they are virtually surrounded by it (some plants 'fix' nitrogen with the help of bacteria living in their root nodules). This is just one other reason for the increasing demand for ammonia.

The Raw Materials Nitrogen and hydrogen are required in the ratio 1:3 by volume. Modern plants use nitrogen from the air and hydrogen obtained by processing natural gas, but the correct ratio of gases can also be obtained from a mixture of 'water gas' and 'producer gas' (*Further Topics*). The carbon monoxide from the mixture is made to react by being passed with excess steam over a hot catalyst consisting of oxides of chromium and iron. The following reaction takes place:

$$N_2(g)+H_2(g)+CO(g)+H_2O(g) \rightarrow CO_2(g)+2H_2(g)+N_2(g)$$

The carbon dioxide is removed by dissolving it in water under pressure, and traces of unreacted carbon monoxide are dissolved in a solution of copper(I) methanoate (formate) under pressure. This process is called the Bosch process and it used to be one of the important methods of hydrogen manufacture. If it is used as part of the synthesis of ammonia the overall process is often called the Haber–Bosch process, but the hydrogen is now usually made by steam-reforming.

The nitrogen and hydrogen gases are compressed and scrubbed to remove as many potential catalyst poisons as possible before they are allowed to enter the converter.

The Conversion The reaction is:

$$N_2(g)+3H_2(g) \rightleftharpoons 2NH_3(g) \qquad \Delta H = -46.2\ \text{kJ mol}^{-1}$$

Study the equation, refer to Le Chatelier's principle, and suggest how you would use temperature and pressure in order to obtain the highest proportion of ammonia at equilibrium.

Pressure You should have realized that high pressures will produce a good yield of ammonia. This is indeed the case as is shown by the following data (by courtesy of ICI). The percentages are the proportions of ammonia at equilibrium.

Temperature K (°C)	Per cent ammonia at equilibrium at pressure (kg cm^{-2}): 155	259	362
623 (350)	46.21	57.46	65.16
723 (450)	22.32	32.87	39.31
823 (550)	9.91	15.58	20.76

The actual pressure used can vary between 150 and 350 atmospheres. Pressures of 1000 atmospheres have been used but such high pressures produce problems. Can you suggest what these might be?

Temperature You should have suggested that low temperature will produce a high yield of ammonia, and the data shows that this is so. Unfortunately, low temperatures would produce such slow rates of reaction that it would take too long to establish equilibrium, even with a positive catalyst. A compromise is made and temperatures of between 623 K (350 °C) and 823 K (550 °C) are normally used. The process of arriving at such a compromise in order to obtain the maximum yield of a product in a given time is called *optimization;* it is very frequently met with in industry. The temperature used in a given Haber process plant is the *optimum* temperature for that plant.

The catalyst The catalyst is usually finely divided iron, mixed with aluminium oxide, potassium hydroxide, and other substances as promoters (page 365). The gases pass over the catalyst beds in large towers called converters (Figure 14.3), of which there are several kinds. The catalyst is poisoned by oxides of carbon, water, and sulphur compounds, and undergoes physical decomposition at temperatures over 933 K (660 °C). The efficiency of the catalyst gradually decreases but it can be used for up to five years in good conditions.

Figure 14.3 Low pressure ammonia plants
(I.C.I. Agricultural Division, Billingham)

The removal of ammonia When the gases leave the converter they are hot and under pressure. Simple cooling by water jackets is enough to liquefy some of the ammonia (why?), and it is removed in the liquid state. Ammonia is also highly soluble in water so that it is possible to remove it from the equilibrium mixture by treatment with water.

Summary

$$N_2(g) + 3H_2(g) \rightleftharpoons 2NH_3(g) \qquad \Delta H \ -ve$$

Theoretical conditions: catalyst, high pressures, low temperatures.
Actual conditions: finely divided iron catalyst and promoters, pressure between 150 and 350 atmospheres, temperature between 623 K (350 °C) and 823 K (550 °C).

The ammonia is removed by liquefaction or by solution in water. Because the operating conditions are not the ideal theoretical ones, only about 10 per cent of the gaseous mixture is converted into ammonia, but this is done quickly and is better than a very slow yield of 90 per cent. In an industrial catalytic process, equilibrium is rarely reached, especially if the process is run continuously.

The Manufacture of Sulphuric Acid by the Contact Process

J. von Liebig once said that it was possible to judge the commercial prosperity of a country by the amount of sulphuric acid it uses. Although sulphuric acid is not quite the industrial indicator it once was, it remains true to say that the acid is one of the fundamentally important industrial chemicals and is always in great demand.

The history of the commercial production of sulphuric acid shows interesting variations. The basic variables have been a suitable source of sulphur dioxide and the actual process by which it is converted into the sulphur trioxide used to make the acid.

Most of the world's supply of sulphuric acid is now produced by the Contact process, a small version of which is used in Experiment 11.9. This is largely because of the demand for higher quality acid and for fuming acid (oleum), neither of which can be obtained by the older Lead Chamber process.

The Raw Materials The source of sulphur(IV) oxide (sulphur dioxide) still varies considerably. Countries with good supplies of natural gas and crude oil extract elemental sulphur in the processing stages. Sulphur is also found as the element in Sicily, Louisiana, and Texas, where large deposits are worked by the Frasch process (page 289). The sulphur is burned in air to produce sulphur(IV) oxide:

$$S(s) + O_2(g) \rightarrow SO_2(g)$$

Sometimes sulphide ores, such as iron pyrites, are roasted in air:

$$4FeS_2(s) + 11O_2(g) \rightarrow 2Fe_2O_3(s) + 8SO_2(g)$$

In Britain the production of sulphur(IV) oxide from naturally occurring calcium sulphate (gypsum or anhydrite) is well established. This method is attractive because cement is a useful by-product, and it makes us more independent of fluctuations in the supply and price of imported sulphur. A mixture of shale (which supplies silicates, aluminates, and iron oxides), coke, and anhydrite is heated in large rotatory kilns over seventy metres in length (Figure 14.4). The charge passes one way, towards the source of heat, and the gases pass in the opposite direction. The gases contain up to

Figure 14.4 Rotatory kilns for roasting anhydrite
(Albright and Wilson, Marchon Division)

9 per cent of sulphur(IV) oxide, and after removal of dust (e.g. by electrostatic precipitation) they are dried and passed on to the converter. The reaction is complicated but an overall summary is:

$$2CaSO_4(s) + C(s) \rightarrow 2CaO(s) + CO_2(g) + 2SO_2(g)$$

↓

Combines with aluminium and silicon compounds to form cement clinker

The Conversion of Sulphur(IV) Oxide to Sulphur(VI) Oxide

The reaction for the Contact process is:

$$2SO_2(g) + O_2(g) \rightleftharpoons 2SO_3(g) \qquad \Delta H = -94.9 \text{ kJ mol}^{-1}$$

The excess oxygen is obtained from the air.

Predict how temperature and pressure may be used to obtain high yields of sulphur(VI) oxide.

Pressure Although high pressures would produce greater yields of sulphur(VI) oxide, atmospheric pressure can be used because the equilibrium position is well to the right.

Temperature Although low temperatures would produce a higher yield of sulphur(VI) oxide, it would take too long to reach equilibrium even when a positive catalyst is used. A temperature of 723 K (450 °C) is found to give optimum yield.

The catalyst Vanadium pentoxide is generally used as a catalyst in the converters (Figure 14.5). Other catalysts such as platinum are more expensive and are very easily poisoned by impurities.

Removal of the sulphur(VI) oxide and the production of the acid The sulphur(VI) oxide is removed from the equilibrium mixture by being dissolved in fairly concentrated sulphuric acid, which is made even more concentrated by the reaction

$$SO_3(g) + H_2O(l) \rightarrow H_2SO_4(aq)$$

Figure 14.5 Converters, sulphuric acid contact plant *(Albright and Wilson, Marchon Division)*

The reaction with water alone is too violent, and even when acid is used heat is generated by the absorption and the vessels have to be cooled.

Some of the very concentrated (often fuming) acid so produced is removed as the product, and the rest is diluted and used to absorb more sulphur(VI) oxide.

Summary

$$2SO_2(g) + O_2(g) \rightleftharpoons 2SO_3(g) \qquad \Delta H \ -\text{ve}$$

$$SO_3(g) + H_2O(l) \rightarrow H_2SO_4(aq)$$

Theoretical conditions: high pressures, low temperatures, positive catalyst.

Actual conditions: atmospheric pressure, 723 K (450 °C), vanadium pentoxide catalyst.

The sulphur(VI) oxide is removed from the equilibrium mixture by absorption in fairly concentrated sulphuric acid.

The Occurrence and Extraction of Metals

Preliminary Treatment

An *ore* is a material from which a metal may be extracted, and it is usually associated with other rocky material from which it must first be removed. This is done by crushing the rock and separating the crude ore by such means as flotation, dissolving the required material in a suitable solvent, removing any magnetic material with magnets, or allowing heavier material to sink to the bottom of the wash liquor.

In Section 5.5 we discovered that the most electropositive metals were found on the left-hand side of the Periodic Table, and that these include sodium and potassium in Group 1, and magnesium and calcium in Group 2. Because of their reactivity they are not found free in nature, and as many of their compounds are soluble in water we should expect to find them in seas, lakes and rivers. Some are found in deposits left when inland seas evaporated, such as the Stassfurt deposits in Germany which contain potassium and magnesium chlorides. The deposits of these metals mainly occur as the chloride and sulphate, or in the case of magnesium and calcium, as the carbonate. Of the other metals dealt with in Chapter 10, zinc, iron, lead and copper occur mainly as the oxide or sulphide ores.

The Extraction of Very Electro–Positive Metals

Very electropositive metals are obtained by electrolysis of the fused chloride. Can you think of two reasons for the choice of chlorides for such electrolyses? In the case of these metals electrolysis is the method best suited for their extraction (page 177). Such a method is only possible if (a) the compound is ionic and thus can be electrolysed, and (b) the metal itself has a fairly low melting point which allows it to be conveniently removed from the cell in the molten state. Metallic chlorides usually fulfil these criteria.

In practice electrolysis of the fused compound in a simple container is unsatisfactory as various other factors must be considered. The extraction of sodium will serve as an example.

Sodium Early methods of extraction, such as the Castner process, used fused sodium hydroxide as the electrolyte instead of fused sodium chloride, but sodium chloride is now preferred as it is cheaper and the

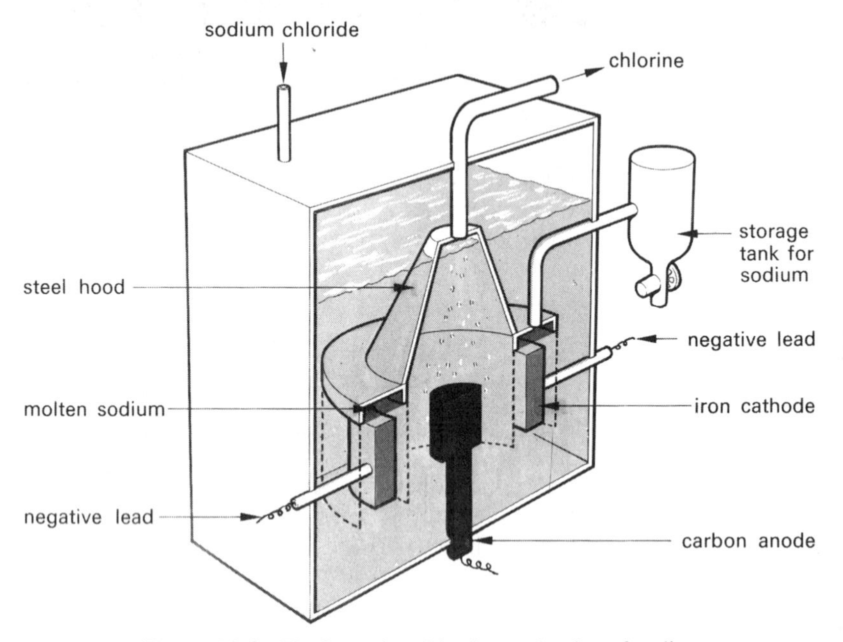

Figure 14.6 The Down's cell for the production of sodium

by-product is more valuable. The modern process utilizes the Down's cell which is illustrated in Figure 14.6. The melting point of sodium chloride is high, 1074 K (801 °C), and this high melting point prevented its use in earlier cells. The addition of up to 60 per cent of calcium chloride lowers the melting point (as does the addition of any impurity to a solid), and the operating temperature of the cell is about 873 K (600 °C). In general, chlorides of the metals under consideration have lower melting points than other compounds and they can function as electrolytes in the fused state because they are ionic solids. In the Down's cell the fused electrolyte contains both calcium and sodium chlorides, but sodium with only a small proportion of calcium is discharged. The calcium crystallizes from the molten alloy on cooling to leave fairly pure liquid sodium. Chlorine is formed at the anode and is a valuable by-product.

At the cathode:

$$Na^+ + e^- \rightarrow Na(l)$$

At the anode:

$$2Cl^- \rightarrow Cl_2(g) + 2e^-$$

Points for Discussion

The design of the cell looks complicated but it is necessary for satisfactory operation.

1. Why is the anode surrounded by a steel hood?

2. The molten sodium is collected in an inverted circular trough and it then runs into an outer storage tank. If the two electrodes were merely placed in the molten electrolyte, the molten sodium would rise and collect at the top of the melt. Apart from the fact that the chlorine evolved would react with the sodium, what other danger could arise?

3. The steel hood collects the chlorine produced and conveys it away from the cell. If the molten sodium collected at the top of a simple cell it would react with air, and if in contact with both electrodes would cause a short circuit.

Aluminium

(a) Purification of the oxide The chief ore is bauxite which is mainly hydrated aluminium oxide, Al_2O_3, $2H_2O$, mixed with small quantities of silica and iron(III) oxide. It is treated with sodium hydroxide solution under pressure and this dissolves the aluminium oxide, which is amphoteric, to form sodium aluminate. After filtration to remove the insoluble impurities, the aluminium solution is seeded with pure aluminium oxide when most of the hydroxide crystallizes in a pure state. On ignition to about 1473 K (1200 °C) almost pure aluminium oxide is obtained.

$$Al_2O_3(s)+3H_2O(l)+2NaOH(aq) \rightarrow 2NaAl(OH)_4(aq)$$

$$2Al(OH)_3(s) \rightarrow Al_2O_3(s)+3H_2O(g)$$

(b) Electrolysis of the oxide This oxide has a high melting point 2323 K (2050 °C) and the molten compound is a very poor conductor. Cryolite (Na_3AlF_6) is added as the oxide dissolves at a lower temperature (1173 K, about 900 °C), in molten cryolite and the melt can be electrolysed. Oxygen is evolved at the anodes, which are made of carbon and slowly burn away to form carbon dioxide so that a significant part of the cost of the process is the replacement of these electrodes. The molten aluminium forms at the cathode and collects on the floor of the cell (Figure 14.7). The cell is lined with graphite which acts as the cathode.

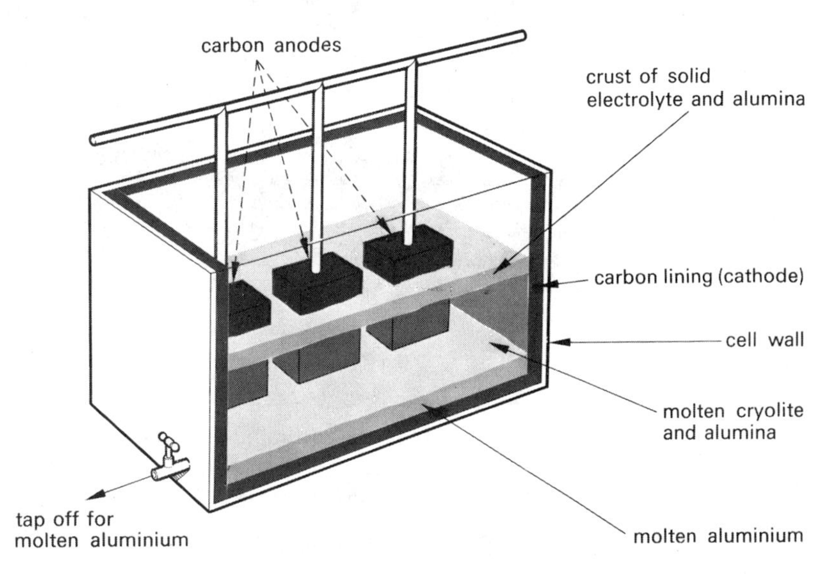

Figure 14.7 The extraction of aluminium by electrolysis

At the cathode:

$$4Al^{3+} + 12e^- \rightarrow 4Al(l)$$

At the anode:

$$6O^{2-} \rightarrow 3O_2(g) + 12e^-$$

Points for Discussion

1. The metals sodium, calcium and magnesium may be obtained by electrolysis of the molten chlorides. Why is fused aluminium chloride not used to extract aluminium? (Refer to the properties of aluminium chloride, Section 10.2.)

2. Three faradays of electricity are needed to liberate 27 g of aluminium, whereas the same quantity of electricity will liberate 69 g of sodium or 60 g of calcium. Why is the metal produced chiefly in highland areas?

The metals so far considered are extracted by electrolysis, but for the metals zinc, iron, lead and copper the method used is *pyrometallurgical*, that is, dependent upon the use of heat in special furnaces. The extraction of iron will serve as an example.

Iron and its conversion into steel The main ores are iron(III) oxide (haematite), and magnetite (tri-iron tetroxide, Fe_3O_4). The treated ore is mixed with coke and limestone and reduced in a blast furnace. Figure 14.8 illustrates a simplified sectional diagram of such a furnace, and Figure 14.9 is a photograph of actual furnaces. The reactions occurring are complex because they are reversible and interrelated, and concentration and temperature are controlling factors. The following is a brief outline of the chemical reactions involved.

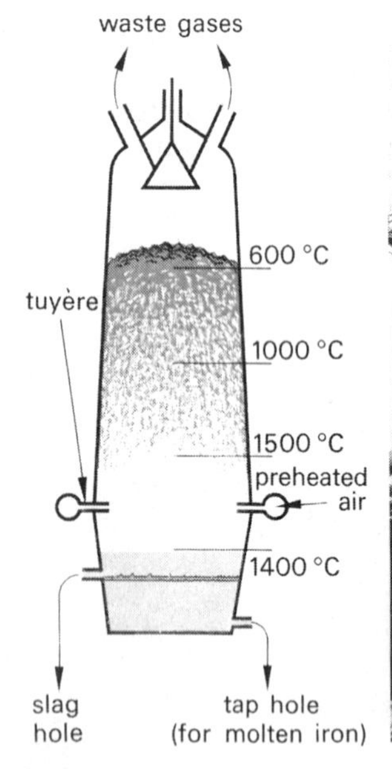

Figure 14.8 The blast furnace

Figure 14.9 The blast furnaces and stoves *(British Steel Corporation, Workington Works)*

The coke burns in the preheated air:

$$C(s) + O_2(g) \rightarrow CO_2(g) \qquad \Delta H = -352\ \text{kJ}$$

The carbon dioxide reacts with more coke to form the monoxide:

$$CO_2(g)+C(s) \rightarrow 2CO(g) \qquad \Delta H = +171\text{ kJ}$$

The carbon monoxide reduces the iron(III) oxide in stages to iron in the upper part of the furnace at a temperature of about 973 K (700 °C):

$$Fe_2O_3(s)+3CO(g) \rightarrow 2Fe(l)+3CO_2(g)$$

Above 1073 K (800 °C) the limestone dissociates into calcium oxide and carbon dioxide:

$$CaCO_3(s) \rightarrow CaO(s)+CO_2(g)$$

The calcium oxide fuses with the silica, present in the ore, and thus removes it in the form of a slag of calcium silicate. This illustrates the acidic nature of silicon dioxide, for it reacts here with a basic oxide.

$$CaO(s)+SiO_2(s) \rightarrow CaSiO_3(l)$$

The dense molten iron forms a layer at the bottom of the furnace, and the molten slag forms a layer above this. The iron is tapped from the furnace at the required time (Figure 14.10). The iron so obtained is called cast iron or pig iron. It is brittle and this means that its use in modern industrial society is extremely limited; most of the cast iron produced is therefore turned immediately into steel. This involves removing impurities such as sulphur and phosphorus which occur in varying proportions in the crude iron, and adding controlled quantities of other elements such as manganese, cobalt, tungsten etc. which give the different varieties of steel their

Figure 14.10 Tapping the blast furnace *(British Steel Corporation, Workington Works)*

distinctive properties. The excess carbon in the pig iron is mainly responsible for its brittle nature, but a controlled proportion of carbon is an essential constituent of steel. (You will remember from Figure 10.2 that the presence of 'foreign atoms' in a metal affects its properties). It is either removed entirely and replaced by a calculated quantity, or the carbon present in the crude iron is reduced to the required proportion.

Figure 14.11 The Open Hearth process *(British Steel Corporation)*

Various methods are used; some of the more important are outlined below.

(a) *The Open Hearth process*

The Open Hearth (Figure 14.11) is a special type of reverberatory furnace in which most of the heat supplied is reflected on to the charge in the furnace from a low roof. The charge consists of molten iron, iron(III) oxide and added scrap steel. Carbon, sulphur and phosphorus are oxidized to carbon monoxide/dioxide, sulphur(IV) oxide and phosphorus(V) oxide respectively. The gases escape via the waste gas flue. The phosphorus(V) oxide combines with the basic lining of the furnace.

It has advantages over the Bessemer process in that the composition of the steel can be more exactly controlled, and the layer of slag formed on the surface protects the steel from oxidation. The charge can also vary widely in the amount of pig iron and scrap iron introduced. Both flue gas and air introduced into the furnace are preheated, and the outgoing hot waste gases are utilized to heat the brickwork of the regenerators. The direction of flow is reversed from time to time so that incoming gases are continually preheated.

Figure 14.12 The Bessemer process—one of the few remaining acid processes *(British Steel Corporation, Workington Works)*

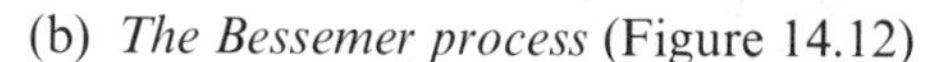

(b) *The Bessemer process* (Figure 14.12)

This is also a batch process and deals with smallish quantities of molten pig iron (about thirty tons). The converter is filled while in a horizontal position, and when swung into a vertical position a blast of air is blown through openings in the bottom at a pressure of about two atmospheres. This oxidizes the impurities to gases which escape, or solids which react with the lining. An oxygen enriched or oxygen/steam mixture reduces the blowing time and also prevents absorption of nitrogen by the molten iron. If phosphorus is absent, or present only in very small quantities, the lining of the converter may be made of silica, but if the percentage is above 0.1 per cent the lining is made of calcined dolomite ($CaCO_3/MgCO_3$) which reacts with the phosphorus oxide formed. As all of the carbon originally present is oxidized to carbon monoxide, a controlled amount of carbon must now be added.

(c) *The LD process* (Figure 14.13)

The process is associated with the steel-producing towns of Linz and Donawitz. The LD converter is tilted to receive the charge of molten iron, and a water-cooled lance made of steel blows oxygen at five to fifteen atmospheres pressure on to the surface. This removes the impurities in the

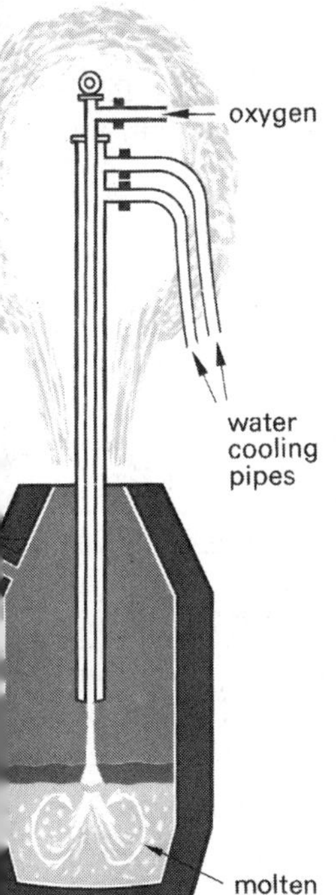

ure 14.13 The LD process *(British Iron and Steel Federation)*

same way as the air blow through the Bessemer. Because of the high temperature produced, it is possible to introduce up to 25 per cent scrap steel without the need for extra heat.

(d) *The Kaldo process*
This is named after the inventor, Kalling, and his workplace, Domnarvet. It is adaptable to a wider variety of steels than the LD process, and is preferred when the phosphorus content is high. Unlike the LD process the converter is maintained at a slight angle to the horizontal and is rotated. The charge consists of molten iron, a little iron(III) oxide and calcium oxide. Oxygen at low pressure is blown on to the surface of the molten metal. Any silica and other impurities form a slag which can be tipped away. More calcium oxide is added and a further oxygen blow carried out.

[Note: the LD, Kaldo and other modern methods are gradually replacing the older ones.]

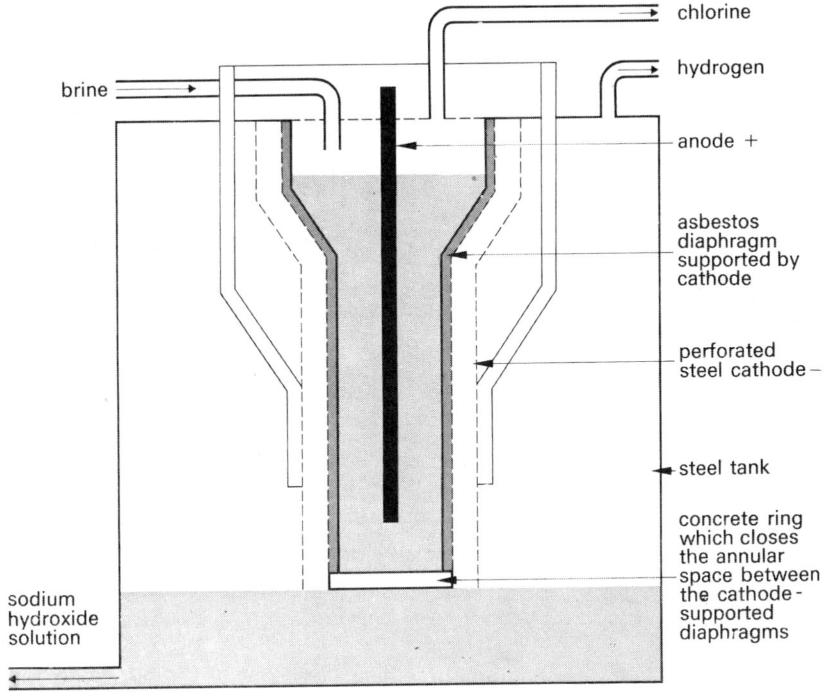

re 14.14 The diaphragm cell

Sodium Hydroxide

The main method of manufacture is by electrolysis of aqueous sodium chloride using either a diaphragm cell or a mercury cell.

A Diaphragm Cell The electrolyte is sodium chloride solution contained in a porous asbestos vessel, which acts as the diaphragm (Figure 14.14). The outside of the diaphragm is supported by a steel mesh which is the cathode. The anode consists of graphite rods suspended in the porous vessel. The diaphragm asbestos vessel is fixed within a steel container. During electrolysis, chlorine is evolved at the anode and passes out of the cell; hydrogen is liberated at the cathode. Sodium ions migrate through the porous diaphragm to the cathode. The sodium ions together with the excess hydroxyl ions at the cathode are removed because the steam, which is blown into the cell, condenses, and dilute sodium hydroxide solution collects on the floor.

In the diaphragm cell the operating temperature is about 358 K (85 °C), which reduces the resistance of the cell. The reactions occurring are:

Cathode: $$2H_3O^+(aq) + 2e^- \rightarrow H_2(g) + 2H_2O(l)$$

Anode: $$2Cl^-(aq) \rightarrow Cl_2(g) + 2e^-$$

Thus, of the four ions present in brine, Na^+, H_3O^+, Cl^- and OH^-, the Na^+ and OH^- are left in solution.

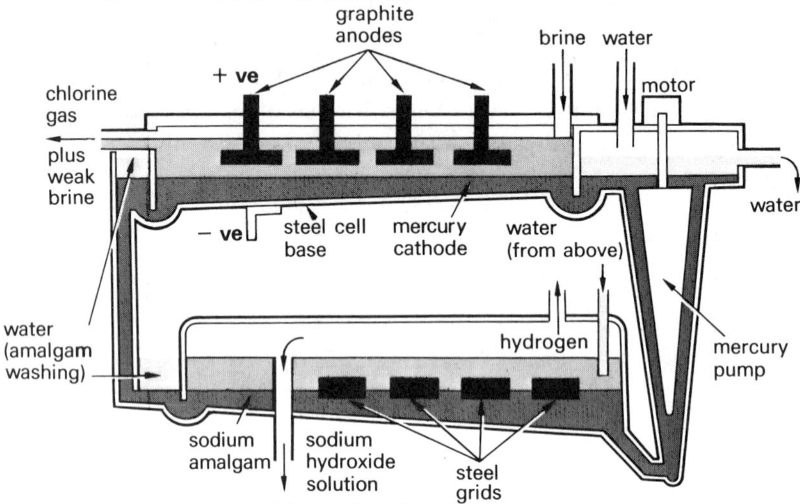

Figure 14.15 The mercury cell

The Mercury Cell This consists of two troughs with opposing slopes and which lie side by side, but are shown one above the other in Figure 14.15. The floor of each cell is covered with a layer of flowing mercury. Sodium chloride solution is the electrolyte in the top compartment, and electrolysis occurs between graphite anodes and the mercury cathode. Chlorine is discharged at the anode:

$$2Cl^-(aq) \rightarrow Cl_2(g) + 2e^-$$

Under these conditions sodium is preferentially discharged at the mercury cathode and dissolves in it to form an amalgam:

$$Na^+(aq) + e^- \rightarrow Na(s)$$

$$Na(s) + Hg(l) \rightarrow Na/Hg(l)$$

The sodium amalgam flows out of the brine compartment into the other trough (the 'caustic compartment') where it is decomposed, in the presence of graphite, by a counter-current of water. The sodium from the amalgam reacts with the water to form sodium hydroxide solution and hydrogen gas is evolved. The mercury, now free of sodium, is returned to the brine compartment. This recycling of the mercury means that the method is an economic one, provided the very high initial capital cost of the mercury and electrolysis plant can be met.

$$2Na/Hg(l) + 2H_2O(l) \rightarrow 2NaOH(aq) + H_2(g) + 2Hg(l)$$

The particular properties of mercury make it eminently suitable for the cathode in the electrolytic preparation of sodium hydroxide. For example, it is a liquid metal and can be run in and out of the cell easily. Can you name two other properties of mercury which are extremely important to the working of the cell? Remember that usually hydrogen is discharged in preference to sodium at a platinum or carbon electrode, and that pure sodium can react with water with explosive violence.

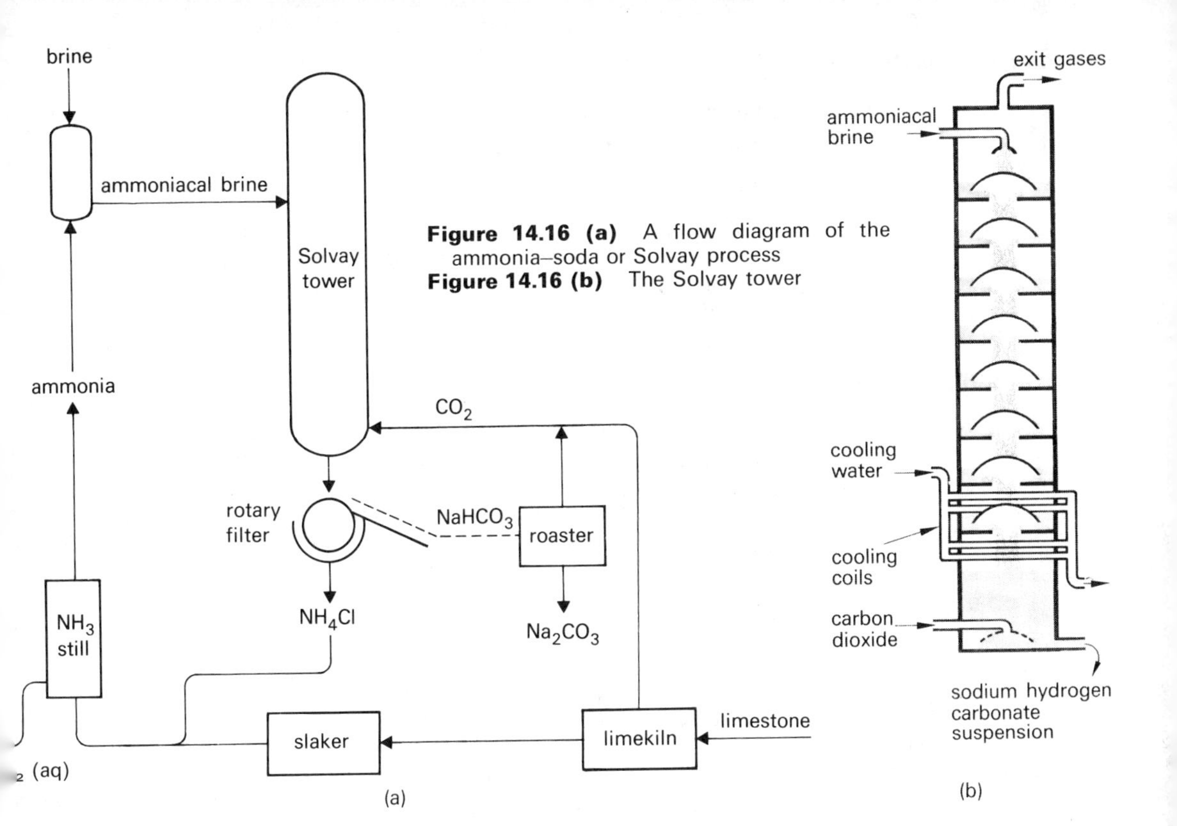

Figure 14.16 (a) A flow diagram of the ammonia–soda or Solvay process
Figure 14.16 (b) The Solvay tower

Sodium Carbonate by the Ammonia–Soda (Solvay) Process

Sodium carbonate is an essential material for a large number of industries. The raw materials are all found in Britain and the process is a very good example of economy in the use of materials. The chemicals required are sodium chloride, limestone, coal and ammonia (as crude ammonia liquor from gas works or pure liquor from ammonia production). Large quantities of water are used for cooling purposes, and to produce steam and brine.

Figure 14.16 (a) is a flow diagram of the process. The limestone is heated in a kiln where it dissociates into calcium oxide and carbon dioxide:

$$CaCO_3(s) \rightarrow CaO(s) + CO_2(g)$$

The carbon dioxide is passed into the Solvay tower (Figure 14.16 (b)). Ammonia is used to saturate the incoming sodium chloride solution, and the resulting 'ammoniacal brine' must be cooled because of the considerable amount of heat liberated during this solution. This 'ammoniacal brine' next passes into the Solvay tower up which the carbon dioxide is blown. The reactions occurring may be summarized as:

$$\underbrace{Na^+(aq) + Cl^-(aq)}_{\text{Brine}} + CO_2(g) + NH_3(g) + H_2O(l) \rightarrow NaHCO_3(s) + NH_4^+(aq) + Cl^-(aq)$$

The precipitated sodium hydrogen carbonate is filtered off on a rotating filter, washed to remove any ammonium chloride, and then heated in a furnace to give sodium carbonate and carbon dioxide:

$$2NaHCO_3(s) \rightarrow Na_2CO_3(s) + CO_2(g) + H_2O(g)$$

The latter gas is passed back into the Solvay tower, and this is only one of the ways in which the gases used in the process are regenerated.

The calcium oxide is slaked in a slaker by addition of water:

$$CaO(s) + H_2O(l) \rightarrow Ca(OH)_2(s)$$

The suspension of calcium hydroxide is heated with the filtrate from the Solvay tower, and ammonia is produced:

$$2NH_4Cl(aq) + Ca(OH)_2(s) \rightarrow 2NH_3(g) + CaCl_2(aq) + 2H_2O(l)$$

If you look carefully at the flow diagram you will see that theoretically there is only one 'waste' product and that is calcium chloride from the ammonia still. The carbon dioxide and the ammonia are recycled into the process, and unavoidable small losses (such as ammonia gas) are made good. This is an excellent illustration of the economical use of by-products.

The Liquefaction of Air

Air consists mainly of oxygen and nitrogen and is the 'raw material' for the commercial production of these two gases. As air is a mixture, its components can be separated by physical means. The principle is basically simple; the air is liquefied and the liquid components separated by fractional distillation. This is possible because they have different boiling points, i.e. liquid nitrogen boils at 78 K (– 195 °C) and liquid oxygen boils at 90 K (– 183 °C).

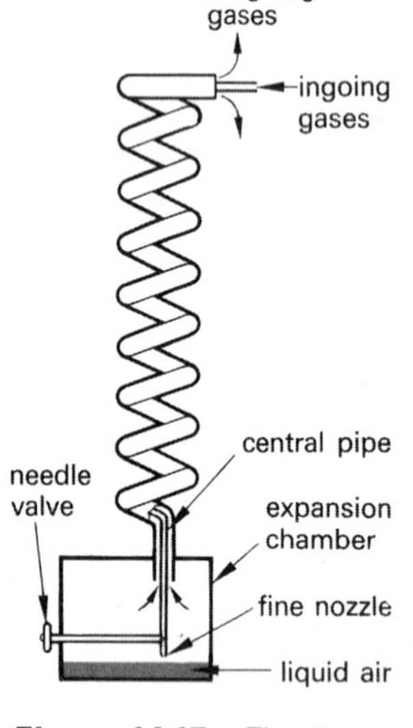

Figure 14.17 The liquefaction of air

The first procedure is to remove both carbon dioxide and water vapour from the air, since at the low temperature of liquefaction, carbon dioxide and water are solids and would block the pipes. The remaining gases (nitrogen, oxygen and noble gases) are compressed at about 200 atmospheres. This produces heat so that the gases emerging from the compressor must next be cooled, and this is done by allowing them to pass through a central pipe surrounded by another pipe containing the emergent cold gases from the expansion chamber (Figure 14.17). The cooled compressed gases emerge from the central pipe via a fine orifice into an expansion chamber (Figure 14.17). As the gas expands it cools (the Joule–Thompson effect). The molecules are close together in the compressed gas, and when they expand into a 'low pressure' chamber they move further apart. To do this they need to overcome the weak intermolecular forces which exist between molecules so close together. The work done in overcoming these forces comes from the intrinsic energy of the gas itself, which is therefore cooled. (A similar phenomenon occurs when air (under pressure) escapes from a bicycle tyre valve.) This cold air now passes out into a pipe surrounding the central pipe, thus cooling the incoming air so that a heat interchange occurs. The gases return to the compressor, and each cycle produces air that is correspondingly colder. The cumulative effect finally liquefies the air.

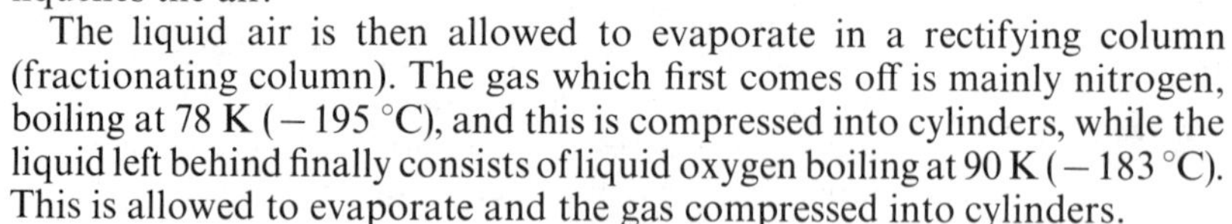

The liquid air is then allowed to evaporate in a rectifying column (fractionating column). The gas which first comes off is mainly nitrogen, boiling at 78 K (– 195 °C), and this is compressed into cylinders, while the liquid left behind finally consists of liquid oxygen boiling at 90 K (– 183 °C). This is allowed to evaporate and the gas compressed into cylinders.

The noble gases are present to a very small extent in air. At the temperature of liquefaction, helium and neon remain as gases and so can be separated from the liquid air. Argon, krypton and xenon are obtained by fractional distillation of liquid oxygen.

The Production of Nitric Acid

Nitric acid can be prepared in the laboratory by heating sodium nitrate with concentrated sulphuric acid (page 284), but although this method was used on an industrial scale, the shortage of sodium nitrate and the increasing demand for nitric acid meant that other raw materials had to be obtained. When ammonia became readily available from the Haber process (page 381) a new route to nitric acid was possible. Although ammonia is an important chemical in its own right, large quantities are

mixture of
1 volume of ammonia +
8 volumes of air

catalyst
layers

gas containing
nitrogen monoxide

re 14.18 The catalyst amber for the produc-n of nitric acid by idation of ammonia

used in the manufacture of nitric acid, which in turn is used in the manufacture of ammonium nitrate, explosives, and in the nitration of benzene and other compounds for the preparation of synthetic dyes, etc.

Ammonia will burn in air enriched with oxygen or, better, in pure oxygen to form nitrogen and steam (Experiment 11.4).

$$4NH_3(g) + 3O_2(g) \rightarrow 2N_2(g) + 6H_2O(g)$$

However, in the presence of platinum as a catalyst nitrogen monoxide is produced instead:

$$4NH_3(g) + 5O_2(g) \rightarrow 4NO(g) + 6H_2O(g)$$

On cooling, the nitrogen monoxide forms nitrogen dioxide by combining with excess oxygen, and this can be converted to nitric acid by dissolving it in water and blowing air through the solution. This series of reactions forms the basis of the industrial manufacture of nitric acid which is summarized below.

(a) Ammonia is mixed with about eight times its volume of air and preheated to about 573 K (300 °C) by passage through a heat interchanger.

(b) The mixture then passes through a catalyst chamber (Figure 14.18) containing several layers of a platinum/rhodium gauze (90 per cent platinum), which acts as the catalyst. The reaction is started by means of an electrically heated resistance wire just above the catalyst:

$$4NH_3(g) + 5O_2(g) \rightarrow 4NO(g) + 6H_2O(g) \qquad \Delta H = -1087 \text{ kJ g-equation}^{-1}$$

The reaction is strongly exothermic; once it has started the heater is turned off and careful control of the reaction mixture keeps the temperature at the level 1173 K (900 °C) which is needed for maximum efficiency.

(c) The gases leaving the catalyst chamber at about 1173 K (900 °C) pass through a heat interchanger, and are cooled to about 523 K (250 °C) by the ammonia/air mixture coming in.

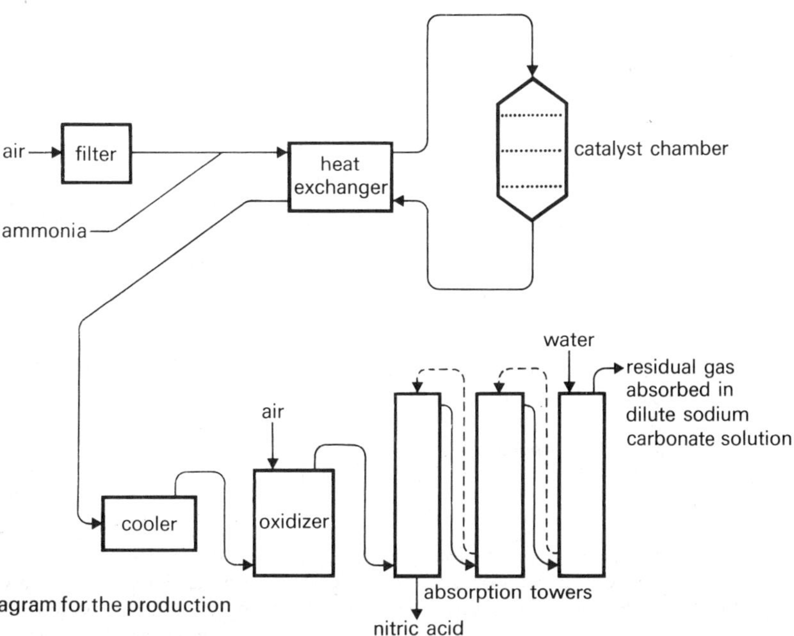

Figure 14.19 Flow diagram for the production of nitric acid

(d) Further cooling of the exit gases takes place in a waste heat boiler. By this time the nitrogen monoxide formed in stage (b) has combined with excess oxygen in the air to form nitrogen dioxide:

$$2NO(g)+O_2(g) \rightarrow 2NO_2(g)$$

(e) The mixture of gases passes to a cooling tower, and then into a series of absorption towers where they meet counter-currents of nitric acid in increasing order of dilution and finally leave a tower through which only water flows (Figure 14.19). The nitrogen dioxide in the gas mixture reacts with the water in the towers:

$$2NO_2(g)+H_2O(l) \rightarrow HNO_2(aq)+HNO_3(aq)$$

The nitrous acid is then oxidized to nitric acid by excess oxygen. The residual oxides of nitrogen present in the acid are expelled by blowing air through the acid, and they are then returned to the first absorption tower.

(f) The concentration of acid obtained is about 55–65 per cent and this is ordinary concentrated nitric acid. If a more concentrated acid is required it may be obtained by distillation of the less concentrated acid with concentrated sulphuric acid.

QUESTIONS CHAPTER 14

1. (a) The formation of methanol (methyl alcohol) from hydrogen and carbon monoxide can be represented by

$$CO(g)+2H_2(g) \rightleftharpoons CH_3OH(l) \qquad \Delta H = +91 \text{ kJ mol}^{-1}$$

What mass of hydrogen would react to cause a heat change of 91 kJ?

(b) What would be the effect on the equilibrium concentration of methanol in this endothermic reaction if (i) the temperature was increased, (ii) the pressure was increased, (iii) the hydrogen concentration was increased? (J. M. B.)

2. Differentiate between the terms 'negative catalyst' and 'positive catalyst'.

3. What is meant by 'the rate of a chemical reaction'?

4. Explain the meaning of (a) promoter, (b) catalyst poison and (c) enzyme.

5. Explain why an increase in the concentration of one or more of the reactants increases the rate of a chemical reaction.

6. An increase in pressure does not noticeably affect the rate of a reaction between solids but increases the rate of a reaction between gases. Explain.

7. Briefly describe three different ways in which the rate of a chemical reaction can be measured.

8. Distinguish between the terms static and dynamic as applied to equilibria.

9. The equation for the reaction by which ammonia is manufactured is:

$$N_2(g)+3H_2(g) \rightleftharpoons 2NH_3(g)$$

(a) What would be the effect on the equilibrium concentration of ammonia of (i) increasing the pressure, (ii) increasing the nitrogen concentration?

(b) The equilibrium concentration of ammonia increases as the temperature is lowered. Is heat evolved or absorbed when ammonia is formed?

(c) Why is a catalyst used in this reaction? (J. M. B.)

10. $2SO_2(g)+O_2(g) \rightleftharpoons 2SO_3(s)+\text{heat evolved} \qquad \Delta H = -189 \text{ kJ mol}^{-1}$

The equation represents a system in equilibrium. State the changes in the equilibrium concentration of sulphur trioxide which would be caused by (a) adding oxygen, (b) heating the mixture, (c) increasing the pressure. (J. M. B.)

11. Each of the following mixtures contains the same number of 'molecules' of both hydrochloric acid and sodium thiosulphate. Which combination would you expect to produce a precipitate of sulphur most quickly?

A 400 cm^3 of M HCl and 400 cm^3 of M $Na_2S_2O_3$
B 200 cm^3 of 2M HCl and 200 cm^3 of 2M $Na_2S_2O_3$
C 100 cm^3 of 4M HCl and 100 cm^3 of 4M $Na_2S_2O_3$
D 200 cm^3 of 2M HCl and 100 cm^3 of 4M $Na_2S_2O_3$
E 400 cm^3 of M HCl and 100 cm^3 of 4M $Na_2S_2O_3$

12. Which of the following statements is *not* true about the reaction between dilute hydrochloric acid and marble chips (calcium carbonate)?

A It is faster after three seconds than it is after ten seconds.
B It slows down with time.
C It eventually stops.
D It proceeds at a constant rate.
E It causes a loss of mass in the reaction vessel contents.

13. In the reaction
$2Y + W \rightleftharpoons Y_2W,\ \Delta H = -8400\ kJ\ mol^{-1}$

which of the following would result in a higher yield of Y_2W?

A The use of a suitable positive catalyst
B Lowering the temperature
C Removal of W
D Reducing the pressure
E Increasing the surface area of W

14. Enzymes are

A substances made by chemists to activate some washing powders
B very active negative catalysts
C catalysts found in living organisms
D synthetic catalysts
E substances which make catalysts more efficient

15. When calcium hydrogen carbonate solution decomposes in nature, an equilibrium is not established because

A the reaction is not reversible
B one or more of the products escapes from the system
C the reaction is too slow
D a negative catalyst is present
E the optimum temperature is too high

The following questions require rather longer answers.

16. Define the principle of Le Chatelier and illustrate it by reference to two reactions which are familiar to you.

17. Explain what is meant by the term catalyst. Describe a simple experiment or experiments to demonstrate the use of a catalyst. Give two industrial reactions in which a catalyst is used, naming the initial substances, the product(s) and the catalyst used as well as any essential conditions. (Long accounts are not wanted, nor are diagrams. The manufacture of nitric acid is excluded.) (C.)

18. Name two important ores of iron. Outline the extraction of iron by the blast furnace process, explaining the essential chemistry. (No diagram is required, nor are technical details.) What differences are there in composition between cast iron and steel? Describe, giving any necessary conditions, the action of (a) hydrochloric acid, (b) nitric acid (one reaction only) and (c) steam on iron. (C.)

19. The following statements are made in a textbook. 'The rates of most chemical reactions are approximately doubled by raising the temperature at which the reactions are carried out by 10 °C.'

'The rate at which a chemical substance reacts is directly proportional to its concentration.'

(a) Describe the experiments you would carry out to test the truth of these two statements when applied to either the reaction between a metal and a dilute acid or the decomposition of hydrogen peroxide catalysed by manganese dioxide (manganese(IV) oxide).

(b) Explain simply, in terms of the ions or molecules present, why the rate of a reaction is increased both by raising the temperature and also by increasing the concentration of the reagents. (C.)

20. Answer the following questions about the manufacture of iron and steel (no diagrams are required).

(a) Give the name and formula of one mineral from which iron is extracted.
(b) Explain how carbon monoxide is formed in the blast furnace.
(c) Write the equation for one reaction by which metallic iron is formed in the furnace.
(d) Explain clearly why limestone (calcium carbonate) is used in the blast furnace and suggest what you think would happen if the limestone were not present.
(e) Name three impurities likely to be present in the 'pig iron' formed in the blast furnace. Give one effect of these impurities on the physical properties of the iron.
(f) Explain how these impurities are removed during the conversion of pig iron into steel. (C.)

21. Describe carefully how sodium carbonate is manufactured by the ammonia–soda (Solvay) process. No details of the plant are required. Give two reasons why this is a very economical process. Explain briefly how glass is manufactured.
(J. M. B.)

22. Give an account of the *chemistry* involved in three of the following:

(a) the production of pig iron,
(b) the manufacture of calcium carbide from limestone,
(c) the electrolytic production of sodium hydroxide,
(d) the extraction of aluminium from pure aluminium oxide.

(Diagrams of commercial plant are not required.)
(A. E. B.)

23. Sulphuric acid is manufactured by converting sulphur dioxide to sulphur trioxide and dissolving this in 95–98 per cent acid, whilst adding water to the appropriate extent.

(a) How is the sulphur dioxide obtained?
(b) State the approximate temperature and relative proportions of gases used in the conversion.
(c) Name one of the catalysts commonly used.
(d) Explain why one uses a moderate temperature and a catalyst.
(e) Why is the sulphur trioxide not dissolved in water directly?
(f) State two important commercial uses of sulphuric acid.

Describe one reaction in each case in which sulphuric acid acts as (i) an acid, (ii) a dehydrating agent, (iii) an oxidizing agent.
(A. E. B.)

24. Draw a diagram of the apparatus which you would use to measure the volume of hydrogen liberated by one gramme of zinc from dilute hydrochloric acid. Describe how you would use the apparatus and state what readings would be necessary if you were required to correct the volume to normal temperature and pressure.

The addition of copper(II) sulphate solution to the zinc and hydrochloric acid increases the rate of the reaction. How would you use the apparatus you have drawn to test whether this is so? What further experiments would you perform to show that it is the copper(II) ions, and not the water or the sulphate ions also present in the copper(II) sulphate solution, which cause the increase in rate?
(J. M. B.)

15 Chemical Measurements. Volumes and Masses

15.1 CHEMICAL FORMULAE. HOW THEY ARE WORKED OUT AND WHAT WE CAN LEARN FROM THEM

A pure chemical compound has a fixed composition and we are therefore justified in representing such a compound by a fixed formula. In this section we will consider what we can learn from a chemical formula and the methods by which it can be determined.

Percentage Composition of Compounds

In a compound the mass of each element can be expressed as a percentage of the total mass of the components and the gravimetric composition is usually given in this way. The method of working is simple and straightforward, as you will see from the following example:

Example 15.1
Calculate the percentage composition by mass of ammonium nitrate.

(a) Write down the formula of the compound: NH_4NO_3.

(b) Find the formula mass by adding together the masses of all the atoms present: $14+(4\times 1)+14+(3\times 16) = 80$.

(c) Express each atomic mass as a percentage of the formula mass. If more than one atom of a particular element is present, then the total mass of these atoms must be used, e.g. as two nitrogen atoms appear in the formula, the mass used is 28 and not 14.

$$\text{Per cent N} = \tfrac{28}{80}\times 100 = 35 \text{ per cent}$$

$$\text{Per cent H} = \tfrac{4}{80}\times 100 = 5 \text{ per cent}$$

$$\text{Per cent O} = \tfrac{48}{80}\times 100 = 60 \text{ per cent}$$

(d) Check that the percentages add up to 100.

Points for Discussion

For all calculations use Table of Atomic Masses and approximate as instructed.

1. Calculate the percentage composition by mass of the following compounds: (a) calcium carbonate, (b) anhydrous copper(II) sulphate, (c) methane, (d) potassium hydrogen carbonate.

2. If a compound contains water of crystallization, the percentage of water should be calculated as a separate unit.

Example 15.2

Calculate the percentage composition by mass of hydrated magnesium chloride, $MgCl_2, 6H_2O$.

Formula mass $= 24+(2\times 35.5)+(6\times 18) = 203$

$$\text{Per cent Mg} = \tfrac{24}{203}\times 100 = 11.82 \text{ per cent}$$

$$\text{Per cent Cl} = \tfrac{71}{203}\times 100 = 34.98 \text{ per cent}$$

$$\text{Per cent } H_2O = \tfrac{108}{203}\times 100 = 53.2 \text{ per cent}$$

Point for Discussion

Determine the percentage composition of each of the following hydrates:

(a) iron(II) sulphate, $FeSO_4, 7H_2O$,
(b) sodium carbonate, $Na_2CO_3, 10H_2O$,
(c) copper(II) sulphate, $CuSO_4, 5H_2O$.

Empirical Formulae

Revise the work you did on finding the formula of magnesium oxide (Section 1.6).

Points for Discussion

1. What is the meaning of an 'empirical' formula?

2. Before you can find the empirical formula of a compound you need to know its composition by mass. What other information do you need?

Calculating the Empirical Formula of a Compound

Except in a few simple cases, such as the formation or reduction of metallic oxides, the quantitative analysis of a compound is too complicated for you to attempt at present, but you will need to know how to calculate the formula of a compound, given the results of its analysis by mass.

Example 15.3

Calculate the simplest (empirical) formula for the compound with the following composition: lead 8.32 g, sulphur 1.28 g, oxygen 2.56 g (Atomic masses: Pb = 207, S = 32, O = 16).

(a) Convert the masses into moles by dividing the mass of the element by the mass of its atomic mass. In this way we are 'counting' the combining atoms by weighing.

(b)

$$Pb = \frac{8.32}{207} = 0.04 \text{ moles}$$

$$S = \frac{1.28}{32} = 0.04 \text{ moles}$$

$$O = \frac{2.56}{16} = 0.16 \text{ moles}$$

The compound contains lead, sulphur, and oxygen atoms combined in the ratio: 0.04 moles of lead: 0.04 moles of sulphur: 0.16 moles of oxygen.

(c) Convert these ratios into whole numbers by dividing throughout by the smallest number (0.04 in this case).

1 mole of lead: 1 mole of sulphur: 4 moles of oxygen

A mole of any substance contains an equal number (6.02×10^{23}) of particles (atoms in this case).

$\therefore$ (6.02×10^{23}) atoms of lead combine with (6.02×10^{23}) atoms of sulphur and with ($4\times 6.02\times 10^{23}$) atoms of oxygen.

∴ 1 atom of lead combines with 1 atom of sulphur and 4 atoms of oxygen.

∴ *Empirical formula is $PbSO_4$*

This type of calculation is conveniently set out in table form.

Element	*Mass (or per cent mass)*	*Molar mass*	*Mass / molar mass*	*Ratio of moles*	*Simplest ratio*
Lead	8.32 g	207 g	8.32/207	0.04	1
Sulphur	1.28 g	32 g	1.28/32	0.04	1
Oxygen	2.56 g	16 g	2.56/16	0.16	4

Ratio of atoms 1:1:4

Empirical formula $PbSO_4$

[Note: if the composition by mass is expressed as percentages the method of working is exactly the same, because then we are considering the mass of each element in 100 g of the compound.]

Points for Discussion

1. Determine the empirical formula of each of the following compounds for which the composition by mass is given:

(a) magnesium 9.5 g, chlorine 28.4 g
(b) copper 40 per cent, sulphur 20 per cent, oxygen 40 per cent [take Cu = 64]
(c) nitrogen 1.40 g, hydrogen 0.40 g, carbon 0.60 g, oxygen 2.40 g
(d) carbon 75 per cent, hydrogen 25 per cent
(e) carbon 40 per cent, hydrogen 6.67 per cent, oxygen 53.3 per cent

2. In the case of hydrated salts, the mass of water of crystallization is taken as one unit and divided by the mass of mole of water.

Example 15.4

A hydrate has the following percentage composition: iron = 20.15 per cent, sulphur = 11.51 per cent, oxygen = 23.02 per cent, water = 45.32 per cent. Determine the empirical formula of the hydrated salt.

Element or group	*Per cent mass*	*Molar mass*	*Per cent mass / molar mass*	*Ratio of moles*	*Simplest ratio*
Fe	20.15	56	20.15/56	0.36	1
S	11.51	32	11.51/32	0.36	1
O	23.02	16	23.02/16	1.44	4
H_2O	45.32	18	45.32/18	2.52	7

Ratio of atoms or groups Fe:S:O:H_2O is 1:1:4:7

Formula is $FeSO_4$, $7H_2O$

Points for Discussion

Calculate the empirical formulae of hydrated salts with the following percentage compositions:

(a) Cu = 25.6 per cent, S = 12.8 per cent, O = 25.6 per cent, H_2O = 36.0 per cent
(b) Na = 16.09 per cent, C = 4.20 per cent, O = 16.78 per cent, H_2O = 62.93 per cent
(c) Mg = 11.82 per cent, Cl = 34.98 per cent, H_2O = 53.20 per cent

Molecular Formulae

The molecular formula gives the *number*, and not just the ratio of the numbers, of each type of atom in one molecule of the compound. A molecular formula can be used only for covalent (molecular) compounds; for ionic substances, the empirical formula is the only one that can be written, as ionic substances have giant structures in which there are no free units such as molecules. Covalent substances can have both empirical and molecular formulae, e.g. the empirical formula for ethane is CH_3, its molecular formula is C_2H_6, but the formula CH_4 for methane is both its empirical and its molecular formula.

Determination of Molecular Formulae

To find the molecular formula of a compound, we need to know both its empirical formula and its molecular mass. The molecular mass of a gas or a volatile liquid can be found experimentally by using the fact that a mole of any gas occupies 22.4 litres at s.t.p. (page 415).

A large container such as a glass globe is weighed when evacuated, and then reweighed when filled with the gas in question. If the volume, temperature, and pressure of the gas are known, its volume at s.t.p. can be calculated and from this the mass of gas that occupies 22.4 litres at s.t.p. can be worked out. This is the mass of a mole of that gas, i.e. its molecular mass.

Example 15.5

Determine the molecular formula of a compound which has the following percentage composition: carbon = 40.00 per cent, hydrogen = 6.66 per cent, oxygen = 53.33 per cent. Molecular mass = 180.

(a) Determine the empirical formula.

Element	*Per cent mass*	*Molar mass*	*Per cent mass / molar mass*	*Ratio of moles*	*Simplest ratio*
Carbon	40	12	40/12	3.33	1
Hydrogen	6.66	1	6.66/1	6.66	2
Oxygen	53.33	16	53.33/16	3.33	1

Ratio of atoms C:H:O = 1:2:1

Empirical formula CH_2O

(b) Use the empirical formula and molecular mass to determine the molecular formula.

Molecular mass = 180

Total mass of atoms in empirical formula = (12+2+16) = 30

Molecular mass = $(30)_n$ where n is the number of empirical formula units in the molecular formula.

$\therefore$ Molecular mass = $(30)_n$ = 180

$\therefore$ n = 6

Molecular formula = $(6 \times CH_2O) = C_6H_{12}O_6$

Point for Discussion

Determine the molecular formulae of compounds having the following percentage composition by mass:

(a) Carbon = 80 per cent, hydrogen = 20 per cent, molecular mass = 30

(b) Hydrogen 5.9 per cent, oxygen 94.1 per cent, molecular mass = 34

(c) Carbon 38.75 per cent, hydrogen 16.1 per cent, nitrogen 45.2 per cent, molecular mass = 31

15.2 VOLUMETRIC ANALYSIS

Chemical analysis is the process by which we can find out the composition of a substance by breaking it down into its constituents. *Qualitative analysis* is used to find the elements present in a given compound or mixture and *quantitative analysis* determines the proportions by mass of these elements.

Quantitative analysis can be carried out gravimetrically; this involves weighing both reactants and products and is often long and tedious. In volumetric analysis the quantitative composition of a substance is worked out by studying reactions which take place between measured volumes of solutions, and as it is easier and more convenient than a gravimetric analysis, it is largely used for routine analyses in the chemical industry. The apparatus is comparatively cheap and simple to use; the two main items are the burette and the pipette (Section 3.4).

Although volumetric analysis can be used for a considerable number of different reactions, in this section we shall deal only with those between acids and bases, and acids and carbonates.

Molar Solutions. A Revision

A standard solution is a solution of known concentration. This could be expressed in grammes per litre of solution etc. but in volumetric work it is usual to express concentration in terms of molarity.

A molar solution is one that contains one mole of a substance per litre of solution.

Revise the work you did in Section 1.6 on moles and molar solutions including the calculations in the text and in the questions at the end of the chapter.

Concentration can also be expressed as multiples or fractions of a molar solution, e.g. a decimolar solution contains one-tenth of a mole per litre of solution, and is written 0.1M. A two molar solution contains two moles per litre and is written 2M. Thus a decimolar solution of hydrochloric acid will contain $36.5/10 = 3.65 \text{ g l}^{-1}$. A 5 molar solution of potassium hydroxide will contain $(5 \times 56) \text{ g} = 280 \text{ g l}^{-1}$.

Conversely, if the concentration is expressed as the mass per given volume of solution, the molarity can easily be calculated.

Example 15.6

2 g of sodium hydroxide are present in 500 cm^3 of solution. What is the molarity of this solution?

2 g in 500 cm^3 of solution
4 g in 1000 cm^3 of solution

The molar mass of sodium hydroxide $= (23+16+1) \text{ g} = 40 \text{ g}$

40 g in 1000 cm^3 is molar

4 g in 1000 cm^3 is 4/40 molar $= 0.1$M

Note:

$$\text{Molarity} = \frac{\text{Mass l}^{-1}}{\text{molar mass}}$$

and mass l^{-1} = molarity × molar mass. Learn these two expressions; they are fundamental to the work in this section.

Points for Discussion

1. (a) Work out the molar masses of: hydrochloric acid, ethanoic (acetic) acid, sodium carbonate.

(b) What is the mass of one mole of: nitric acid, potassium hydroxide, calcium carbonate?

2. Calculate the concentration in g l^{-1} of the following solutions: 2M sulphuric acid, 0.5M nitric acid, 0.1M potassium hydroxide.

3. Calculate the molarity of each of the following:

(a) 10.6 g of sodium carbonate in 500 cm^3 solution

(b) 6.9 g of potassium carbonate in 100 cm^3 solution

(c) 9.8 g of sulphuric acid in 250 cm^3 solution, 0.7 g of potassium hydroxide in 50 cm^3 solution.

The Use of Molarities in Neutralization Reactions

In volumetric neutralization reactions between acids and bases, a solution of known concentration is added to a solution whose concentration is unknown until the colour change of an indicator shows that the end point has been reached and the reaction is complete. This process is known as titration (Section 3.4). From the volumes of the solutions used in the titration, and the molarity of one solution, the molarity of the other solution can be calculated.

Example 15.7
Find the molarity of a solution of hydrochloric acid, 15.0 cm^3 of which will exactly neutralize 10.0 cm^3 of a 0.2M solution of sodium hydroxide.
The first step is to write the equation for the reaction:

$$NaOH(aq) + HCl(aq) \rightarrow NaCl(aq) + H_2O(l)$$

From the equation we see that:

1 mole of sodium hydroxide reacts with 1 mole of hydrochloric acid
∴ 1 litre of M NaOH reacts with 1 litre of M HCl
10 cm^3 of M NaOH react with 10 cm^3 of M HCl
10 cm^3 of 0.2M NaOH react with 10 cm^3 of 0.2M HCl
*But the volume of hydrochloric acid used in the titration is 15.0 cm^3, i.e. the acid is *less* concentrated and the molarity must be only 10/15 of 0.2M
∴ 10 cm^3 of 0.2M NaOH react with 15.0 cm^3 of $(0.2 \times 10/15)$ M HCl $= 0.133$M

Example 15.8
(In this example we give an alternative method of working the calculation from first principles and also a method for using the formula when the number of reacting molecules is not the same.)
What is the molarity of a solution of sulphuric acid, 10.0 cm^3 of which require 25.0 cm^3 of a solution of 4M potassium hydroxide for complete neutralization?

$$2KOH(aq) + H_2SO_4(aq) \rightarrow K_2SO_4(aq) + H_2O(l)$$

2 moles of KOH react with 1 mole of H_2SO_4
25 cm^3 of 4M KOH contains $(\frac{25}{1000} \times 4)$ moles
∴ 10 cm^3 of the H_2SO_4 must contain *half* this number of moles (see equation)
10 cm^3 of the H_2SO_4 must contain $\frac{1}{2}(\frac{25}{1000} \times 4)$ moles
1000 cm^3 of the H_2SO_4 contain $\frac{1}{2}(\frac{25}{1000} \times 4) \times \frac{1000}{10} = 5$ moles
∴ Acid is 5M.

*If you find difficulty in following this part of the calculation, you can use the formula $M_1 \times V_1 = M_2 \times V_2$ (where M_1, M_2 are the molarities of the two solutions and V_1, V_2 the respective volumes), *but* this only applies when the equation for the reaction shows that the number of molecules of acid and base that react together are the same, i.e. one of each in this example.

Using the Formula We cannot use $M_1V_1 = M_2V_2$ directly because in this case the numbers of reacting molecules is *not* the same. 2 moles of KOH are used for each mole of H_2SO_4. However, there is a way of getting round this difficulty and that is to *reduce* the volume of the reactant involving *most* moles by the ratio in which the reacting molecules appear in the equation.
e.g.

$$H_2SO_4 + 2KOH$$

(Most molecules)

Ratio is 1:2

The volume of the reactant involving most molecules (KOH) = 25 cm^3. This volume is now reduced by $\frac{1}{2}$ and becomes 12.5 cm^3. We now use this volume in the formula:

$$\begin{aligned} M_1V_1 &= M_2V_2 \\ 12.5 \times 4 &= M_2 \times 10 \\ M_2 &= (12.5 \times 4)/10 \\ &= 5 \end{aligned}$$

Although this method may produce the correct answer, you will be a better chemist if you can *explain* the calculations by one of the two other alternatives which involve basic chemical principles.

Point for Discussion

Determine the molarity of the following solutions:

(a) hydrochloric acid, 25 cm^3 of which neutralize 20 cm^3 of 1.5M sodium hydroxide solution;
(b) sulphuric acid, 20 cm^3 of which neutralize 30 cm^3 of 0.1M potassium hydroxide solution;
(c) sodium hydroxide, 10 cm^3 of which neutralize 15 cm^3 of 2.5M ethanoic (acetic) acid;
(d) nitric acid, 10 cm^3 of which react with 25 cm^3 of 0.4M ammonia solution.

Experimental Work on Volumetric Analysis

You should now be familiar with the method of calculating molarities from the results of titrations of acids and bases, and will be ready to do some practical work. Before starting the next experiment, look back to page 84 and thoroughly revise the work on the use of a burette and pipette which you used in Experiment 3.11 to prepare a neutral solution.

Experiment 15.1
To Determine the Molarity of a Solution of Hydrochloric Acid *

Using the apparatus and procedure of Experiment 3.11 with phenolphthalein as the indicator, titrate the given solution of hydrochloric acid, of unknown concentration, with the solution of sodium hydroxide of known molarity. Record your results as in Experiment 3.11 and use them, with the given molarity of the sodium hydroxide solution, to calculate the molarity of the hydrochloric acid.

Further Experiments Involving Neutralization Reactions In order to familiarize yourself with the techniques of volumetric work and also to obtain more practice in doing the calculations, we suggest that you repeat Experiment 15.1 using solutions of other acids and bases, e.g. find the molarity of a solution of potassium hydroxide using 0.1M sulphuric acid, and the molarity of a solution of nitric acid using 0.2M sodium hydroxide solution.

Care must be taken over the choice of indicator, the suitability of which depends on the *strength* of the acid and base used (Section 3.3). The following table will be a useful guide:

Titration	*Indicator*
Strong acid–strong base	Any indicator
Weak acid–strong base	Phenolphthalein
Strong acid–weak base	Methyl orange or methyl red
Weak acid–weak base	No indicator is satisfactory

Reactions Between Acids and Soluble Carbonates

These experiments are carried out in exactly the same manner as the titrations between acids and bases, methyl orange being used as the indicator. The calculations are similar to those for acid–base reactions.

Example 15.9
2.12 g of anhydrous sodium carbonate were dissolved in water and the solution made up of 200 cm^3. 25 cm^3 of this solution, on titrating with nitric acid, was found to neutralize 15 cm^3 of the acid. Calculate the molarity of the nitric acid.

First find the molarity of the sodium carbonate solution.
2.12 g of Na_2CO_3 in 200 cm^3 of solution
$\therefore$ 10.6 g of Na_2CO_3 in 1000 cm^3 of solution
Molar mass of Na_2CO_3 is $(46+12+48)$ g $= 106$ g

Solution of Na_2CO_3 is $(10.6/106)$M $=$ *0.1M*

The equation for the reaction is:

$$Na_2CO_3(aq)+2HNO_3(aq) = 2NaNO_3(aq)+CO_2(g)+H_2O(l)$$

1 mole of sodium carbonate reacts with 2 moles of nitric acid
25 cm^3 of 0.1M Na_2CO_3 contains $(\frac{25}{1000}\times 0.1)$ moles
$\therefore$ 15 cm^3 of given HNO_3 contains $2(\frac{25}{1000}\times 0.1)$ moles
1000 cm^3 of given HNO_3 contains $1000/15\times 2(\frac{25}{1000}\times 0.1)$ moles $=$ 0.33 moles

Nitric acid is 0.33M.

Using the formula: $M_1V_1 = M_2V_2$

$Na_2CO_3+2HNO_3$
↑
(Most molecules) Ratio is 1:2

Volume of reactant involving most molecules (nitric acid) $=$ 15 cm^3. This volume is reduced by half and becomes 7.5 cm^3. We now use this volume in the formula.

$$0.1\times 25 = M_2\times 7.5$$
$$M_2 = 0.33$$

The acid is 0.33M.

Experiment 15.2

†(a) To Prepare a Standard Solution of Sodium Carbonate and to Calculate its Molarity. (b) To Determine the Molarity of a Solution of Hydrochloric Acid Using the Prepared Standard Sodium Carbonate Solution *

Apparatus

Watch-glass, 250 cm³ beaker, 250 cm³ graduated flask, burette and stand, white tile, pipette, spatula, stirring rod, wash-bottle, filter funnel, access to balance sensitive to 0.001 g.

Pure anhydrous sodium carbonate, dilute hydrochloric acid, tepid deionized water, methyl orange.

Procedure

(a) The aim of this stage is to transfer *all* of an accurately weighed quantity of sodium carbonate into deionized water and make up the solution to exactly 250 cm³. Great care is needed to ensure that this is done.

(b) Weigh the watch-glass and record its mass (correct to 0.001 g). Add spatula measures of pure anhydrous sodium carbonate until about 2.5 g have been added. Record the accurate total mass and find the mass of the sodium carbonate by subtraction.

(c) Carefully transfer the sodium carbonate to the beaker and using the wash-bottle of deionized water, wash the watch-glass with a stream of water so that the washings run into the beaker (Figure 15.1). Add about 100 cm³ of deionized water and stir gently. When all the powder has dissolved, pour the solution extremely carefully through the filter funnel into the graduated flask. Wash the stirring rod so that the washings run into the beaker and wash out the beaker with deionized water, pouring the washings into the flask. Add deionized water to the flask to within 2 cm of the etched graduation mark, and then add the water slowly from a pipette until the bottom of the meniscus corresponds with the plane of the graduation mark when viewed at eye level. Do not allow the drops to run down the inside neck of the flask. Put in the stopper and invert the flask several times so that the solution is thoroughly mixed.

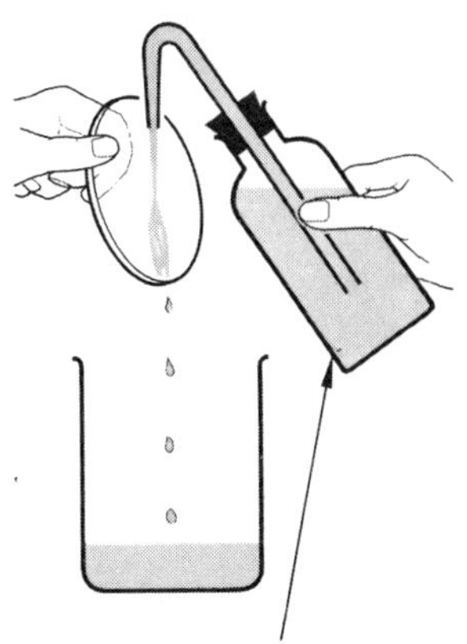

Figure 15.1 Washing the last traces of sodium carbonate from the watch-glass and into the beaker

(d) Titrate 10 cm³ portions of the solution with the given hydrochloric acid, using methyl orange as indicator. Make a table of your results in the usual way.

Points for Discussion

1. From the mass of sodium carbonate dissolved in 250 cm³ of solution, calculate the molarity of your sodium carbonate solution.

2. From the volume of acid used, calculate the molarity of the hydrochloric acid. Look back at the last example if you are not sure of the working.

3. Pure anhydrous sodium carbonate is made by heating sodium hydrogen carbonate ($NaHCO_3$) to constant mass; write the equation for this reaction.

15.3 VOLUME CHANGES IN GASES

Alterations in the volume of a gas may be due to physical changes or chemical reactions.

Physical Factors

Physical factors which affect the volume of a gas are temperature and pressure, and you have probably learnt of these changes and the laws that govern them in your physics lessons. As these factors must always be taken into account when dealing with gas volumes in chemistry, we will consider them briefly now.

Effect of Change of Pressure

This was first investigated by Robert Boyle in 1662. From his experimental results the law that bears his name was formulated. This states that 'the volume of a fixed mass of gas is inversely proportional to the pressure, if the temperature remains constant'.

The law can be expressed mathematically as $V \propto 1/P$ or $PV = $ a constant and thus $P_1V_1 = P_2V_2$ (where P_1 and V_1 are the initial pressure and volume, and P_2 and V_2 the final pressure and volume).

Effect of Change of Temperature

This is given by Charles' Law. 'The volume of a fixed mass of gas is directly proportional to its temperature (on the Kelvin scale), if the pressure remains constant.'

The law can be expressed mathematically as $V \propto T$, from which $V/T = $ a constant and $V_1/T_1 = V_2/T_2$.

[Note: T_1 and T_2 are on the Kelvin scale. A temperature expressed in Kelvin is equal to the temperature expressed in degrees Celsius plus 273, e.g. 30 °C = (30 + 273) K = 303 K.]

Effect of Change of Both Pressure and Temperature

The mathematical interpretations of the two laws can be combined to give an expression for the volume of a gas when *both* pressure and temperature are changed.

$$\frac{P_1V_1}{T_1} = \frac{P_2V_2}{T_2}$$

This is known as the *general gas equation*. By substituting in the equation any five of the variables, we can find the value of the sixth variable.

Example 15.10

The volume of a fixed mass of gas is 300 cm³ at 0 °C and 760 mm pressure. What will be the volume at 273 °C and 380 mm pressure?

Substituting in the general gas equation, with the temperatures on the Kelvin scale:

$$\frac{760 \times 300}{273} = \frac{380 \times V_2}{546}$$

$$V_2 = \frac{760 \times 300 \times 546}{380 \times 273} = 1200$$

$$V_2 = 1200 \text{ cm}^3$$

[Note: it is a simple matter to make a rough check of your answer. Look at the two variables T and P. If T increases then V increases. If P increases also, then V decreases.]

In some cases it may be necessary to find the new pressure of a gas which undergoes a change in temperature and volume.

Example 15.11
The volume of a given mass of gas is 400 cm^3 at 680 mm pressure and 57 °C. What will be the pressure if the volume decreases to 300 cm^3 and the temperature to 24 °C?

As before, substitute in the general gas equation:

$$\frac{P_1V_1}{T_1} = \frac{P_2V_2}{T_2}$$

$$\frac{680 \times 400}{330} = \frac{P_2 \times 300}{297}$$

$$P_2 = \frac{680 \times 400 \times 297}{300 \times 330}$$

$$= 816 \text{ mm}$$

Pressure is 816 mm.

Standard Temperature and Pressure

As small changes in temperature and pressure make considerable differences to the volume of a gas, these volumes are compared under a standard set of conditions known as Standard Temperature and Standard Pressure (s.t.p.). Standard Temperature is 273 K (0 °C) and Standard Pressure is 760 mm of mercury (one atmosphere).

Points for Discussion

1. A gas has a volume of 190 cm^3 at 27 °C and 840 mm pressure. What will be its volume at s.t.p.?

2. A certain mass of gas has a volume of 200 cm^3 at 31 °C and 608 mm pressure. If the volume of the gas becomes 160 cm^3 at a temperature of 47 °C, what will be the new pressure?

3. Revise the work you did on the kinetic theory in Chapter 1 and explain Boyle's and Charles' Laws in terms of this theory.

Chemical Factors

Reactions Involving Gases

You will already be familiar with the dramatic changes in the volumes of gases during chemical reactions. Obvious examples are the copious production of hydrogen from the reaction between zinc and hydrochloric acid or the large volume of carbon dioxide evolved when copper(II) carbonate is heated. There are other cases where a chemical reaction brings about a *decrease* in the total volume of gas as when oxygen is passed over heated copper, and there are also examples of reactions in which there is no change in volume between gaseous reactants and products, e.g. when chlorine and hydrogen react to form hydrogen chloride.

Special apparatus has been designed for measuring the volume changes in gases during chemical reactions. One such piece of apparatus is a *eudiometer*, which is used for reactions which need heat or an electric discharge to start them or which take place with explosive violence. If there is a eudiometer in the laboratory you may be allowed to examine it. Simpler methods involve the use of gas syringes as the following experiment demonstrates.

Experiment 15.3

†To Determine the Volume Changes in the Reaction between Nitrogen Monoxide and Oxygen ***

Apparatus

Two glass syringes (100 cm^3), two stands and syringe holders, three-way stopcock, rubber tubing.
Oxygen cylinder, apparatus for producing pure nitrogen monoxide.

Procedure

(a) Connect the two syringes and the stopcock by rubber tubing as shown in Figure 15.2.

(b) Fill one syringe with nitrogen monoxide and push the gas out again. Do this several times, finally leaving 40 cm^3 of the gas in the syringe. Repeat the procedure using oxygen in the other syringe so that 50 cm^3 of oxygen remain in the syringe after the preliminary 'washing out'.

(c) Turn the stopcock so that the two syringes are connected and push 5 cm^3 of oxygen into the syringe containing the nitrogen monoxide. Close the stopcock, wait until there is no further change in volume and note the volume of gas in each syringe.

(d) Repeat the process with further 5 cm^3 portions of oxygen until there is no change in the *total* volume of gas in the two syringes.

Points for Discussion

1. Write down the name and formula of the brown gas formed in the reaction.

2. When the reaction finished, what volume of oxygen had been used?

3. What was the volume of nitrogen monoxide at the start of the experiment? Did all this gas react with oxygen? Give a reason for your answer.

4. Write down the volumes of oxygen and nitrogen monoxide that reacted together.

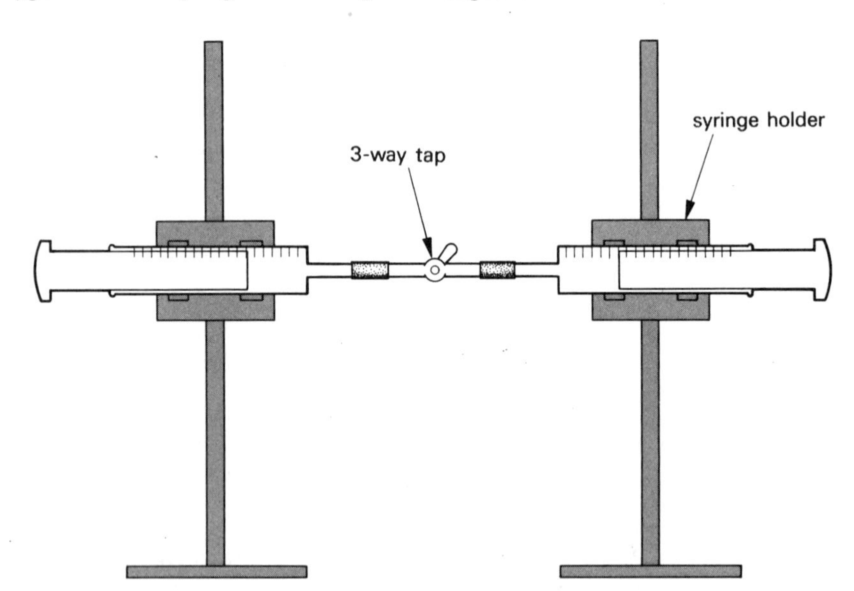

Figure 15.2 Investigating the volume relationship between nitrogen monoxide and oxygen

The Relationship Between Volumes of Combining Gases

The following are the results of some experiments in which gas syringes or a eudiometer have been used to find the volume changes in gases during chemical reactions.

(a) 100 cm^3 of hydrogen react with 100 cm^3 of chlorine to produce 200 cm^3 of hydrogen chloride.

(b) 200 cm^3 of hydrogen react with 100 cm^3 of oxygen to produce 200 cm^3 of steam.

(c) 100 cm^3 of carbon monoxide react with 50 cm^3 of oxygen to produce 100 cm^3 of carbon dioxide.

(d) 50 cm^3 of nitrogen react with 150 cm^3 of hydrogen to form 100 cm^3 of ammonia.

The gases in each reaction were measured under the same conditions of temperature and pressure.

Points for Discussion

1. You will notice that in some cases there is a decrease in the total volume of gas during the reaction. This does not mean that matter has been destroyed but that it has been rearranged so as to occupy a smaller volume. This will be explained in detail later in the section.

2. Look carefully at the volumes of the reacting gases and the volumes of the products. Which of the following would be the best way of expressing the relationship between these volumes?

A. The total volume of the reactants is the same as the total volume of the products.

B The volumes of the reactants and the volumes of the products are in a simple whole number ratio.

C There is always a decrease in the volume in a reaction involving gases.

D There is always an increase in volume in a reaction involving gases.

E There is no obvious relationship between the volumes involved in reactions between gases.

Gay-Lussac's law At the beginning of the nineteenth century, the French scientist, Joseph Gay-Lussac, was investigating the nature of air and trying to find out if it was a mixture or a compound. He collected a great variety of air samples and analysed them. (He even made a perilous journey in a balloon to collect some air at high altitudes.) He did this by exploding the oxygen in a sample of air, with hydrogen in a eudiometer. He found that, in all cases, two volumes of hydrogen combined with one of oxygen, and this constant but simple ratio led him to experiment with other gases. In 1808 he summarized his results as follows. 'When gases combine, they do so in volumes which bear a simple ratio to one another and to the volume of the product if gaseous. All volumes must be measured at, or corrected to, the same temperature and pressure.'

Did your answer to Points for Discussion 2 agree with Gay-Lussac's law?

Example 15.12

48 cm^3 of methane were mixed with 212 cm^3 of oxygen and the mixture was exploded. The product, after cooling and reverting to the initial temperature and pressure, occupied 164 cm^3 of which 116 cm^3 were unused oxygen. Show that these results illustrate Gay-Lussac's law.

Methane + oxygen → carbon dioxide + steam (condenses to negligible volume of water)

Volume of oxygen used $= (212-116)\ cm^3 = 96\ cm^3$

Volume of carbon dioxide formed $= (164-116)\ cm^3 = 48\ cm^3$

48 cm^3 methane react with 96 cm^3 oxygen to form 48 cm^3 carbon dioxide

1 volume methane reacts with 2 volumes oxygen to form 1 volume carbon dioxide

Ratio of reacting and final volumes is 1:2:1 which is in accordance with Gay-Lussac's law.

Points for Discussion

1. 60 cm^3 oxygen were added to 24 cm^3 of carbon monoxide and the mixture was ignited. The product occupied 72 cm^3, of which 48 cm^3 were unused oxygen. Show that this is an illustration of Gay-Lussac's law.

2. 24 cm^3 of methane and 96 cm^3 of oxygen were exploded together. The final volume, measured under the original conditions, was 72 cm^3, neglecting the water formed. 48 cm^3 of this was unused oxygen. Show that this illustrates Gay-Lussac's law.

3. 60 cm^3 of hydrogen were mixed with 40 cm^3 of chlorine and exposed to bright sunlight. The total volume of gas was unchanged, but when shaken with water the volume was reduced to 20 cm^3 of gas, which was shown to be pure hydrogen. Are these results in accordance with Gay-Lussac's law?

Gay-Lussac. Dalton and Avogadro

The Connection Between Gay-Lussac and Dalton

Gay-Lussac's work was published in 1808. Four years earlier Dalton had propounded his theory that atoms combined in the ratio of *small whole numbers*. The obvious question for scientists of the time was, 'What is the connection between Dalton's 'small whole numbers' of combining atoms and Gay-Lussac's 'small whole numbers' of combining volumes of gases'?

Avogadro

The theories of Dalton and Gay-Lussac were integrated in 1811 when the Italian scientist, Amedeo Avogadro, by what might be called a brilliant guess, propounded his now famous hypothesis. This stated that, 'equal volumes of gases, under the same conditions of temperature and pressure, contain equal numbers of *molecules*'. The really important part of Avogadro's work was his introduction of the idea that gases consist of molecules. You now take this for granted, but at the time of Avogadro scientists were aware of the existence of only one fundamental particle, the atom. They knew that *different* atoms could make up a unit (a compound atom) but had no idea that atoms of the same element could join to make another kind of basic particle. Avogadro considered that molecules were groups of atoms, usually pairs, which formed the fundamental units of a gaseous element. Once this idea was conceived, his hypothesis was the next obvious and logical step, and the findings of both Dalton and Gay-Lussac became compatible.

Let us now see what happens when we apply Avogadro's hypothesis to the combination of hydrogen and chlorine.

1 volume of hydrogen + 1 volume of chlorine → 2 volumes of hydrogen chloride

Applying Avogadro's hypothesis:

X *molecules* of hydrogen + X *molecules* of chlorine → 2X *molecules* of hydrogen chloride

∴ 2 molecules of hydrogen chloride contain 1 molecule of hydrogen and 1 molecule of chlorine
∴ 1 molecule of hydrogen chloride contains $\frac{1}{2}$ molecule of hydrogen and $\frac{1}{2}$ molecule of chlorine

Now if, as Avogadro suggested, the molecules of most gaseous elements are diatomic, i.e. contain two atoms, half a molecule of hydrogen will be one atom; the molecule of hydrogen chloride will consist of one atom of hydrogen and one atom of chlorine and will have the formula HCl.

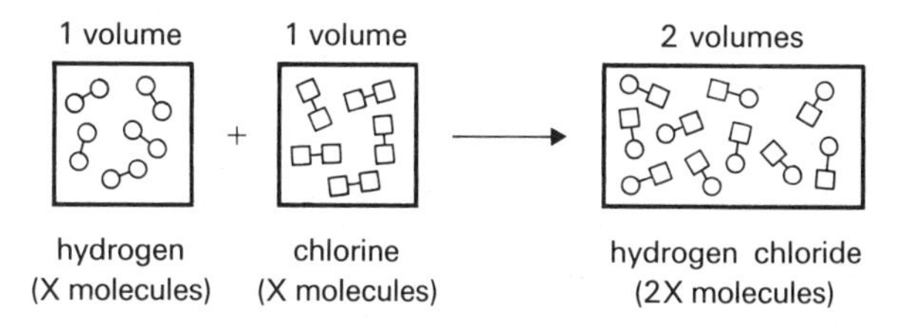

Thus Avogadro was able to use the ideas of both Dalton and Gay-Lussac and couple them with his own inspired theory of the gaseous molecule to produce what is now acknowledged to be one of the corner stones of chemistry. In spite of this, Avogadro's work was neglected for many years and it was not until 1858 that, due to the efforts of S. Cannizzaro, it was accepted and its importance realized. It was still called a hypothesis because, although it could be used to explain a number of known facts, there was no practical method of proving it to be true. However its assumption forms the basis of all our ideas on gases and the pressures they exert, it explains Gay-Lussac's law, enables us to find the formulae of gases, explains why the moles of a balanced equation are in the same simple ratio as the volumes of gases taking part in the reaction, and provides the basis for the standard method of finding the molecular masses of gases. The vast accumulation of quantitative data on mass and volume relationships which we have at our disposal today leaves no doubt as to the validity of Avogadro's hypothesis, which is now more correctly termed Avogadro's law. As a somewhat belated recognition of his work, the number of particles in a mole is known as Avogadro's Number: approximately 6×10^{23} (Section 1.6).

Gay-Lussac and Avogadro

We can use Avogadro's law to explain the volume relationships in reactions involving gases which were first summarized by Gay-Lussac. For example, in the reaction between hydrogen and oxygen:

$2H_2(g)$	$+O_2(g)$	$\rightarrow 2H_2O(g)$
2 molecules	1 molecule	2 molecules
2x molecules	x molecules	2x molecules

Using the converse of Avogadro's law, i.e. that equal numbers of molecules of gases which are subjected to the same pressure and temperature conditions will occupy equal volumes, we can write:

2 volumes hydrogen react with 1 volume oxygen to form 2 volumes of steam

We can show this relationship diagrammatically as:

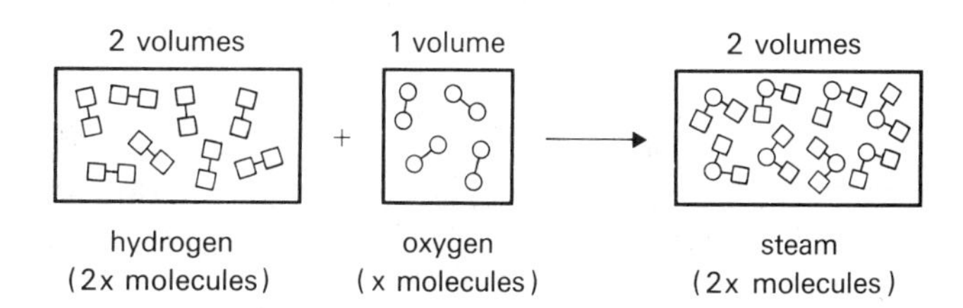

This also shows clearly how the rearrangement of the molecules of hydrogen and oxygen to form molecules of steam produces a decrease in the number of particles and so a decrease in the total volume of gas. (Can you explain this decrease in volume in terms of the kinetic theory? Remember that the volumes are measured at constant pressure.)

Point for Discussion

In a similar manner, show the relationship between numbers of molecules and the ratio of the reacting volumes in the following reactions: (a) carbon monoxide and oxygen, (b) nitrogen and hydrogen, (c) carbon monoxide and steam, (d) sulphur(IV) oxide (sulphur dioxide) and oxygen.

Diatomic Molecules

We have said that Avogadro's work was more in the nature of an inspired guess than logical reasoning from observed experimental facts. The statement that gases existed as clusters of atoms and, in particular, that the molecules of gaseous elements were diatomic, was incapable of verification at the time. All the same it did form a convenient link between the work of Dalton and Gay-Lussac and became an important working hypothesis, forming the basis of much important chemical theory and experimental work in the years that followed its acceptance. Today there is such an overwhelming mass of evidence that gaseous elements, except for the noble gases, usually exist as diatomic molecules that this fact is universally accepted by scientists.

Some Important Consequences of Avogadro's Law

In Chapter 1 you learned that chemists can compare the masses of atoms and molecules even though an individual atom or molecule cannot be weighed directly. You may have wondered how such comparisons can be made. You will learn how the use of Avogadro's law enables this to be done.

The Relative Masses of Molecules

Suppose we consider two gases, A and B, and we weigh equal volumes of each under the same conditions. We find that the mass of gas A is exactly twice the mass of gas B. Now as we have weighed the same volume of both gases, we have by Avogadro's law, weighed equal numbers of molecules of each. Let this number be x.

$$\therefore \frac{\text{Mass of x molecules of gas A}}{\text{Mass of x molecules of gas B}} = \frac{2}{1}$$

Dividing by x

$$\frac{\text{Mass of 1 molecule of gas A}}{\text{Mass of 1 molecule of gas B}} = \frac{2}{1}$$

This means that the ratio of the masses of *single* molecules of any two gases is the same as the ratio of the masses of equal volumes of these two gases measured under the same conditions.

Thus if we take as a unit, the mass of one kind of molecule, the mass of any other molecule can be compared to it and so relative molecular masses can be obtained, simply by comparing equal volumes of gases under the same conditions.

The Volume of a Mole of Gas

You will remember that as individual particles such as atoms, ions and molecules are so small, chemists use a large number of particles as a basic unit when comparing the amounts of different substances taking part in chemical reactions. This practical unit is the mole, or Avogadro's Number and it consists of 6.02×10^{23} particles.

As the particles which make up most gases are molecules, a mole of any gas will contain 6.02×10^{23} molecules. By Avogadro's law all gases containing this number of molecules will occupy the same volume, if measured under the same conditions. This volume is *22.4 litres at s.t.p.* and is known as the *molar volume* or *gramme-molecular volume*. Thus one mole of any gas at s.t.p. will occupy approximately 22.4 litres. (Under normal laboratory conditions the molar volume is approximately 24 litres.)

Points for Discussion

1. What is the volume occupied at s.t.p. by, (a) a mole of hydrogen, (b) 2 moles of chlorine, (c) 0.5 moles of carbon dioxide, (d) 0.1 mole of hydrogen sulphide?

2. At a temperature of 323K (50 °C) and a pressure of 760 mm, will a mole of oxygen contain (a) more than, (b) less than, (c) the same number as, the Avogadro Number of molecules?

The Mass of a Mole of Gas

Although the *volume* occupied by a mole is the same for all gases under the same conditions, the molar masses of different gases vary because the masses of individual molecules vary (Section 1.6). The mass of a mole of oxygen is the mass of 6.02×10^{23} oxygen molecules. This can be shown experimentally to be 32 g. Now the atomic mass of oxygen is 16 and as the molecule of oxygen is diatomic, its molecular mass must be 32. The mass of a mole of oxygen is therefore its molecular mass expressed in grammes. This value is also known as the *gramme-molecular mass*. The same reasoning can be applied to other gases, so you can write the mass of a mole (or the gramme-molecular mass) of any gas simply by expressing its molecular mass in grammes.

Thus the mass of a mole of hydrogen (H_2) = (2×1) g = 2 g. The gramme-molecular mass of carbon dioxide, CO_2 = $(12+32)$ g = 44 g. Remember the gramme-molecular mass of a gas occupies the gramme-molecular volume, which is 22.4 litres at s.t.p.

Points for Discussion

1. What is the volume, measured at s.t.p. of, (a) 16 g of oxygen, (b) 71 g of chlorine, (c) 10 g of hydrogen, (d) 4.4 g of carbon dioxide, (e) 16 g of sulphur(IV) oxide (sulphur dioxide)?

2. What is the mass of (a) 22.4 litres of ammonia, (b) 112 cm^3 of carbon monoxide, (c) 5.6 litres of hydrogen sulphide, (d) 4480 litres of nitrogen? (All volumes measured at s.t.p.)

Vapour Density

The relative vapour density of a gas is defined as the mass of a given volume of the gas compared with the mass of the same volume of hydrogen, measured under the same conditions of temperature and pressure.

This can be expressed as:

$$\text{Vapour density of any gas G} = \frac{\text{Mass of any volume of G}}{\text{Mass of the same volume of hydrogen}}$$

(Volumes measured at same temperature and pressure.)

In the case where the volume is the *molar volume* at s.t.p., then the mass of the molar volume of the gas G will be its gramme-molecular mass and the mass of the molar volume of hydrogen will be 2 g.

$$\therefore \text{ vapour density of gas G} = \frac{\text{molar mass of G}}{2}$$

$$\therefore\ 2 \times \text{vapour density of any gas} = \text{molar mass of the gas.}$$

This is a most important relationship because, as it is relatively easy to determine the vapour density of a gas, it provides a simple means of finding its molecular mass. The task of comparing the masses of particles too small to be seen is not as difficult as it may first appear. Learn the relationship between vapour density and molar mass and make sure that you fully understand how the concepts in this chapter enable such molecular masses to be determined.

Example 15.13

An evacuated flask has a mass of 80.050 g. When filled with hydrogen its mass is 80.052 g and filled with chlorine at the same temperature and pressure its mass is 80.120 g. Calculate from these results, the molecular mass of chlorine.

Mass of hydrogen = (80.052 − 80.050 g) = 0.002 g

Mass of same volume of chlorine = (80.120 − 80.050) g = 0.070 g

$$\text{Vapour density of chlorine} = \frac{0.070}{0.002} = 35$$

Molecular mass of chlorine = *35 × 2* = *70*

Points for Discussion

1. Calculate from the molecular mass, the vapour density of each of the following gases: carbon monoxide, hydrogen chloride, nitrogen, sulphur(IV) oxide (sulphur dioxide).

2. What are the molecular masses of gases having the following vapour densities: (a) 8.5, (b) 22, (c) 16, (d) 17?

3. A gas cylinder filled with hydrogen holds 10 g of the gas. The same cylinder holds 160 g of gas A or 220 g of gas B under the same temperature and pressure conditions. Calculate the molecular masses of gas A and gas B.

The Molecular Formula of a Gas

We have already shown how, if we accept Avogadro's law, we can prove that the molecular formula of hydrogen chloride can be worked out from the volume changes which occur during the reaction between hydrogen and chlorine. In a similar manner and by using specialized apparatus to follow the volume changes, the molecular formula of a number of gaseous compounds can be obtained.

Points for Discussion

[Note: in all the following cases, the volumes given are measured at the same temperature and pressure.]

1. A measured volume of an oxide of nitrogen is decomposed by heated nickel; the volume of the gas is reduced by half and the gas remaining is nitrogen. Deduce the formula for the gas. (Vapour density of gas is 23. Atomic masses: N = 14, O = 16).

2. A mixture of 125 cm^3 of hydrogen and 125 cm^3 of chlorine is exposed to bright sunshine in a suitable container. The resulting gas, which is found to be pure hydrogen chloride, has a volume of 250 cm^3. Deduce the formula for hydrogen chloride.

3. When excess charcoal was burnt in a known volume of pure oxygen, and the apparatus cooled, it was found that there was no change in volume but the residual gas was pure carbon dioxide. What is the formula for the gas? (Molecular mass = 44; C = 12, O = 16).

15.4 CHEMICAL EQUATIONS

A chemical equation has been described as a type of chemical shorthand but it is much more than this. As well as telling the chemist what reactants are used and which products are formed, it indicates to him the mass and volume relationships between these substances and may also tell him the degree of energy change during the reaction. If state symbols are used they provide additional information as to whether the substances concerned in the reaction are solids, liquids, gases or free ions in solution.

Writing Equations

When we write a chemical equation we take for granted two concepts in Dalton's atomic theory, (a) that atoms cannot be created or destroyed, (b) that the atoms, molecules or ions which appear in an equation are representative of the much larger masses of the substances used in the reaction.

Experiment 15.4

†To Compare the Total Mass of the Reactants with the Total Mass of the Products in a Chemical Reaction **

Apparatus
Divided flask, two teat pipettes, access to balance. Solutions of sodium chloride and silver nitrate.

Procedure
(a) Half fill one of the compartments of the divided flask with sodium chloride solution by means of the teat pipette. Using the other pipette, add about the same volume of silver nitrate solution to the other compartment. Take care that the solutions do not mix during this process (Figure 15.3).

(b) Weigh the flask and contents and record the mass.

(c) Swirl the solutions in the flask so that they mix together, but ensure that no liquid is spilt.

(d) Reweigh the flask and contents.

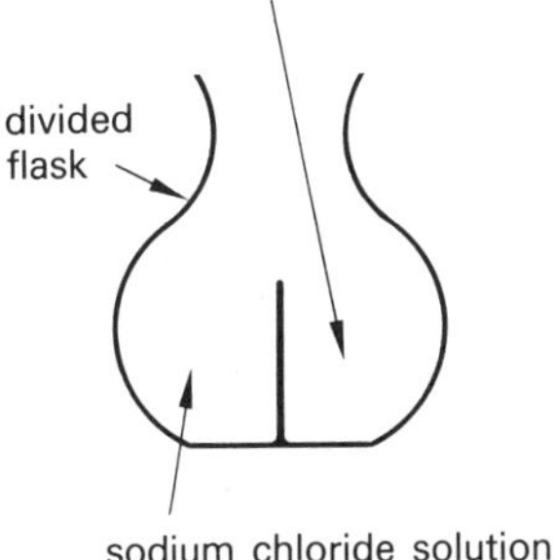

Figure 15.3 To prove the Law of Conservation of Mass

Points for Discussion
1. What proof have you that a chemical reaction took place in the flask?

2. Write down the name of any new substance formed during the reaction.

3. Was there any difference between the total mass of the reactants and the total mass of the products?

Can Atoms be Created or Destroyed?

The reaction that took place in the divided flask illustrates a fundamental chemical law, the *Law of Conservation of Matter* which states that, 'matter cannot be created or destroyed, but can be changed into different forms'. As atoms are units of matter, the law should also apply to them. However, you probably know that atoms are not quite so permanent as Dalton had supposed and that matter can be changed into *energy* according to Einstein's equation, $E = mc^2$. However as this type of reaction only occurs to any measurable degree in such circumstances as the explosion of an atomic bomb or the working of a nuclear power station, it is not likely to be encountered during the normal work of a school chemistry laboratory.

The Law of Conservation of Matter, therefore, applies to all the chemical reactions that you have to deal with and so when writing the equation for any reaction you must be sure that the exact number of atoms which make up the chemical formulae of the reactants on the left-hand side of the equation appear on the right-hand side of the equation, to make up the chemical formulae of the products.

Many different equations have been used throughout this book and you should by now be able to write most of the simpler equations without difficulty. The work in this section will be, therefore, mainly a summary of what you have already learnt.

A Word of Warning

It is a waste of time to try to write chemical equations if you do not know the symbols and valencies of the atoms of the more common elements.

Rules for Writing Equations

1. Be sure that you know the names of all the reactants and products.
2. Write the equation in words. (As you become more skilled in the use of formulae, this step can be omitted.)
3. Write down a 'skeleton' equation, using the exact formulae for all the substances taking part and leaving a space in front of each formula for adding numbers.
4. Balance the equation.
5. Add the state symbols behind each reagent.

Example 15.14
Write the equation for the reaction between sodium and water.

1. Sodium reacts with water to produce a solution of sodium hydroxide; hydrogen is evolved.
2. Sodium + Water → Sodium hydroxide + Hydrogen
3. $Na + H_2O \rightarrow NaOH + H_2$
   (Remember that most of the gaseous elements are diatomic.)
4. There are two atoms of hydrogen on the L.H.S. (left-hand side) and three atoms on the R.H.S. (right-hand side). We must first make the number of hydrogen atoms on the R.H.S. an even number, because we are dealing in 'units of two' in the molecule H_2O. This can be achieved by having two NaOH units.

$$Na + H_2O \rightarrow 2NaOH + H_2$$

On the R.H.S. we now have 'in excess' one Na, two H, and one O and the equation can be balanced by adding these units to the L.H.S.

$$2Na + 2H_2O \rightarrow 2NaOH + H_2$$

5. Finally we add the state symbols.

$$2Na(s) + 2H_2O(l) \rightarrow 2NaOH(aq) + H_2(g)$$

The whole operation is, of course, carried out in one line.

Example 15.15
Write the equation for the combustion of butane in oxygen.

1. Butane (C_4H_{10}) will react with oxygen to form carbon dioxide and steam.
2. Butane + Oxygen → Carbon dioxide and Steam
3. $C_4H_{10} + O_2 \rightarrow CO_2 + H_2O$
4. To balance an equation in which rather large numbers of atoms are involved, consider first the substance which has the largest numbers of

atoms and 'balance' the equation with respect to that substance.

$$C_4H_{10}+O_2 \rightarrow 4CO_2+5H_2O$$

We could now balance this equation by putting '$6\frac{1}{2}$' in front of the oxygen molecule but as we cannot have half molecules, we complete the equation by multiplying throughout by two.

$$2C_4H_{10}(g)+13O_2(g) \rightarrow 8CO_2(g)+10H_2O(g)$$

Point for Discussion

Write equations for (a) the reaction between zinc and dilute sulphuric acid, (b) the reaction between lead monoxide and dilute nitric acid, (c) the reaction between chlorine and ammonia, (d) the reaction between copper and nitric acid to produce nitrogen monoxide, (e) the reaction between chlorine and hot concentrated potassium hydroxide, (f) the reaction between tri-lead tetroxide and dilute nitric acid.

Ionic Equations

In an ionic equation we only include those ions which are formed or changed in the course of the reaction. This is important as there are many reactions in which there is no change in some of the ions but the ordinary stoichiometric equation does not show this.

Consider the reaction between sodium chloride solution and silver nitrate solution (Experiment 15.4).

$$NaCl(aq)+AgNO_3(aq) \rightarrow AgCl(s)+NaNO_3(aq)$$

is a sound balanced chemical equation but the ionic equation

$$Ag^+(aq)+Cl^-(aq) \rightarrow AgCl(s)$$

is a clearer representation of what has happened. The Na^+ and NO_3^- ions are *not* joined in solution, as the first equation suggests. They were free ions at the start of the reaction and they are free at the end, so are not included in the ionic equation. Such ions are known as 'spectator ions' as they just 'watch from the side-lines'.

An ionic equation can be obtained as follows:

1. Determine the stoichiometric equation as on page 418.
2. From your knowledge of bonding and of solubilities, decide which substances are ionic, and which ions are free in solution.
3. Rewrite the equation in ionic form.
4. Eliminate any species which appear on both sides of the equation as free ions (i.e. the spectator ions).

As you gain experience in dealing with ionic equations you will be able to telescope these separate steps into one operation.

For the reaction between $AgNO_3(aq)$ and $NaCl(aq)$:

1. $AgNO_3(aq)+NaCl(aq) \rightarrow AgCl(s)+NaNO_3(aq)$
2. and 3. $Ag^+(aq)+NO_3^-(aq)+Na^+(aq)+Cl^-(aq) \rightarrow$
$AgCl(s)+Na^+(aq)+Cl^-(aq)$
4. $Ag^+(aq)+Cl^-(aq) \rightarrow AgCl(s)$

[Note: it is useless to attempt this type of equation unless you have a sound knowledge of bonding, types of reaction and solubilities.]

Point for Discussion
Write ionic equations for the reactions between:

(a) aqueous solutions of barium chloride and magnesium sulphate,

(b) aqueous solutions of calcium nitrate and sodium carbonate,

(c) an aqueous solution of sodium hydroxide and dilute nitric acid,

(d) zinc and an aqueous solution of copper(II) sulphate.

Reversible Reactions

The arrow → between two sides of an equation signifies that the reaction goes practically to completion. The sign ⇌ means that the reaction is reversible and both forward and back reactions are going on at the same time. You have learnt about this kind of reaction in Section 14.2. Here we would only remind you to use the sign ⇌ when writing an equation for a reversible reaction.

Energy Changes

Energy changes are shown in equations as heat given out or taken in (Chapter 13). This is expressed in joules per gramme-equation or more conveniently in kilojoules per gramme-equation, ΔH. Remember in an exothermic reaction where heat is given out and the system loses energy, ΔH is negative. In endothermic reactions ΔH is positive.

Completion of Unbalanced Equations

It is a useful exercise to complete unbalanced equations which represent actual reactions but where the symbol for one of the atoms is replaced by X.

Example 15.16
Balance the following equation:

$$XO_2 + HCl \rightarrow XCl_2 + Cl_2 + H_2O$$

1. Balance the oxygen atoms:

$$XO_2 + HCl \rightarrow XCl_2 + Cl_2 + 2H_2O$$

2. Balance the hydrogen and chlorine atoms:

$$XO_2 + 4HCl \rightarrow XCl_2 + Cl_2 + 2H_2O$$

Point for Discussion
Balance the following equations:

(a) $NH_3 + H_2O + X_2SO_4 \rightarrow (NH_4)_2SO_4 + XOH$

(b) $XOH + Cl_2 \rightarrow XCl + XClO_3 + H_2O$

(c) $X_2O_5 + H_2O \rightarrow H_3XO_4$

(d) $H_2XO_4 + C \rightarrow H_2O + XO_2 + CO_2$

Interpretation of Equations

Example 15.17
What is meant by the equation:

$$CaO(s) + H_2O(l) \rightarrow Ca(OH)_2(s) \qquad \Delta H = -67.2\ kJ$$

Solid calcium oxide and liquid water react together in the ratio one unit: one unit (formula units and molecules respectively) to form one unit of solid calcium hydroxide (formula unit). This is in itself not particularly useful in the practical sense; we wish to deal with *measurable* quantities of the substances. It is more useful to interpret the equation in terms of units X times as large as a formula unit or a molecule, where X is Avogadro's Number, i.e. 6.02×10^{23}. These are measurable quantities.

∴ X formula units of solid calcium oxide react with X molecules of water to form X formula units of solid calcium hydroxide.
∴ 1 mole of calcium oxide reacts with 1 mole of water to form 1 mole of calcium hydroxide.

This is the practical interpretation of a typical equation.

The equation also shows the energy change involved and its nature and this is normally quoted for the quantities shown in the equation measured in moles. Thus when 1 mole of calcium oxide reacts with 1 mole of water, 67.2 kJ are liberated.

Point for Discussion

Interpret the following equations:

(a) $3Fe(s) + 4H_2O(g) \rightleftharpoons Fe_3O_4(s) + 4H_2(g)$

(b) $CH_4(g) + 2O_2(g) \rightarrow CO_2(g) + 2H_2O(l) \quad \Delta H = -1142\ kJ$

(c) $Pb^{2+}(aq) + 2Cl^-(aq) \rightarrow PbCl_2(s)$

(d) $Ca^{2+}(aq) + CO_3^{2-}(aq) \rightarrow CaCO_3(s) \quad \Delta H = +12.6\ kJ$

An Important Use for Equations. Reacting Weights and Volumes

A chemical equation provides the information we need to calculate the masses and in some cases the volumes, of the reactants and products. This is particularly important in industry. An engineer must be able to compare the actual yield of a product with the theoretical one so as to assess the efficiency of a process.

Example 15.18

What is the mass of magnesium oxide produced by the complete combustion of 9.6 g magnesium in pure oxygen?

1. Write down the balanced equation for the reaction.

$$2Mg(s) + O_2(g) \rightarrow 2MgO(s)$$

2. Write down the number of moles of each substance mentioned in the problem

$$\underset{(2\ \text{moles})}{2Mg} + O_2 \rightarrow \underset{(2\ \text{moles})}{2MgO}$$

3. Using the table of atomic masses work out the molar masses:

2 moles Mg:(2 × 24) g → 2 moles MgO:2(24 + 16) g
48 g magnesium produce 80 g magnesium oxide
∴ 9.6 g magnesium produce (80 × 9.6/48) g magnesium oxide
= 16 g magnesium oxide

Points for Discussion

1. Calculate the mass of calcium oxide produced when 20.0 g of calcium carbonate is decomposed by heat.

2. A solution containing 8.0 g of sodium hydroxide is neutralized by hydrochloric acid. What mass of sodium chloride will be produced if the solution is evaporated to dryness?

3. What mass of copper will remain if 3.2 g of copper(II) oxide are completely reduced to the metal? (Take the atomic mass of copper as 64.)

4. If the mass of more than one substance is required you must calculate each mass separately.

Example 15.19
Calculate the mass of calcium hydroxide that will react with 21.4 g of ammonium chloride. What mass of ammonia gas will be produced?

$2NH_4Cl(s)$	$+Ca(OH)_2(s)$	$\rightarrow 2NH_3(g) + CaCl_2(s) + 2H_2O(g)$
2 moles	1 mole	1 mole
107 g	74 g	34 g

(a) 107 g ammonium chloride react with 74 g calcium hydroxide
$\therefore$ 21.4 g ammonium chloride react with $(74 \times 21.4/107)$ g calcium hydroxide = *14.8 g calcium hydroxide*

(b) 107 g ammonium chloride produce 34 g ammonia
$\therefore$ 21.4 g ammonium chloride produce $(34 \times 21.4/107)$ g ammonia = *6.8 g ammonia*

Point for Discussion

Calculate the mass of hydrogen chloride used, and the mass of calcium chloride formed, when 10 g of calcium carbonate reacts with excess hydrochloric acid.

Reacting Volumes

We normally measure the amount of a gaseous product by quoting its volume. To do this we first determine the mass of the gas as in the last few examples and then convert the mass into the volume it would have at s.t.p., using the fact that the gramme-molecular volume of any gas at s.t.p. is 22.4 litres.

Example 15.20
Calculate the volume of carbon dioxide (measured at s.t.p.) evolved when 10.0 g of potassium hydrogen carbonate are completely decomposed by heating.

$2KHCO_3(s)$	$\rightarrow K_2CO_3(s) + CO_2(g) + H_2O(g)$
2 moles	1 mole
$2(39+1+12+48)$ g	$(12+32)$ g
200 g	44 g

A mole of carbon dioxide is 44 g.

$\therefore$ 44 g will occupy 22.4 litres at s.t.p.
200 g of $KHCO_3$ evolve 22.4 litres of CO_2 at s.t.p.
$\therefore$ 10 g of $KHCO_3$ evolve *1.12 litres of CO_2 at s.t.p.*

Points for Discussion

1. What volume of hydrogen (at s.t.p.) will be produced by the action of excess dilute hydrochloric acid on 6.5 g of zinc?

2. What would be the volume of carbon dioxide produced, at s.t.p., if 2.5 g of pure calcium carbonate reacted completely with dilute nitric acid?

Mass and Volume Relationships under Conditions Other than s.t.p.

In some cases the volume of gas required in the calculation is not measured at s.t.p. The basic calculation is carried out as with the foregoing examples and then the volume that would be formed at s.t.p. is converted to the volume at the temperature and pressure required.

Example 15.21
Calculate the volume of hydrogen, measured at 364 K (91 °C) and 800 mm, which is evolved when 32.5 g of zinc are treated with excess dilute sulphuric acid.

$$Zn(s) + H_2SO_4(aq) \rightarrow ZnSO_4(aq) + H_2(g)$$

1 mole — 1 mole
65 g — 2 g
1 g-molecular volume
22.4 litres at s.t.p.

65 g zinc produce 22.4 litres of hydrogen at s.t.p.
∴ 32.5 g zinc produce 11.2 litres of hydrogen at s.t.p.

Use the formula $\frac{P_1V_1}{T_1} = \frac{P_2V_2}{T_2}$ to convert to required conditions.

$$\frac{760 \times 11\,200}{273} = \frac{800 \times V_2}{(273+91)}$$

$$V_2 = \frac{760 \times 11.2 \times 364}{273 \times 800}$$

$$= \textit{14.18 litres}$$

Points for Discussion

1. Calculate the volume of ammonia, measured at s.t.p., which will reduce 60 g of copper(II) oxide to copper. If the temperature is reduced to 266 K (−7 °C) and the pressure to 720 mm, what will be the new volume of ammonia? (Take the atomic mass of copper to be 64.)

2. Find the volume of sulphur(IV) oxide measured at 312 K (39 °C) and four atmospheres pressure, which would be obtained by burning 4.0 g of sulphur in excess oxygen.

3. What will be the volume of ethyne obtained, at 312 K (39 °C) and 600 mm pressure, by the action of excess water on 4.0 g of calcium carbide?

4. What volume of carbon dioxide, at 351 K (78 °C) and 684 mm pressure, will be produced by the action of heat on 4.4 g of sodium hydrogen carbonate?

Using Standard Solutions

Example 15.22
100 cm³ of a solution of hydrochloric acid dissolved 3.0 g of magnesium ribbon. Calculate the molarity of the acid.

$$Mg(s) + 2HCl(aq) \rightarrow MgCl_2(aq) + H_2(g)$$

1 mole — 2 moles
24 g — 2 moles

24 g magnesium react with 2 moles hydrochloric acid
∴ 3 g magnesium react with (2 × 3/24) moles hydrochloric acid = 0.25 moles
0.25 moles of hydrochloric acid in 100 cm³ solution
∴ 2.50 moles of hydrochloric acid in 1000 cm³ solution
Acid is 2.50 molar.

Points for Discussion

1. Find the mass of pure iron that would be dissolved by 500 cm³ of 0.1 M sulphuric acid.

2. A solution of sodium carbonate, containing 5.3 g of the dissolved salt, will exactly neutralize 200 cm³ of hydrochloric acid. Calculate the molarity of the acid.

Volume–Volume Relationships

In reactions that concern only gases, it is only necessary to consider the relative volumes of the gases involved.

Example 15.23

What is the volume of oxygen that will combine with 100 cm^3 of carbon monoxide at the same temperature and pressure?

$$\begin{array}{lll} 2CO(g) & +O_2(g) & \rightarrow 2CO_2(g) \\ 2 \text{ molar vol.} & 1 \text{ molar vol.} & \end{array}$$

2 molar volumes carbon monoxide react with 1 molar volume oxygen
100 cm^3 of carbon monoxide react with *50 cm^3 of oxygen*

Example 15.24

A mixture of 150 cm^3 of oxygen and 50 cm^3 of hydrogen is exploded in a suitable apparatus. Give the names and volumes of the gases remaining, (a) all volumes measured at 760 mm and 393 K (120 °C), (b) all volumes measured at 760 mm and 293 K (20 °C).

$$\begin{array}{lll} 2H_2(g) & +O_2(g) & \rightarrow 2H_2O(\text{l or g}) \\ 2 \text{ molar vol.} & 1 \text{ molar vol.} & 2 \text{ molar vol.} \end{array}$$

2 volumes hydrogen react with 1 volume oxygen to give 2 volumes steam
50 cm^3 hydrogen react with 25 cm^3 oxygen to give 50 cm^3 steam

(a) Volume of oxygen remaining = $(150-25)$ cm^3 = 125 cm^3
Gases remaining: 125 cm^3 oxygen, 50 cm^3 steam

(b) *Gases remaining: 125 cm^3 oxygen.* (The steam will have condensed to a negligible volume of water at 293 K (20 °C).)

Points for Discussion

1. A mixture of 500 cm^3 hydrogen and 125 cm^3 chlorine is exploded in bright sunshine. Give the names and volumes of the gases remaining after the reaction, measured at the same temperature and pressure.

2. Give the names and volumes of the remaining gases when a mixture of 50 cm^3 oxygen and 40 cm^3 carbon monoxide is ignited (all volumes measured at room temperature and pressure).

You should now be able to work out, from an equation, the masses and/or volumes of products formed from known amounts of starting materials, under any conditions of temperature and pressure. It should be obvious that the ability to obtain correct results for the calculations depends on a sound understanding of basic chemical reactions, their equations, the gramme-mole concept and Avogadro's law. Make sure that you fully understand these basic ideas as once you have grasped them you should have no difficulty with the calculations.

QUESTIONS CHAPTER 15

For questions involving atomic masses, use the Table of Approximate Atomic Masses given in the back endpaper.

Assume that 1 gramme-molecule of a gas occupies 22.4 litres at s.t.p.

1. Calculate the empirical formula of a compound that has the composition: 52.0 per cent zinc, 9.6 per cent carbon, 38.4 per cent oxygen. (A. E. B.)

2. An organic compound was found to contain 12.8 per cent carbon, 2.1 per cent hydrogen and 85.1 per cent bromine. The vapour density of the compound was 94.

(a) What is the empirical formula of the compound?
(b) What is the molecular mass of the compound?
(c) What is the molecular formula of the compound? (J. M. B.)

3. A gas has a vapour density of 13.5, and contains 3.70 per cent hydrogen and 44.44 per cent carbon by mass. Analysis shows the gas to contain hydrogen, carbon and nitrogen only. Calculate the molecular formula of this gas. (J. M. B.)

4. Calculate the formula of a hydrocarbon containing 82.8 per cent by mass of carbon and having a molecular mass of 58. (C.)

5. In a reaction between a solution of a metallic hydroxide (formula MOH) and dilute hydrochloric acid, 20.0 cm^3 of the alkali reacted with 25.0 cm^3 of the acid. The acid concentration was 4.00 g per litre and that of the alkali was 7.67 g per litre. Calculate the formula mass of the alkali and hence the atomic mass of the metal.

Describe in detail how you would determine the volume of the acid and alkali which exactly neutralize each other. (J. M. B.)

6. 10 cm^3 of a solution of sodium hydroxide required 25 cm^3 of 0.4M hydrochloric acid for neutralization. Calculate the molarity of the alkali solution. (A. E. B.)

7. The formula mass of lead nitrate is 331. 6.62 g of lead nitrate were dissolved in water and the volume of the solution made up accurately to 500 cm^3. What was the molarity of this solution? (W. J. E. C.)

8. Describe in detail how you would prepare 250 cm^3 of an exactly 0.05M solution of sodium carbonate and use it to determine the exact concentration of an approximately 0.05M solution of sulphuric acid. You are provided with the usual apparatus for volumetric analysis. Name the indicator and state the colour change. Explain how you would work out the result.

$$Na_2CO_3(aq) + H_2SO_4(aq) \rightarrow Na_2SO_4(aq) + CO_2(g) + H_2O(l)$$

$$2Na^+ + CO_3^{2-} + 2H^+ + SO_4^{2-} \rightarrow 2Na^+ + SO_4^{2-} + CO_2 + H_2O$$

(C.)

9. (a) Find the ratio of the number of atoms in 16 g of sulphur to the number of atoms in 46 g of sodium.

(b) Find the ratio of the number of molecules in 32 g of methane, CH_4, to the number of molecules in 5.6 litres of hydrogen at s.t.p.

(c) One oxide of manganese, Mn, contains 2.4 g of oxygen combined with 0.1 mole of manganese atoms. What is the simplest formula of this oxide? (W. J. E. C.)

10. State (a) Gay-Lussac's law of gaseous volumes, (b) Avogadro's law.

Describe any one experiment you choose which fully illustrates Gay-Lussac's law.

Four grammes of sulphur on heating in oxygen give 2.8 litres of sulphur dioxide. Calculate the molecular mass of sulphur dioxide. (A. E. B.)

11. State Gay-Lussac's law of combining volumes.

200 cm^3 of a gaseous element X_2 reacted with 650 cm^2 of a gaseous element Y_2 to form 450 cm^3 of a mixture of XY_3 and Y_2. It was later found that 50 cm^3 of excess of Y_2 remained unused. All volumes were measured under the same conditions of temperature and pressure.

(a) What volume of XY_3 was formed in the reaction?
(b) Write a statement to show the relationship between the volumes of X_2 and Y_2 used and the volume of XY_3 formed.
(c) Give a balanced molecular equation for the reaction. (J. M. B.)

12. One volume of oxygen combines with chlorine to form two volumes of a gaseous oxide of chlorine which has a vapour density of 43.5. What is its formula? (J. M. B.)

13. When exploded with excess oxygen, ethane (C_2H_6) reacts as in the equation:

$$2C_2H_6(g) + 7O_2(g) \rightarrow 4CO_2(g) + 6H_2O(l)$$

If 20 cm^3 of ethane are exploded with 100 cm^3 of oxygen and the gaseous products are reduced to the original laboratory temperature and pressure, what will be the volume and composition of the resulting mixture of gases? (C.)

14. Molar solutions (solutions containing 1 gramme-formula per litre) of a metal sulphate and barium chloride were prepared. The following mixtures were made in small tubes.

Volume of sulphate solution (cm^3)	*Volume of chloride solution* (cm^3)	*Volume of water* (cm^3)
(a) 2	2	8
(b) 2	4	6
(c) 2	6	4
(d) 2	8	2
(e) 2	10	nil

The tubes were centrifuged and the depths of the precipitates formed were measured and recorded as shown.

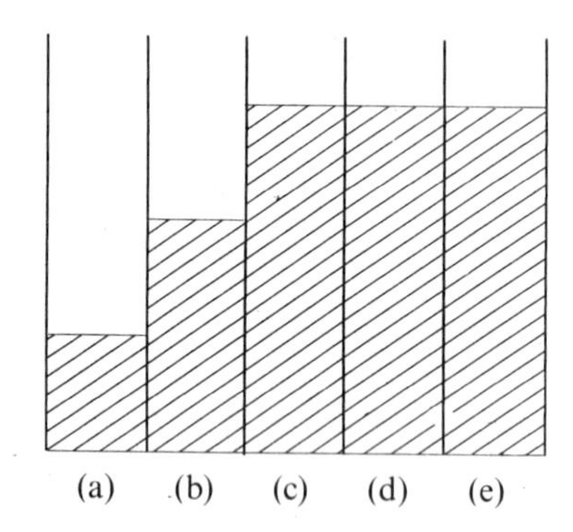

(i) Write the ionic equation for the reaction.
(ii) What is the mass of barium chloride in 1 litre of its solution?
(iii) Why did the volume of the precipitate stay constant in (c), (d) and (e)?
(iv) What volume of barium chloride solution reacts with 1 litre of the sulphate solution?
(vi) How many moles (gramme-formulae) of barium chloride react with 1 mole (1 gramme-formula) of the sulphate?
(vii) Using the symbol M for the unknown metal, write the formula for the sulphate and hence the equation for the reaction.
(viii) 1 mole (1 gramme-formula) of the metal sulphate has a mass of 342 g. What is the atomic mass of M?
(ix) How and why was the total volume of the mixture kept constant?
(x) What was the purpose of centrifuging the tubes? What other process might have been used? (J. M. B.)

15. When 3.1 g of a carbonate MCO_3 are heated to constant mass, 2.0 g of the oxide of the metal are formed. On reduction of this oxide, 1.6 g of pure metal remains.

(a) Describe experiments by which these figures could be determined in the laboratory.
(b) Give equations for the two reactions.
(c) Calculate the atomic mass of M.
(d) Calculate the volume of carbon dioxide evolvėd during the heating of the carbonate. (A. E. B.)

16. An experiment showed that when 0.1 formula mass of sodium hydrogen carbonate was heated to constant mass, 5.3 g of solid remained. Show that these results agree with the following equation:

$$2NaHCO_3(s) \rightarrow Na_2CO_3(s) + H_2O(g) + CO_2(g)$$

What fraction of a mole of carbon dioxide molecules should have been given off in the above experiment?

What volume would the carbon dioxide occupy at s.t.p? (W. J. E. C.)

17. The formula of basic copper carbonate is $CuCO_3$, $Cu(OH)_2$.

(a) What mass of copper(II) oxide can be obtained by the action of heat on 4.42 g of basic copper carbonate?
(b) If 4.42 g of basic copper carbonate reacted with excess dilute sulphuric acid, what volume of carbon dioxide, measured at s.t.p., would be evolved? (J. M. B.)

18. When aqueous solutions of silver nitrate and calcium chloride are mixed, solid silver chloride is precipitated, the equation for the reaction being

$$2AgNO_3(aq) + CaCl_2(aq) \rightarrow 2AgCl(s) + Ca(NO_3)_2(aq)$$

(a) What volume of 0.5 molar calcium chloride would you add to 100 cm^3 of molar silver nitrate just to complete the above reaction?
(b) What mass of solid silver chloride (formula mass = 143.5) would be formed? (W. J. E. C.)

19. 4.76 g of the hydrated chloride of a divalent metal M, of atomic weight 59, contain 2.60 g of the anhydrous salt. Find the number of molecules of water of crystallization in each molecule of the hydrated salt. (W. J. E. C.)

20. When heated in a current of nitrogen, magnesium reacts with it to form magnesium nitride, Mg_3N_2.

Magnesium nitride reacts with water to form magnesium hydroxide and ammonia.

(a) Write the equation for the first reaction and calculate the volume of nitrogen, measured at s.t.p., required to react with 8.0 g of magnesium.

(b) Write the equation for the second reaction and calculate the volume of ammonia produced, measured at s.t.p., if the magnesium nitride formed in (a) reacts completely with water. (Assume that all the ammonia can be driven off and collected.)

(J. M. B.)

21. 'Calcite is pure calcium carbonate'. Is this statement borne out by the following experimental results?

A large crystal of calcite weighing 3.753 g was placed in 25.0 cm^3 of a 2M solution of hydrochloric acid and left there until action ceased. After washing and drying, the crystal was found to weigh 1.253 g.

What volume of carbon dioxide at s.t.p., was formed during the reaction? (C.)

Index

ACETALDEHYDE *see* Ethanal
Acetic acid, *see* Ethanoic acid
Acetone, *see* Propanone
Acetylene, *see* Ethyne
Acetylides, 321
Acheson process, 269
Acidity, 80
Acids, 70
  action on metals, 78, 91
  anhydride, 118
  and bases, 75–6
  carboxylic, 327
  common laboratory, 70
  effect on indicators, 73
  general properties, 70–5
  oxides, 118, 275, 277
  reaction with insoluble bases, 93
  reactions with carbonates, 75, 92
  salts, 87–8
  strong, 79
  weak, 80
  see also names of individual acids
Activated charcoal, 45, 302
Activation energy, 344
Addition, polymerization, 329, 330
  reaction, 313
Adsorption, 45, 54
Adsorbed layer, 178
Aerobic respiration, 112
After damp, 309
Air, composition of, 108
  liquefaction of, 394
  reactions of metal with, 172
  solubility in water, 110
Alcohol, absolute, 325
  aliphatic, 321
  denatured, 329
  dihydric, 321
  monohydric, 321
  polyhydric, 321
  tests for, 329
  trihydric, 321
Alcoholic beverages, 323
Alkali, 76
  caustic, 78
  general properties, 77–9
  reaction with ammonium compounds, 79
  reaction with metals, 78
Alkanes, 306
  combustion of, 308
  decomposition of, 309
  properties of, 306–10
Alkenes, 310
  physical constants, 311
  properties of, 313–5
  reactions with halogens, 313
  test for, 315
  uses of, 315
Alkynes, 316
  distinction from alkenes, 321
  preparation of, 316
  properties of, 318–20
Allotropes, 193
  metastable, 280
  of phosphorus, 198
  of sulphur, 195, 196, 197
Allotropy, 192, 193, 195–8
  of oxygen, 288
  of phosphorus, 279
  of sulphur, 195–8, 288
Alloys, 256
Aluminium (the element), 252, 253
Aluminium compounds
  chloride, 253
    anhydrous, 214
  hydroxide, 253
  oxide, 138, 252, 382, 387
  sulphate, 253
Alums, 253
Ammonia, molecular shape, 200
  preparation of, 280
  properties of, 282
  soda process, 393
  solution, 283
  synthesis of, 381
Ammonium, chloride, 56, 279, 284
  ion, 206
  nitrate, 241, 285
  nitrite, 279
  phosphate, 280
  salts, properties of, 284, 285
Amphoteric, hydroxide, 255
  oxide, 118, 138, 252, 254, 261
Analysis, qualitative, 403
  quantitative, 403
  simple chemical, 56
  volumetric, 403
Anhydrite, 217, 383
Anions, 156, 159
Annealing of glass, 276
Anode, 155
  sacrificial, 179
Anodizing, 252
Anti-knock, 365
Antimony pentachloride, 329
Argon, 109, 394
Aromatics, 305
Arsenic, 131, 268
Aspirator, 109
  Atmospheric nitrogen, 278
Atom, 2, 5
Atomic, diameter, 9
  hydrogen, 124
  lattice, 192
  mass, chemical, 20, 22, 186
  mass, physical, 20, 22
  number, 19, 20, 133
  size, 9
  structure, 19, 20
  theory, 16, 129
  volume, 186
Attraction, forces of, 15
  intermolecular, 347
Avogadro, Amadeo, 25, 28, 412, 413
  number, 25, 26, 413

BACTERIA, 278
Baeyer's reagent, 312, 314
Bakelite, 333
Baking powder, 273
Bases, 76
Basic carbonates, 238
  iron(II) nitrate, 241
  oxide, 118
Battery, 152
Bauxite, 387
Beer, 323
Benzene, 287
Berzelius, 18
Bessemer process, 390
Bismuth oxychloride, 375
Blank test, 177
Bleaching, 293
Boiling, effect of pressure on, 111
  effect on hard water, 222
  point, 346
  point of a mixture of liquids, 61
  point of a pure liquid, 60
Bomb calorimeter, 351
Bond, breaking, 346
  energies, 377
  making, 346
Bonding, covalent, 35
  ionic, 30
Boyle, R., 408
Bragg, W. H. and W. L., 133
Brass, 263
Bromide, ion, 155
Bromine, 292, 297
  molecule, 155
  movement of molecules in air, 9, 10
  reaction with ethene, 313
  reaction with methane, 308
  reaction with steel wool, 144–5
Bronze, 261
  Age, 175
Brownian movement, 11
Burette, 4, 85
Butanol, 322
But-1-ene, 311
Butene isomers, 315

CALCITE CRYSTAL, 188
Calcium (the element), 142, 248–9
  ion in water, 217
  ion, test for, 251
Calcium compounds, carbide, 248, 316, 317
  carbonate, 27, 220, 221, 238, 250, 362
  chloride, 250, 386
  hydrogen carbonate, 238
  hydroxide, 217, 237, 250
  oxide, 249, 250, 309
  phosphate, 279
  silicate, 275, 389
  sulphate, 213, 217, 221, 250
Calgon, 223, 228, 233
Calorie, 344
Calorific values, 352
Camphor, 10, 58
Cane sugar, 45, 295, 323
Cannizzaro, S., 323, 413
Carbon, 268, 269, 270
  'amorphous', 268, 269
  cycle, 270
  fibre, 269
Carbon dioxide, molecular shape, 199
  preparation of, 271
  properties of, 247, 272
  reaction with calcium carbonate, 217
  test for, 75
  uses of, 273
Carbon monoxide, 273
  uses of, 274
Carbonates, basic, 238, 251, 261
  effect of heat on, 233
  reaction of acids with, 74
  test for, 90, 237, 239
Carbonic acid, 237
Carnallite, 251
Carnivorous animals, 277
Carboxylic acid, 308
  group, 327
Carothers, W. H., 331
Cast iron, 256, 389
Castner process, 386
Catalyst, 364
  in reversible reactions, 374
  negative, 365
  positive, 365
Cathode, 155, 157, 161
  rays, 133
Cathodic protection, 179
Cation, 156, 159
Cell, simple, 177
  voltaic, 177
Cellulose, 336
Cement, 251
Centrifuge, 44
Chamber process, 383
Change, of concentration in reversible reactions, 375
  of temperature in reactions, 363
Charcoal, activated, 45, 269, 302
Charles, J. A., 14, 408
Chatelier, H. L. Le, 379, 380
Chemical, energy, 343
  equilibrium, 367
  formulae, 399
Chloride ion, 32, 157, 160
  properties of, 243
  test for, 90
Chlorine, 128, 297, 386
  action on alkalis, 301
  as an oxidizing agent, 300
  compounds of, 301
  isotopes, 21, 23
  preparation of, 298
  properties of, 299 et seq
  reaction with steel wool, 144
  water, 300
Chloroform, *see* Trichloromethane
Chromatography, gas, 54
  separation of ballpoint ink by, 55
  separation of chlorophyll, 55
  separation of colourless materials by, 56
  separation of sugars by, 338
Chromatogram, 55
Chromium(III) ion, 293
Clarke's process, 222
Clay, 271
Cleavage plane, 188
Clock, iodine, 361
Coal, 343
Collection of gases, 100–5
Colorimeter, 82
Combination, direct, 92
Combustion, 112
Complex salts, 253
Composition, percentage, of compounds, 399
Compounds, 62, 64, 66
Concentration, changes of, 361
Condensation polymerization, 330, 331
Conduction of electricity, 151, 153
Conductor, 154
Conservation of energy, law of, 356
Contact process, 383, 384
Converter, ammonia, 382
Copolymer, 332
Copper (the element), 263
  effect of heating in air, 63, 106, 108
  ion, test for, 265
  purification of by electrolysis, 167
  reaction with acids, 74
Copper(I) compounds, 265
  acetylide, 321
  chloride (ammoniacal), 318
  oxide, 265
Copper(II) compounds, carbonate (basic), 264
    effect of heat on, 232
  chloride, 264, 365
  hydroxide, 232, 264
  nitrate, 232, 264
  oxide, 93, 264
  pyrites, 263
  sulphate, 264
    effect of heat on, 62
    electrolysis of, 162, 164, 165
  sulphide, 265
Corundum, 252
Coulomb, 179
Covalent, bond, 35
  compounds, 154
Criteria of purity, 58
Crude oil, 52
Cryolite, 387
Crystal, 188
  growth, 187
  molecular, 194
Crystallization, 44, 45
  of a salt, 91–2
  water of, 215, 233
Cuprite, 263
Cycle, carbon, 270
DECOLORIZATION, 45
Decomposition, double, 93–8, 235, 238
  thermal, 286
Dehydrating agent, 295
Deliquescence, 215, 216
Deliquescent, 215
Density, vapour, 415
Depolymerization, 336
Desiccator, 250
Detergents, 219, 224–8
Deuterium, 122
1,6-Diaminohexane, 335
Diamond, 186, 193, 194, 269, 344
Diaphragm cell, 391
Diatomic molecules, 414
Diffusion, 11
1,2-dihydroxyethane, 314
Direct combination, 231
Dirt removal, 226
Displacement, 172
  of halogens, 297
Dissociation, thermal, 286
Distillate, 47, 49
Distillation, 46, 47, 48
  fractional, 46–53
  simple, 46, 48
Dolomite, 390
Double layer, 170
Double salt, 253
Downs cell, 386
Downward delivery, 102
Drechsel bottle, 102
Drikold, 272
Drying, agent, 102, 295, 323
  of gases, 102, 280
Ductile, 256

EFFLORESCENT, 215–6, 246, 248
Electricity, conduction of by elements, 64, 151
  Faraday's laws, 179–82
  from chemical reactions, 168–9
  from reactions in voltaic cells, 177–8
  nature of, 152
  units of, 179–80
Electrochemical series, 167–77
  related to decomposition of metallic compounds, 233
  related to reactivity of oxides, 235
Electrode, 155
  effect of change of electrode on electrolysis of copper(II) sulphate solution, 164–5
  standard hydrogen, 170–1
  use of mercury, 392
Electrode potential, 170
  in predicting chemical behaviour, 170–4
Electrolysis, 151–82
  Faraday's laws of, 179–82
  in anodizing aluminium, 252
  of sodium hydroxide in industry, 391–2
  role of water in, 157–9
  to prepare aluminium, 387–8
  to prepare sodium, 386–7
  to produce 'heavy water', 122
Electrolyte, 154–5
Electron(s), 18
  arrangement in atoms, 19
  calculating the number in an atom, 20–1

ease of acceptance by non-metals, 145, 147
ease of removal from metal atoms, 142, 147
electronic structure and Faraday's laws, 181
Faraday as a mole of electrons, 181
in electric currents, 152
in electrode potentials and reactivity, 169–73
in electrolysis, 155–6
in halogen displacements, 298
in metal displacement reactions, 168–70, 172
in oxidation and reduction, 205–10, 259–60
in producing X-rays, 133
significance in bonding, 30–9
significance in chemical properties, 134–47
significance in Periodic Table classification, 132–47
Electronegativity, 146–7
in halogen displacements, 298
in metals and the electrochemical series, 169–70, 175
Electronic structure, 19–21
and bonding, 30–9
and the Periodic Table, 132–47
significance of structure of noble gases, 30
unique case of hydrogen, 134
Electroplating, 166
Electrovalent bond, *see* Ionic bond
Elements, 1–2, 62–4
changes down metallic groups, 140–2, 147
changes down non-metallic groups, 143–7
determination of oxidation numbers, 205–7
differences between elements, compounds and mixtures, 65–8
displacement reactions, 172
division into metals and non-metals, 63
electrode potentials of, 170–1
individual metals and their compounds, 231–66
individual non-metals and their compounds, 268–302
in nature, 114
in Periodic classification, 127–47
some physical properties, 136
some structural information, 140
Empirical formula, 28–9, 400–1
Endothermic, 344–6, 348
in reactions attaining equilibrium, 376–9
Energy, in combustion, photosynthesis, respiration and rusting, 112–4
in other chemical reactions, 343–57
Enthalpy, 347
Enzymes, 366
in detergents, 227
in hydrolysis of starch, 336–40
from yeast, 323
Epsom salts, 252
Equations,
how to write them, interpret them, 417–24
Equilibrium, 367–80
dynamic and static, 371–4
effect of temperature, pressure and concentration changes, 374–80
in the Contact process, 383–5
in the Haber process, 381–3
le Chatelier's principle, 379–80
Esters, 327–9
in making soap, 219, 329
polyesters, 332–3
Etching, 276
Ethanal, 326–7
Ethane, 306–10
from petroleum, 53
strength of C—C bond in, 357
Ethanoic acid, 70, 71, 327–8
heat of neutralization of, 353–4
Ethanol, 321–9
as an isomer of methoxymethane, 200–1
determination of boiling point, 60
determination of heat of combustion, 350–1
distillation of aqueous solution, 48
Ethene, 310–6, 330–1
distinction from ethyne, 321
electronic structure and shape of molecule, 199
in manufacture of ethanol, 325
strength of C=C bond in, 357
Ether, 3, 5, 327
Ethylene, *see* Ethene
Ethylene glycol, *see* 1,2-dihydroxyethane
Ethylethanoate (acetate), 328
Ethyne, 316–21
oxy-acetylene blowpipe, 117
Eudiometer, 409
Evaporation, 45
Exothermic, 344–6, 348
in equilibria, 376–9
Extraction (of metals), 167, 175, 177, 385–91
(*see also* Individual metals)

FARADAY, laws of electrolysis, 179–82
unit of electricity, 181
Fats, hydrogenation of, for margarine, 125
in soap manufacture, 219, 227
Fehling's test, 336
Fermentation, 322–5
Ferric, *see* Iron compounds
Ferrous, *see* Iron compounds
Fertilizers, 278, 280, 282, 285, 381
Fibre-glass, 277
Filler, 334
Filtrate, 44
Filtration, 43–4
Fixed nitrogen, 278
Fluorine, *see* Halogens
Foods, 112–3, (*see also* Glucose, Fehling's test, etc.)
Formulae, determination of from knowledge of bonding, 34–9
empirical, 28, 29, 400–1
experimental determination, 27–30
of gases, 410–7
of salts, 88–9
quantitative interpretation etc., 399–402
Fountain experiment, 281
Fractional distillation, 48, 50
of crude oil, 52–3
of ethanol and water, 323
of liquid air, 394
Fractionating column, 49, 51, 52–3
Fractions, 51–3
Frasch process, 289
Freezing point, 58
effect of impurities on, 58–60
of gases, data, 105
of solids, data, 136, 139, 141, 142
Fructose, 323, 336
Fuels, 112, 310, 343, 351–2
Functional group, 305
Fusion, electrolysis of fused compounds, 154
in sun, 119

GALENA, 261
Galvanizing, 178, 255
Gas chromatography, 56
Gases, Boyle's law, 15, 408
change of state, 5, 185
Charles' law, 14, 408
data on, 105
diffusion of, 9
Gay-Lussac's law, 411–4
general gas equation, 408
liquefaction and freezing for collection of, 104
liquefaction of, 15
preparation, collection, purification, etc., 100–5, 295
speed and movement of molecules, 11–4
temperature and volume changes, 14
volume changes and calculations on, 407–16
Gas pressure, 13–4
Gay-Lussac's law, 411–4
Geiger-Müller tube, 373–4
Generators, gas, 100–1
Germanium, 131, 147
Giant molecules (macromolecules), 198
Giant structures, 34, 138, 185–6, 190–2
and the Periodic Table, 138
data on, 139, 140
Glass, 275–7
Glucose, 323, 336
product of hydrolysis of starch, 338–40
Glycogen, 340
Gold, 173, 175, 176, 177
Gossage process, 247
Gramme-atoms, 24–6
to find formulae, 27–30, 400–1
Gramme-formulae, 26
Gramme-molecular volume, 415
Gramme-molecules, 26–7
Graphite, 186, 190, 193–4, 196, 269–70
Groups (in Periodic Table), 127–45
elements of Group 4, 268–77
elements of Group 5, 277–88
elements of Group 6, 288–97
elements of Group 7, 297–302
Gypsum, 249, 251–2

HABER PROCESS, 381–3
Haematite, 256, 388

Halogens, 132, 143–7, 297–302
  data on physical properties, 143
  reaction with ammonia, 282
  reaction with hydrocarbons, 309–10, 312–14, 318–19
  reaction with hydrogen, 145
  reaction with iron wool, 144
  reaction with water, 128, 143
Hardness of water, 216–28
  action of soaps on, 200
Heat changes (in chemical reactions), 343–57
  for heat of reaction, combustion, etc., 348–57
Heavy water, 122
Heterogeneous, 363
Hexane (from petroleum), 53
Hexamethylene diamine, 335
Hofmann voltameter, 161–2
Homogeneous, 363
Homologous series, 305
Hydrated (hydrates), 215, 253
Hydration energy, 93–5, 353–6
Hydrides, 139, 140
  of silicon, 271
  see also Hydrogen
Hydrocarbons, 305–21
  in crude oil, 52–3
  in structure of soaps and detergents, 225–7
  isomerism in, 201–2
Hydrochloric acid, 70, 71, 73, 302
  action on carbonates, 74–5
  action on metals, 73, 74, 175
  as 'parent' of the chloride salts, 87, 88
  as typical dilute acid, 96
  electrolysis of, 162
  heat of neutralization of, 352–4
  oxidation to produce chlorine, 298
  test for, 90
  to prepare sodium chloride, 84–6
Hydrogen, 119–25, 145
  addition reactions, 314, 319
  data on, 105
  difficulty of placing in Periodic Table, 134
  displacement from acids by metals (electrochemical series), 175
  early use as standard for atomic masses, 22
  from electrolysis, 157–9
  in Haber process, 381–3
  in oxidation and reduction, 204, 210
  ions from water and influence in electrolysis, 157–9
  overvoltage in simple cell, 177–8
  reactions showing affinity of chlorine for, 299–300
  speed of molecules, 13
  standard hydrogen electrode, 170–1
  test for, 74
  uniqueness of atomic structure and relationship to bonding, 38, 134
Hydrogen carbonates, action on acids, 74–5
  as salts, 88
  distinction from carbonates, 238–9
  preparation from carbon dioxide, 237
  test for, 90
  use in precipitating 'pure' carbonates, 238
Hydrogen chloride, 301–2
  addition reactions of, 319–20
  in preparing anhydrous metallic chlorides, 253, 257
  speed and movement of molecules, 12, 13
  see also Hydrochloric acid
Hydrogen peroxide, 228–30
  as an oxidizing/reducing agent, 209–10, 229, 291, 361
  catalytic decomposition of, 114
  in converting iron(II) to iron(III), 259–60
  use of negative catalysts in keeping solution stable, 365
Hydrogen sulphates, 88, 241–2
Hydrogen sulphide, 245, 290
  as reducing agent, 209–10, 260, 287, 291–2
  data on, 105
  drying of, 102
  effect on paints, 255
  in precipitating metallic sulphides, 244–5
  test for, 229, 245
Hydrogen sulphites, 296
Hydrolysis, 214–15, 253, 257
  in organic chemistry, 328
  of bismuth trichloride in investigating equilibria, 375–6
  of natural polymers, 336–7
Hydronium ions, 158–9
Hydrophilic, 225, 322
Hydrophobic, 225, 322
Hydrosphere, 114, 119
Hydroxide ions (from water, and in electrolysis), 157–9, 162
Hydroxides, 235–7
Hydroxyl group (in organic chemistry), 321–9
Hygroscopic, 215–16, 248
Hypochlorites, 301
Hypochlorous acid, 300–1

IMMISCIBLE LIQUIDS, 47
Impurities, effect on boiling point of liquids, 61
  effect on melting point of solids, 58–60
  removal during salt preparations, 92
  removal of in gas preparations, 100–2
  removal of from solids and liquids, 43–57
Indicators, 71, 72, 73, 77, 81, 83
Industrial processes, 381–96
  see also individual processes
Ink, 47, 54
Insoluble bases, *see* Bases
Insoluble salts, 93–6
Intermolecular collisions, 13
  in rates of reactions, 362, 363
Intermolecular forces, 36
  and surface tension of water, 224–6
  in Joule-Thompson effect, 394
  in latent heat of fusion, 346–7
  in molecular structures, 190–2
Intramolecular forces, 36
Intrinsic energy, 344
Iodine, 143, 144, 297–302
  clock, 361
  difficulty of placing in Periodic Table, 132–3
  diffusion of in illustrating equilibria, 369–70
  in test for starches, 336
  see also Halogens
'Iodoform' test, 327
Ion exchange, 223–4
Ionic bonding, 30–5
  and electrolysis, 153–6
  giant ionic structures, 190–2
  ionic hydrides, 123
Ionic equations, 151–72, 205, 419–20
Ionic radii, 140–2, 143
Ions, 5, 31–2, 39
  attraction between, lattice energies, 93–5
  hydration energies of, 93–5
  in double decomposition, 93–5
  in electrolysis, 154–6
  preferential discharge of in electrolysis, 158–60
  see also Ionic bonding
Iron (the element), 65, 73–4, 116, 173–4, 175
  as a catalyst in the Haber process, 381–3
  conversion to steel, 389–91
  electronegativity and electrode potential, 169–72
  for general reactions of metallic compounds, 231–45
  in producing iron(II) sulphate and other salts, 90–1
  production of in blast furnace, 388–9
  properties of the metal and its compounds, 256–61
  rendering passive, 240
  rusting and protection of, 178–9
Iron compounds, properties, 256–61
  interconversions of iron(II) and iron(III) compounds, 258–60
  iron(II) nitrate, special preparation, 239
  iron(II) sulphide, 65–6
  iron(III) chloride, hydrolysis of, 215
Isomers, 200–2
Isomorphism, 253, 255
Isotopes, 21–3
  of hydrogen, 122
  radioactive isotopes in investigating chemical equilibria, 372–4

JOULE, 344
Joule-Thompson effect, 394

KALDO PROCESS, 391
Kinetic energy, 13, 14, 192, 343
  and allotropes of sulphur, 196–8
Kinetic theory, 13, 15, 16
  and change of state, 4, 5, 15, 185
  and diffusion, 3, 9, 10, 11
  and gas pressure, 13, 14
  and intermolecular collisions, 13
  and speeds and movement of gas molecules, 11–14
  see also Brownian movement
Kipp's generator, 100–1
Kurchatovium, 132

LD PROCESS, 390–1
Large molecules, 329–40
Latent heat, of fusion, 60
of vaporization, 346–7
Lattice energy, 93–5, 355
Lavoisier, Antoine, 114, 119, 127
Law of, Conservation of Matter, 417–18
Constant Proportions, 17
Multiple Proportions, 17
Lead (the element), action with acids, 175
action on steam, 173–4
as an element of Group 4, 147
electronegativity and electrode potential, 169–72
general reactions and properties of the metal and its compounds, 261–3
Lead(II) compounds, 261–3
general reactions of metallic compounds, 231–45
bromide, electrolysis when fused, 154
chloride in investigating equilibria, 373–4
compounds in salt preparations, 92, 95
ethanoate (acetate), in test for hydrogen sulphide, 229
oxide, amphoteric nature, 118
oxide, reduction of by hydrogen, 124
sulphide, 229
Le Chatelier's Principle, 379–80
in the Contact process, 383–5
in the Haber process, 381–3
Light, influence on reaction rates, 366
influence on substitution reactions in organic chemistry, 309
Limekiln, 271
Limelight, 250
Limestone,
in limekiln, 271
in preparing calcium oxide in the laboratory, 251
in producing hard water, 217–18
in the Solvay process, 393–4
Limewater, 75
Liquefaction, of air, 394
of other gases, 15
Liquids, change of state of, 5, 185
determination of boiling point of, 60
diffusion in, 10
effect of impurities on boiling point of, 61
movement of molecules of, 13, 16
separation if immiscible, 47
separation if miscible, 46, 48, 49, 50–3
Lithium (the element), 141–2
lithium chloride, electrolysis when dissolved or fused, 152, 156–7
Lithosphere, 114, 119
Litmus, 72, 73, 77

MACROMOLECULES, 198, 329
Magnesium (the element), 251
reaction with water, 135
Magnesium compounds, carbonate, 251
chloride, 251
hydroxide, 251
oxide, 251
determination of formula of, 28
reaction with water, 138
sulphate, 252
uses of, 252
Magnesium ion, 252
Maltose, 338, 339
Mass (*see* Atomic mass, Molecular mass etc.)
Mass number, 20
Matter, divisibility of, 6
Melting points, determination of, 58–9
Mendeléev, Dmitri, 130
Mercury cell, 392
Metal, carbonates, 237–9
chlorides, 242–4
hydroxides, 235–7
nitrates, 239–41
oxides, 231–4
sulphates, 241–2
sulphides, 244–5
Metals, 231 et seq
extraction of, 167, 177, 385 et seq
general properties of, 63–4
reaction of with acids, 73–4, 175
reaction of with air, 172
reaction of with chlorine, 299
relative chemical reactivity of, 171
Methane, laboratory preparation, 307
reactions of, 308–9
Methylated spirit, 329
Mica, 194
Miscible liquids, 47
Mixtures, 64–6
separation of, 44 et seq
Molar solutions, 27, 403
Molarity, 403
determination of, 405, 407
use of, 404
Molar volume (Gramme-molecular volume), 415
Mole, the, 27
Mole of gas, mass of, 415
volume of, 414
Molecular, crystals, 194
formulae, 402
mass, 24
shapes, 198–200
structures, 190–1
Molecule, 5
Molecules, large, 329
Monoclinic sulphur, preparation of, 195
Monomer, 329
Moseley, H., 133
Movement of particles, 9–13

NATURAL POLYMERS, 336–40
Neutral oxides, 118
Neutralization, 82–6, 404
heat of, 348, 352–4
Neutron, 18
Newlands, John, 129
Nitrates, heat on, 241
preparation of, 239
test for, 240
Nitric acid, 284–8
acidic properties, 286
as a nitrating agent, 286
as an oxidizing agent, 287–8
industrial production of, 394
physical properties, 286
preparation of (laboratory), 284
thermal decomposition of, 286
uses of, 288
Nitric oxide, *see* Nitrogen monoxide
Nitrogen (the element), 277
compounds of, 280 et seq
cycle, 277
fixed, 278
oxides of, 288
preparation of, 109, 278
properties of, 277, 279
uses of, 279
Nitrogen monoxide, volume changes in reaction with oxygen, 410
Noble gases, 30
Non-metals, 63–4, 268 et seq
Nucleus, 18
Nylon, 332
laboratory preparation of, 335

OIL, spreading of, 6
Oil drop experiment, 8
Olefines, *see* Alkenes
Open Hearth Process, 390
Organic acids, 327
Organic chemistry, 305–42
Oxidation, 204 et seq
in electrolysis, 167
in terms of electron transfer, 205
in terms of hydrogen and oxygen, 204
in terms of oxidation number, 205
number, 205–6
of alcohol, 326
of iron(II) sulphate, 259
Oxides, of metals, 231–4
types of, 118–9
Oxidizing agents, 208–10
Oxygen, bonding in, 36
chemical properties of, 116
physical properties of, 115–6, 288
preparation (laboratory) of, 114–5
test for, 117
uses of, 117

pH, 81–2
meter, 82
Packing in solids, 189
Paraffins, *see* Alkanes
Particles, 2
movement of in gaseous state, 9–13
movement of in liquid state, 10–11
movement of in solid state, 10
packing together of, 186
size of, 5 et seq
Percentage composition of compounds, 399
Periodic Classification of elements, 127–34
Periods, 132
trends across, 135
Permanently hard water, 220
Peroxides, 119
Perspex, 330
Petroleum, fractional distillation of, 52–3

Phosphate treatment, 224
Phosphorescence, 280
Phosphorus (the element), 277, 279 et seq.
chemical properties of, 280
manufacture of, 279
oxide, 138
physical properties of, 277, 279
reaction with water, 137
uses of, 280
Photography, 366
Photosynthesis, 113
Physical properties of gases, 105
Pigments, separation of, 55
Plastic sulphur, 196
Plasticizer, 334
Polyester, 328, 332
Polyethylene (Polythene), 330
Polymerization, 314, 320, 329 et seq.
addition, 329–31
condensation, 330, 331–3
of ethene, 314, 330
of methylmethacrylate, 330
of styrene, 330, 334
of vinyl chloride, 330
Polymers, natural, 336–40
synthetic, 329–35
Polymorphism, 192
Polystyrene, 330
preparation (laboratory) of, 334
Polyvinyl chloride (P.V.C.), 320, 330
Potassium, reaction with water, 127
chlorate(V), thermal decomposition of, 364
manganate(VII). 2. 6
Potential energy, 343
Precipitation, 93–4
heat of, 348, 349–50
Preparation of gases, 100
Pressure (gas), effect on boiling point, 111
effect of temperature and volume changes on, 14–15, 408
standard, 409
Promoters, 365
Proof spirit, 324
Propanone, 61, 72
Proton, 18
Purification, of compounds, 43 et seq
of copper, 167
of gases, 100
Purity, criteria of, 58

QUALITATIVE ANALYSIS, 403
Quantitative analysis, 403
Quartz crystals, 188

RADICALS, 39
Ramsay, Sir William, 109
Rate of movement of gases, 12–13
Rate of reaction, 360–6
effect of catalysts on, 364–5, 366
effect of concentration on, 360–2, 366
effect of light on, 366
effect of surface area on, 363, 366
effect of temperature on, 363, 366
iodine clock, 361
measurement of rate, 360
of calcium carbonate and hydrochloric acid, 361, 363
of sodium thiosulphate and hydrochloric acid, 362
Rayleigh, Lord, 109
Reacting volumes, 422
Reaction trend, in metallic groups, 142
in non-metallic groups, 145
Rectified spirit, 323
Redox reactions, 208
of sulphur dioxide in aqueous solution, 291–2
Reducing agents, 208–10
Reduction, 204 et seq
in electrolysis, 167
in terms of electron transfer, 205
in terms of hydrogen and oxygen, 204
in terms of oxidation number, 205
of iron(III) sulphate, 260
Respiration, 112
Reversible reactions, 367 et seq, 420
effect of catalyst on, 379
effect of concentration changes on, 375
effect of pressure changes on, 378
effect of temperature changes on, 376
Rhombic sulphur, preparation of, 195
Rochelle salt, 364
Röntgen, W. K., 133
Rusting, 114, 178
conditions needed for, 178
prevention of, 178

S.T.P., 409
Sacrificial anode, 179
Saliva, enzymes in, 337
Salt, separation from sand, 44
Salts, 86–97
acid, 87, 88
complex, 253
definition of, 87
double, 253
insoluble, 93
normal, 87, 88
preparation of, 90–7
Saponification, 329
Saturated compounds, 310
Saturated solution, equilibrium with solid, 374
Scheele, C. W., 114, 277
Scrubbing, 381
Sebacoyl dichloride, 335
Seeding, 387
Separation of mixtures, 43–57
Series, electrochemical, 167–77
homologous, 305
Shapes of molecules, 198–201
Silica, 274
Silicates, 275
Silicon (the element), as in Period 3, 136, 139, 147
compared with carbon, 268
manufacture, 270
occurrence, 271
preparation of, 271
properties of, 271
Silicon compounds, dioxide, 118, 138, 274, 275
hydrides, physical properties, 139
tetrachloride, 139
Silver, chloride, heat of precipitation, 349
halides, 366
nitrate, 90
oxide, 236
Soap, composition, 219
preparation of, 219
reaction with hard water, 200
Soapless detergents, 227, 228
Soda lime, 307
Sodamide, 246
Sodium (the element), 127, 129, 245–8
bonding with chlorine, 32
chemical properties, 245, 246
combustion in oxygen, 116
physical properties, 136, 141, 245
occurrence, 245
reaction with chlorine, 299
reaction with ethanol, 326
reaction with water, 127, 135
Sodium compounds, aluminate, 236
carbonate, 238, 248, 355, 393
chlorate(V), 301
chloride, 139, 160, 163, 248
ethanoate, 307
ethoxide, 326
hydride, 123, 139
hydrogen carbonate, 238, 248, 393
hydrogen sulphate, 88, 242
hydrogen sulphite, 296
hydroxide, 77–9, 84, 164, 235, 236, 246, 391–2
hypochlorite, 301
monoxide, 246
nitrate, 248
nitrite, 279
peroxide, 246
potassium tartrate, 364
sulphate, 248
sulphite, 296
thiosulphate, 296, 362
Solids, in equilibrium with a saturated solution, 374
movement of particles in, 10
packing in, 189
structure of, 187
Solution, as a method of obtaining pure substances, 43
heat of, 349
molar, 403
standard, 403
Solvay process, 393
Solvent, definition of, 44
water as a, 214
Specific heat of water, 213
Spectator ions, 94, 95, 419
Spectrograph, mass, 24
Stalactites, 218, 367
Stalagmites, 218, 367
Standard, electrode potential, 170, 171
pressure, 409
solution, 403
of sodium carbonate, 407, 423
temperature and pressure (s.t.p.), 409
Starch, as a natural polymer, 329
hydrolysis of, 337
structure of, 339
test for, 336
Static equilibrium, 371
States of matter, 5
Steam, reactions with metals, 136, 173
reactions with non-metals, 136
Stearic acid, 219

Steel, composition of, 256
manufacture, 389–91
properties of, 261
Stoichiometric equation, 419
Strong acids, comparison of strengths of, 79
definition of, 353
indicator for, 406
Structural formulae, 201
Structure, 185–202
atomic, 18, 20
crystal, 187–9
diamond, 194
giant, 138, 140, 185, 190, 191
giant ionic, 140, 191, 192
graphite, 193
molecular, 138, 140, 185, 190, 191
of elements in Period 3, 140
starch, 339
sodium chloride, 192
sulphur, 195
Sublimation, 56, 57
Substitution reaction, 309
Sugar, cane, 295, 325
reducing, identification of, 338
test for, 336
Sulphates, as salts, 88
preparation, 241–2
properties, 241
test for, 89
Sulphides, preparation, 244
properties, 245
test for, 245
Sulphites, 296
Sulphur, combustion in oxygen, 117
compounds, 289–96
extraction, 288
liquid, 196
monoclinic, 195
plastic, 196
properties of, 289
rhombic, 195
uses of, 289
Sulphur(IV) oxide, as an acidic oxide, 118
as an oxide of an element of Period 3, 138, 139
chemical properties, 290–3
manufacture, 383
physical properties of, 105
preparation of, 290
Sulphur(VI) oxide, in Contact process, 384
preparation of, 293
properties of, 294
Sulphuric acid, as a common laboratory acid, 71
as a drying agent, 102
dehydration of sugar cane, 295
electrolysis of, 163
manufacture, 383–5
preparation in laboratory, 294
properties of, 294–6
reactions with metals, 175
test for, 90
uses of, 296
Sulphurous acid, 290, 296
Superphosphate of lime, 280
Surface, area, effect on rate of reaction, 363
tension, 224, 225
Symbols for atoms, 17, 18
Synthetic, ammonia, manufacture, 381–3
rubber, 315, 321
Syringe method of collecting gases, 104

TANTALUM, 231
Tartaric acid, 70
Tellurium, 132
Temperature, effect on equilibrium, 376, 377, 380
effect on gas pressure, 14
effect on rate of reaction, 363
effect on sulphur, 198
standard, 409
transition, 196
volume changes with, 408
Temporary hardness of water, 220, 221, 367
Terylene, 314, 328, 332, 333
Tetrabromomethane, 309
Tetrachloroethane, 319
Tetrachloromethane, molar heat of vaporization, 347
solvent, 335
Thermal decomposition, 115
definition of, 284
of ammonium compounds, 284
of nitric acid, 286
of potassium chlorate, 364
Thermal dissociation, 284
Theory, atomic, 16
kinetic, 13
Thermo-plastic material, 332
Thermo-setting material, 332
Thompson, S. G., 132
Tin, 147
Tin-plating, 178
Titanium, 231
Titration, 85, 242, 404
Transition elements,
as an addition to Mendeléev's table, 132
the special case of, 140
Triads, Döbereiner's, 129
Tribromomethane, 309
Trichloromethane, 329, 370
Triiodomethane, 327
Tritium, 122
Turpentine, reaction with chlorine, 299

UNIVERSAL INDICATOR, 81
Unsaturated compounds, 310, 313
Urea, 282
Urea-formaldehyde resins, 332, 333

VALENCY, 39, 159
Vanadium pentoxide, 384
Vaporization, heat of, 186
Vapour density, 415
Velocity of gaseous molecules, 13, 14
Vinegar, 71, 324, 327
Vinyl chloride, 315, 320, 330
Voltaic cells, 177
Voltameter, 161
Hofmann's, 162, 163
Volume, atomic, 26, 186
changes with temperature, 14, 407
of a mole of gas, 414
strength of hydrogen peroxide, 228
–volume relationships, 424
Volumetric analysis, 403–7

WASHING SODA, as a salt, 89
effect on hard water, 222
Water, 213–28
air dissolved in, 110
as a solvent, 214
boiling point, 213, 346
effect of pressure on, 111
chemical properties, 214
composition of, 164
electrolysis of, 163
energy changes during synthesis, 346
hardness, 216–24
heat of vaporization, 213, 347
heavy, 122
hydrogen bonding in, 213
hydrolysis by, 214
molecule, shape of, 200
of crystallization, 215, 233
physical properties, 213
reaction with magnesium, 135
reaction with metals, 173
reaction with metallic oxides, 246, 250
reaction with non-metallic oxides, 272, 290
reaction with sodium, 127
softening, 221–4
surface tension of, 224
synthesis of, 346
test for, 215
Water gas, 273, 381
Weak acids,
comparison of strengths of, 79
heat of neutralization of, 353
indicator for, 406
Weiner, C., 11
Wrought iron, 261

XANTHOPHYLL, 55
Xenon, 108, 394
X-ray, analysis, 193
diffraction, 189

YEAST, fermentation by, 323

ZIRCONIUM, 231
Ziegler, K., 331
Zinc (the element), 254–5
Zinc compounds, carbonate, 255
chloride, 255
–copper couple, 168
hydroxide, 255
nitrate, 255
oxide, 118, 254
sulphate, 255
sulphide, 255
Zincates, 236, 255

Table of Atomic Masses

Name	*Symbol*	*Atomic Number*	*Chemi Atom Mas*
Actinium	Ac	89	227
Aluminium	Al	13	27
Americium	Am	95	243
Antimony	Sb	51	122
Argon	Ar	18	40
Arsenic	As	33	75
Astatine	At	85	210
Barium	Ba	56	137
Berkelium	Bk	97	247
Beryllium	Be	4	9
Bismuth	Bi	83	209
Boron	B	5	11
Bromine	Br	35	80
Cadmium	Cd	48	112
Caesium	Cs	55	133
Calcium	Ca	20	40
Californium	Cf	98	251
Carbon	C	6	12
Cerium	Ce	58	140
Chlorine	Cl	17	35
Chromium	Cr	24	52
Cobalt	Co	27	59
Copper	Cu	29	64
Curium	Cm	96	247
Dysprosium	Dy	66	162
Einsteinium	Es	99	25
Erbium	Er	68	16
Europium	Eu	63	15
Fermium	Fm	100	25
Fluorine	F	9	1
Francium	Fr	87	22
Gadolinium	Gd	64	15
Gallium	Ga	31	7
Germanium	Ge	32	7
Gold	Au	79	19